Embodied Carbon in Buildings

Francesco Pomponi • Catherine De Wolf
Alice Moncaster

Editors

Embodied Carbon in Buildings

Measurement, Management, and Mitigation

 Springer

Editors
Francesco Pomponi
REBEL (Resource Efficient
Built Environment Lab)
Edinburgh Napier University
Edinburgh, UK

Alice Moncaster
School of Engineering and Innovation
Open University
Milton Keynes, UK

Catherine De Wolf
School of Architecture, Civil and
Environmental Engineering
Ecole Polytechnique Fédérale de Lausanne
(EPFL)
Lausanne, Switzerland

ISBN 978-3-319-72795-0   ISBN 978-3-319-72796-7   (eBook)
https://doi.org/10.1007/978-3-319-72796-7

Library of Congress Control Number: 2018930100

Printed on acid-free paper

This Springer imprint is published by Springer Nature
The registered company is Springer International Publishing AG
The registered company address is: Gewerbestrasse 11, 6330 Cham, Switzerland

Irina Bolshakova and Gerardo and Franca Pomponi

Lieve Vanhee and Philippe and Daniel De Wolf

Richard Bridgeman and Poppy and Cicely Moncaster Bridgeman

Foreword

The significance of embodied carbon in buildings has been recognised only relatively recently. Early attempts to quantify the carbon implications of specific construction materials were noble and served to identify the very real challenges in determining with any confidence the true figures of embodied carbon. The number of variables was daunting, and the postulation of generalised figures for individual materials served only to demonstrate the unreliability of depending on the values for any particular setting.

More recently, there has been a drive internationally to improve the understanding of embodied carbon in materials and in the whole cycle of construction and operation. As buildings become more energy-efficient, the embodied carbon component becomes more significant in the decision-making process for environmentally responsible design.

The publication of this book on embodied carbon in buildings could not be more timely. As research proliferates into the many different facets of the topic, the authors have brought together contributions which together cover the key questions to be addressed in a digestible and accessible form. Tackling management, measurement and mitigation in the three sections of the book is most impressive, and by dealing with uncertainty, much of the scepticism of life cycle analysis and its past in confusion and unreliable data can be dispelled. The geographical coverage globally is invaluable, and the authors have been assiduous in bringing such a diverse collection of contributions together in a single volume.

Pomponi, De Wolf and Moncaster, who are all eminent in this field, can be very proud of this seminal publication.

Professor Peter Guthrie

Peter Guthrie, OBE, FREng, was appointed the UK's first professor of engineering for sustainable development at the University of Cambridge in 2000. A civil engineer with geotechnical specialisation by background, for the first half of his career, Peter worked as a practising engineer on infrastructure projects, working extensively in Africa and Asia as well as on major UK projects such as the Channel Tunnel Rail Link and London 2012 Olympic Park. In 1980, Peter founded RedR Engineers for Disaster Relief, and in 1994, he was awarded an OBE. With a passion for integrating social and environmental considerations into engineering design, since 2000 he has led research at Cambridge that enables engineers to deliver more sustainable outcomes. In recent years, a particular focus of his has been on improving the understanding of energy efficiency in buildings and the challenges (financial, social and environmental) faced in delivering significantly lower carbon emissions from building stock. His other main focus is on resilience in infrastructure.

The Editors

Dr Francesco Pomponi is the Vice Chancellor's Fellow at Edinburgh Napier University. Francesco leads the Resource Efficient Built Environment Lab (REBEL) in Edinburgh, a group of scientists addressing the issue of resource scarcity and their efficient use from a multi-disciplinary point of view. Francesco's expertise lies with life cycle assessment, embodied carbon and circular economy, and he moved to academia after 6 years in industry as engineer and project manager. He is part of the newly launched Annex 72 of the International Energy Agency and has recently chaired the 'Life Cycle Assessment and Carbon Accounting' Forum of the International Passive and Low Energy Architecture Conference and the 'Design for Sustainability' Track of the International Sustainable Development Research Society Conference. He is a fellow of the RSA, a member of the IET and a fellow of the HEA. An associate member of St Edmund's College University of Cambridge and of Cambridge Architectural Research (CAR), Francesco regularly collaborates with practitioners and researchers from South and Central America, Africa, Europe and of course the UK.

Dr Catherine De Wolf is a Marie Sklodowska-Curie postdoctoral fellow from the European Commission and a Swiss Government Excellence Scholar at the Swiss Federal Institute of Technology in Lausanne (Ecole Polytechnique Fédérale de Lausanne, EPFL) where she works on low-carbon structural design within the Structural Xploration Lab (SXL). She also worked as a researcher at the University of Cambridge while obtaining her PhD in building technology at the Massachusetts Institute of Technology (MIT), after studying both civil engineering and architecture at the Vrije Universiteit Brussel and Université Libre de Bruxelles. She closely collaborated with leading engineering firms including Arup, Ney & Partners and Thornton Tomasetti on embodied carbon assessment in buildings. This led to her nomination on the board of the Carbon Leadership Forum and the launch of the Structural Engineers 2050 Commitment Initiative.

Dr Alice Moncaster is a senior lecturer in engineering at the Open University. She remains a visiting fellow at the University of Cambridge, where she was previously a lecturer in engineering and director of the IDBE masters course, and a fellow of Newnham College. The move to academia followed 10 years in industry as a civil/ structural engineer, during which time she became increasingly concerned about the responsibility of the construction sector for climate change. Alice's research focuses on reducing the ecological impacts of the built environment. She led the research group Cambridge University Built Environment Sustainability (CUBES) as part of the Centre for Sustainable Development at Cambridge between 2010 and 2017 and has been the UK participating expert on the International Energy Agency Annex 57, and now Annex 72, since 2012.

Introduction

Embodied carbon is, to some extent, an odd beast. Its importance is evident and the beneficial consequences of its reduction undeniable. We know that the built environment is a major source of our carbon excesses, yet most policies focus only on part of the picture by capping operational energy consumption, for the use of buildings. We also know that the Intergovernmental Panel on Climate Change (IPCC) has warned that carbon reductions are needed now, not in 30 years' time. Lowering the immediate emissions related to current building construction and demolition, the embodied carbon, is an obvious way to do so. In recent years, research on embodied carbon has therefore increased.

Many fields of research develop steadily over the years, led by a small and coherent community of experts. Others quietly die, as the world moves on. Yet, for a very few topics, a moment comes when the world suddenly wakes up to their importance, and interest and attention start to snowball. This is such a moment for the subject of this book, the greenhouse gas emissions resulting from the construction of buildings. Within this snowballing, of industry consultancies producing tools, of manufacturers benchmarking their products, of academics working together on major projects and even of the rumblings of political and regulatory change, there is, however, a real danger that the knowledge will become so dispersed that any real progress will be lost. Instead of forming a coherent body of work to inform policy and evoke real change in how we construct our built environment, we run the real risk of finding ourselves in a meaningless avalanche of disconnected ideas. This book, therefore, sets out to perform a vital task – to extract coherence, not chaos, from this outpouring of intellectual endeavour.

Following the Paris Agreement, many nations have revamped their carbon plans, climate change drafts and carbon reduction targets. However, most governments remain stuck on the same single track of promoting operational energy efficiency in buildings, seemingly reluctant to acknowledge that this ignores an essential part of the picture. More energy-efficient buildings may reduce energy use and carbon emissions in the long term, but without a parallel focus on embodied energy and carbon, the real savings that could be made right now are lost, often instead resulting in an increase in short-term impact. Without a holistic understanding of the data, a sincere estimate of the uncertainties and an appreciation of the impact of human behaviour – both of occupiers and of constructors – this is a gamble with the future of our environment.

We hope, therefore, that we have succeeded in representing, within this one volume, a persuasive argument for the importance of including embodied emissions in all aspects of construction. The argument is constructed over the first three sections through the main areas of debate over the *measurement* of embodied carbon, the key concepts of its *management* and a comprehensive overview of the *mitigation* strategies being proposed and enacted. The final section acknowledges that there are geographical differences in both context and approach, providing an overview of the state of knowledge and practice across regions of the world.

Correct understanding of estimates is an essential starting point in the embodied carbon debate. If we cannot agree on our numbers, the conversation is prevented from moving forward. The first section, therefore, includes three chapters dedicated to uncertainty analysis, each of which offers novel and diverse points of view on the topic. The section also features chapters on the embodied carbon of different structural materials as well as the inclusion of some uncommon variables in embodied carbon assessments, such as surface albedo.

The management section is perhaps the most diverse in the book and the one with the greater interdisciplinary outlook. It features chapters looking at early design tools, others aimed at bridging the current gap between research and practice and some looking at the significance of life cycle stages often neglected in embodied carbon assessments as well as the identification of carbon hotspots.

The third section on mitigation is the natural conclusion of the 'embodied carbon journey' offered in the book. In other words, now that we know how to quantify embodied carbon and that we have also learned how to manage it, how can we actually reduce it? The section features a diverse set of chapters, looking at novel opportunities offered by the principles of a circular economy, sustainable technologies and optimisation strategies at both material and building levels.

Views from different regions of the world conclude the book, and we are very proud of the broad coverage we managed to achieve. This section includes contribution from Australia, a world leader in embodied carbon, Africa, North and South America, Europe and China. We strongly believe all chapters offer a stimulating learning opportunity for all those interested.

We hope that this book will succeed in its aims: to educate and enthuse both practitioners and scholars, to provide a comprehensive starting point for the novel researcher in the field and to act as an essential reference source for everyone working on this topic. Most of all, we hope to have created a document that collates, connects and makes sense of the current state of knowledge and that identifies clearly the questions still to be answered.

We believe that bringing together key researchers in this area has already started the process of creating a virtual global community highlighting and validating their different views while acknowledging the similarity of the challenges we are facing. We hope that both readers of and contributors to this book will return to their work with renewed spirit and positivity, in the recognition that together we form a strong, passionate community working together to create real change towards a low-carbon future.

Contents

Contributors

A. Akbarnezhad School of Civil and Environmental Engineering, UNSW, Sydney, NSW, Australia

I. Andresen Norwegian University for Science and Technology (NTNU), Department of Architecture, Trondheim, Norway

H. Birgisdottir Danish Building Research Institute, Aalborg University, Copenhagen, Denmark

L. M. Campos Universidad Colegio Mayor de Cundinamarca, Bogotá, Colombia

J. Connaughton School of Construction Management and Engineering, University of Reading, Reading, UK

S. Cooper-Searle Department of Engineering, University of Cambridge, Cambridge, UK

B. Cousins-Jenvey Expedition Engineering/Useful Projects Ltd, London, UK

Department of Architecture and Civil Engineering, University of Bath, Bath, UK

Systems Centre, University of Bristol, Bristol, UK

R. H. Crawford Faculty of Architecture, Building and Planning, The University of Melbourne, Parkville, VIC, Australia

D. Davies Magnusson Klemencic Associates, Seattle, WA, USA

C. De Wolf School of Architecture, Civil and Environmental Engineering, Ecole Polytechnique Fédérale de Lausanne (EPFL), Lausanne, Switzerland

D. Densley Tingley Civil and Structural Engineering Department, University of Sheffield, Sheffield, UK

G. K. C. Ding School of the Built Environment, Faculty of Design, Architecture & Building, University of Technology Sydney, Ultimo, NSW, Australia

B. Doepker Magnusson Klemencic Associates, Seattle, WA, USA

I. Ellingham Cambridge Architectural Research Ltd, Cambridge, UK

W. Fawcett Cambridge Architectural Research Ltd, Cambridge, UK

S. Finnegan Liverpool School of Architecture (LSA), University of Liverpool, Liverpool, UK

S. M. Fufa SINTEF Building and Infrastructure, Oslo, Norway

J. Gantner Department of Lifecycle Assessment, Fraunhofer Institute for Building Physics, Stuttgart, Germany

J. Giesekam Sustainability Research Institute, School of Earth and Environment, University of Leeds, Leeds, UK

J. B. Guinée Department of Industrial Ecology, Institute of Environmental Sciences (CML), Leiden University, Leiden, The Netherlands

A. W. A. Hammad School of Civil and Environmental Engineering, UNSW, Sydney, NSW, Australia

M. Hedlund Magnusson Klemencic Associates, Seattle, WA, USA

R. Heijungs Department of Industrial Ecology, Institute of Environmental Sciences (CML), Leiden University, Leiden, The Netherlands

Department of Econometrics and Operations Research, Vrije Universiteit, Amsterdam, The Netherlands

J. Hong School of Construction Management and Real Estate, Chongqing University, Chongqing, China

A. Houlihan Wiberg National Technical University of Norway (NTNU), Trondheim, Norway

K. Hyde School of Construction Management and Engineering, University of Reading, Reading, UK

L. Johnson Magnusson Klemencic Associates, Seattle, WA, USA

N. C. Kayaçetin Middle East Technical University, Ankara, Turkey

N. Kibwami School of the Built Environment, Department of Construction Economics and Management, Makerere University, Kampala, Uganda

A. B. Liel Department of Civil, Environmental, and Architectural Engineering, University of Colorado Boulder, Boulder, CO, USA

T. Malmqvist KTH Royal Institute of Technology, Stockholm, Sweden

R. Marsh Danish Building Research Institute, Aalborg University, Copenhagen, Denmark

M. A. Mendoza Beltran Department of Industrial Ecology, Institute of Environmental Sciences (CML), Leiden University, Leiden, The Netherlands

A. M. Moncaster School of Engineering and Innovation, Open University, Milton Keynes, UK

F. Nygaard Rasmussen Danish Building Research Institute, Aalborg University, Copenhagen, Denmark

J. Ochsendorf Department of Architecture and Civil and Environmental Engineering, Massachusetts Institute of Technology (MIT), Cambridge, MA, USA

S. Perera School of Computing, Engineering and Mathematics, Western Sydney University, Penrith, Australia

F. Pomponi REBEL (Resource Efficient Built Environment Lab), Edinburgh Napier University, Edinburgh, UK

D. Rey School of Civil and Environmental Engineering, UNSW, Sydney, NSW, Australia

S. Richardson TSBE Centre, University of Reading, Reading, UK

M. Schmidt Faculty of Architecture, Building and Planning, The University of Melbourne, Parkville, VIC, Australia

G. Q. Shen Department of Building and Real Estate, The Hong Kong Polytechnic University, Hung Hom, Hong Kong

K. Simonen Department of Architecture, University of Washington, Seattle, WA, USA

E. Soulti Building Research Establishment, Watford, UK

A. Souto-Martinez Department of Civil, Environmental, and Architectural Engineering, University of Colorado Boulder, Boulder, CO, USA

W. V. Srubar III Department of Civil, Environmental, and Architectural Engineering, University of Colorado Boulder, Boulder, CO, USA

A. Stephan Faculty of Architecture, Building and Planning, The University of Melbourne, Parkville, VIC, Australia

T. Susca Bundesanstalt für Materialforschung und –prüfung (BAM), Berlin, Germany

E. J. Sutley Department of Civil, Environmental, and Architectural Engineering, University of Kansas, Lawrence, KS, USA

M. Tang School of Construction Management and Real Estate, Chongqing University, Chongqing, China

A. M. Tanyer Middle East Technical University, Ankara, Turkey

D. Trabucco Council on Tall Buildings and Urban Habitat/IUAV University of Venice, Venice, Italy

A. Tutesigensi School of Civil Engineering, University of Leeds, Leeds, UK

M. Victoria Scott Sutherland School of Architecture and Built Environment, Robert Gordon University, Aberdeen, UK

M. K. Wiik SINTEF Building and Infrastructure, Oslo, Norway

Part I
Measurement

Chapter 1
Uncertainty Analysis in Embodied Carbon Assessments: What Are the Implications of Its Omission?

M. A. Mendoza Beltran, Francesco Pomponi, J. B. Guinée, and R. Heijungs

Introduction

Embodied carbon assessments of buildings are methodologically similar to the more well-known and standardized life cycle assessment (LCA) focused on the quantification of carbon emissions throughout the life cycle of buildings. Generally, this type of studies results in single-point estimates, based on deterministic data, which in many cases represents an average numerical output which embeds no information on the likelihood, significance or variability of that value. In comparative studies, for instance, when the performance of two buildings is compared, the LCA point value results are superposed and directly compared. The allegedly less environmentally detrimental alternative is chosen without considering the risk of making a wrong decision.

Such deterministic assessments have many associated uncertainties. For instance, inventory data for complex systems such as buildings is variable and sometimes non-existent, undetermined and ambiguous. Also, methodological choices are made during the different phases of LCA, introducing uncertainty in the results particularly in comparative contexts. Yet, many LCAs and embodied

M. A. Mendoza Beltran (✉) · J. B. Guinée
Department of Industrial Ecology, Institute of Environmental Sciences (CML), Leiden University, Leiden, The Netherlands
e-mail: mendoza@cml.leidenuniv.nl

F. Pomponi
REBEL (Resource Efficient Built Environment Lab), Edinburgh Napier University, Edinburgh, UK

R. Heijungs
Department of Industrial Ecology, Institute of Environmental Sciences (CML), Leiden University, Leiden, The Netherlands

Department of Econometrics and Operations Research, Vrije Universiteit, Amsterdam, The Netherlands

© Springer International Publishing AG 2018
F. Pomponi et al. (eds.), *Embodied Carbon in Buildings*,
https://doi.org/10.1007/978-3-319-72796-7_1

carbon assessments of buildings lack uncertainty analysis that address the presence of these sources of uncertainty and that accompany and enrich the interpretation of results.

This chapter will provide the reader with an overview of uncertainty analysis in LCA, from the rationale to the methodological challenges through to the increased usefulness of the results in comparison with point value assessments. This chapter will reflect on the consequences of disregarding uncertainty in LCA and attempt to explain how uncertainty analysis interacts with decision-making and how it can benefit and facilitate environmentally conscious decisions.

Meanwhile, a large body of literature is available on how to conduct an uncertainty analysis in LCA. Diverse techniques and methods that can be suited to address uncertainty in different applications and contexts are already part of the existing literature. Although we will provide reference to seminal literature and key studies for the less experienced reader, this chapter will not focus on the technical details for implementing an uncertainty analysis. Rather, we investigate the consequences of abstaining in the practice of uncertainty analysis in LCAs of buildings. Thus, we focus in the comparison of a deterministic LCA and an LCA with uncertainty analysis.

Uncertainty in LCA

Certainty is the idea of confidence, assurance and accuracy about our knowledge of the truth. Certainty and truth exist, evading discussions on philosophical scepticism that are self-defeating as denying their existence is already accepting a truth with certainty (Briggs 2016). The idea of uncertainty is based upon the existence of truth by acknowledging there is something that is but cannot be fully known. Uncertainty does not exist in objects themselves, aside from the sense of existence, but only in our mind or intellect (Briggs 2016). Therefore, it is our incapacity to know the truth that underlines uncertainty. In fact, there are many ways to treat uncertainty, but probability is one of the most used ones. Probability is the language of uncertainty that explains the limitations in our knowledge of the truth (Briggs 2016). This is why many fields of knowledge have relied on probability to help treat this limitation, and the field of LCA is no exception, as will be shown below.

Uncertainty has been researched for about 30 years in LCA. The increased attention that LCA received during the 1990s as a tool to describe environmental impacts of products in the broad sense came along with criticism about the drawbacks of this decision support framework used by governments and companies (Udo de Haes 1993). One of the major limitations are uncertainties around it (Finnveden 2000; Ross et al. 2002), which threaten the reliability of decision-makers on the results and recommendations from LCAs. Guinée et al. (1993) mentioned that: "A valuation of environmental profiles without an assessment of the reliability and validity of the results, is of little value".

Some of the first dedicated research to uncertainty treatment in LCA appeared during the 1990s. Uncertainty analysis in LCA was defined by Heijungs (1996) as "the study of the propagation of unintentional deviations" in order to understand "those areas where product and process improvement lead to the highest environmental gain". Similarly, Huijbregts (1998a, b) identified the usefulness of uncertainty analysis in LCA to help decision-makers judge the significance of the differences in product comparisons, options for products improvements or the assignment of eco-labels. Weidema and Wesnæs (1996) were the first to describe and apply data quality indicators (DQIs), semi-quantitative numbers providing information about the quality of the data, and data quality goals (DQGs), the desired quality of the data, in an LCA context. This methodological development known as the "pedigree-matrix" in LCA jargon, inspired by the purely qualitative proposal of Funtowicz and Ravetz (1990), is one of the most widely applied techniques to semi-quantitatively address uncertainty of data in LCA. This method was later incorporated in the ecoinvent database (Frischknecht et al. 2007). DQIs enabled early probabilistic approaches to account for data uncertainties and LCA models evolved from deterministic models to stochastic models characterized by probability distributions (Kennedy et al. 1996).

Yet only until the end of the 1990s and beginning of the twenty-first century, a general framework that distinguished various types of uncertainty and variability in LCA was proposed and further studied (Huijbregts 1998a; Björklund 2002). These frameworks are of particular importance as they differentiate various types of uncertainty and variability in LCA as well as recognize that different types of uncertainty and variability might require different treatment (Huijbregts 1998b). The types of uncertainty and variability are (according to a combination of Huijbregts 1998a; Björklund 2002):

• Parameter uncertainty: data inaccuracy, data gaps and unrepresentative data
• Uncertainty due to methodological choices
• Model uncertainty
• Epistemological uncertainty
• Spatial variability
• Temporal variability
• Sources and objects variability
• Mistakes

While uncertainty refers to lack of knowledge about the truth (Briggs 2016), variability makes reference to inherent differences within a population attributable to natural heterogeneity of values (Björklund 2002). Therefore, while uncertainty can be reduced, variability cannot be reduced but only better estimated, for instance, with better sampling (Björklund 2002). In the interest of brevity, from here on we use uncertainty to refer to both uncertainty and variability types together.

Types of Uncertainty

Although a detailed description of each type of uncertainty falls beyond the scope of this work, a very brief explanation of some uncertainty types is provided to exemplify what they entail. We ask readers to consult Björklund (2002) for a detailed description of each type of uncertainty.

Parameter uncertainty has been associated to data inaccuracy (Huijbregts et al. 2001), unavailability and to unrepresentative data (Björklund 2002). This is uncertainty due to, for example, wrong inventory data, missing data or data that refers to different technologies, places or temporal resolutions than the intended one. *Methodological choice uncertainty* is due to the unavoidable choices of practitioners along the phases of LCA on topics like functional units, system boundaries (Tillman et al. 1994), allocation methods (Weidema 2000; Guinée and Heijungs 2007), impact categories and characterization methods and factors (Huijbregts 1998b; Finnveden 1999). *Model uncertainty* refers to simplification aspects of LCA such as aggregation and the modelling aspect of LCA, for example, linear and non-linear models (Heijungs and Suh 2002), derivation of characterization factors (Björklund 2002) or estimation of emissions with exogenous specialized models. *Variability* refers to intrinsic fluctuations of a numerical property (Björklund 2002) such as the yield of a hectare of arable land. *Epistemological uncertainty* emerges from the lack of knowledge on system behaviour, for instance, when modelling future systems (Björklund 2002).

Approaches to Deal with Uncertainties in LCA

Different types of uncertainty in LCA may require different types of treatment. There are different approaches to deal with uncertainties in LCA. In certain cases, the aim is to reduce uncertainty in order to generate a more reliable assessment and therefore, better support for decision-making. In other cases, the aim is to reflect the uncertainty of the result as an extra piece of information to the decision-maker. In general, the main approaches to different types of uncertainty are (Heijungs and Huijbregts 2004) the scientific, the constructivist, the legal and the statistical approaches. These approaches use additional research, consensus or agreement, authority and probability and statistics to deal with uncertainty. From these approaches, only the statistical approach explicitly incorporates uncertainty in the outcomes of LCA (Heijungs and Huijbregts 2004).

Statistical approaches to parameter uncertainty have led to sophisticated methods to quantify input uncertainties (Heijungs and Frischknecht 2005; Bojacá and Schrevens 2010; Henriksson et al. 2013; Ciroth et al. 2013; Muller et al. 2014; Qin and Suh 2016), to propagate such uncertainties through the LCA model (Imbeault-Tétreault et al. 2013; Groen et al. 2014; Heijungs and Lenzen 2014), to interpret outputs with uncertainty (Heijungs and Kleijn 2001; Prado-Lopez et al.

2014, 2015; Henriksson et al. 2015; Cucurachi et al. 2016) as well as to approaches that deal with all the above (Hung and Ma 2009; Andrianandraina et al. 2015; Gregory et al. 2016; Wei et al. 2016).

Dealing with uncertainty due to methodological choices in LCA has mostly been approached by the constructivist and legal approach. Consensus among stakeholders on the choices or predefining (ISO 2006) or mandating the choices reduces uncertainty in the outcomes (Heijungs and Huijbregts 2004) and increases comparability of studies. Environmental Product Declaration (EPD) schemes as well as Product Category Rules (PCRs) are examples of such approaches to deal with uncertainty due to choices (Del Borghi 2013). More recently, statistical and mathematical approaches to treat choice uncertainty have been proposed too (Jung et al. 2013; Cruze et al. 2014; Mendoza Beltran et al. 2015; Hanes et al. 2015). These incorporate the effects of uncertainty due to the different choices on the outcomes.

For model uncertainties, statistical approaches have been published (Padey et al. 2013; Andrianandraina et al. 2015). Typically, these treat parameter and model uncertainty simultaneously.

Within the different approaches, a large number of tools to deal with uncertainty in LCA are available (Table 1.1).

Regardless of the availability of the different tools and the widely agreed recommendation that dealing with different sources of uncertainty in LCA is a vital step to increase reliability on LCA results, few studies apply any method for such purpose (Ross et al. 2002; Lloyd and Ries 2007). It appears that the latest literature review of LCAs including uncertainty was performed by Lloyd and Ries (2007), which despite being about a decade ago is the best available. This study reviewed 400 journal publications and 2000 websites resulting from the search terms uncertainty or variability and LCA and published up to 2004. They narrowed down their review to 24 studies, from which about half contained applications in specific sectors and the rest focused on method development. From the case study articles, only two focus on the building sector (Chevalier and Tfino 1996; Huijbregts et al. 2003). Since the review of Lloyd and Ries (2007), the trends have not changed (Pomponi and Moncaster 2016). Despite of the addition of the DQI pedigree and its associated uncertainty estimation to the ecoinvent data, and the incorporation of methods for uncertainty propagation in mainstream software, such as SimaPro and OpenLCA, doing such an analysis is still more an exception than a rule. Although, including a preliminary uncertainty analysis in LCA studies is nowadays simpler than ever.

LCAs of Buildings with Uncertainty Analysis

A recent review by Pomponi and Moncaster (2016) focused on LCAs and embodied carbon studies of buildings. This review classified 77 studies according to the life cycle stages taken into account, among other characteristics. The stages were specified as in the BS EN 15978 framework (Fig. 1.1). Pomponi and Moncaster

Table 1.1 Main sources of uncertainty in LCA and some techniques and methods to treat them

Type / Tools	Parameter uncertainty	Model uncertainty	Uncertainty due to choices	Spatial variability	Temporal variability	Sources and objects variability
Scientific approach						
Additional measurements	x					x
Scenario modelling		x	x	x	x	x
Non-linear modelling		x				
Multimedia modelling		x		x		
Constructivist approach						
Expert judgements/peer review	x		x			x
Rules of thumb	x					
Legal approach						
Standardization	x		x			
Prescription of specific methods (e.g. ILCD)	x		x			
Statistical approach						
Probabilistic simulation	x		x			x
Data quality indicators	x			x	x	x
Uncertainty importance analysis (global sensitivity analysis)	x	x	x	x	x	x
Classical statistical analysis	x			x	x	x
Bayesian statistical analysis	x			x	x	x
Sensitivity analysis	x	x	x	x	x	x
Interval arithmetic	x			x	x	x
Correlation and regression analysis	x					x

Adapted from Huijbregts (1998a), Björklund (2002), and Heijungs and Huijbregts (2004)

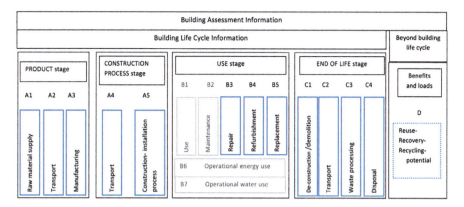

Fig. 1.1 BS EN 15978 framework

(2016) showed that most studies exclude or ignore uncertainty analysis and have a short-sighted approach to the life cycle of buildings, i.e. many only include manufacturing stages and few include impacts during occupation (use stage) or end-of-life.

There are, however, a few recent studies that include some treatment of uncertainty. These are studies where mostly a large number of scenarios has been assessed, therefore producing a range of results that convey a level of uncertainty (Pomponi et al. 2015, 2016). Alternatively, they use Monte Carlo simulations to propagate parameter uncertainty (Blengini and Di Carlo 2010; Heeren et al. 2015; Hoxha et al. 2017; Pomponi et al. 2017). In the case of Heeren et al. (2015), a comprehensive sensitivity analysis has been enriched with information about the assumed distributions of parameters, as well as correlation between parameters and outputs, allowing a much deeper interpretation of combined effects of materials used and energy demand of buildings, as well as of trade-offs in the results. The authors modelled a high-number of combinations resulting from different samplings of certain parameters per life cycle stage. Their findings are therefore in the form of a range of results with statistical description (e.g. mean, standard deviation, quartiles), and this certainly helps a more informed decision-making process.

Despite the existence of these more recent studies that include some handling of uncertainty, it is atypical for LCAs in the building sector to formally consider uncertainty. The first study which accounted for parameter, methodological choices and model uncertainties in the building sector is that by Huijbregts et al. (2003). The authors focused on two insulation alternatives for a Dutch one-family dwelling and developed and tested a methodology that takes into account parameter uncertainty through Monte Carlo (MC) simulations and scenario and model uncertainty by means of resampling different scenarios and model assumptions iteratively. Their results indicated that all types of uncertainty influence the outcomes of their study, thereby showing that the three sources of uncertainty should be evaluated and accounted for simultaneously.

The other example is the study by Vieira and Horvath (2008). The authors approached uncertainty differently, in a more qualitative fashion. However, their work also aims at reducing uncertainty in LCAs of building by eliminating some of the value judgements used in common approaches for allocation. The authors map differences in uncertainty between attributional and consequential LCAs. They tested their approach by comparing the outcomes of both approaches applied to concrete in a typical US building frame, concluding that neither appeared to yield more complete results.

The Inclusion of Uncertainty Analysis in LCAs

As shown in the literature review, studies taking into account uncertainty in LCAs of buildings are scarce. Those studies do treat uncertainty by means of scenarios and statistical approaches and display a range of different outcomes. Scenarios are mainly applied to show the effect of parameter, methodological choices and model uncertainty. Otherwise, MC simulation is used as a propagation method for uncertainty in model parameters and for inventory data (Huijbregts et al. 2003; Heeren et al. 2015).

Given that the main aim of this study is to understand what uncertainty analysis adds to LCA, we illustrate the difference between a deterministic LCA and an LCA including uncertainty analysis. For such purpose, we complemented a deterministic LCA of a household in the UK (Monahan and Powell 2011) to include uncertainty of life cycle inventory data by means of the protocol of Henriksson et al. (2013) using additional secondary data. We further propagate this source of uncertainty in the inputs to the outputs using MC simulations. Finally, both deterministic and uncertainty analysis LCA results are presented for comparing the outcomes of both approaches for the same household. Below we present the implementation of the illustrative case.

Illustration Case Implementation

We build on the case and inventory data published by Monahan and Powell (2011). The choice for this specific study follows mainly two reasons: the clear availability of inventory data and the simplified approach to implement the carbon emissions, which is typical for embodied carbon assessments of buildings. Monahan and Powell (2011) implemented carbon intensities for the full supply chains of materials used in the construction phase without an actual representation of the supply chain processes. A more classic approach to implement the life cycle of products, used in other LCAs, connects the inputs of foreground processes to background processes from LCA databases or from other secondary data sources which do include varied and interconnected processes in the supply chains. The study by

Monahan and Powell (2011) is an embodied carbon assessment for a household in the UK. The functional unit for this study is the external, thermal envelope of a three bedroom, semi-detached house with a total footprint area of 45 m^2 and a total internal volume of 220 m^3. The phases of the life cycle included in the system are product and construction stages according to Fig. 1.1.

From Monahan and Powell (2011), we use the quantity of materials used in construction as well as we calculate the average embodied carbon coefficient (ECC) for each material (Table 1.2). Using three data points, i.e. the average, the low estimate and the high estimate for the ECC per material, the weighted average, overall dispersion parameter phi (Heijungs and Frischknecht 2005) and the assumed distribution were calculated following the protocol and decision tree from Henriksson et al. (2013). Using these parameters, we implemented the system in the CMLCA software (http://www.cmlca.eu/). We used 1000 MC simulations to propagate uncertainties in the inputs to the outputs. Only carbon emissions are included in the inventory. Results are presented at the inventory level for carbon emissions to air.

Results

Figure 1.2 shows the embodied carbon content per material used in the construction in tons of carbon dioxide. Figure 1.3 shows the percentage of the contribution of each material to the total embodied carbon per household. These two figures correspond to our implementation of the system of Monahan and Powell (2011). Both results are very similar as expected.

Results including inventory data uncertainty estimates for the total embodied carbon content per household are shown in Fig. 1.4. Not only do these outcomes show there is variability in the total embodied carbon (i.e. from 18 to 67 tons of CO_2/household), but they also show the frequency in which different results are likely to be obtained. The average embodied carbon is still around the result for the deterministic LCA, i.e. 35 tons of CO_2/household.

The contribution of each group of materials to the total embodied carbon was calculated. Figure 1.5 shows these results for each group of materials and 1000 MC simulations. This analysis enables improved identification of processes that require data refinement. Alternatively, if data is considered already as the best available as possible, technological improvements can be identified that would lead to higher mitigation of emissions. For example, Fig. 1.3 shows that waste treatment is responsible for 12% of carbon emissions; however, Fig. 1.5 shows that this could be as high as 30%.

Table 1.2 Inventory data for household constructed in the UK

Category	Material	Quantity (kg) Monahan and Powell (2011)	Emissions (kgCO$_2$) Monahan and Powell (2011)	Average ECC (kgCO$_2$/kg [a]) Monahan and Powell (2011)	Average ECC pedigree-matrix Weidema and Wesnæs (1996)	Low estimate of ECC[a]	Low ECC pedigree-matrix Weidema and Wesnæs (1996)	High estimate of ECC[a]	High ECC pedigree-matrix Weidema and Wesnæs (1996)	Weighted mean ECC [overall dispersion] Henriksson et al. (2015)	Distribution
Metals	Aluminium	260	2140	8.23	[1,1,1,1,1,4]	2.12	[2,3,2,2,3,5]	10.93	[2,3,2,2,3,5]	7.7 [0.58]	Lognormal
	Steel	251	956	3.81	[1,1,1,1,1,4]	1.54	[2,3,2,2,3,5]	3.98	[2,3,2,2,3,5]	3.48 [0.42]	Lognormal
Minerals	Brick	2264	1175	0.52	[1,1,1,1,1,4]	0.18	[2,3,2,2,3,5]	0.55	[2,3,2,2,3,5]	0.471 [0.47]	Lognormal
	Cement	2023	798	0.39	[1,1,1,1,1,4]	0.222	[2,3,2,2,3,5]	0.86	[2,3,2,2,3,5]	0.44 [0.62]	Lognormal
	Concrete	56,651	9863	0.17	[1,1,1,1,1,4]	0.05	[2,3,2,2,3,5]	0.22	[2,3,2,2,3,5]	0.16 [0.55]	Lognormal
	Gypsum plaster	1349	413	0.31	[1,1,1,1,1,4]	0.22	[2,3,2,2,3,5]	0.43	[2,3,2,2,3,5]	0.32 [0.33]	Lognormal
Openings	Windows	1277	1996	1.56	[1,1,1,1,1,4]	1.384	[2,3,2,2,3,5]	3.42	[2,3,2,2,3,5]	1.82 [0.5]	Lognormal
	Doors	142	246	1.73	[1,1,1,1,1,4]	1.23	[2,3,2,2,3,5]	2.69	[2,3,2,2,3,5]	1.8 [0.39]	Lognormal
Plastics	HD polyethylene	56	90	1.61	[1,1,1,1,1,4]	1.58	[2,3,2,2,3,5]	2.11	[2,3,2,2,3,5]	1.68 [0.18]	Lognormal
	LDPE	29	72	2.48	[1,1,1,1,1,4]	1.97	[2,3,2,2,3,5]	3.1	[2,3,2,2,3,5]	2.5 [0.23]	Lognormal
	Poly-isocya-nate insulation	187	561	3.00	[1,1,1,1,1,4]	1.384	[2,3,2,2,3,5]	3.18	[2,3,2,2,3,5]	2.78 [0.38]	Lognormal
	Polythene	146	285	1.95	[1,1,1,1,1,4]	1.89	[2,3,2,2,3,5]	2.58	[2,3,2,2,3,5]	2.04 [0.19]	Lognormal
	PUR insulation	195	585	3.00	[1,1,1,1,1,4]	1.23	[2,3,2,2,3,5]	3.11	[2,3,2,2,3,5]	2.74 [0.42]	Lognormal

Timber	Composite board products	4330	3462	0.80	[1,1,1,1,1,4]	0.69	[2,3,2,2,3,5]	1.475	[2,3,2,2,3,5]	0.89 [0.47]	Lognormal
	Larch	1315	1421	1.08	[1,1,1,1,1,4]	1.007	[2,3,2,2,3,5]	1.61	[2,3,2,2,3,5]	1.15 [0.27]	Lognormal
	Engineered timber	222	152	0.68	[1,1,1,1,1,4]	0.41	[2,3,2,2,3,5]	0.72	[2,3,2,2,3,5]	0.64 [0.28]	Lognormal
	Softwood	6792	3056	0.45	[1,1,1,1,1,4]	0.3	[2,3,2,2,3,5]	0.541	[2,3,2,2,3,5]	0.44 [0.29]	Lognormal
Fuel	Mains gas (kWh)	1107	226	0.20	[1,1,1,1,1,4]	0.18	[2,3,2,2,3,5]	0.25	[2,3,2,2,3,5]	0.21 [0.18]	Lognormal
	UK grid electricity (kWh)	11,106	948	0.09	[1,1,1,1,1,4]	0.07	[2,3,2,2,3,5]	0.48	[2,3,2,2,3,5]	0.15 [0.88]	Lognormal
	Diesel (l)	2070	363	0.18	[1,1,1,1,1,4]	0.14	[2,3,2,2,3,5]	0.305	[2,3,2,2,3,5]	0.19 [0.4]	Lognormal
Waste	(tkm)	5350	4934	0.92	[1,1,1,1,1,4]	0.116	[2,3,2,2,3,5]	1.07	[2,3,2,2,3,5]	0.82 [0.66]	Lognormal
Transport	(tkm)	9372	883	0.09	[1,1,1,1,1,4]	0.033	[2,3,2,2,3,5]	0.193	[2,3,2,2,3,5]	0.097 [0.68]	Lognormal
Total		34,625									

Data from Monahan and Powell (2011) and other secondary data as referenced on the table. Final inputs used in the LCA calculations correspond to the weighted mean and overall dispersion parameter (Phi) calculated based on Henriksson et al. (2015)

[a]The values used for the minima and maxima of ECCs have been taken from the following sources: Gabi LCA Databases (2017), Swiss Centre For Life Cycle Inventories (2007), JRC-IES (2010), Hammond and Jones (2011), Hill et al. (2011), Zabalza Bribián et al. (2011), Bin and Parker (2012), Akbarnezhad et al. (2014), García-Segura et al. (2014), Atmaca and Atmaca (2015), Heeren et al. (2015), Zhang and Wang (2015), Hill and Dibdiakova (2016), Ökobaudat (2016), and Gan et al. (2017)

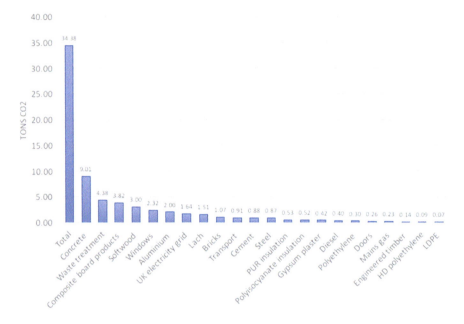

Fig. 1.2 Tons of carbon dioxide emissions to air from each material input

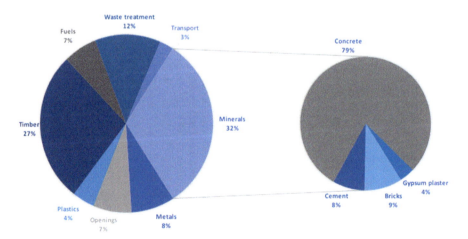

Fig. 1.3 Percentage of embodied carbon per group of inputs to the total embodied carbon per household

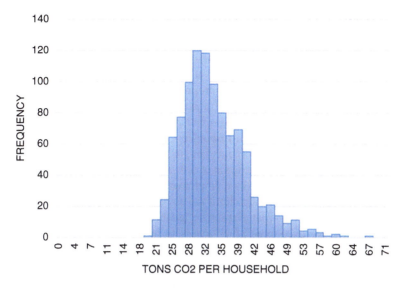

Fig. 1.4 Histogram of total embodied carbon content per household accounting for uncertainty in parameter uncertainty (using 1000 MC simulations)

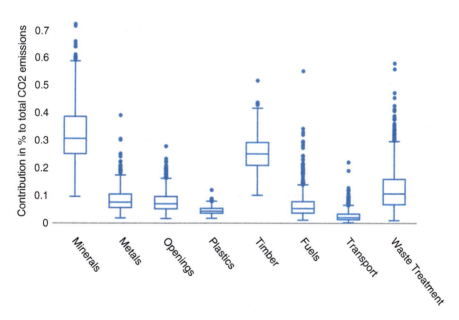

Fig. 1.5 Boxplots of contribution of different groups of inputs to the embodied carbon per household including parameter uncertainty (using 1000 MC simulations)

Discussion

Our illustrative case study exemplifies some of the limitations of deterministic LCAs particularly those with an aggregated representations of supply chains. In relation to the later characteristic, i.e. an implementation without an actual representation of the supply chain of the production of each material/input, the identification of hotspots and possible mitigation activities can be hampered. Despite the identification of materials or input responsible for a large proportion of the emissions (in the contribution analysis), it is not possible to identify where in the supply chain of such material those emissions are occurring; thus, mitigation options cannot be specified. This simplified implementation of the embodied carbon in the life cycle of buildings can also limit the assessment of innovative technologies such as the use of recycled materials as no feedbacks into upstream supply chains are represented. We encourage practitioners to include as much as possible full representations of supply chains following the life cycle thinking, particularly if environmental hotspots and mitigation potentials of innovative technologies or practices are to be identified. This is of utmost relevance in assessments following circular economy principles.

Moreover, deterministic LCAs can lead to biased recommendations to decision-makers. As our case study shows, average/point value results do not cover outcomes for all the possibilities of combinations including variability of the data (not to mention methodological choices and model uncertainty that we did not account for). From such results, recommendations that can omit likely outcomes can be made. For instance, in our case, technological changes to improve waste treatment might not be considered as a priority recommendation, as is the improvement in the use of minerals and timber. However, uncertainty analysis results showed that waste treatment can still have a large potential for reducing the carbon emissions of the system.

The most important limitation of deterministic LCAs becomes more evident for comparative LCAs. When comparing two systems, the embodied carbon results correspond to two distributions (or two sets of MC simulations) and not two point values. Although the interpretation of results including uncertainty in comparative LCAs is more complex than that of two point values, results including uncertainty are much more informative of the ranges of the outcomes and more robust as they add likelihood information and can serve as a base for statistical significance quantification (Henriksson et al. 2015). Moreover, trade-offs between two alternatives can also be calculated if the scope of the assessment goes beyond carbon emissions. Although we do not elaborate an example of results of a comparative LCA with uncertainty, methods for the interpretation of such results are discussed in Gregory et al. (2016), Wei et al. (2016), and Mendoza Beltran et al. (2017).

We showed that deterministic LCAs are broadly used in embodied carbon assessment of buildings. These assessments can lead to conclusions like "Material X is responsible for the Y% of embodied carbon emissions of a household" or "stage Z of the life cycle of buildings are responsible for most of the carbon emissions of a building". We showed that the statistical treatment of various sources of uncertainty in LCA can enrich the assessment by opening possibilities overlooked in deterministic type of studies. Other treatments of uncertainty such as standardization could serve the purpose of reducing uncertainty among different studies. However, this treatment also misses the point of studying a complex system accounting for variability and uncertainty which could lead to identification of improvement options not identifiable when the results are calculated with a fixed (consented, standard) set of data, choices, etc. We encourage practitioners to include uncertainty, where possible, and particularly for the purpose of identifying innovation options and emission reduction possibilities in the system.

Finally, we are aware of the challenges that availability of data might impose for practitioners willing to include and treat uncertainty in LCA. However, as shown in this chapter, there are techniques and a large pool of methods to help practitioners with such undertaking. An increasing number of such techniques and methods are being implemented in software for LCA. It is, nonetheless, important to emphasize that in any case and as written by Briggs (2016): "mistaking models as reality, or as certain, [...]" should be abolished as a practice and science should return to its true pursuit: "a proper understanding of essence and cause" (Briggs 2016).

Conclusion

In this chapter, we provided an overview of uncertainty analysis in LCA. We presented a literature review that covered from the rationale to the methodological challenges through to the increased usefulness of the results in comparison with point value assessments.

Uncertainty and variability are present in many aspects of environmental assessments particularly in LCAs. Main sources of uncertainty include parameter uncertainty, model uncertainty and uncertainty due to methodological choice and variability, among others. Disregarding uncertainty in LCA can have consequences. For instance, *single-value or deterministic LCAs lead to oversimplified outcomes or unjustified confidence in outcomes.* Thus, deterministic LCAs can lead to biased recommendations to decision-makers particularly in the case of comparative LCA.

Uncertainty analysis provides additional information for the decision-making process. *Addressing uncertainty and variability provides a large added value for practitioners, as it adds information about the significance and robustness of the results.* This information can benefit and facilitate environmentally conscious decisions as innovation opportunities can be overlooked when not addressing uncertainty.

In the specific case of buildings, which are characterized by extremely long lifespans where elements interact dynamically in both temporal and spatial dimensions, accounting for uncertainty and variability is vital to produce reliable assessments. Few assessments exist that include uncertainty and variability estimates. The oversimplified implementation of the ECCs of inputs to buildings makes it impossible to assess hot-spots in the supply chains and to have better assessments of circular alternatives. It is therefore fundamental that both academics and practitioners in the construction sector align with best practice within the LCA community to ensure they produce assessments that are accurate, reliable and most importantly useful.

References

Akbarnezhad, A., Ong, K. C. G., & Chandra, L. R. (2014). Economic and environmental assessment of deconstruction strategies using building information modeling. *Automation in Construction, 37*, 131–144.

Andrianandraina, Ventura, A., Senga Kiessé, T., et al. (2015). Sensitivity analysis of environmental process modeling in a life cycle context: A case study of hemp crop production. *Journal of Industrial Ecology.* https://doi.org/10.1111/jiec.12228.

Atmaca, A., & Atmaca, N. (2015). Life cycle energy (LCEA) and carbon dioxide emissions (LCCO2A) assessment of two residential buildings in Gaziantep, Turkey. *Energy and Buildings, 102*, 417–431.

Bin, G., & Parker, P. (2012). Measuring buildings for sustainability: Comparing the initial and retrofit ecological footprint of a century home–the REEP house. *Applied Energy, 93*, 24–32.

Björklund, A. E. (2002). Survey of approaches to improve reliability in LCA. *International Journal Life Cycle, 7*, 64–72.

Blengini, G. A., & Di Carlo, T. (2010). The changing role of life cycle phases, subsystems and materials in the LCA of low energy buildings. *Energy and Buildings, 42*, 869–880. https://doi.org/10.1016/j.enbuild.2009.12.009.

Bojacá, C. R., & Schrevens, E. (2010). Parameter uncertainty in LCA: Stochastic sampling under correlation. *International Journal of Life Cycle Assessment, 15*, 238–246. https://doi.org/10.1007/s11367-010-0150-0.

Briggs, W. (2016). *Uncertainty the soul of modeling, probability and statistics.* New York: Springer International Publishing.

Chevalier, J. L., & Tfino, J. F. L. E. (1996). Requirements for an LCA-based model for the evaluation of the environmental quality of building products. *Building and Environment, 31*, 487–491. https://doi.org/10.1016/0360-1323(96)00016-9.

Ciroth, A., Muller, S., Weidema, B., & Lesage, P. (2013). Empirically based uncertainty factors for the pedigree matrix in ecoinvent. *International Journal of Life Cycle Assessment.* https://doi.org/10.1007/s11367-013-0670-5.

Cruze, N. B., Goel, P. K., & Bakshi, B. R. (2014). Allocation in life cycle inventory: Partial set of solutions to an ill-posed problem. *International Journal of Life Cycle Assessment, 19*, 1854–1865. https://doi.org/10.1007/s11367-014-0785-3.

Cucurachi, S., Borgonovo, E., & Heijungs, R. (2016). A protocol for the global sensitivity analysis of impact assessment models in life cycle assessment. *Risk Analysis, 36*, 357–377. https://doi.org/10.1111/risa.12443.

Del Borghi, A. (2013). LCA and communication: Environmental product declaration. *International Journal of Life Cycle Assessment, 18*, 293–295. https://doi.org/10.1007/s11367-012-0513-9.

Finnveden, G. (1999). Methodological aspects of life cycle assessment of integrated solid waste management systems. *Resources, Conservation and Recycling, 26*, 173–187. https://doi.org/10.1016/S0921-3449(99)00005-1.

Finnveden, G. (2000). On the limitations of life cycle assessment and environmental systems analysis tools in general. *International Journal of Life Cycle Assessment, 5*, 229–238.

Frischknecht, R., Jungbluth, N., Hans-Jörg, A., et al. (2007). Overview and methodology – ecoinvent report no.1. Swiss Centre for Life Cycle Inventories, Dübendorf.

Funtowicz, S., & Ravetz, J. (1990). *Uncertainty and quality in science for policy.* Dordrecht: Kluwer Academic Publishers.

Gabi LCA Databases. (2017). In. http://www.gabi-software.com/databases/gabi-databases/

Gan, V. J., Cheng, J. C., Lo, I. M., & Chan, C. (2017). Developing a CO 2-e accounting method for quantification and analysis of embodied carbon in high-rise buildings. *Journal of Cleaner Production, 141,* 825–836.

Garcia-Segura, T., Yepes, V., & Alcala, J. (2014). Life cycle greenhouse gas emissions of blended cement concrete including carbonation and durability. *International Journal of Life Cycle Assessment, 19,* 3–12.

Gregory, J., Noshadravan, A., Olivetti, E., & Kirchain, R. (2016). A methodology for robust comparative life cycle assessments incorporating uncertainty. *Environmental Science & Technology.* acs.est.5b04969. https://doi.org/10.1021/acs.est.5b04969.

Groen, E. A., Heijungs, R., Bokkers, E. A. M., & de Boer, I. J. M. (2014). Methods for uncertainty propagation in life cycle assessment. *Environmental Modelling and Software, 62,* 316–325. https://doi.org/10.1016/j.envsoft.2014.10.006.

Guinée, J. B., & Heijungs, R. (2007). Calculating the influence of alternative allocation scenarios in fossil fuel chains. *International Journal of Life Cycle Assessment, 12,* 173–180. https://doi.org/10.1007/s11367-006-0253-9.

Guinée, J. B., Heijungs, R., Udo de Haes, H. A., & Huppes, G. (1993). Quantitative life cycle assessment of products: Classification, valuation and improvement analysis. *Journal of Cleaner Production, 1,* 81–91. https://doi.org/10.1016/0959-6526(93)90046-E.

Hammond, G., & Jones, C. (2011). *Embodied carbon: The inventory of carbon and energy (ICE).* University of Bath. Edited by BSRIA and printed by ImageData Ltd. ISBN 978 0 86022 703 8.

Hanes, R. J., Cruze, N. B., Goel, P. K., & Bakshi, B. R. (2015). Allocation games: Addressing the ill-posed nature of allocation in life-cycle inventories. *Environmental Science & Technology, 49,* 7996–8003. https://doi.org/10.1021/acs.est.5b01192.

Heeren, N., Mutel, C. L., Steubing, B., et al. (2015). Environmental impact of buildings – what matters? *Environmental Science & Technology, 49,* 9832–9841. https://doi.org/10.1021/acs.est.5b01735.

Heijungs, R. (1996). Identification of key issues for further investigation in improving the reliability of life cycle assessments. *Journal of Cleaner Production, 4,* 159–166.

Heijungs, R., & Frischknecht, R. (2005). Representing statistical distributions for uncertain parameters in LCA. *International Journal of Life Cycle Assessment, 10,* 248–254. https://doi.org/10.1065/lca2004.09.177.

Heijungs, R., & Huijbregts, M.A.J. (2004). A review of approaches to treat uncertainty in LCA.

Heijungs, R., & Kleijn, R. (2001). Numerical approaches towards life cycle interpretation five examples. *International Journal of Life Cycle Assessment, 6,* 141–148. https://doi.org/10.1007/BF02978732.

Heijungs, R., & Lenzen, M. (2014). Error propagation methods for LCA – A comparison. *International Journal of Life Cycle Assessment, 19,* 1445–1461. https://doi.org/10.1007/s11367-014-0751-0.

Heijungs, R., & Suh, S. (2002). The computational structure of life cycle assessment. *International Journal of Life Cycle Assessment, 7,* 314–314.

Henriksson, P. G. J., Guinée, J., Heijungs, R., et al. (2013). A protocol for horizontal averaging of unit process data – Including estimates for uncertainty. *International Journal of Life Cycle Assessment, 19,* 429–436.

Henriksson, P. J. G., Heijungs, R., Dao, H. M., et al. (2015). Product carbon footprints and their uncertainties in comparative decision contexts. *PLoS One, 10,* e0121221. https://doi.org/10.1371/journal.pone.0121221.

Hill, C., & Dibdiakova, J. (2016). The environmental impact of wood compared to other building materials. *International Wood Production Journal, 7,* 215–219. https://doi.org/10.1080/20426445.2016.1190166.

Hill, N., Walker, H., Beevor, J., & James, K. (2011). *Guidelines to Defra/DECC's GHG conversion factors for company reporting: Methodology paper for emission factors*. London: Department for Environment, Food and Rural Affairs.

Hoxha, E., Habert, G., Lasvaux, S., et al. (2017). Influence of construction material uncertainties on residential building LCA reliability. *Journal of Cleaner Production, 144*, 33–47. https://doi.org/10.1016/j.jclepro.2016.12.068.

Huijbregts, M. A. J. (1998a). LCA methodology application of uncertainty and variability in LCA part I : A general framework for the analysis of uncertainty and variability in life cycle assessment. *International Journal of Life Cycle Assessment, 3*, 273–280.

Huijbregts, M. A. J. (1998b). Application of uncertainty and variability in LCA part II : Dealing with parameter uncertainty and uncertainty due to choices in life cycle assessment. *International Journal of Life Cycle Assessment, 3*, 343–351.

Huijbregts, M., Norris, G., & Bretz, R. (2001). Framework for modelling data uncertainty in life cycle inventories. *Journal Life Cycle, 6*, 127–132.

Huijbregts, M. A. J., Gilijamse, W., Ragas, A. M. J., & Reijnders, L. (2003). Evaluating uncertainty in environmental life-cycle assessment. A case study comparing two insulation options for a Dutch one-family dwelling evaluating uncertainty in environmental life-cycle assessment. A case study comparing two insulation options for. *Environmental Science & Technology, 37*, 2600–2608. https://doi.org/10.1021/es020971.

Hung, M. L., & Ma, H. W. (2009). Quantifying system uncertainty of life cycle assessment based on Monte Carlo simulation. *International Journal of Life Cycle Assessment, 14*, 19–27. https://doi.org/10.1007/s11367-008-0034-8.

Imbeault-Tétreault, H., Jolliet, O., Deschênes, L., & Rosenbaum, R. K. (2013). Analytical propagation of uncertainty in life cycle assessment using matrix formulation. *Journal of Industrial Ecology, 17*, 485–492. https://doi.org/10.1111/jiec.12001.

ISO. (2006). *Environmental management – Life cycle assessment – Requirements and guidelines*. Switzerland: International Organization for Standardization. ISO 14044:2006.

JRC-IES. (2010). *International Reference Life Cycle Data System (ILCD) Handbook - General guide for Life Cycle Assessment – Detailed guidance*. Luxembourg: Publications Office of the European Union. First edition.

Jung, J., Assen, N., & Bardow, A. (2013). Sensitivity coefficient-based uncertainty analysis for multi-functionality in LCA. *International Journal of Life Cycle Assessment, 19*, 661–676. https://doi.org/10.1007/s11367-013-0655-4.

Kennedy, D. J., Montgomery, D. C., & Quay, B. H. (1996). Data quality stochastic environmental life cycle assessment modeling: A probabilistic approach to incorporating variable input data quality. *International Journal of Life Cycle Assessment, 1*, 199–207.

Lloyd, S. M., & Ries, R. (2007). Characterizing, propagating, and analyzing uncertainty in life-cycle assessment. *Journal of Industrial Ecology, 11*, 161–181. https://doi.org/10.1162/jiec.2007.1136.

Mendoza Beltran, A., Heijungs, R., Guinée, J., & Tukker, A. (2015). A pseudo-statistical approach to treat choice uncertainty: The example of partitioning allocation methods. *International Journal of Life Cycle Assessment*. https://doi.org/10.1007/s11367-015-0994-4.

Mendoza Beltran, A., Prado-Lopez, V., Font Vivanco, D., et al. (2017). Quantified uncertainties in comparative LCAs: What can be concluded?

Monahan, J., & Powell, J. C. (2011). An embodied carbon and energy analysis of modern methods of construction in housing: A case study using a lifecycle assessment framework. *Energy and Buildings, 43*, 179–188. https://doi.org/10.1016/j.enbuild.2010.09.005.

Muller, S., Lesage, P., Ciroth, A., et al. (2014). The application of the pedigree approach to the distributions foreseen in ecoinvent v3. *International Journal of Life Cycle Assessment*. https://doi.org/10.1007/s11367-014-0759-5.

Ökobaudat. (2016). Ökobaudat, Informationsportal nachhaltiges bauen. www.oekobaudat.de. Accessed 1 Apr 2016.

Padey, P., Girard, R., le Boulch, D., & Blanc, I. (2013). From LCAs to simplified models: A generic methodology applied to wind power electricity. *Environmental Science & Technology, 47*, 1231–1238. https://doi.org/10.1021/es303435e.

Pomponi, F., & Moncaster, A. (2016). Embodied carbon mitigation and reduction in the built environment: What does the evidence say? *Journal of Environmental Management, 181*, 687–700. https://doi.org/10.1016/j.jenvman.2016.08.036.

Pomponi, F., Piroozfar, P. A. E., Southall, R., et al. (2015). Life cycle energy and carbon assessment of double skin façades for office refurbishments. *Energy and Buildings, 109*, 143–156. https://doi.org/10.1016/j.enbuild.2015.09.051.

Pomponi, F., Piroozfar, P. A. E., & Farr, E. R. P. (2016). An investigation into GHG and non-GHG impacts of double skin façades in office refurbishments. *Journal of Industrial Ecology, 20*, 234–248. https://doi.org/10.1111/jiec.12368.

Pomponi, F., D'Amico, B., & Moncaster, A. (2017). A method to facilitate uncertainty analysis in LCAs of buildings. *Energies, 10*, 524. https://doi.org/10.3390/en10040524.

Prado-Lopez, V., Seager, T. P., Chester, M., et al. (2014). Stochastic multi-attribute analysis (SMAA) as an interpretation method for comparative life-cycle assessment (LCA). *International Journal of Life Cycle Assessment, 19*, 405–416. https://doi.org/10.1007/s11367-013-0641-x.

Prado-lopez, V., Wender, B. A., Seager, T. P., et al. (2015). Tradeoff evaluation improves a photovoltaic case study. *Journal of Industrial Ecology, 00*, 1–9. https://doi.org/10.1111/jiec.12292.

Qin, Y., & Suh, S. (2016). What distribution function do life cycle inventories follow? *International Journal of Life Cycle Assessment.* https://doi.org/10.1007/s11367-016-1224-4.

Ross, S., Evans, D., & Webber, M. (2002). How LCA studies deal with uncertainty. *International Journal of Life Cycle Assessment, 7*, 47–52. https://doi.org/10.1007/BF02978909.

Swiss Centre For Life Cycle Inventories. (2007). Ecoinvent database 2.2. Ecoinvent Cent. 2.0.

Tillman, A. M., Ekvall, T., Boumann, H., & Rydberg, T. (1994). Choice of system boundaries in life cycle assessment. *Journal of Cleaner Production, 2*, 21–29.

Udo de Haes, H. A. (1993). Applications of life cycle assessment: Expectations, drawbacks and perspectives. *Journal of Cleaner Production, 1*, 131–137. https://doi.org/10.1016/0959-6526(93)90002-S.

Vieira, P. S., & Horvath, A. (2008). Assessing the end-of-life impacts of buildings. *Environmental Science & Technology, 42*, 4663–4669. https://doi.org/10.1021/es0713451.

Wei, W., Larrey-Lassalle, P., Faure, T., et al. (2016). Using the reliability theory for assessing the decision confidence probability for comparative life cycle assessments. *Environmental Science & Technology, 50*, 2272–2280. https://doi.org/10.1021/acs.est.5b03683.

Weidema, B. P. (2000). Avoiding co-product allocation in life-cycle assessment. *Journal of Industrial Ecology, 4*, 11–33. https://doi.org/10.1162/108819800300106366.

Weidema, B., & Wesnæs, M. (1996). Data quality management for life cycle inventories – An example of using data quality indicators. *Journal of Cleaner Production, 4*, 167–174.

Zabalza Bribián, I., Valero Capilla, A., & Aranda Usón, A. (2011). Life cycle assessment of building materials: Comparative analysis of energy and environmental impacts and evaluation of the eco-efficiency improvement potential. *Building and Environment, 46*, 1133–1140. https://doi.org/10.1016/j.buildenv.2010.12.002.

Zhang, X., & Wang, F. (2015). Life-cycle assessment and control measures for carbon emissions of typical buildings in China. *Building and Environment, 86*, 89–97. https://doi.org/10.1016/j.buildenv.2015.01.003.

Chapter 2
Probabilistic Approaches to the Measurement of Embodied Carbon in Buildings

J. Gantner, W. Fawcett, and I. Ellingham

Introduction

Most decisions in everyday life and also in professional life must be taken on the basis of imperfect information. The lack of certainty is generally perceived as a deficiency. In the seventeenth century Rene Descartes described uncertainty as the antithesis of knowledge (Cottingham 2013). The ambition of early scientists was to replace ignorance and uncertainty with knowledge and certainty, but the development of mathematical, physical and social sciences the nineteenth and twentieth centuries showed that uncertainty is itself a phenomenon to be studied and understood (Bernstein 1996). A large body of theory and expertise has been built up in such fields as probability theory and decision-making under uncertainty.

Any attempt to measure the embodied carbon in buildings or building components encounters numerous issues of uncertainty. We do not find a simple dichotomy between exact knowledge of embodied carbon, or total ignorance. Instead, there is a spectrum that runs from precise or near-exact knowledge to total or near-total ignorance. Estimates of embodied carbon are rarely if ever at the extreme ends of the spectrum, but somewhere in between. The position in the spectrum is a significant attribute of our estimates and will have a bearing on how we use the estimates (Fig. 2.1).

If we treat uncertain estimates as if they were exact, we are throwing away valuable information that can help us make better decisions. To make full use of the knowledge that we have about embodied carbon, we should take account of the

J. Gantner
Department of Lifecycle Assessment, Fraunhofer Institute for Building Physics, Stuttgart, Germany

W. Fawcett (✉) · I. Ellingham
Cambridge Architectural Research Ltd, Cambridge, UK
e-mail: william.fawcett@carltd.com

© Springer International Publishing AG 2018
F. Pomponi et al. (eds.), *Embodied Carbon in Buildings*,
https://doi.org/10.1007/978-3-319-72796-7_2

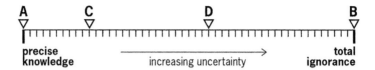

Fig. 2.1 Diagram of the spectrum from precise knowledge of embodied carbon to total ignorance. All estimates fall somewhere on the spectrum. In case A, the embodied carbon can be precisely measured; in case B, there is absolutely no information about embodied carbon. Cases C and D are more typical, with some uncertainty in the estimate for case C and considerably more for case D

degree of uncertainty in estimates. This chapter puts forward suggestions about how this can be done.

Sources of Uncertainty

Uncertainty manifests itself in different ways. It can include incomplete or contradictory information, as well as linguistic inaccuracies, variability, and errors. Three types of uncertainty can be identified (tagged with Donald Rumsfeld's aphorisms of 2002):

I. Uncertainty where known parameters follow a pattern of statistical randomness ('things we know we know')
II. Uncertainty where known parameters influence a system but their effect is not clear ('things we know we don't know')
III. Uncertainty from parameters that influence a system, but we are not even aware that they exist ('things we don't know we don't know').

Relative to the above three forms of uncertainty, this chapter focuses primarily on case I, where we have some knowledge of the statistical variation of the uncertain factors affecting embodied carbon. There is much less that can be done with case II in estimating embodied carbon, but for designers it can indicate where flexibility should be prioritised, to reduce the risk of damaging outcomes. Nothing can be said about the factors in case III, by definition.

In the context of construction projects, architects, planners and other decision-makers typically struggle with the following problems of uncertainty which affect all aspects of design, construction and management:

(A) Insufficient and incomplete information (e.g. lack of data on construction materials, geometry of the building, etc.)
(B) Inaccuracies in simulation models (e.g. energy simulation software, building utilisation forecasts, etc.)
(C) Unpredictability of future events (e.g. changes of use, duration of service life, technological changes, socio-economic changes, etc.).

Looking specifically at the challenge of measuring or estimating embodied carbon, uncertainties can be viewed under four headings (Gantner 2017a):

1. *Uncertainty about the current embodied carbon of construction materials, components and whole buildings*

   We know that every building component and construction process that makes up a current construction project must have a specific quantity of embodied carbon, but it is very difficult to measure. A building component, say, a steel beam, does not come with an embodied carbon history that would say what proportion was new or recycled steel, whether the steel plant was efficient or inefficient and whether it used energy from fossil or renewable sources, how far it travelled and by what means of transport, and so on. Instead of measuring the embodied carbon in that specific steel beam, we have to rely on estimates of the typical embodied carbon of similar steel beams taken from databases. Estimating is difficult when we see a specific steel beam, but even harder at design stage when a steel beam is specified long before its sourcing is known. As embodied carbon is given greater priority, we can expect better record keeping and larger and more complete databases, which will reduce the uncertainty in estimates. But there will always be uncertainty, which should be acknowledged and preferably quantified.

This uncertainty applies to estimates of the embodied carbon in today's construction projects. If the estimate of embodied carbon is extended to a life cycle view and includes not just the construction of buildings but also their life and end of life, the problems of uncertainty are amplified.

2. *Uncertainty about the future embodied carbon of construction materials and components, including technological innovation*

   It is certain that the extractive, industrial and transportation process that are the source of the embodied carbon in construction materials and components will change over time. Therefore identical materials and components will have a different embodied carbon measurement in the future compared to today. Given the current concern with environmental issues and the pressure for decarbonisation, it is possible that future embodied carbon will be lower, but the direction and amount of change will vary for different construction materials and components. Detailed predictions are impossible, but it may be possible to forecast general trends.

The following source of uncertainty applies to all aspects of construction when taking the life cycle view – including the measurement of embodied carbon.

3. *Uncertainty about future events in the service life of built assets, including length of component life, component replacement or substitution, changes of use, end of life*

   Because they are so long lasting, buildings are exposed to especially high uncertainties over their life cycle compared to other short-life manufactured products. During their service lives most buildings will be exposed to uncertain events such as changes in technology, society, regulations, energy sources, and changes in use. These factors can seriously compromise any analysis that projects the current situation into the future.

The final source of uncertainty applies whenever we wish to make a valid comparison between two measurements:

4. *Uncertainty about system boundaries and methods of measurement*
    The measurement of embodied carbon depends on the boundaries set for the system, and the ranking of alternatives may vary for different boundaries. Standards that define boundaries, such as EN 15804 (2012), EN 15978 (2012), ISO 14040 (2009), ISO 14044 (2006), the ILCD handbook (2010a, b, c, 2012), and EeBGuide (Gantner et al. 2015a, b), aim to ensure comparability between measurements. However, for particular cases there is always a question about what are actually the appropriate boundaries to use.

Representation of Uncertainty

Current Measurement

When considering a building or building component that has already been constructed or manufactured, it is reasonable to suppose that there is in fact an exact value for its embodied carbon, but in almost every case we do not know precisely what it is: our knowledge lies somewhere between exact knowledge and total ignorance. How can this uncertain knowledge be represented and quantified?

Estimates of embodied carbon are the outcome of a calculation process involving many variables; in a simple example, the embodied carbon in a steel beam is the product of two variables, the mass of the beam in kilograms, and the embodied carbon per kilogram of steel. The variables may have different uncertainties; here, the knowledge of the mass of the steel beam may be exact but knowledge of the embodied carbon per kilogram uncertain. The uncertainty in our knowledge of the embodied carbon in the beam derives from uncertainty in one or more of the underlying variables. Some of the variables may not be specifically carbon-related, for example, the mass of a building component or the distance travelled from factory gate to construction site, but uncertainty in these variables contributes to the overall uncertainty in the measurement of embodied carbon.

In considering the variables that are used in measuring embodied carbon, it is useful to distinguish between discrete (or category) variables, and continuous (or numerical) variables. In the discrete case the value sought is one of two or more mutually exclusive categories, but we are uncertain which applies; for example, we may be uncertain whether the aluminium in a window falls into the category 'new' or the category 'recycled'. Other variables can vary continuously; for example, the size of a window, or how many kilometres it is transported from the factory gate to the building site.

The representation of uncertain knowledge is simpler for discrete variables. Each of the categories is assigned a probability, such that the sum of probabilities for all categories is 1, or 100%. We might know, for example, that there is 90% probability that the aluminium in a window is recycled and 10% that it is new. This is not the same as saying that it is made of 90% recycled aluminium and 10% new aluminium – it is either one or the other. Often our knowledge about discrete variables is non-numerical; for example, we may know that 'most' or 'almost all' aluminium windows use recycled material. However, in order to quantify uncertain knowledge, we have to assign numerical probabilities to the categories. If we have no knowledge about which category is likely to be correct, it is usual to assign equal probabilities to all categories (Bernstein 1996, p. 50).

There are several ways of representing uncertain knowledge of continuous variables. The following list starts with the weakest (Fig. 2.2):

1. A single number that represents the best estimate; this is a single-point estimate – or the 'best guess'.
2. A range from the lowest to the highest possible values, with no indication of where the correct value lies within the range. This defines a uniform distribution between the lowest and highest possible values.
3. A three-point estimate (3PE) based on the lowest conceivable value, the most likely value and the highest conceivable value. This defines a triangular probability distribution. Three-point estimates are a good way of capturing the knowledge of experienced practitioners, who can offer their wisdom on the most likely, the lowest possible and the highest possible values for an uncertain variable.
4. An empirical distribution of data from comparable cases where the embodied carbon was measured. If the cases form a representative sample of the range of possible values, they can be used as a probability distribution.
5. A mathematical probability distribution, such as a binomial, normal or lognormal distribution, that describes the range of values that could occur and their probabilities. Such a distribution is usually calibrated with empirical data. This approach is only valid when the mathematical distribution truly reflects reality; for example, with a defined annual probability of component failure, expected component life follows a mathematical distribution. It can be tempting to use an arbitrary distribution for an uncertain variable for the sake of mathematical elegance; however, this achieves little and its spurious precision may be misleading.

Single-point estimates (case 1) are widely used for measurements of embodied carbon. They are simple and familiar, but they are the most basic representation and provide no information about the uncertainty surrounding the measurement. Perhaps uncertainty is ignored because few people know how to handle it and prefer to avoid the issue.

Some information sources on embodied carbon give ranges (case 2), for example some entries in the Bath database (Hammond and Jones 2011). Probability distributions are seldom used (cases 3–5), although three-point estimates are sometimes

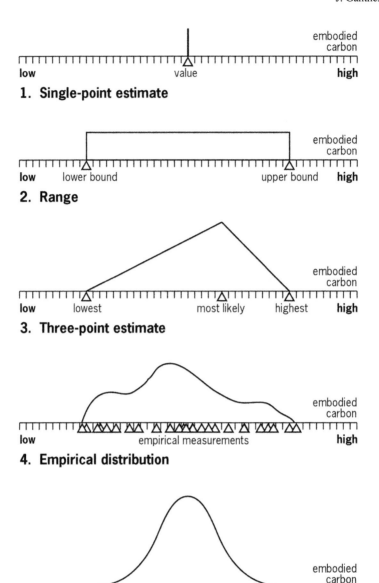

1. Single-point estimate

2. Range

3. Three-point estimate

4. Empirical distribution

5. Mathematical distribution

Fig. 2.2 Diagram showing five ways of representing the measurement of an uncertain variable, starting with the weakest

used for recording expert opinion (BCIS (Building Cost Information Service) 2006). With any of these representations an important parameter is the average value of the distribution, but by itself this is no more than a single-point estimate; it

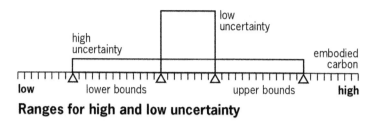

Ranges for high and low uncertainty

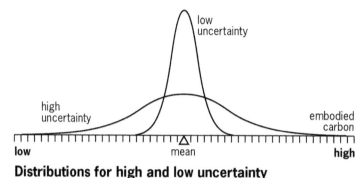

Distributions for high and low uncertainty

Fig. 2.3 Ranges (case 2) for low and high uncertainty (above); and mathematical distributions (case 5) for low and high uncertainty (below)

excludes information about uncertainty. A range (case 2) is narrow if there is a small amount of uncertainty, but wide if there is a lot of uncertainty; and probability distributions (cases 3–5) are narrow and spiky if there is a small amount of uncertainty, but wide and flat if there is high uncertainty. The shape of the distribution varies with the amount of uncertainty, even when the average is unchanged (Fig. 2.3).

The uncertainty of a distribution can be measured by its standard deviation, with larger values indicating higher uncertainty. Another approach is to slice the probability distribution into percentile segments, typically for the 10th, 20th, 30th, etc., percentiles. With low uncertainty, these percentile values are tightly clustered around the average, but with high uncertainty the percentile values diverge from the average (Table 2.1). Probability distributions can be visualised using probability density graphs (as in Fig. 2.3), or as cumulative probabilities or S-curves, making easier to read off percentile values (Fig. 2.4).

Life Cycle Measurement

So far, we have been concerned with the initial state of buildings or building components, where there is assumed to be an exact measurement for embodied carbon about which we have uncertain knowledge. Moving to a lifecycle view the

Table 2.1 Table showing units of embodied carbon at percentile intervals for four alternative materials with varying levels of uncertainty, all with an average value of 100 units

	Material A	Material B	Material C	Material D
	Level of uncertainty			
Percentiles	Very low	Low	High	Very high
100%	112	135	158	170
90%	106	119	132	145
80%	104	113	121	129
70%	103	108	113	118
60%	101	104	106	109
50%	100	100	100	100
40%	99	96	94	91
30%	97	92	87	82
20%	96	87	79	71
10%	94	81	68	55
0	88	65	42	19

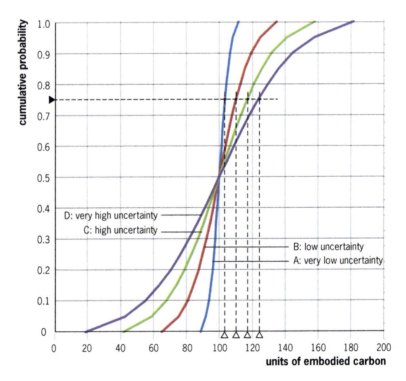

Fig. 2.4 Cumulative probability distribution for the embodied carbon of four alternative materials with varying levels of uncertainty (as described in Table 2.1), all with an average value of 100 units

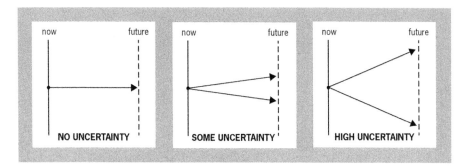

Fig. 2.5 Fans of uncertainty

situation is different, because the future embodied carbon from maintenance, refurbishment or replacement has not yet been incurred and so cannot have an exact value. In order to carry out life cycle evaluation we have to estimate this future embodied carbon.

Future estimates of embodied carbon are more uncertain than current measurements, because critical variables, such as the embodied carbon in a kilogram of steel, are likely to change over time from today's values (Ellingham and Fawcett 2006). This factor is often ignored and today's values are carried over into life cycle measurements; very little reliance can be placed on life cycle measurements of this type.

The range of possible future values for a variable can be visualised with a fan of uncertainty (Fig. 2.5). Time moves from left to right, with today on the left-hand side. The current value of a variable under consideration is a point on the left and the width of the fan represents the extent of possible future variation – the wider the fan the greater the expected variability over time and therefore the greater the uncertainty about future values.

Tree Representations

The fan of uncertainty can be quantified as a tree of possible future values branching from an initial value. Over time the variable will in fact follow a single path through the tree, but there are many possible paths and we do not know in advance which one will be followed. A binomial tree has two-way branches at each generation, typically a year apart, with probabilities (which may be equal or unequal) attached to the upward and downward branches (Fig. 2.6). At each generation, the range of possible values increases and the probability attached to each one diminishes. The tree establishes the probability distributions for the variable at each generation, which get progressively wider and flatter to reflect increasing uncertainty.

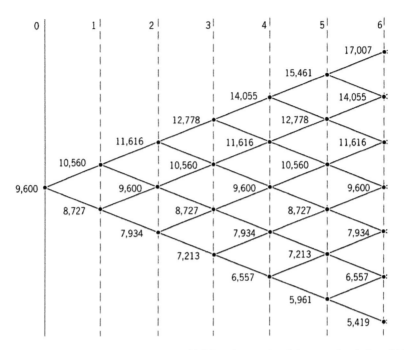

Fig. 2.6 Binomial tree with a starting value of 9600, and upward and downward variation of 10% per year; the first six generations are shown. The range of possible values increases at each generation

Monte Carlo Simulation

Tree representations work well for a single variable with consistent evolution from generation to generation. However, the measurement of embodied carbon usually involves a calculation with several variables. In this situation, a second method of estimating future values is more appropriate: Monte Carlo simulation (it can also be used for simple situations as an alternative to trees). With Monte Carlo simulation there can be many variables, each with its own probability distribution that can vary from generation to generation.

Monte Carlo simulation arrives at the probabilistic measurement of embodied carbon by running a large number of simulated scenarios, typically thousands. In each scenario, the variables that determine the measurement of embodied carbon begin at their known initial values and at each time step values are randomly sampled from the variables' pre-defined probability distributions; the embodied carbon for the scenario is calculated with these sampled values. Each scenario can be thought of as one path through a very large tree of future possibilities that is too large for all the branches to be enumerated. In aggregate, the values from many scenarios establish a probability distribution for the embodied carbon of the system being modelled. It is comparable to case (4) above (Fig. 2.2), a distribution formed from a representative sample of instances. As with a tree representation, there is increasing uncertainty at each generation.

Decision-Making Under Uncertainty

A primary use of measurements of the embodied carbon in buildings is for comparing alternative strategies or materials at design stage and deciding which to adopt. In the rare cases where there is exact knowledge of the embodied carbon of the alternatives, decision making is easy – select the alternative with the lowest embodied carbon.

When the measurements of embodied carbon are uncertain the situation changes, because the decision maker's subjective attitude to uncertainty comes into play. A consequence of subjectivity is that different decision makers may evaluate the same uncertain measurements differently, and choose different alternatives.

Two important factors in the subjective evaluation of uncertain data are risk aversion and time preference (Gollier 2001; Jordaan 2005). Risk aversion applies to both current measurements of embodied carbon and life cycle analysis, and requires a probabilistic approach. Time preference only applies to life cycle analysis, but is relevant to both deterministic and probabilistic approaches.

Risk Aversion

Most people are risk averse. Faced with two alternatives, one having a certain measurement of embodied carbon and the other having an uncertain measurement with an average value equal to the certain measurement, most people prefer the alternative with the certain measurement. This is a well-researched empirical fact, and can be rationalised by proposing that when an uncertain process is equally likely to produce favourable and unfavourable outcomes, people are more distressed by the unfavourable cases than they are delighted by the favourable ones (Kahneman and Tversky 2000, Chap. 1). This asymmetry gives more psychological weight to unfavourable outcomes, so the subjective balance between favourable and unfavourable outcomes is offset to the unfavourable side. The subjective balance point can be called the certainty equivalent of the uncertain process, and is less favourable than its average value.

When comparing alternatives that have uncertain measurements, it is the certainty equivalents that should be compared, not the average values. The impact of risk aversion increases with the level of uncertainty, widening in the offset between the average and the certainty equivalent. When comparing alternatives with different levels of uncertainty, a more uncertain alternative with a low average measurement could be rejected in favour of one with a higher average measurement but lower certainty equivalent (as in the example below).

Table 2.2 Percentile values for probabilistic distributions of the embodied carbon measurement for alternatives A with lower uncertainty and B with higher uncertainty

Percentile or confidence level	Alternative A (lower uncertainty)	Alternative B (higher uncertainty)
100%	810	940
90%	760	860
75%	710	755
50%	640	580
25%	535	400
10%	460	280
0	400	200

If we only know the average of an uncertain measurement of embodied carbon it is impossible to take account of risk aversion and unsatisfactory decisions may be made.

Risk aversion can be analysed with the percentile chart. The example compares alternatives with different levels of uncertainty in the measurement of embodied carbon (Table 2.2). Alternative A has lower uncertainty and Alternative B higher uncertainty. The average values at the 50th percentile favour Alternative B (580 compared to 640 for Alternative A). This would be the basis for decision making by someone who was risk neutral, i.e. equally willing to accept the risk of good or poor outcomes (or someone who had average values only). However, a risk averse decision maker is disproportionately influenced by the high embodied carbon cases and has a certainty equivalent that is offset from the average measurement. A way of quantifying the certainty equivalent offsets is to look at the percentiles above the risk-neutral 50th percentile. Suppose that a given decision maker's level of risk aversion is represented by the offset from the 50th to the 75th percentiles: then Alternative A is favoured (710 compared to 755 for Alternative B). The percentile offset to establish certainty equivalents varies for different decision makers in accordance with their level of risk aversion. When used in this way, the percentiles can be interpreted as confidence levels.

Most decision makers are not consciously aware of their level of risk aversion and cannot express it as a percentile confidence level. If they could, then it would be a purely mechanical operation to determine their preference by comparing uncertain alternatives at the appropriate confidence level. Because most decision makers are unable to quantify their level of risk aversion, it is more helpful to provide them with data about the alternatives and their uncertainties (as in Table 2.2), so that they can apply intuitive trade-offs in arriving at decisions.

Time Preference

In life cycle analysis, the measurements of embodied carbon are assembled for each time step in the study period, derived from a deterministic model, a probabilistic tree, or Monte Carlo simulation. The uncertainty in these measurements will increase at each time step.

The set of measurements for all time steps is unwieldy, and for decision making they must be integrated into a single life cycle measurement. The simplest way of doing this is to add together the measurements for all time steps, and this is the normal procedure in life cycle assessment (LCA). It gives equal weight to all measurements from the first to the last time steps, which may be decades apart. The simple addition approach is questionable because of time preference. Everyday experience and numerous experiments have shown that people attach more weight events that occur now or in the near future as compared to distant events, and this effect increases as events become more distant. It is hard to separate pure time preference from the increasing uncertainty of future events, but researchers have tried to do this and suggest that a typical rate of pure time preference is between 1% and 2% per year (Oxera Consulting 2002). The rate varies between decision makers: those who have a long-term view and attach considerable weight to events far in the future have a lower rate of time preference; others who have a short-term view and care more about current events or those in the near future have higher rate of time preference.

An example shows the application of a percentage rate of time preference to the life cycle embodied carbon measurements of two hypothetical alternatives over a 20-year study period (Table 2.3). The time preference ratio is applied at each year of the study period and the adjusted values are summed to give a present equivalent of the life cycle measurements. Without time preference, Alternative B has the lower value for cumulative embodied carbon, but when a time preference rate of 1.5% is applied the ranking reverses and Alternative A has the lower value. This is because the estimated embodied carbon in Alternative B is mostly incurred early in the study period and the values are relatively little affected by applying time preference; where in Alternative A the estimated embodied carbon is mostly incurred late in the study period and the values are much more affected by applying time preference. The application of time preference implies a short-term view, and the higher the rate of time preference the more pronounced this is.

The mathematical procedure for applying a percentage time preference ratio to life cycle estimates of embodied carbon is exactly the same as for discounting in financial life cycle costing, but the time preference ratio is normally much smaller, because financial discount rates include factors for uncertainty and rising wealth. Discounting is often rejected in environmental life cycle assessment because the future of the planet cannot be regarded as being of diminishing importance (Stern 2008, Chap. 2; Hellweg et al. 2003). But the purpose of life cycle analysis is not to make predictions (or judgments) about the future, but to make current decisions efficiently with the knowledge that is currently available. From this perspective, it

Table 2.3 An example of a time preference rate of 1.5% per year applied to the life cycle embodied carbon measurements for two alternatives over a 20-year study period. The cumulative life cycle measurements without and with time preference are shown. Without time preference Alternative B has the lower cumulative value, but with time preference Alternative A has the lower cumulative value

Year	Alternative A	Alternative B	Time Pref = 1.5%	Alt A * TP	Alt B * TP
1	85	270	100.0%	85	270
2	95	260	98.5%	94	256
3	105	250	97.0%	102	243
4	115	240	95.6%	110	229
5	125	230	94.1%	118	217
6	135	220	92.7%	125	204
7	145	210	91.3%	132	192
8	155	200	90.0%	139	180
9	165	190	88.6%	146	168
10	175	180	87.3%	153	157
11	185	170	86.0%	159	146
12	195	160	84.7%	165	135
13	205	150	83.4%	171	125
14	215	140	82.2%	177	115
15	225	130	80.9%	182	105
16	235	120	79.7%	187	96
17	245	110	78.5%	192	86
18	255	100	77.3%	197	77
19	265	90	76.2%	202	69
20	275	80	75.0%	206	60
Cumulative	3600	3500		3043	3131

is arguably inefficient to attach equal weight to measurements for the first time steps and last time steps in the study period, when the later estimates are much more uncertain.

Just as for risk aversion, time preference is subjective and varies between decision makers, but most decision makers are not consciously aware of their level of time preference and cannot express it as a percentage ratio. It can be helpful to present decision makers with information about the impact of the time preference ratio and allow them make intuitive judgments in arriving at their decisions.

Sensitivity Analysis

The measurement of embodied carbon involves calculations with many uncertain parameters. In a deterministic approach, fixed values are assigned to all the parameters, and in a probabilistic approach uncertainty is explicitly modelled for some of the parameters but for others fixed values are used. Sensitivity analysis is a way of investigating the uncertainty surrounding the fixed values. For chosen parameters, alternative fixed values are assigned in a systematic way and the measurement of embodied carbon repeated. Comparison of the alternative parameter values and the corresponding variation in the measurement of embodied carbon shows whether or not the parameter value is critical for the decision between alternatives, and whether there are threshold values. Threshold values simplify decision making; for example, suppose that locally-sourced material A is preferable to remotely-sourced material B if material B has to be transported more than 150 km, but material B is preferable if it is transported less than 150 km – with this threshold value it is not necessary to know the exact transport distance, only whether it is more or less than 150 km.

Sensitivity analysis gives decision makers a better understanding of the solution space of the decision they have to make. It increases confidence compared to simply accepting the output of black box calculations. A critical issue is whether changing the parameter values in sensitivity analysis alters the ranking of the alternatives being compared. In decision making the ranking of alternatives is often critical, rather than the actual measurements of the alternatives. Sensitivity analysis can help the decision maker understand which of the following categories applies to their decision:

- One of the alternatives is clearly superior to any others and can be selected with a high degree of confidence, without the need for applying more detailed analysis
- Two or more of the alternatives are so close that the analysis methods being used cannot establish a preferred solution; the selection can be made on the basis of other criteria
- Two or more alternatives are close and more detailed analysis is justifiable in order to identify the preferred alternative.

This understanding can avoid wasting time in detailed analysis when the first two cases apply.

Measurement and Decision Making

It is worth emphasising the difference between measuring embodied carbon and decision making on the basis of the measurements. Decision making involves the decision maker's subjective value system, and the ranking of alternatives with the same measurements can vary for different decision makers. The methods described in this section do not alter the measurements, but aim to map them onto the decision maker's value system to reveal the ranking of alternatives in that value system.

The objective is to make the optimal decision, taking account of the uncertainty of measurements and the decision maker's value system. However, making an optimal decision does not guarantee a good outcome. With uncertainty poor outcomes can occur even when good decisions have been made – the point is that good decisions maximise the likelihood of a good outcome without guaranteeing that it will occur. At the time of decision, it is impossible to know whether or not the outcome will turn out well or badly, because it depends on uncertain future circumstances. The outcome of a decision is only known with certainty in hindsight.

Flexible Strategies

Life cycle measurement of embodied carbon aims to help current decision makers maximise the expected embodied carbon performance over the life cycle or study period. However, during the life cycle new decisions will be taken, for example decisions about the replacement of components that have a shorter service life than the study period.

In conventional life cycle models it is assumed that the first decision is repeated like-for-like throughout the study period (Fig. 2.7a). This is unrealistic. At the time of component replacement there is a new decision and previously rejected alternatives can be reconsidered (Fig. 2.7b). Also, some alternatives will disappear during the service life, and new ones become available (Fig. 2.7c). Although future decisions cannot be predicted, they improve life cycle performance by responding to new information that is not available to today's decision makers. For example, in a particular project we may be confident that timber windows are a good decision today compared to, say, aluminium or plastic alternatives, but we cannot know whether like-for-like replacement with timber windows will be a good decision in 2050. But the decision makers in 2050 will have new information which they will use when choosing between alternatives. Decision making that uses new knowledge can be expected to improve life cycle performance compared to being locked into like-for-like decision making.

To maximise the benefit from future knowledge, current design strategies should incorporate many opportunities for future decision making – that is, they should be flexible. This is not a new idea, but recent developments transform it from an attractive but elusive idea into a precise and quantifiable attribute.

The new approach to modelling flexibility is based on real options ideas (Ellingham and Fawcett 2006; Mun 2006; CILECCTA 2013), and applies when future decision makers can choose between currently defined alternatives (case (b) in Fig. 2.7). The alternatives and the rules for deciding which to select have to be specified. The model uses Monte Carlo simulation to assign values to uncertain variables, and in each simulation run the most favourable of the defined alternative is chosen at each replacement cycle or decision point: flexibility is only used when it is advantageous (Fig. 2.8). Over multiple Monte Carlo simulation runs the probabilistic measurement of embodied carbon in a flexible strategy is established and can be compared to the measurement of other, non-flexible strategies. An example is given in Fawcett et al. (2012).

Some issues when modelling and measuring flexibility:

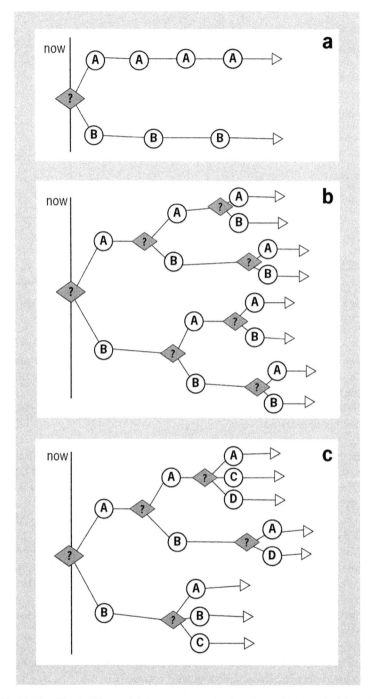

Fig. 2.7 (a) The like-for-like model for component replacement. (b) New decisions between alternatives can be made at each replacement cycle. (c) New alternatives will appear, and old alternatives disappear, when future replacement decisions are made

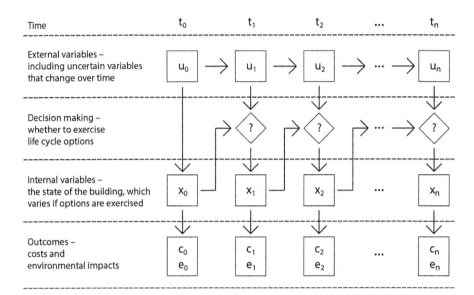

Fig. 2.8 In a life cycle model of a flexible strategy the alternatives are referred to as options, and changing from one alternative to another is referred to as exercising an option

- Because a flexible strategy adopts the most favourable alternative at each step in the life cycle, its life cycle measurement is likely to be superior to alternatives without flexibility. It is likely to avoid extreme outcomes and therefore have a probability distribution with a narrower spread and lower uncertainty than non-flexible alternatives.

- If there is an up-front embodied carbon penalty for a flexible strategy, the flexibility benefit over the life-cycle must be compared to the initial investment to determine whether it is should be adopted; for example, in deciding whether to construct a roof with separate framing, insulation and waterproofing elements, which is more expensive but provides greater life cycle flexibility than integrated panels combining structure, insulation and waterproofing.
- Flexibility cannot be modelled or measured with deterministic approaches to embodied carbon, only with a probabilistic approach. Ignoring flexibility makes life cycle measurement less realistic, because future decision makers will take advantage of flexibility even if it is ignored in a life cycle model.
- If we wish to reduce embodied carbon over the life cycle, not just measure it, we should prioritise flexibility in current design strategies.
- Flexibility can be quantified in many situations, but the principles can also be applied qualitatively – indeed 'flexible thinking' is perhaps a greater contribution to good decision making than quantification.

- Flexibility that allows buildings and infrastructure to respond to future climate change is now referred to as adaptation, and the approach described here is highly applicable to design for resilience.

Worked Example

Introduction

The worked example compares deterministic and probabilistic approaches to the measurement of embodied carbon in an innovative construction system employing structural steel modules. The modules can be disassembled at end of life and re-used in a follow-on building. The system involves some structural redundancy which brings an up-front embodied carbon penalty, but re-use of the steel modules gives embodied carbon benefits in follow-on buildings, potentially giving an overall reduction in embodied carbon. Given that the take-up of the system's re-usability is uncertain when the building is first erected, is the innovative system a good choice in embodied carbon terms?

This question is investigated by comparing the life cycle embodied carbon of a case study office building when built with either the innovative steel module system or conventional reinforced concrete construction. The data for the case study is based on a commercial research for the manufacturer of the steel module system (Gantner 2017b), but the probabilistic modelling is a research exercise by the authors. The case study office building has three floors of length 63.4 m and width 14.9 m, with an internal floor area of 2,593m^2. It is assumed that the secondary elements, finishes, etc., are identical for the two structural systems, as is the heat loss of 104 kWh/m^2/year, so the comparison of embodied carbon is restricted to the structure only. The measurement is on a life cycle basis, for a study period of 60 years.

Basis for Measurement

The measurement of embodied carbon in the worked example uses the method of life cycle assessment (LCA), which is a recognised basis for the quantification of the environmental impacts of processes, products or services, including embodied carbon. Standardised methods of LCA are defined in ISO 14040 (2009) and 14044 (2006), to ensure a consistent and transparent approach to evaluating environmental impacts, and the comparability of results.

System Boundaries and Assumptions

The following LCA lifecycle stages were considered (as defined in EN 15978 2012):

- A1–A3: Raw material supply of building materials, transport to the manufacturer and production of built-in building materials
- C+D: End of life and credits – the end of life of the building materials is modelled according to DGNB (German Green Building Council) (2009). For the end of life of the modular steel system the pattern of re-use is modelled.

Framing the Life Cycle Model

The LCA model is based on the following life cycle scenario.

- A new building is built in 2017, constructed using either the modular steel system or reinforced concrete.
- When the building reaches end of its life, the reinforced concrete building's structure is demolished and the steel building's structure is either demolished or its modules are re-used. Materials or components recovered at end of life are re-used in a follow-on building, which need not be on the same site as the one demolished or for the same purpose.
- If the first-generation building used reinforced concrete, the follow-on building also uses reinforced concrete. If the first-generation building used steel modules, either the follow-on building uses steel and incorporates the re-used modules, or if the modules are not re-used the follow-on building uses reinforced concrete.
- The same cycle repeats for successive generations, up to the end of the study period.
- When a follow-on building is constructed, its embodied carbon measurement does not use today's values but takes account of the expected decarbonisation of energy and construction processes up to the year when it is built. The measurement of embodied carbon credits for any materials or components that are re-used is also adjusted for decarbonisation up to the date of re-use, rather than using the measurements that applied when they were first produced.

The life cycle study period was 60 years and could cover more than one generation of building construction and replacement.

Data

Data values are given in Table 2.4, 2.5, and 2.6.

Table 2.4 Embodied carbon in construction, and rate of decarbonisation of construction

	Steel modules	Reinforced concrete
Embodied carbon for initial construction (kg CO_2e/m^2, 2017 values)	344.39	301.1
Embodied carbon at end-of-life if building demolished (kg CO_2e/m^2, 2017 values – negative values means that embodied carbon in follow-on building is reduced)	−1.06	0.58
Reduction in embodied carbon of follow-on building if steel modules are re-used at end-of-life of predecessor building (%)	50%	n/a
Expected rate of decarbonisation of construction up to 2030 (% per year)	1.50%	0.53%
Expected rate of decarbonisation of construction after 2030 (% per year)	0.50%	0.50%

Table 2.5 Service life

Expected service life for deterministic measurement	28 years
Expected service life for probabilistic measurement	Three-point estimate with values:
	Lowest conceivable value = 14 years,
	Most probable value = 28 years,
	Longest conceivable value = 44 years

Table 2.6 Re-use of steel modules

Probability of re-use of steel modules at end of service life, 2017 value	80%
Decline in probability of re-use of steel modules at end of service life, per year after 2017	0.5%

There are two uncertain factors in the worked example: first, the service life of the first generation building and follow-on buildings, represented by a three-point distribution; and second, the probability that steel modules will be re-used in a follow-on building, represented as a discrete variable for the probability of re-use. Limiting the uncertainty to two factors is a great simplification because many other factors in the example are in reality uncertain, but it allows the significant differences between deterministic and probabilistic measurement to be illustrated.

Method of Measurement: Deterministic Analysis

A deterministic calculation is relatively easy as all variables are assigned fixed values, some varying over time in a specified way, as for decarbonisation. For deterministic analysis, the service life was fixed at 28 years, approximately the average of the values generated by the three-point distribution; and the probability of re-use of the steel modules was handled by taking a weighted average of the embodied carbon for the two cases (re-use or no re-use) in accordance with the probability of re-use. The calculation moved through the 60-year study period, calculating the embodied carbon incurred by the steel and concrete alternatives in each year. These values were summed to give the life cycle measurement of embodied carbon, without any adjustment for time preference. Because the measurement was deterministic, it was not possible to apply any adjustment for risk aversion.

The deterministic approach gives single values for the life cycle measurement of embodied carbon in the steel module and reinforce concrete alternatives, with no information about the uncertainty surrounding the measurements.

Method of Measurement: Probabilistic Analysis

The probabilistic measurement of embodied carbon used Monte Carlo simulation with multiple runs. In each run, values for the two uncertain variables were sampled from the probability distributions described above, and with these values the life cycle embodied carbon of the two alternative construction systems were measured. Each Monte Carlo run generates a different result, and when the measurements from all the runs are assembled they give probability distributions for the life cycle embodied carbon of the two construction systems over the 60-year study period.

The Monte Carlo simulation was carried out for 1000 runs using Excel. In Monte Carlo simulations, the largest possible number of runs is desirable, but in this case 1000 runs provide a reasonably stable solution – more runs would give a smoother probability distribution, but might fall into the trap of spurious accuracy, when an apparently precise result obscures the fact that is in fact subject to uncertainty.

In each Monte Carlo run the two construction systems were modelled in parallel with the same sampled values for the service life of the original and follow-on buildings. This made the comparison of the structural systems easier by revealing how many of the runs favoured each alternative. This indicates the confidence that the decision-maker can have, depending on whether, for example, one alternative performed best in 95% of the runs or just 52% of the runs.

The base case was a 60-year study period, but the model was built with 100 annual time steps, so the results for other study periods could be examined.

The same calculation model was used for the deterministic calculation, with fixed values replacing the sampling of uncertain variables in probabilistic Monte Carlo simulation. In deterministic mode only one model run was needed.

Results: Deterministic Analysis

The results for deterministic analysis over a 60-year study period show that the steel module construction system has lower life cycle embodied carbon than the

Table 2.7 Results for deterministic analysis over the 60-year study period and other study periods, showing the aggregate embodied carbon for the alternative constuction systems

Study period (years)	25	40	50	60	100
Concrete embodied carbon (kg CO_2e/m^2)	301.1	579.2	579.2	820.9	1029.94
Steel embodied carbon (kg CO_2e/m^2)	344.4	527.2	527.2	723.0	909.8
Steel advantage	−14.4%	9.9%	9.9%	13.5%	13.2%

reinforced concrete alternative (Table 2.7). There is no information about the uncertainty surrounding these measurements.

This table also shows the deterministic results for different study periods. The 25-year study period is shorter than the deterministic building life, 28 years, so the building is never replaced and the reinforced concrete building has lower embodied carbon. The building is replaced once in the 40 and 50-year periods, twice in the 60-year period, and three times in the 100-year study period, and in all these cases the steel system has lower embodied carbon due to the re-use of the steel modules.

Results: Probabilistic Analysis

The average of the measurements of embodied carbon from the 1000 Monte Carlo runs over a 60-year study period also show that the steel module construction system has lower life cycle embodied carbon than the reinforced concrete alternative (Table 2.8). Information about the uncertainty surrounding these measurements is discussed below.

This table also shows the probabilistic results for different study periods, which are generally consistent with the deterministic results for the same study periods.

The table includes data about divergent outcomes, i.e. the percentage of 1000 Monte Carlo runs for which the construction system that was preferred overall was out-performed by the alternative. The figures are quite high – a quarter to a third of runs for 50 and 60-year study periods – showing that even a good decision may lead to a poor outcome. It shows that apparent certainty suggested by the deterministic measurements is not justified.

The probability distribution of results from the 1000 Monte Carlo runs for the 60-year study period only is shown in a histogram (Fig. 2.9). It is a complex distribution with multiple peaks, corresponding to the number of rebuildings in the 60-year Monte Carlo runs. In some runs there is one rebuilding; in others there are two, giving higher measurements for embodied carbon; and in a very few runs there are three rebuildings, with the highest measurements for embodied carbon. Looking at the clusters of results for one and two rebuildings, there is a wider spread of values for the steel system, depending on whether the modules are re-used or not, compared to the concrete system where the values are more tightly grouped. This shows that there is greater uncertainty about the measurement of embodied carbon for the steel module system.

Table 2.8 Results for probabilistic analysis, showing the average of the aggregate embodied carbon in the initial and follow-on buildings, for 1000 Monte Carlo runs

Study period (years)	25	40	50	60	100
Concrete embodied carbon (kg CO_2e/m^2)	384.4	558.6	615.9	710.3	979.2
Steel embodied carbon (kg CO_2e/m^2)	402.9	524.7	565.5	639.4	858.0
Steel advantage	−4.8%	6.5%	8.9%	11.1%	14.1%
Divergent outcomes	18.0%	41.0%	34.4%	27.2%	14.2%

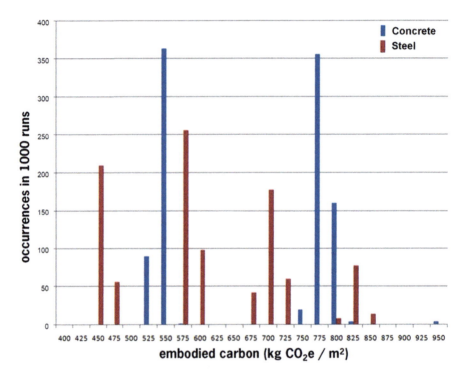

Fig. 2.9 Distribution of embodied carbon measurements for 1000 Monte Carlo runs over a 60-year study period, in 25 kg CO_2e/m^2 bands. The clusters of results correspond to Monte Carlo runs with one rebuilding with lower embodied carbon measurements (to the left), and two rebuildings with higher embodied carbon measurements (to the right)

Further information about the probabilistic results can be derived from the cumulative distribution of results over the 1000 Monte Carlo runs, when the embodied carbon measurements are sorted in order from the lowest to the highest (Fig. 2.10). The cumulative distribution for the concrete system has two distinct near-vertical segments, corresponding to the two peaks in the histogram. The line for the steel system is more continuous, corresponding to the wider range of values seen in the histogram. Although the steel system has greater uncertainty, only an extremely risk averse decision maker, requiring a confidence level over 90%, would prefer concrete.

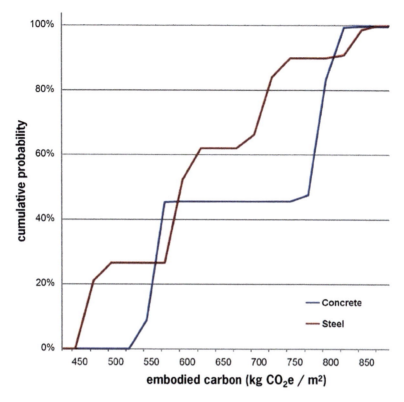

Fig. 2.10 Cumulative distribution of embodied carbon measurement for 1000 Monte Carlo runs over a 60-year study period

Sensitivity Analysis

The measurements given above are based on the input values in Tables 2.1, 2.2, and 2.3. Most of these values are in fact uncertain estimates. Sensitivity analysis with alternative values for key input parameters provides additional information. A single example is used for illustration: the probability of re-use of the steel modules. This is a critical parameter because it is the embodied carbon benefit from re-use that causes the steel module system to out-perform the concrete system. In the model, an initial probability of re-use of 80% is set, which declines by 0.5% per year (Table 2.7). The sensitivity chart shows the impact of re-running the Monte Carlo simulation using alternative values for the initial probability between 15% and 95%, with all other parameters in the model unchanged (Fig. 2.11). With a very low probability of re-use, the steel module system's embodied carbon benefit disappears and the concrete system is preferred.

The threshold value of about 40% is important: it is difficult to determine a precise value for the probability of re-use, but for decision making it is only

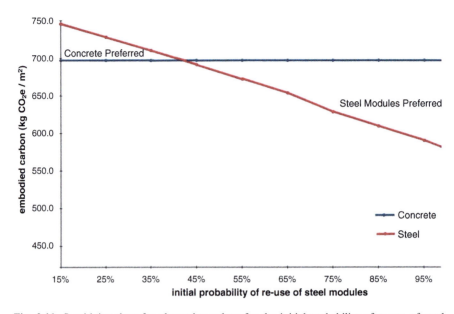

Fig. 2.11 Sensitivity chart for alternative values for the initial probability of re-use of steel modules between 15% and 95% – average of 1000 Monte Carlo runs over a 60-year study period for each value. The performance of the concrete system is unaffected by changes in the value of this parameter

necessary to determine whether it is above or below the threshold value, which is a much simpler task.

Sensitivity analysis is very useful for decision-makers, but the number of possible sensitivity analysis exercises is enormous. The focus should be on critical parameters.

Conclusions

The most fundamental argument for a probabilistic approach is that uncertainty is pervasive in the measurement of embodied carbon in buildings, and is accentuated in life cycle analysis. Uncertainty and its implications are overlooked in conventional deterministic measurement. The move to probabilistic analysis can build on techniques that have been developed in other disciplines for representing uncertainty and for decision making under uncertainty.

Probabilistic analysis provides additional quantified information about embodied carbon measurements. It can, in some cases, change the embodied carbon ranking of alternative strategies. As well as adding to quantified data, the insights from probabilistic analysis enhance the decision maker's depth of understanding of embodied carbon, allowing judgment to be exercised more effectively.

With probabilistic analysis, the impact of subjective risk aversion can be studied; it varies between decision makers and is an important factor in the choice between alternative strategies. Ignoring risk aversion detaches embodied carbon analysis from real world decision making.

Time preference, which is another important subjective factor in decision making, applies to life cycle measurement of embodied carbon in both deterministic and probabilistic approaches, but there are unresolved debates about the appropriate way of handling the time dimension in life cycle measurement of embodied carbon.

Compared to conventional deterministic approaches to the measurement of embodied carbon, probabilistic approaches do not make decision making *easier*. On the contrary, deterministic approaches offer an easy and superficially conclusive basis for embodied carbon measurement and decisions, but modelling the uncertainty that is prevalent in any study of embodied carbon will enhance understanding and should lead to *better* decisions.

Acknowledgments Software for probabilistic modelling of life cycle costing (LCC) and life cycle assessment (LCA), including sensitivity analysis and flexibility, was developed in the collaborative CILECCTA project (Construction Industry Life-cycle Costing and Assessment, 2009–13) (see CILECCTA 2013). It was funded by the European Community Programme FP7/2007–2013 for Research, Technological Development and Demonstration Activities under EC Grant Agreement no. 229061.

References

BCIS (Building Cost Information Service). (2006). *Life expectancy of building components: Surveyors' experiences of buildings in use*. London: Royal Institution of Chartered Surveyors.

Bernstein, P. L. (1996). *Against the gods: The remarkable story of risk*. New York: Wiley.

CILECCTA. (2013). Sustainability within the construction sector: CILECCTA – Life cycle costing and assessment (e-handbook for EC-funded collaborative research project).

Cottingham, J. (Ed.). (2013). *Descartes: Meditations on first philosophy*. Cambridge: Cambridge University Press.

Deutsche Gesellschaft für Nachhaltiges Bauen (DGNB). (2009). *DGNB Handbuch – Neubau Büro- und Verwaltungsgebäude (DGNB handbook – New construction of office and administration buildings)*. Stuttgart: Kohlhammer Druck.

Ellingham, I., & Fawcett, W. (2006). *New generation whole-life costing: Property and construction decision-making under uncertainty*. Oxford: Taylor & Francis.

EN 15804. (2012). *Sustainability of construction works – Environmental product declarations – Core rules for the product category of building products*. Berlin: Beuth Verlag.

EN 15978. (2012). *Sustainability of construction works – Assessment of environmental performance of buildings – Calculation method (German version EN 15978:2011 (2012))*. Berlin: Beuth Verlag.

Fawcett, W., Hughes, M., Krieg, H., Albrecht, S., & Vennström, A. (2012). Flexible strategies for long-term sustainability under uncertainty. *Building Research and Information, 40*(5), 545–557.

Gantner, J. (2017a). Wahrscheinlichkeitsbasierte Methode zum Erfassen von Unsicherheiten in der Ökobilanz bei Entscheidungen unter Berücksichtigung zukünftiger Ereignisse

(Probability-based methods of capturing uncertainty in life cycle assessment by considering future events in decision making). PhD thesis, University of Stuttgart (to be published).

Gantner, J. (2017b). *Ressourceneffizienz durch Wiederverwendung am Beispiel Modulbau* (Resource efficiency through reuse using the example of modular construction). Presentation at Bau 17. Munich.

Gantner, J., Wittstock, B., et al. (2015a). *EeBGuide guidance document – Part A: Products. Operational guidance for life cycle assessment studies in of the energy efficient buildings initiative.* Stuttgart: Fraunhofer Verlag.

Gantner, J., Wittstock, B., et al. (2015b). *EeBGuide guidance document – Part B: Buildings. Operational guidance for life cycle assessment studies in of the energy efficient buildings initiative.* Stuttgart: Fraunhofer Verlag.

Gollier, C. (2001). *Economics of risk and time*. Cambridge, MA: MIT Press.

Hammond, G., & Jones, C. (2011). *Embodied carbon: The inventory of carbon and energy*. Bath: University of Bath/BSRIA.

Hellweg, S., Hofstetter, T. B., & Hungerbuhler, K. (2003). Discounting and the environment. Should current impacts be weighted differently than impacts harming future generations? *International Journal of Life Cycle Assessment, 8*, 8–18.

International Reference Life Cycle Data System (ILCD). (2010a). *Handbook – General guide for life cycle assessment – Detailed guidance*. Luxembourg: European Commission.

International Reference Life Cycle Data System (ILCD). (2010b). *Handbook – Analysis of existing environmental impact assessment methodologies for use in life cycle assessment*. Luxembourg: European Commission.

International Reference Life Cycle Data System (ILCD). (2010c). *Handbook – Framework and requirements for life cycle impact assessment models and indicators*. Luxembourg: European Commission.

International Reference Life Cycle Data System (ILCD). (2012). *Handbook – Characterization factors of the ILCD, recommended life cycle impact assessment methods, database and supporting information*. Luxembourg: European Commission.

ISO 14040. (2009). *Environmental management – Life cycle assessment – Principles and framework*. Berlin: Beuth Verlag.

ISO 14044. (2006). *Environmental management – Life cycle assessment – Requirements and guidelines*. Berlin: Beuth Verlag.

Jordaan, I. (2005). *Decisions under uncertainty: Probabilistic analysis for engineering decisions*. New York: Cambridge University Press.

Kahneman, D., & Tversky, A. (Eds.). (2000). *Choices, values and frames*. New York: Cambridge University Press.

Mun, J. (2006). *Real options analysis* (2nd ed.). Hoboken: Wiley.

Oxera Consulting. (2002). *Social time preference rate for use in long-term discounting*. London: UK Government: Office of the Deputy Prime Minister and Department of Environment Food and Rural Affairs.

Stern, N. H. (2008). *The economics of climate change*. London: The Stationery Office.

Chapter 3
Uncertainty Assessment of Comparative Design Stage Embodied Carbon Assessments

S. Richardson, K. Hyde, and J. Connaughton

Introduction

Embodied carbon assessment is subject to theoretical, methodological and practical limitations. Emissions associated with a particular material or product may occur at different times, often decades and possibly even centuries apart. Emissions of all greenhouse gases must be converted into a common metric to allow comparison.

Furthermore, production processes are complex and subject to many variations including sources of raw material, production volumes, batching and so forth. Data are therefore often based on averages.

Many production processes result in two or more co-products, and emissions from the process must then be assigned to each product in some meaningful way to avoid double counting.

Unless the assessment is conducted retrospectively, at the end of the building life cycle, assumptions must be made about future life cycle stages in order to account for all relevant emissions. Practically, data collection can be very resource intensive, meaning simplifications may be necessary such as the use of secondary and proxy data or estimates, the exclusion from the study of certain parts of the building or stages in the building life cycle and the use of cut-off criteria to exclude those materials or processes deemed to contribute less than a specified threshold value.

The opportunity to reduce embodied carbon is greatest in the early stages of design when the precise nature and quantum of different material inputs are uncertain.

S. Richardson (✉)
TSBE Centre, University of Reading, Reading, UK
e-mail: stephentrichardson@googlemail.com

K. Hyde · J. Connaughton
School of Construction Management and Engineering, University of Reading, Reading, UK

© Springer International Publishing AG 2018   51
F. Pomponi et al. (eds.), *Embodied Carbon in Buildings*,
https://doi.org/10.1007/978-3-319-72796-7_3

Different studies deal with these limitations in different ways, which has led to the variability in results from different assessments of embodied carbon in buildings (Darby 2014; Dixit et al. 2010; Moncaster and Song 2012). In this work, which has been carried out in collaboration with Sainsbury's Supermarkets Ltd., we focus on the process analysis method of embodied carbon assessments. Yet similar problems of uncertainty also affect input-output and hybrid methods (Yang and Heijungs 2017), and it is anticipated that some aspects of the approach to addressing uncertainty that is proposed in this chapter may also be usefully applied to these.

One of main purposes of a design stage embodied carbon assessment is as a decision support tool. The method may be used by designers to identify which elements of the building are responsible for the largest contribution to embodied carbon. They may also evaluate the effect of adopting alternative design options on the embodied carbon of the building. Such assessments can then be used to support the choice of design alternatives to reduce embodied carbon. Uncertainty about the results of the embodied carbon assessment leads to uncertainty in such decision-making processes.

It is important to recognize that uncertainty is subjective and is viewed differently by people who have different roles (Heijungs and Huijbregts 2004). When considering uncertainty assessment methods, it is therefore important to determine the context in which uncertainty is being considered. For Sainsbury's the rationale for undertaking this work was primarily to support their efforts to reduce embodied carbon. Therefore, it is the effects of uncertainty on decision-making in relation to embodied carbon reduction measures that might be considered for supermarkets that were of primary interest. The uncertainty assessment methods that were developed in this research were applied to a case study of two design alternatives for an element of a supermarket building, where one design was expected to reduce embodied carbon. Taking this approach allowed the scope of the uncertainty assessment to be limited to just the materials relevant to the case study.

Uncertainty assessment is an established technique which 'identifies, analyses and manages uncertainties' (Skinner 2012). Uncertainty assessment has been developed and applied in a range of different fields including environmental modelling, of which embodied carbon assessment is a form. A model 'represents a compromise between desired functionality, plausibility, and tractability' (Walker et al. 2003). It is this process of compromise and simplification of reality that leads to uncertainty about the outcomes. Uncertainty in the outcomes of the model is one consequence of uncertainty in the model formulation.

Review of Uncertainty Literature

In the context of environmental modelling, a number of uncertainty typologies have been proposed to distinguish between uncertainties with different characteristics. Walker et al. (2003) distinguish different uncertainties using a conceptual framework, which the authors refer to as the three dimensions of uncertainty (Fig. 3.1). These dimensions are the location, level and nature. *Location* specifies whether

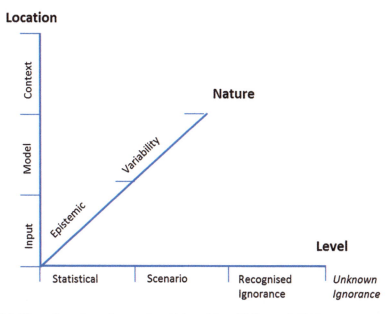

Fig. 3.1 Three dimensions of uncertainty (Adapted from Walker et al. 2003)

uncertainty is in the input data (input uncertainty), in the model's relationships or equations (model uncertainty) or in the assumptions and choices on which the model is based (context uncertainty). The *level* of uncertainty is a scale which is intended to codify how certain or uncertain something is. The *level* of uncertainty therefore ranges from determinism or absolute certainty at one end of the scale to absolute uncertainty at the other. Between these extremes, Walker et al. differentiate between statistical uncertainty, where probabilities can be estimated; scenario uncertainty, where different outcomes are anticipated but the likelihood or probability of each is not known; and recognized ignorance, or 'known unknowns' where 'the scientific basis for developing scenarios is weak' (Walker et al. 2003). The *nature* dimension further distinguishes between epistemic uncertainties that can be reduced through improved knowledge and uncertainties that are due to natural variabilities of the system to be modelled. This typology was found to be widely cited and applied in the literature. Its main advantage over other typologies which have been proposed is that it recognizes that uncertainties may have multiple characteristics. For example, uncertainty in a data input may be either epistemic or variable in nature and at the same time be relatively more or less uncertain than other data. Sources of uncertainty can thus be categorized in matrix form according to each of the dimensions.

Ross et al. (2002) suggest that uncertainty has been widely discussed within the field of LCA since the late 1990s, linked to its inclusion in the ISO 14040 series standards. Within ISO 14040 there is a recommendation to consider uncertainty effects, a recommendation which becomes mandatory if the results are to be published for the purpose of comparative assertions (British Standards Institution

2006a, b). Despite this, in their review of 30 LCA reports published between 1997 and 2002, Ross et al. (2002) found that only 14 of these discussed uncertainties in the results. Of these, only three conducted an uncertainty assessment, either qualitative or quantitative.

In a more recent review, Lloyd and Ries (2007) found that, from a selection of 100 studies identified in an initial screening of online publications, just 24 journal or conference papers included a quantitative assessment of uncertainty in a life cycle impact category. They point to a lack of available information for characterizing uncertainty as being one key reason why it is not more widely applied.

In contrast to the ISO 14040 LCA standards, the more recently developed EU standards for assessing the environmental sustainability of buildings (EN 15978) and construction products (EN 15804) make no reference to uncertainty at all (British Standards Institution, 2010, 2011, 2014a). It is therefore unsurprising that in published LCA and embodied carbon studies of buildings, little attention has been paid to the effects of uncertainty on results (Pomponi et al. 2017; Pomponi and Moncaster 2016; Cousins-Jenvey et al. 2014).

In order to ascertain the extent to which uncertainty has been addressed in the field of building LCA and embodied carbon assessments and the methods that have been applied, a review following the approach of Ross et al. (2002) was conducted. The starting point for this analysis was to collate a list of all papers cited in three relevant literature reviews (Sartori and Hestnes 2007; Ramesh et al. 2010; Cabeza et al. 2014), and these were supplemented with further searches in online journal catalogues for relevant keywords.

After initial selection, papers were divided into four types of study depending on whether they were an assessment of a whole building or single component or building element and those that were full LCA and those which were life cycle or embodied carbon assessments. Each paper was then searched for a number of terms that might indicate consideration of uncertainty. The terms were *uncertain* (including singular and plural references and uncertainty), *sensitivity*[1] (and sensitive), *data quality*, *error* and *variability* (and variation, variance, etc.). These search terms were chosen based on a preliminary review of papers that were known to address uncertainty either explicitly or implicitly. In this preliminary sample, these were the terms most frequently used in the discussion of issues relevant to uncertainty. A summary of the results of this review process is shown in Table 3.1.

Less than one quarter of the studies reviewed were found to contain a quantitative or qualitative assessment of uncertainty.

---

[1] Saltelli and Annoni (2010) define uncertainty analysis (or assessment) and sensitivity analysis in terms of the questions they seek to answer. They suggest that the former seeks to answer 'How uncertain is this inference?', whilst the latter aims to determine 'Where is this uncertainty coming from?' (p. 1508). However, they acknowledge that the two terms are sometimes used interchangeably to encompass either or both objectives.

Table 3.1 Results of the systematic literature review showing which of 63 peer-reviewed LCA or embodied/life cycle carbon assessment of buildings or building elements/materials included reference to uncertainty

Type of study	Total positive search results[a]	Search terms[b]						No relevant use of search term[c]
		Uncertainty[d]	Sensitivity[d]	Data quality	Error	Variation[d]		
Building LCA	10	9	10	4	5	5		9
Building embodied/life cycle carbon assessment	16	4	5	4	8	9		4
Material/component LCA	8	5	5	4	3	6		12
Material/component embodied/life cycle carbon assessment	0	0	0	0	0	0		1
Totals	34	18	20	12	16	20		26

[a]Indicates number of studies of each type which contained relevant use of one or more of the search terms
[b]An asterisk denotes that alternative variations of the search terms such as singular/plural and adjectives/nouns were included in the search
[c]A relevant use was one which related to uncertainty in the inputs or results of the study. Only the error and variation
[d]Search terms yielded nonrelevant instances in 4 of the 33 papers listed here as having no relevant use

Identifying Relevant Sources of Uncertainty

The first step in the uncertainty assessment method is to identify sources of uncertainty. This was undertaken through a systematic literature review. The sources of uncertainty identified were categorized using Walker et al.'s (2003) uncertainty matrix. The uncertainty matrix including examples of sources of uncertainty identified in the literature on embodied carbon is reproduced in Fig. 3.2. Note that *recognized ignorance*, a level of uncertainty defined in Walker et al.'s matrix, has been excluded here. By definition, recognized ignorance is something that cannot be evaluated or quantified directly. However, Funtowicz and Ravetz (1990) proposed that evaluating data quality can provide an indication of which parts of a model are most subject to recognized ignorance. Whilst this was undertaken as part of the wider research here, it is beyond the scope of the current study. So too is a full discussion of all the sources of uncertainty listed in Fig. 3.2. However, those sources that are relevant to the case studies assessed here are discussed in greater detail in the results section.

It is important to note that there are multiple sources of uncertainty at both the statistical and scenario levels. Sources of statistical uncertainty are defined as those for which quantitative ranges can be determined (Walker et al. 2003; Warmink et al. 2010; Refsgaard et al. 2007). All the sources of statistical uncertainty identified here are located in the model inputs because they relate to uncertainties in carbon factors or material and energy quantities or both. Scenario uncertainties are characterized

		Level *Nature*			
		Statistical		Scenario	
		Epistemic	*Variability*	*Epistemic*	*Variability*
Location	Model Context			• Allocation rules • System boundary • Time horizon of emissions • Level of data aggregation • Specification • Other methodological choices	
	Model Structure			• Functional unit	
	Model Inputs	• Measurement error • Transport • Energy Mix	• Process variation • Transport • Service life length • Maintenance / Refurbishment Frequency • Energy Mix • End of Life processing	• Geographic representative-ness • End of life processing	

Fig. 3.2 Uncertainty matrix (After (Walker et al. 2003)) categorizing sources of uncertainty in embodied carbon assessments according to their nature, level and location

by multiple alternative options where the likelihood of each occurring is unknown or where the options are not subject to probabilities but represent subjective choices (Skinner 2012).

Method

Van Asset and Rotmans (2002) hold that 'adequate uncertainty treatment implies that uncertainties salient for the decision-making process are identified, characterized and communicated'.

These three steps are reflected in the approach promoted by Heijungs and Huijbregts (2004) and thus form the basis of the approach applied here. Firstly, potential sources of uncertainty were identified and categorized using the uncertainty matrix. The identification step is important for selecting an appropriate method for the subsequent characterization of uncertainty, since different methods have strengths and weaknesses in terms of the types of uncertainty that they can account for (van der Sluijs et al. 2004). The sources of uncertainty identified indicate that a combination of quantitative and qualitative techniques is most appropriate. The final step, communication, takes the form of a case study, chosen to illustrate the impact that uncertainty can have on design stage decisions for reducing embodied carbon.

The Dutch National Institute for Public Health and the Environment (RIVM) has produced a tool catalogue for uncertainty assessment (van der Sluijs et al. 2004). The guide describes different uncertainty assessment methods in detail and also provides an indication of which types of uncertainty the different methods are most suited to dealing with. A similar set of recommended methods can be found in Refsgaard et al. (2007), Table 3.5.

Two methods that have wide applicability for a range of different types and levels of uncertainty are scenario analysis and expert elicitation (van der Sluijs et al. 2004). In a scenario analysis, the different scenarios modelled are defined qualitatively rather than according to probabilities or statistical ranges (Walker et al. 2003). Similarly, expert elicitation is typically used where system complexity or a lack of data means that quantitative uncertainty ranges cannot be determined statistically – a situation often encountered in embodied carbon assessments (Pomponi et al. 2017). Expert judgements may be elicited on both quantitative and qualitative aspects of uncertainty.

Step 1: Identification

- Literature review
- Uncertainty typology/uncertainty matrix

Step 2: Characterization

- Expert elicitation
- Scenario analysis

Step 3: Communication

- Case studies
- Scenario results for case studies

Scope of the Uncertainty Assessment

The context for this study is the use of embodied carbon assessment to support design decisions taken to reduce embodied carbon. A case study comparing the embodied carbon of steel and glulam design alternatives for the structural frame of a supermarket is used to illustrate the uncertainty assessment. As such, the scope of the uncertainty assessment is restricted to the materials relevant for this case study example. Material substitution is a commonly cited approach to reducing the embodied carbon of buildings (Pomponi and Moncaster 2016), and the case study used here represents a typical example of such a material substitution which has been employed on past supermarket buildings in the UK. To further simplify the analysis, the scope of the embodied carbon assessments for the two alternatives was limited to the building frame. Qualitative and quantitative data on uncertainty, obtained through expert elicitation, were then applied to the embodied carbon assessments of both design alternatives in order to model the effect that uncertainty has on each and, in particular, on the predicted reduction in embodied carbon that can be achieved by choosing one design over another.

The sources of uncertainty identified in the uncertainty matrix were assessed qualitatively and quantitatively for steel and timber through expert elicitation and subsequent scenario analysis.

Eliciting Expert Judgements of Uncertainty for Steel and Timber

Expertise 'refers to performance in a particular domain [...] that is superior to the performance of a number of other people within that same domain' (Schvaneveldt et al. 1985). It is this expertise, rather than representativeness or sample size, that supports the validity and defensibility of the outcomes of expert elicitation. Hoffman states that 'the accumulation of skill based on experience and practice are key' to the development of expertise. The criteria for selecting experts were adapted from Cooke and Goossens (1999) and included:

- Reputation in the field of interest
- Experience in the field of interest
- Number of publications in the field of interest
- Relevant qualifications

- Diversity in background
- Interest and availability to participate

The elicitations conducted for this study were based on a formal protocol proposed by Slottje et al. (2008) and further developed by Knol et al. (2010). In addition to the requirements of this protocol, experts were asked to consider scenario and statistical uncertainties separately. The objectives of the elicitations were:

- To elicit qualitative judgements of the types of scenario uncertainty affecting specific materials
- To elicit quantitative judgements of the statistical uncertainty affecting specific materials

Different experts were consulted for the two materials, steel and glulam, on the basis of their field of expertise, particularly their familiarity with the production processes for a given material. The results presented here are based on an initial group elicitation conducted with ten experts and a further five individual elicitations, three assessing steel and two assessing timber.

The sources of scenario uncertainty specific to each material identified during the expert elicitations are represented in the form of a decision tree diagram. This uncertainty is then propagated through the embodied carbon model using scenario analysis. The total number of scenarios for the scenario analysis of each material is determined by the number of possible paths through the decision tree. This is shown beneath each decision tree diagram in the subsequent sections. Each scenario represents one possible combination of assumptions and choices that affect the embodied carbon of that material. Together, these scenarios represent the scenario uncertainty for that material.

For each material, carbon factors were then sought or derived from published sources to match each of the scenarios represented by the decision trees. Wherever possible, the data were taken directly or adapted from relevant environmental product declarations (EPDs). Where EPDs were not available, data from other sources such as industry bodies, national carbon databases or academic studies were used. By identifying or deriving a carbon factor which corresponds to each set of assumptions and choices, it was then possible to calculate the resultant embodied carbon for the case study material for all scenarios. In this way, the scenario uncertainties are propagated through the embodied carbon model so that their effect on the results can be evaluated.

After assessing scenario uncertainties, each expert was then asked to estimate the effect of statistical uncertainties using an iterative procedure, based on an approach known as the interval method (Slottje et al. 2008). The results of this part of the elicitation procedure for all materials are displayed in Table 3.2 to allow comparison. The results for each individual material and the combined impact of both statistical and scenario uncertainties on the carbon factors for that material are then presented.

Table 3.2 Estimates of the statistical uncertainty ranges for embodied carbon factors of selected construction materials elicited by semi-structured interview

Material	Statistical uncertainty estimate (%)	
	Lower bound	Upper bound
Steel	−15	30
Timber	−15	20

Elicited Statistical Uncertainties

The elicited statistical uncertainty ranges are reported as upper and lower bounds in Table 3.2. Where estimates differed between experts, the values with the greatest magnitude or modulus have been used.

There is limited possibility to validate the experts' estimates since few uncertainty assessments have been undertaken for embodied carbon and no examples have been found that treat scenario and statistical uncertainty separately, as has been done here. The Inventory of Carbon and Energy (ICE) (Hammond and Jones 2011), which is one of the most comprehensive reviews of embodied energy and carbon factors published to date, includes indicative uncertainty ranges for some materials. However, the ranges given are for embodied energy rather than embodied carbon and do not separate scenario and statistical uncertainty. The ranges from the ICE data are all higher than those elicited for this research. This would be expected given that the ICE ranges are expected to include the effects of both statistical and scenario uncertainties. For timber, the statistical uncertainty ranges estimated in this research were compared to those presented by Rüter and Diederichs (2012) for the variability in GWP of different types of timber. Since their ranges are based on their own inventory data and results, it may be assumed that common assumptions and methods have been used. If this is valid, then the ranges they present would represent sources of statistical uncertainty and would exclude most sources of scenario uncertainties. The uncertainty ranges in Rüter and Diederichs (2012) were found to be greater than the ranges elicited here which suggests that the values in Table 3.2 may underestimate the actual variation due to statistical uncertainty. These should therefore be seen as preliminary estimates, and this serves to highlight the need for greater availability and transparency of uncertainty data for the embodied carbon factors of different materials.

Elicited Scenario Uncertainties

Steel

Steel is manufactured in one of three ways, the basic oxygen furnace (BOF) route, the electric arc furnace (EAF) route and the open hearth furnace (OHF) route. The OHF route accounted for just 1.3% of production in 2010 and continues to decline due to poor economics compared to the other two routes (World Steel Association

2011). Countries that are known to still have operating OHF plants represent less than 3% of total UK steel imports (UK Steel 2016). Therefore, the OHF production route has also been excluded from the analysis conducted here.

Scrap steel is recycled via both the BOF and EAF routes, however, whilst scrap steel content in the BOF process is generally between 10% and 30% (World Steel Association 2011, p. 72); the EAF process typically has much higher scrap content, up to 100% (Darby 2014). This means that energy consumption per tonne of EAF steel is much lower than for BOF steel since the additional processing stage to reduce the iron ore is either not required (in the case of 100% scrap content) or is greatly reduced. Carbon emissions published for the EAF route are typically 60–85% lower than for the BOF route (Darby 2014, Table 3–32, 31 Material Profile: Steel).

Three sources of scenario uncertainty were identified:

- System boundary with respect to the treatment of recycling (here the options are the recycled content approach, which gives credit to recycled material inputs, or the system expansion approach, which credits recyclability based on the net scrap, that is, the percentage scrap recovered at end of life minus the scrap input)
- Level of data aggregation with respect to different production routes for steel
- Geographic representativeness with respect to EU-produced steel or steel from non-EU countries (rest of the world)

Through the expert elicitations, the possible alternative choices or assumptions relating to each of these were identified, as shown in the decision tree in Fig. 3.3, and lead to a total of 21 possible scenarios for the embodied carbon factor of steel.

Since the different scenarios relate to the product stages of the life cycle, it was necessary to identify carbon factors for production of steel sections and sheet (module A1 according to BS EN 15804 (British Standards Institution 2014a)) and transportation to the fabricator or secondary manufacturer in the UK (module A2). The emissions from fabrication (A3), transport to site (A4), construction (A5) and demolition and disposal (C1–4) are the same for all scenarios.

These carbon factors are then combined with the experts' assessments of statistical uncertainty presented in Table 3.2. This gives an upper bound of +30% and a lower bound of −15%.

Table 3.3 presents the carbon factors for the supply of un-fabricated structural sections (module A1) derived for each of the 21 scenarios along with details of which sources were used. The carbon factors for the remaining modules A2–4 and C1–4 are presented in Table 3.4.

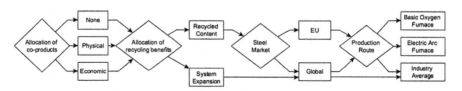

Fig. 3.3 Decision tree representing the methodological choices and assumptions that give rise to scenario uncertainty in embodied carbon factors for structural steel

Table 3.3 Carbon factors for production of structural steel for each of the 21 scenarios identified through expert elicitation of uncertainty and depicted in Fig. 3.3

Scenario	Allocation to coproducts	System boundary (allocation of recycling)	Level of aggregation	Production route	Module A1 carbon factor (kgCO$_2$e/kg)	Sources[a]
Steel-1	None	Recycled content	EU	BOF[b]	2.652	Ruukki (2014)
Steel-2				EAF[c]	0.687	Ruukki (2014), Institut Bauen and Umwelt e.V. (2014)
Steel-3				Avg.	1.886	Ruukki (2014), Institut Bauen and Umwelt e.V. (2013)
Steel-4			Global	BOF	As EU	
Steel-5				EAF	As EU	
Steel-6				Avg.	2.331	Darby (2014)
Steel-7		System expansion	Global	Avg.	1.230	Ruukki (2014), European Commission Joint Research Centre (2013)
Steel-8	Physical	Recycled content	EU	BOF	2.440	Ruukki (2014)
Steel-9				EAF	0.632	Institut Bauen and Umwelt e.V. (2014)
Steel-10				Avg.	1.735	Institut Bauen and Umwelt e.V. (2013)
Steel-11			Global	BOF	As EU	
Steel-12				EAF	As EU	
Steel-13				Avg.	2.030	Darby (2014)
Steel-14		System expansion	Global	Avg.	1.130	(European Commission Joint Research Centre (2013)
Steel-15	Economic	Recycled content	EU	BOF	2.541	Darby (2014), Ruukki (2014)
Steel-16				EAF	0.653	Darby (2014), Ruukki (2014), Institut Bauen and

(continued)

Table 3.3 (continued)

Scenario	Allocation to coproducts	System boundary (allocation of recycling)	Level of aggregation	Production route	Module A1 carbon factor (kgCO₂e/kg)	Sources[a]
						Umwelt e.V. (2014)
Steel-17				Avg.	1.807	Darby (2014), Ruukki (2014), Institut Bauen and Umwelt e.V. (2013)
Steel-18			Global	BOF	As EU	
Steel-19				EAF	As EU	
Steel-20				Avg.	2.233	Darby (2014)
Steel-21		System expansion	Global	Avg.	1.177	Darby (2014), Ruukki (2014), European Commission Joint Research Centre (2013)

[a]Where multiple sources are listed, data have been combined to derive a carbon factor matching the relevant set of assumptions or choices
[b]BOF: basic oxygen furnace route for steel production
[c]EAF: electric arc furnace route for steel production

Table 3.4 Carbon factors selected for transport to the fabricator (module A2), fabrication (A3), transport to site (A4) and end-of-life processing (C1–4) for structural steel

Module	Carbon factor (kgCO₂e/kg)	Sources
A2	EU: 0.042	These figures represent a weighted average of transport emissions for steel sourced from the EU and steel sourced globally. They are based on factors originally derived in Darby (2014) but using updated industry data for UK steel imports (UK Steel 2016) and the most recent UK emissions factors for road and sea transportation (Department for Environment, Food and Rural Affairs 2015)
	Global: 0.107	
A3	0.450	Darby (2014)
A4	0.0146	Based on the average transport distance for UK metals in 2015 (Department for Transport 2016) and the average emissions per tonne/km of HGV freight in the UK (Department for Environment, Food and Rural Affairs 2015)
C1–4	0.06	Steel Construction Institute End of life LCA and embodied carbon data for common framing materials (n.d.)

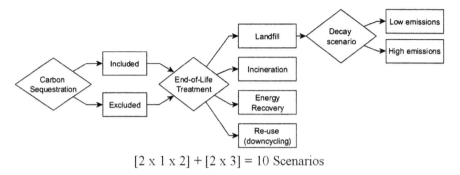

$$[2 \times 1 \times 2] + [2 \times 3] = 10 \text{ Scenarios}$$

Fig. 3.4 Decision tree depicting the main sources of scenario uncertainty for the embodied carbon of timber

Timber

Timber is used in construction in a wide variety of applications. In the material comparison case study, its use is in the form of glued laminated (glulam) timber, an engineered timber typically used for structural applications.

The sources of scenario uncertainty identified for timber depicted in Fig. 3.4 give a total of ten scenarios to be modelled. These comprise the methodological choice of whether to include the carbon sequestered during tree growth and the end-of-life processing options. For the end of life, four options of landfill, incineration, energy recovery and reuse in lower-grade timber or *down-cycling* were identified. For landfill, a further decision node was included to reflect different assumptions about the rate of decay of timber once sent to landfill. The rate of decay is measured in terms of the percentage of carbon which is released as either CO_2 or methane, and this is known as the degradable organic carbon fraction (DOCf) (Hogg et al. 2011). The two scenarios included here are for a low decay rate (DOCf = 0.1%) and a high decay rate (DOCf = 38.5%), and these have been selected based on available data (Wood for Good 2013a).

Since the different scenarios relate to the product stages and the end-of-life stage, carbon factors were derived for modules A1–3, modules C1–4 and module D. These carbon factors are then combined with the experts' assessments of statistical uncertainty presented in Table 3.2.

Case Study: Steel Vs. Glulam Structural Frame

This case study compares the embodied carbon of a glulam timber frame with an alternative steel design. The design data were provided by Sainsbury's. In order to simplify the analysis, only a sub-section of the glulam timber frame (as depicted in Fig. 3.5) for the sales area is assessed and compared to a steel frame design for a sub-section of the same dimensions.

Height to
beams:
7.200 m

24.000 m

18.000 m

Fig. 3.5 Sub-section of the structural frame for a supermarket for which the embodied carbon of two alternative materials is assessed in this case study

Glulam has been used for the structural frame of an increasing number of supermarkets and large retail stores in the UK. It is often associated with flagship projects intended to showcase environmental sustainability (Ijeh 2014). In an unpublished assessment of Sainsbury's Dartmouth store, the glulam frame was estimated to have 57% less embodied carbon than a steel alternative, leading to a 27% saving on the total embodied carbon for the building (dCarbon8 2008). Therefore, this case study illustrates how uncertainty affects the estimation of potential embodied carbon savings that are achievable by using glulam as an alternative material to steel. The quantities of steel and glulam in each design are shown in Table 3.6.

Scope and Boundaries

In this case study, it was assumed that neither of the structural frames requires maintenance or replacement during the life cycle of the store, which is modelled as 25 years (see Richardson et al. (2014) for justification of this assumption). Therefore, modules B1–5, relating to recurring embodied carbon during the use phase, have been omitted from the comparison. In practice, either frame option may require some maintenance in the form of paint or preservative treatment; however, prior studies have shown the impact of maintenance to be negligible compared to other life cycle stages (Kofoworola and Gheewala 2009), and its exclusion here is therefore not expected to impact the validity of the results.

The carbon factors for steel listed in Table 3.3 and those for glulam in Table 3.5 were used to model each of the scenarios for the two frame options. Since the

Table 3.5 Carbon factors derived for assessing scenario uncertainty of the embodied carbon of glulam timber

Scenario	Credit for sequestration	End of life	A1	A2-3	A1-3	Source	A4	Source	C1-4	D	Source
1	Seq	LF1	-730.20	242.20	-488.00	1	37.20	2	90.60	-0.22	3
2	Seq	LF2	-730.20	242.20	-488.00		37.20		934.00	-79.10	4
3	Seq	I	-730.20	242.20	-488.00		37.20		846.00	0.00	4
4	Seq	IER	-730.20	242.20	-488.00		37.20		846.00	-593.0	4
5	Seq	RU	-730.20	242.20	-488.00		37.20		819.00	-8.41	4
6	No Seq	LF1	57.80	242.20	300.00		37.20		90.60	-0.22	3
7	No Seq	LF2	57.80	242.20	300.00		37.20		934.00	-79.10	4
8	No Seq	I	57.80	242.20	300.00		37.20		846.00	0.00	4
9	No Seq	IER	57.80	242.20	300.00		37.20		846.00	-593.0	4
10	No Seq	RU	57.80	242.20	300.00		37.20		31.00	-8.41	5

Sources:

1. Wood for Good (2013b)

2. Sequestered carbon estimated using Eq. 1 in British Standards Institution (2014b)

3. Calculated based on 1400 km by road from Central Europe (plus return) and 100 km by sea and vehicle emission factors from Department for Environment, Food and Rural Affairs (2015)

4. Wood Solutions Australia (2015) and includes estimate of transport for waste from Steel Construction Institute End of life LCA and embodied carbon data for common framing materials (n.d.), Department for Transport (2016)

5. Calculated by subtracting the estimated sequestered or biogenic carbon (Eq. 1 in British Standards Institution (2014b)) from the value for reuse in Wood for Good (2013b)

Seq credit given for carbon sequestered in the growth cycle; *No Seq* no credit for sequestered carbon; *LF1* and *LF2* landfill with low and high decay rate, respectively; *I* incineration without energy recovery; *IER* incineration with energy recovery; *RU* reuse

Table 3.6 Material quantities for two alternative structural frames matching the dimensions shown in Fig. 3.5

| Material | Material quantity | | | |
| | Glulam frame | | Steel frame | |
	Volume (m^3)	Mass (tonnes)	Volume (m^3)	Mass (tonnes)
Steel (7850 kg/m^3)	0.17	1.33	1.38	10.83
Glulam (490 kg/m^3)	33.72	16.52	0	0

glulam frame option contains both steel and glulam, it would be necessary to model 210 different scenarios, 10 different glulam scenarios for each of the 21 steel scenarios ($21 \times 10 = 210$ scenarios in total). However, the comparison of the two options was simplified by omitting the steel from the glulam option and reducing the total amount of steel in the steel frame option by the equivalent amount. In reality, the life cycle impacts for steel joints for a timber frame may not be exactly equal to those of an equivalent quantity of steel for a pure steel frame. However, the simplification is not expected to have a significant effect on the results and means it is only necessary to model 10 scenarios for the embodied carbon of the glulam in the timber frame option and compare these to the 21 scenarios for the steel frame option.

Case Study Results

The results of the uncertainty assessment of the comparative embodied carbon assessment of the two frame options are shown in Fig. 3.6. The results for the steel frame have been grouped together by the assumption about production route, since this has the greatest impact on the results. For the results based on industry average production, a further sub-grouping has been applied based on whether the system expansion or recycled content approach has been used.

The results of the uncertainty assessment show that the glulam frame has a greater overall range of possible values than the steel frame. The extremes of the range for the glulam frame are -14.6 (Glulam-1) and 52.6 tCO$_2$e (Glulam-7) when both scenario and statistical uncertainties are considered, whilst for steel they are 14.4 and 49.6 tCO$_2$e (Steel-9 and Steel-4, respectively).

By breaking these ranges down into specific scenarios, each based on a defined set of assumptions as has been done here, it is possible to see which sources of uncertainty are causing this variability and also the relative magnitude of the effect of one source of uncertainty compared to another. For example, it can be seen that in general, within each of the scenarios grouped together in Fig. 3.7, there is relatively little variation compared to the difference between one group and another. The steel scenarios are grouped according to the assumption made about the production route. This indicates clearly how important this assumption is when comparing the steel frame to the glulam alternative. The choice of production route

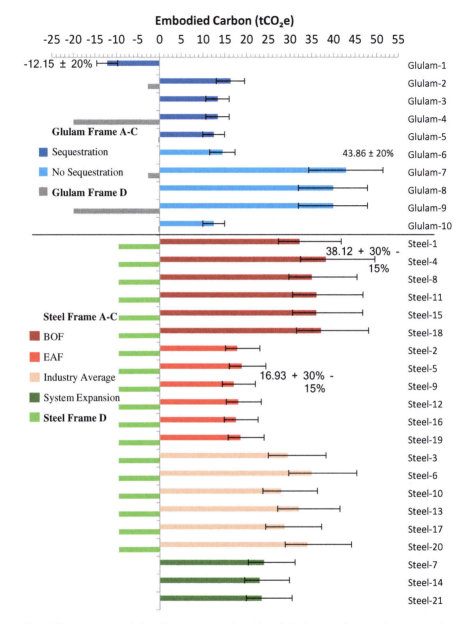

Fig. 3.6 Assessment of the effects of scenario and statistical uncertainty on the comparative embodied carbon assessment of glulam and steel structural frame options for a supermarket. The bars indicate variation due to sources of scenario uncertainty, whilst the error lines indicate additional variation due to sources of statistical uncertainty (see notes overleaf). Notes (scenario classifications): Steel-1-7, no allocation to coproducts; Steel-8-14, physical allocation; Steel-15-21, economic allocation; Steel-1, 8, and 15, EU BOF Steel; Steel-4, 11, and 18, Global BOF Steel; Steel-2, 9, and 16, EU EAF Steel; Steel-5, 12, and 19, Global EAF Steel; Steel-7, 14, and 21, industry average with system expansion (for recycling); Steel-1, 4, 8, 11, 15, 18, recycled content. Glulam-1 and 6, landfill with low decay; Glulam-2 and 7, landfill with high decay;

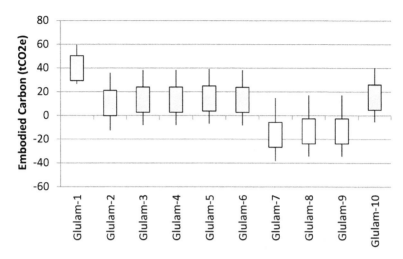

Fig. 3.7 Difference in embodied carbon between steel and glulam alternatives for each combination of scenarios modelled as part of the uncertainty assessment. The chart shows cradle to grave embodied carbon (modules A–C) only. Note that values greater than 0 indicate a saving achieved when using glulam instead of steel. Values below 0 indicate that the use of glulam leads to an increase in embodied carbon.

has a greater effect on the difference between the embodied carbon of the steel and the glulam for a given glulam scenario than either the choice of allocation method (for coproducts) or the choice of production location.

Similarly, for the glulam frame, the choice of whether to credit sequestration can be seen to have a much greater effect on embodied carbon than most of the different end-of-life assumptions. The exceptions to this are the two alternative assumptions about landfill. If the timber is assumed to be sent to landfill, the assumed decay rate is a very important source of uncertainty. Assuming a low decay rate leads to a scenario where carbon sequestered in the timber is effectively stored underground and is not released, or in the case that no credit is given for sequestration, then the end-of-life emissions are nevertheless greatly reduced when compared to other waste treatment options. By contrast, if a high rate of decay is assumed, then the landfill option is seen to lead to slightly higher emissions since a proportion of the carbon is assumed to be emitted as methane which has a higher global warming potential the CO_2.

When considering the comparison of the two frame options, the effect of scenario uncertainties is particularly stark. The box and whisker plot in Fig. 3.7 depicts the uncertainty in the estimated embodied carbon saving that can be

---

Fig. 3.6 (continued) Glulam-3 and 8, incineration; Glulam-4 and 9, energy recovery; Glulam-5 and 10, recycling, life cycle modules according to BS EN 15804 (British Standards Institution 2014a): A–C: raw material supply, transport, manufacture, transport to site, demolition and disposal. D benefits beyond the life cycle; construction (A5) and use (B) are excluded

achieved when using glulam timber instead of steel. Positive values mean that the glulam frame has lower embodied carbon than the steel frame. The box shows the effect of scenario uncertainty (i.e. the maximum and minimum savings when comparing a given timber scenario with all of the steel scenarios), whilst statistical uncertainty has been represented by error bars or whiskers. For each of the different timber scenarios, the column in the chart represents the range of savings when compared to the steel scenarios. The upward error bars represent the greatest possible increase in the difference between the two options when estimates of statistical uncertainty are included by comparing the estimated upper limit for glulam (+20%) with the estimated lower limit for steel (-15%). Conversely, the downward error bars are the comparison of the lower limit for glulam (-20%) with the upper limit for steel (+30%).

In the most optimistic case, and considering only the scenario, the use of timber saves between 30 and 50 tCO_2e compared to steel. At the other extreme, comparing the steel with the least favourable scenario for glulam leads to an increase in embodied carbon of between around 5 and 25 tCO_2e. This gives an overall variability of approximately 75 tCO_2e. In relative terms, the two extremes represent either a saving of embodied carbon of up to 170% or an increase in embodied carbon of 160% through replacing steel with timber.

In Fig. 3.7, the 21 data points for each glulam scenario are not depicted individually but are shown amalgamated into a single box plot which depicts the overall spread of the results. For just under 70% of scenario combinations modelled, the use of glulam for the frame results in an embodied carbon saving versus the steel frame. It can be seen that the steel is estimated to have lower embodied carbon when compared to timber scenarios where sequestration is not credited and the timber is assumed to be either incinerated, or sent to energy recovery, or sent to landfill where a high rate of decay is assumed.

It can be seen that glulam scenarios 2–6 and 10 all have error bars which extend below the x-axis. This indicates that if estimates of statistical uncertainty are included, then more of the scenario combinations could lead to results where the glulam frame has higher embodied carbon than the steel. In these cases, even with the inclusion of sequestered carbon in timber, inefficiencies in the timber production, or other factors leading to statistical uncertainty, could negate any carbon saving.

The very wide range of possible outcomes highlights that modelling and considering each of the individual scenarios are important to fully understand the effects and the sources of uncertainty. This approach allows designers and decision-makers to understand the relative impact of different sources of uncertainty when making design and material choices, for example, between glulam and steel. The implications and justifications for making assumptions about a given material can therefore be reviewed and evaluated as part of the decision-making process.

Discussion

The results of the research show that for each material considered, sources of scenario level uncertainties have a greater effect on embodied carbon results than sources of statistical level uncertainty. Moreover, for the two materials assessed, a small number of specific sources of scenario uncertainty produce the dominant influences upon uncertainty. For example, aggregating data for the whole steel industry masks the relatively higher embodied carbon impacts arising from BOF steel when compared to EAF steel. The rationale for aggregating data from these two processes relates to the constrained market for scrap steel. The BOF process has an upper limit on the amount of scrap that can be introduced to the feedstock of around 30%, whereas EAF feedstock may be up to 100% scrap. Yet recovery rates for scrap steel are already very high, and hence, in discussions on this matter in academic research (e.g. Jones 2009) and industry reports (e.g. World Steel Association 2011), it has been argued that introducing a requirement for the steel in a building to have a high recycled content has no net benefit and could in extreme cases lead to greater inefficiencies in steel production. This view is based on the assumption that, given the constrained market for scrap steel, attempting to increase the recycled content in the steel for one building would only reduce availability of recycled steel elsewhere. On this basis, there is no benefit to be realized from penalizing BOF steel with a high carbon factor and crediting EAF steel with a lower carbon factor. The two production routes are both part of a single system, which is already achieving near-maximal efficiency in terms of the rate of steel recovery and recycling (World Steel Association 2011). As long as the total demand for steel continues to grow, the supply of scrap will not be able to meet demand.

Yet there is a flaw in the argument, since it is based on the assumption that the only alternative to EAF steel, with its high recycled content and relatively low carbon factor, is BOF steel. Whilst this may be true in some cases, there are many applications in construction where viable alternatives to steel exist. The case study presented here, which assessed the replacement of steel with an alternative timber solution for a supermarket's structural frame, is one such example. If more emphasis were placed on ensuring the steel used in buildings has a high recycled content, the constrained availability of this steel could conceivably lead to increased use of alternative materials. In this case, incentivizing the use of high recycled content EAF steel through the use of carbon factors that reflect the true carbon emissions of this process could lead to a net benefit.

For timber, the choice of whether or not to include sequestered carbon in the calculation has been shown to have the greatest effect on the embodied carbon factor. A straightforward way to make its inclusion more defensible is to specify that timber is from certified sustainable sources (Darby 2014) that are managed to ensure that there is no net deforestation. Thus, as timber is harvested, it is replaced with new growth, which increases the amount of carbon sequestered overall.

The results of the scenario analyses have also highlighted the significant uncertainty related to decisions regarding the end-of-life treatment of timber. One of the scenarios in which timber is sent to landfill at the end of its life has shown lower cradle to grave emissions than the alternative energy recovery or material reuse scenarios. The embodied carbon of glulam assessed in this scenario was found to be -0.36 tCO_2e per cubic metre if sequestration in the growth cycle was credited and $+0.428$ tCO_2e/m^3 if it was not. By contrast, energy recovery and reuse were found to give embodied carbon factors of $+0.395$ and $+0.368$ tCO_2e/m^3, respectively (1.183 and 0.368 tCO_2e/m^3 without sequestration). The alternative landfill scenario, which assumes a higher rate of decay, was shown to result in of 0.483 tCO_2e/m^3 (1.271 tCO_2e/m^3 without sequestration). Thus these two alternatives produced both the highest and lowest values for embodied carbon. This serves to illustrate how great the effect of scenario uncertainties can be on the embodied carbon results. The low rates of decay have been demonstrated in controlled tests where bulk timber is sent to landfill, whereas the tests that produced high rates of decay were based on ground timber (Wood Solutions Australia 2015).

On the evidence of this assessment alone, the most carbon efficient approach to dealing with timber at the end of life appears to be to send it to landfill with as little processing as possible in order to minimize rates of decay. This can be seen as a form of carbon capture and storage. Carbon dioxide is sequestered from the atmosphere whilst the tree is growing and is stored as biogenic carbon in the timber. When the timber is harvested, this carbon remains stored in the timber product. If the low decay rate assumption is valid, when the timber product is disposed of, the carbon is released back to the atmosphere very slowly, in contrast to the peak emissions from burning timber for energy.

Thus, the apparent end-of-life emissions from reuse or recycling of timber are often misleading. These emissions are included for carbon accounting purposes since, otherwise, double counting can occur by including credits at both stages of use. These calculations of carbon emissions for accounting purposes do not reflect the actual emissions of carbon dioxide to the atmosphere. Thus, users of such data must take this into consideration when basing design, investment or policy decisions on such results.

The waste hierarchy approach set out in the EU Waste Framework Directive (European Union 2008) sets the order of priority or preference for waste treatment. Prevention of waste is the most preferable approach followed by reuse, then recycling and then energy recovery. Disposal to landfill is the least favoured waste treatment option, and this is reflected in UK Government initiatives which have explored policy measures to prevent timber being sent to landfill (Department for Environment, Food and Rural Affairs 2013). The results of an embodied carbon assessment should therefore not be considered in isolation but in conjunction with other indicators and guidance for achieving environmental sustainability across the building life cycle.

Conclusions and Further Work

Sources of uncertainty that affect the embodied carbon factors of building materials can be divided into two distinct groups based on the level of uncertainty that they introduce. There are uncertainties related to different assumptions and methodological choices which can be made when estimating carbon factors for a material. These introduce levels of uncertainty which are characterized by alternative scenarios where the probability of each scenario is not known or cannot be meaningfully determined. The second group of sources of uncertainty are characterized by levels of uncertainty where it is possible to establish quantitative ranges for a given carbon factor.

The effects of both statistical and scenario uncertainty on the carbon factors for steel and glulam timber were assessed. The results showed that the levels of uncertainty are so great that the glulam timber commonly considered to be a low carbon material may, under certain combinations of assumptions or conditions, result in higher embodied carbon emissions than steel. In around 70% of the scenarios modelled, timber was found to have lower embodied carbon. For steel, whether data are aggregated across both major production routes or not has been shown to be a key source of uncertainty. For timber, the choice of whether sequestered carbon is credited to the product and the assumptions made about the end-of-life treatment were found to be the most important sources of scenario level uncertainty.

Given that timber appears preferable in a greater percentage of scenarios, the results could be used in practice to identify ways in which to make these scenarios more reasonable or probable. For instance, the use of sustainably certified timber may make claiming credits for sequestered carbon more justifiable. Furthermore, a design team might explore measures that can be taken at the design stage which make a particular end-of-life outcome more likely. The use of scenario assessment demonstrated here could thus support decision-making for lower embodied carbon at the level of individual projects and at an organizational or national policy level.

Importantly, these results also serve to highlight that that recently developed European standards do not provide the necessary clarity to mitigate these sources of uncertainty. Indeed, with respect to the sequestered carbon in timber, rules intended to prevent double counting arising through reuse, the guidance is positively misleading.

The scenario assessment technique applied here works well when assessing small numbers of materials. Yet it starts to become more impractical as more materials are included because the number of scenarios increases significantly and becomes unmanageable. Further research initiatives to develop the method should look for ways to assess the impacts of scenario level uncertainties using computer simulations so that their effects on the embodied carbon of whole buildings may be evaluated.

Acknowledgements This work has been funded by EPSRC and Sainsbury's Supermarkets Ltd. as part of an Engineering Doctorate. The authors wish to acknowledge the support of Sainsbury's in providing some of the data for this work. We also gratefully acknowledge the generous contributions of time and knowledge of those who participated in the various rounds of expert elicitations conducted during the course of this study and the wider research of which it is a part.

References

Asselt, M. B. A., & Rotmans, J. (2002). Uncertainty in integrated assessment modelling. *Climatic Change, 54*, 75–105.

British Standards Institution. (2006a). *ISO 14040:2006 environmental management. Life cycle assessment. Principles and framework*. London: BSI Standards Ltd..

British Standards Institution. (2006b). *ISO 14044:2006 environmental management – Life cycle assessment – Requirements and guide-lines*. London: BSI Standards Ltd..

British Standards Institution. (2010). *BS EN 15643–1:2010 sustainability of construction works – Sustainability assessment of build-ings part 1: General framework*. London: BSI Standards Ltd..

British Standards Institution. (2011). *BS EN 15978:2011 sustainability of construction works – Assessment of environmental performance of buildings – Calculation method*. London: BSI Standards Ltd..

British Standards Institution. (2014). *BS EN 15804:2012+A1 sustainability of construction works – Environmental product declarations – Core rules for the product category of construction products*. London: BSI Standards Ltd..

British Standards Institution. (2014b). BS EN 16449:2014 wood and wood-based products: Calculation of the biogenic carbon content of wood and conversion to carbon dioxide.

Cabeza, L. F., Rincón, L., Vilariño, V., Pérez, G., & Castell, A. (2014). Life cycle assessment (LCA) and life cycle energy analysis (LCEA) of buildings and the building sector: A review. *Renewable and Sustainable Energy Reviews, 29*, 394–416.

dCarbon8. (2008). Sainsbury's Dartmouth – Carbon footprint assessment. Unpublished Internal Document, Sainsbury's Supermar-kets Ltd.

Cooke, R. M., & Goossens, L. J. H. (1999). Procedures guide for structured ex-pert judgment. European Commission – Directorate-General for Re-search.

Cousins-Jenvey, B., Walker, P., Shea, A., Sykes, J., & Johansson, A. (2014). Comparing deterministic and probabilistic non-operational building energy modelling. *Sustainable Habitat for Developing Societies, 30*, 135–142.

Darby, H. (2014). Investigation of the relative impacts on global warming of embodied and operational carbon emissions from buildings. EngD, University of Reading.

Department for Environment, Food and Rural Affairs. (2013). Wood waste landfill restrictions in England: Call for evidence – Analysis.

Department for Environment, Food and Rural Affairs. (2015). DEFRA carbon factors. In: CarbonSmart UK conversion factors. http://www.ukconversionfactorscarbonsmart.co.uk/. Accessed 18 May 2016.

Department for Transport. (2016). Road freight statistics: Table RFS0112 – Average length of haul by commodity: Annual 2004–2015.

Dixit, M. K., Fernández-Solís, J. L., Lavy, S., & Culp, C. H. (2010). Identification of parameters for embodied energy measurement: A literature review. *Energy and Buildings, 42*, 1238–1247.

European Commission Joint Research Centre. (2013). European Life Cycle Database (ELCD). In: European platform on life cycle assessment. http://eplca.jrc.ec.europa.eu/ELCD3/. Accessed 26 Sept 2016.

European Union. (2008). Directive 2008/98/EC of the European parliament and of the council on waste and repealing certain directives.

Funtowicz, S. O., & Ravetz, J. R. (1990). *Uncertainty and quality in science for policy*. Netherlands: Springer.

Hammond, G., & Jones, C. (2011). The inventory of carbon and energy (ICE). BSRIA

Heijungs, R., & Huijbregts, M. A. (2004). A review of approaches to treat uncertainty in LCA. Proceedings of the IEMSS conference.

Hogg, D., Ballinger, A., & Oonk, H. (2011). Inventory improvement project – UK landfill methane emissions model final report to Defra and DECC. Eunomia Research and Consulting.

Ijeh, I. (2014). The eco-supermarket pioneers. In: Building. http://www.building.co.uk/buildings/technical-case-studies/the-eco-supermarket-pioneers/5068185.article. Accessed 16 Feb 2017.

Institut Bauen und Umwelt e.V. (2013). Environmental product declaration – Structural steel: Sections and plates – BauforumStahl e.V.

Institut Bauen und Umwelt e.V. (2014). Environmental product declaration – structural section steel CELSA Group.

Jones, C. (2009). Embodied impact assessment: The methodological challenge of recycling at the end of building lifetime. *Construction Information Quarterly, 11*, 140–146.

Knol, A. B., Slottje, P., van der Sluijs, J. P., & Lebret, E. (2010). The use of expert elicitation in environmental health impact assessment: A seven step procedure. *Environmental Health, 9*, 19.

Kofoworola, O. F., & Gheewala, S. H. (2009). Life cycle energy assessment of a typical office building in Thailand. *Energy and Buildings, 41*, 1076–1083.

Lloyd, S. M., & Ries, R. (2007). Characterizing, propagating, and analyzing uncertainty in life-cycle assessment: A survey of quantitative approaches. *Journal of Industrial Ecology, 11*, 161–179.

Moncaster, A. M., & Song, J.-Y. (2012). A comparative review of existing data and methodologies for calculating embodied energy and carbon of buildings. *International Journal of Sustainable Building Technology and Urban Development, 3*, 26–36.

Pomponi, F., & Moncaster, A. (2016). Embodied carbon mitigation and reduction in the built environment – What does the evidence say? *Journal of Environmental Management, 181*, 687–700.

Pomponi, F., D'Amico, B., & Moncaster, A. (2017). A method to facilitate uncertainty analysis in LCAs of buildings. *Energies, 10*, 524.

Ramesh, T., Prakash, R., & Shukla, K. K. (2010). Life cycle energy analysis of buildings: An overview. *Energy and Buildings, 42*, 1592–1600.

Refsgaard, J. C., van der Sluijs, J. P., Højberg, A. L., & Vanrolleghem, P. A. (2007). Uncertainty in the environmental modelling process–a framework and guidance. *Environmental Modelling & Software, 22*, 1543–1556.

Richardson, S., Hyde, K., Connaughton, J., & Merefield, D. (2014). Service life of UK supermarkets: Origins of assumptions and their impact on embodied carbon estimates. World Sustainable Building Conference, Barcelona.

Ross, S., Evans, D., & Webber, M. (2002). How LCA studies deal with uncertainty. *Int J LCA, 7*, 47–52.

Rüter, S., & Diederichs, S. (2012). Ökobilanz-Basisdaten für Bauprodukte aus Holz [LCA-data for Timber Construction Products]. Bundes-forschungsinstitut für Ländliche Räume, Wald und Fischerei Institut für Holztechnologie und Holzbiologie, Hamburg.

Ruukki. (2014). Environmental product declaration – Structural steel construction products.

Saltelli, A., & Annoni, P. (2010). How to avoid a perfunctory sensitivity analysis. *Environmental Modelling & Software, 25*, 1508–1517.

Sartori, I., & Hestnes, A. G. (2007). Energy use in the life cycle of conven-tional and low-energy buildings: A review article. *Energy and Buildings, 39*, 249–257.

Schvaneveldt, R. W., Durso, F. T., Goldsmith, T. E., Breen, T. J., Cooke, N. M., Tucker, R. G., & De Maio, J. C. (1985). Measuring the structure of expertise. *International Journal of Man-Machine Studies, 23*, 699–728.

Skinner, D. J. C. (2012). A novel approach for identifying uncertainties within environmental risk assessments. Ph.D., Cranfield University.

Slottje, P., van der Sluijs, J. P., & Knol, A. B. (2008). Expert elicitation: Methodological suggestions for its use in environmental health impact assessments. National Institute for Public Health and the Environment.

Steel Construction Institute End of life LCA and embodied carbon data for common framing materials. (n.d.). In: Steelconstruction.info. http://www.steelconstruction.info/End_of_life_LCA_and_embodied_carbon_data_for_common_framing_materials. Accessed 5 May 2016.

UK Steel. (2016). Key statistics 2016. London.

van der Sluijs, J. P., Janssen, P. H. M., Petersen, A. C., Kloprogge, P., Risbey, J. S., Tuinstra, W., & Ravetz, J. R. (2004). RIVM/MNP guidance for uncertainty assessment and communication: Tool catalogue for uncertainty assessment. RIVM.

Walker, W. E., Harremoës, P., Rotmans, J., van der Sluijs, J. P., van Asselt, M. B. A., Janssen, P., & von Krauss, M. P. K. (2003). Defining uncertainty: A conceptual basis for uncertainty management in model-based decision support. *Integrated Assessment, 4*, 5–17.

Warmink, J. J., Janssen, J. A. E. B., Booij, M. J., & Krol, M. S. (2010). Identification and classification of uncertainties in the application of environmen-tal models. *Environmental Modelling & Software, 25*, 1518–1527.

Wood for Good. (2013a). Wood for good life cycle database – Kiln dried softwood.

Wood for Good. (2013b). Wood for good life cycle database – Glued laminated timber.

Wood Solutions Australia. (2015). Environmental product declaration – Softwood timber – Wood solutions. Forrest and wood products Australia Ltd.

World Steel Association. (2011). Life cycle assessment methodology report. Brussels.

Yang, Y., & Heijungs, R. (2017). On the use of different models for consequential life cycle assessment. *The International Journal of Life Cycle Assessment*. https://doi.org/10.1007/s11367-017-1337-4.

Chapter 4
Embodied Carbon of Wood and Reinforced Concrete Structures Under Chronic and Acute Hazards

A. Souto-Martinez, E. J. Sutley, A. B. Liel, and W. V. Srubar III

Introduction

One of the primary challenges in calculating total life cycle embodied carbon (EC) in buildings is the prediction of expected service life of materials and components, especially in the face of both chronic (e.g., chloride-induced corrosion, carbonation, climate change) and acute (e.g., seismic, wind, flood) hazards. Current whole-building life cycle assessment (WBLCA) methodologies often neglect these in-service impacts in the environmental accounting by assuming lifespans equal to the expected life of the building or by assuming materials will incur no damage or degradation during service. For example, Peuportier (2001) compared the environmental impacts of different building designs by conducting detailed WBLCAs. For the use phase, the study assumed that building materials and components had similar durability, thereby excluding any impacts due to damage, maintenance, or replacement. Likewise, Junnila and Horvath (2003) quantified the environmental impacts of an office building in Finland assuming a lifespan of 50 years without accounting for in-service impacts. Similar assumptions were made in Basbagill et al. (2013), Bribian et al. (2009), Cabeza et al. (2014), and Khrasreen et al. (2009).

Omission of in-service impacts is due, in large part, to the complexity and uncertainty of predicting in-use performance and the lack of simple models and methodologies that can be easily incorporated into LCA frameworks. This omission

A. Souto-Martinez · A. B. Liel · W. V. Srubar III (✉)
Department of Civil, Environmental, and Architectural Engineering, University of Colorado Boulder, Boulder, CO, USA
e-mail: wsrubar@colorado.edu

E. J. Sutley
Department of Civil, Environmental, and Architectural Engineering, University of Kansas, Lawrence, KS, USA

© Springer International Publishing AG 2018
F. Pomponi et al. (eds.), *Embodied Carbon in Buildings*,
https://doi.org/10.1007/978-3-319-72796-7_4

has been identified in the literature as one of the principal WBLCA challenges and sources of uncertainty (see Khrasreen et al. 2009; Leicester 2001; Simonen 2014; Souto-Martinez et al. 2017; Srubar 2015b; Stambaugh 2017; Welsh-Huggins and Liel 2017a).

In recognition of this challenge, this chapter serves as a technical resource for those seeking to integrate service-life prediction models and loss-estimation methodologies into WBLCA frameworks for building materials, components, and structures. The primary focus of this chapter will center on (1) EC accounting methodologies for wood and concrete materials and structures, (2) current guidelines for incorporating service-life estimates into WBLCA, (3) existing service-life prediction models of wood and concrete materials and components, and (4) a review of methodologies that link loss assessments from acute hazard events to EC accounting for reinforced concrete- and wood-frame building typologies.

EC Accounting Methodologies

Life cycle assessment (LCA) is a standardized methodology used to track and quantify the environmental impacts of a product or process throughout its life cycle (International Standards ISO-14025, −14040, −14044). LCA is a powerful approach that can be used to not only quantify environmental impacts but also, through contributory analyses, identify areas of improvement or, via comparative assertions, compare material (e.g., concrete mixture) or building design alternatives. In recent years, several literature reviews have been published regarding LCA in the building sector. These reviews have addressed potential users, drivers, challenges, and limitations of LCA (see, e.g., Bribian et al. 2009; Chau et al. 2015; Erlandsson and Borg 2003; Sharma et al. 2011).

In general, LCA approaches can be classified into three main categories: (1) - process-based LCA, (2) economic input-output LCA (EIO-LCA), and (3) hybrid LCA. Process-based LCAs calculate energy inputs and potential outputs by utilizing a process flow diagram for each life cycle phase. For buildings, this would include process, manufacturing, and transportation energy for all building products, construction energy, use-phase operational energy, and any post-use energy expended or reclaimed. A process-based LCA approach has the advantage of specificity and high-precision results. However, disadvantages include limited access to data, time, and cost, making accurate, replicable analyses difficult. EIO-LCA calculates environmental impacts of an entire economic sector (e.g., construction) and links economic impacts of that economic sector's supply chain to emissions and energy consumption. Disadvantages of EIO-LCA include high variability for specific individual materials, a lag in data collection, and inherent low-fidelity data measurements. The hybrid LCA approach was recently developed to address the disadvantages of both process-based and EIO-LCA methods. Hybrid LCAs strike a balance by estimating initial embodied impacts of materials using an economy-wide scope,

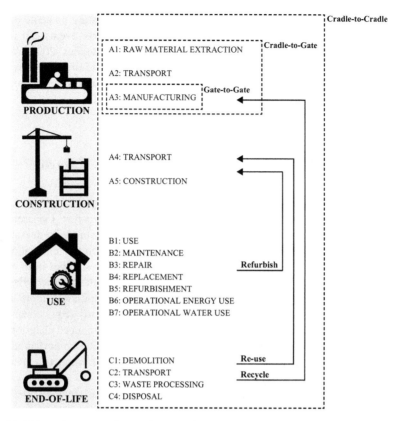

Fig. 4.1 Life cycle stages and example system boundaries of building materials, components, and structures, noting refurbishment, reuse, or recycling

while impacts in the operation and end-of-life phases are specifically analyzed using higher fidelity, process-based models.

LCA results depend—among other things—on system boundary definitions. Depending on the stages included in the analysis, environmental impacts can be reported "gate-to-gate," "cradle-to-gate," "cradle-to-grave," or "cradle-to-cradle." A description of these is provided in Fig. 4.1, as well as the phases of a building product life cycle that can be included in the system boundary. Of particular interest are the system boundaries that include or exclude the use phase. The use phase includes activities, such as scheduled or recurring maintenance and repair, and impacts due to continual operations (i.e., lifetime energy use).

The importance of chronic-hazard durability assumptions on building LCA is demonstrated in the literature by Dixit et al. (2012), Grant and Ries (2013), Grant et al. (2014), Strand and Hovde (1999), and Srubar (2015b). While buildings for LCA are usually modeled using service-life assumptions (e.g., 50 years), Grant et al. (2013) demonstrates the importance of assessing building lifetime variability, since total impacts will depend highly on operational energy, maintenance, and replacement during the use phase. In addition, a growing literature on acute hazards in WBLCA has shown that considering earthquakes and other events can impact design decisions (Junnila and Horvath 2003; Mileti and Gailus 2005; Welsh-Huggins and Liel 2017b). Thus, the ability to predict service-life performance is increasingly critical for improved accuracy in measuring total lifetime environmental impacts.

Quantifying the EC of Buildings Under Chronic Hazards

Anticipated service life is a critical factor when conducting accurate LCAs that include the use phase of a building material, component, or structure. Concrete- and wood-based building materials and components are especially prone to degradation from decades-long exposure to chronic hazards. Depending on the application, for example, reinforced concrete materials are exposed to chlorides, either from marine or de-icing salts, which exacerbate the risk of chloride-induced corrosion. Exposed concrete materials must also resist deterioration due to freeze-thaw, sulfate attack, carbonation, and multiple mechanical degradation mechanisms (e.g., creep, fracture). Wood-based materials must resist years of moisture-induced deterioration, including mold growth, fungal attack, termite attack, and marine borer degradation. Furthermore, the duration of loads and application can lead wood to experience mechanical degradation via creep and relaxation.

A wealth of experimental studies conducted over the past few decades have highlighted durability concerns in concrete- and wood-based construction, which have led to equation-based modeling of material durability to estimate time-to-damage, plan maintenance activities, and facilitate material design (Otieno et al. 2011). As shown in Fig. 4.2, these models can be used in LCA methodology to predict service lifetime, given preferred performance criteria and limiting damage states (i.e., chloride concentration that initiates corrosion). These predictions can be used (1) to evaluate whether candidate concrete- or wood-based materials or components will last for intended target design lifetime as illustrated in Fig. 4.2 and (2) to account for expected replacements during the use phase in calculating total life cycle EC.

In the next two sections, a review of current equation-based service-life prediction models for reinforced concrete- and wood-based materials, respectively, is presented. Strengths and limitations of these models are discussed along with more advanced models, review articles, and commercially available software. For some models, example results are shown to illustrate how the results can be incorporated into a holistic LCA framework for quantifying total life cycle EC.

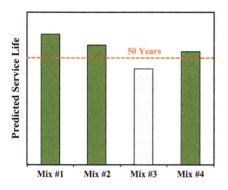 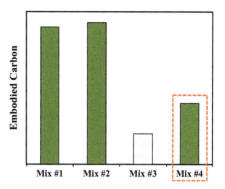

Fig. 4.2 Illustrative example showing the expected service life of potential candidate mixture designs and corresponding embodied carbon. Expected service life can be defined as part of the functional unit. If 50 years is a target design service life, Mix #3 does not meet the functional unit requirements. Therefore, it is eliminated as a possible mixture design candidate in the LCA, and Mix #4 would be the most environmentally preferred

Reinforced Concrete

Chloride-Induced Corrosion

For reinforced concrete in chloride-laden environments, the predicted time to corrosion-induced cracking has often been defined as a principal failure limit state. Time to corrosion-induced cracking has been estimated using a corrosion damage model first proposed by Tuutti (1982). The model, shown in Fig. 4.3, assumes that the progression of concrete damage increases linearly with time after corrosion initiation. The total service life, t_s, of reinforced concrete is considered the sum of two successive time periods, namely, (1) the time to corrosion initiation, t_i, which is governed by the diffusion of chlorides throughout the concrete media, and (2) the time to corrosion cracking, t_c.

Time to corrosion initiation has been most often predicted using diffusion models that are analytical or numerical solutions to Fick's second law of diffusion. The following analytical solution to Fick's second law of diffusion has been used extensively to model steady-state chloride diffusion and to predict time to corrosion initiation in reinforced concrete (Tanaka et al. 2006; Srubar 2015a):

$$c_x(x,t) = c_o \left[1 - \mathrm{erf}\left(\frac{x}{2\sqrt{Dt}} \right) \right]$$

where c_x is the chloride concentration (kg/m^3) in the reinforced concrete at a distance from the surface, x (m), at time, t (s), D is the apparent chloride diffusion coefficient (m^2/s), and c_o is the chloride concentration (kg/m^3) at the surface of the concrete. The time to corrosion initiation is defined as the time at which the chloride concentration at a distance from the concrete surface to the reinforcement (i.e., cover depth) is equal to the critical chloride concentration threshold required to initiate rebar

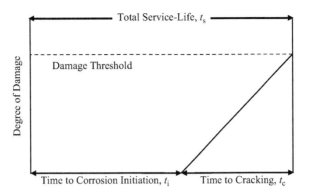

Fig. 4.3 Two-phase model for total service life of reinforced concrete structures

corrosion (0.7–1.2 kg/m^3) (Sfikas and Ingham 2016; Srubar 2015a; Tuutti 1982). Models that relate critical concrete mixture properties (e.g., water-to-cement ratio) to apparent chloride diffusion coefficients of concrete pastes, mortars, and concretes have been proposed, including models that incorporate effects of supplementary cementitious materials, like fly ash, slag, silica fume, and metakaolin (Riding et al. 2014; Stambaugh 2018).

While simple and useful, the error function solution to the diffusion equation has notable limitations, such as assuming steady-state conditions and ignoring chemical reactivity of chloride with metastable minerals in cementitious binders. Researchers have addressed some of these limitations over the past few decades by improving the accuracy of chloride diffusion models. For example, Srubar (2015a) extended the error function solution to account for the use of both contaminated and uncontaminated recycled aggregates. Additional improvements have been made by researchers employing 1-, 2-, and 3-D numerical finite difference solutions to solve the diffusion equation, which address limitations of the error function solution by accounting for nonsteady-state boundary conditions and time and temperature dependence of chloride diffusion coefficients. More sophisticated physics-based transport models have also been developed that account for reactive transport of chloride and existence of cracks (Takewaka et al. 2003). These physics-based models, while more accurate, do sacrifice simplicity and computational expense.

Several commercially available software tools can be used to estimate the anticipated service life of steel-reinforced concrete structures exposed to chlorides. *Life-365*,[1] for example, calculates the initiation period and total service life of reinforced concrete structures, as well as providing life cycle cost analyses of alternate corrosion protection systems. *EUCON*[2] is able to design a concrete mixture that meet

---

[1]Life-365 Software: http://www.life-365.org
[2]EUCON Software: http://avid.force.com/pkb/articles/en_US/download/EUCON-PRO-Software-version-3-1-2

requirements on strength and service life, which are inputted by the user, not only considering chlorides ingress but also other degradation phenomena (e.g., carbonation) of reinforced concrete structures. *DURACON*[3] is another tool that predicts durability of concrete structures under different environmental conditions. Its simulation is based on Fick's second law of diffusion, the same finite difference calculations employed by *Life-365* and *EUCON*. *STADIUM,*[4] however, is a finite element (FE) analysis software that predicts concrete service life. *STADIUM* bases its calculations on chemical reactive transport modeling in concrete, providing the user with information regarding time to corrosion initiation.

Freeze-Thaw Deterioration

As micro- and macroporous brittle materials, cement pastes, mortars, and concretes are prone to surface and internal damage due to water ingress and subsequent freeze-thaw cycling. Models that predict freeze-thaw durability are less common than models for chloride ingress, since damage caused by chloride-induced corrosion is more common and freeze-thaw degradation mechanisms are complex. Predictive models are available in the literature, but most are based solely on isolated experimental studies. Penttala (2006), for example, proposed a degradation model based on experimental surface scaling and internal freeze-thaw damage of concrete samples in saline and nonsaline environments:

$$D = a + b \left[\left(\frac{W}{C} \right)^c / A^d / B^e \right]$$

where W/C is the water-to-cement ratio in decimal form, A is the air content (%), B is curing time (days), D is the relative dynamic modulus (%), and a, b, c, d, and e are regression-derived coefficients shown in Table 4.1. According to ASTM C666 (2015), concrete specimens pass freeze-thaw durability tests if the dynamic modulus is 60% or higher after 300 freeze-thaw cycles, which could be used as a limit state in the service-life design of frost-resistant concrete if the average freeze-thaw cycles are known for the geographic region of interest.

Limitations of this model include the prediction of dynamic modulus rather than time-to-damage, limiting its generalizability. To address these limitations, other, albeit more numerically complex, models that predict freeze-thaw degradation have been proposed by several authors (Attiogbe 1996; Bažant et al. 1988; Cho 2007; Tikalsky et al. 2004; Wenting et al. 2012). The reader is referred to these studies for more details.

---

[3]DURACON Software: http://duracon.software.informer.com
[4]STADIUM Software: https://www.simcotechnologies.com/what-we-do/stadium-technology-portfolio

Table 4.1 Freeze-thaw deterioration modeling coefficients

Damage type	a	b	c	d	e
Scaling	−57.80	8679	3.356	0.4957	0.2142
Internal damage	−831.9	835	−0.08153	−0.03092	0.001866

Sulfate Attack

The chemistry of cementitious materials imparts a chemical vulnerability of hydration products to internal (e.g., unreacted gypsum, gypsum-containing aggregates) or external (e.g., groundwater) sources of sulfate. Like freeze-thaw deterioration, models to predict time-to-damage due to sulfate attack are also primarily based on experimental results. Two empirical models proposed by Kurtis et al. (2000) predict expansion of concrete materials caused by sulfate attack as a function of W/C, exposure duration (i.e., time), and tricalcium aluminate (C_3A) content in the cement. The experimental program tested the effects of these variables on the sulfate resistance of hundreds of different concrete mixtures in a non-accelerated testing program. For concretes made with cements with low (<8%) C_3A contents, expansion can be predicted by:

$$E = 0.0246 + 0.0180 \left(t \times \frac{W}{C} \right) + 0.00016 (t \times c_a)$$

and for concretes made with cements with high (>8%) C_3A contents, expansion can be predicted by:

$$\ln(E) = -3.753 + 0.930t + 0.0998 \ln(t \times c_a)$$

where E is expansion (%), c_a is the C_3A content in the cement (%), W/C is the water-to-cement ratio (%), and t is time in years. Other models for sulfate attack found in the literature have been proposed by Glasser et al. (2008), Monteiro and Kurtis (2003), Roziere et al. (2009), and Tixier and Mobasher (2003).

Carbonation-Induced Corrosion and Carbon Sequestration

The process of carbonation is a chemical reaction that primarily occurs between calcium hydroxide in hydrated cement and atmospheric carbon dioxide (CO_2), resulting in calcium carbonate. This chemical reaction is of recent interest for two primary reasons. First, the reaction depletes hydroxide ions in the concrete pore solution, which effectively lowers the pH from approximately 12.5 to 9.0. This reduction can destabilize the passive oxide layer of mild steel reinforcement and lead to carbonation-induced reinforcement corrosion in the presence of sufficient oxygen and water. Therefore, similar to chloride resistance, sufficient cover depth is required to protect steel reinforcement, especially in severe exposure conditions. Second, while concrete carbonation can lead to rebar corrosion, carbonation is a carbon sequestration process, which is a positive environmental benefit (Souto-Martinez et al. 2017).

Several carbonation studies have resulted in models to calculate carbonation depth at any given time and, inversely, predict when corrosion due to carbonation might initiate. One such model has been proposed by the Portuguese National Laboratory of Civil Engineering (Monteiro et al. 2012):

$$x = \sqrt{\left(\frac{2 \cdot c_c \cdot t}{R}\right) \cdot \left[\sqrt{k_0 k_1 k_2}\left(\frac{1}{t}\right)^n\right]}$$

where x is the corrosion depth (m), c is the environmental CO_2 concentration (kg/m^3), t is exposure time (years), k_0 is equal to 3.0, k_2 is equal to 1.0 for standard curing, and R is the carbonation resistance coefficient (kg year/m^5) that is calculated for Type I and Type II cement according to:

$$R = 0.0016 \cdot f_c^{3.106}$$

and for Types III–V and white cement according to:

$$R = 0.0018 \cdot f_c^{2.862}$$

where f_c is the compressive strength of the concrete (MPa). The factors k_1 and n, shown in Table 4.2, are dependent upon exposure classifications.

Other models that predict carbonation depth and time to carbonation-induced corrosion that may be of interest include Ahmad (2003), Ann et al. (2010), Jiang et al. (2000), Kashef-Haghighi et al. (2015), Papadakis et al. (1991), Saetta and Vitaliani (2004), Steffens et al. (2002), and Wang and Lee (2009).

In recognition that carbonation is a carbon sequestration process that is often left out of use-phase LCAs, some studies have attempted to quantify the amount of CO_2 sequestered during the service life of ordinary portland cement (OPC) concrete structures (Collins 2010; García-Segura et al. 2014; Kosmatka et al. 2002; Lee et al. 2013; Mounanga et al. 2004; Pade and Guimaraes 2007; Yang 2014).

Souto-Martinez et al. (2017) derived a general mathematical expression to predict C_s, the total theoretical mass of sequesterable CO_2 (kg CO_2), in a structural concrete element that accounts for cement type, cement quantity, addition of supplementary cementitious material (SCMs), degree of carbonation, and time:

$$C_s = \phi_c \, C_m \cdot [V_c \cdot m]$$

where ϕ_c is the degree of carbonation, which has been shown empirically to range from 0.4 to 0.8, m is the total mass of cement per unit volume of concrete (kg/m^3) obtained from the concrete batch mixture proportions, V_c is the total carbonated volume, and C_m is the carbon sequestration potential, which can be calculated by:

Table 4.2 Parameter values for k_1 and n based on exposure classification

Parameter	XC1 Dry or permanently humid	XC2 Humid, rarely dry	XC3 Moderately humid	XC4 Cyclically humid and dry
k_1	1.0	0.20	0.77	0.41
n	0	0.183	0.02	0.08

Table 4.3 Parameter values for k_1 and n based on exposure classification

Cement type	α	Supplementary cementitious material (SCM)	Average σ % SiO_2	β
Type I	0.165	Fly ash (class F)	50%	0.55
Type II	0.163	Fly ash (class C)	25%	0.27
Type III	0.166	Slag	35%	0.38
Type IV	0.135	Silica fume	90%	0.99
Type V	0.161	Metakaolin	50%	0.55
White	0.203			

$$C_m = \alpha - \beta \cdot y$$

where y is the percent replacement (by mass of cement) by SCMs in decimal form. Table 4.3 lists values for the coefficient α, which accounts for variation in cement type, and β, which accounts for the chemical reactivity of SCMs.

The total carbonated volume at any point in time can be calculated by multiplying the total carbonation depth, x, by the total surface area, A, of exposed concrete members:

$$V_c = A \cdot x$$

The model used to calculate carbonation depth was discussed previously (Monteiro et al. 2012). Figure 4.4 illustrates example results achieved by the carbon sequestration model. For more details and a thorough discussion of the utility of the model, please see Souto-Martinez et al. (2017).

Shrinkage and Creep

There are several models to predict shrinkage and creep. However, these models often involve complex numerical calculations, and inclusion of such models herein is beyond the scope of this review. Instead, the reader is referred to Shah et al. (1998), who presented a fracture mechanics model to quantitatively predict the shrinkage cracking behavior of restrained concrete ring specimens, and Goel et al. (2007) who developed a comprehensive review of models that predict creep and shrinkage in concrete.

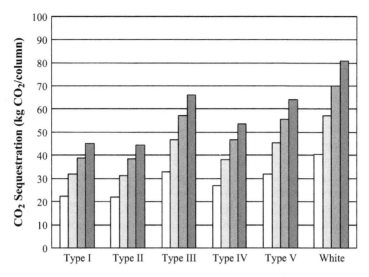

Fig. 4.4 Illustrative results from the carbon sequestration model, showing the effect of cement type on carbon sequestration potential in a high-concentration (800 ppm) CO_2 environments at 25 (○), 50 (○), 75 (○), and 100 (●) years for a 40 MPa concrete column (0.1 × 0.1 × 3 m)

Wood and Engineered Wood

Previous literature reviews have identified tools and methods for service-life prediction of wood-based materials. For example, Leicester (2001) published a review of existing models to predict durability of timber. Williams et al. (2000) published an overview of wood properties and their effect on long-term durability. Brischke et al. (2006) presented a comprehensive review on decay-influencing factors of wood and identified approaches to implement methods and tools for life prediction of wood-based materials and components.

Compared to concrete, substantially fewer models have been developed for predicting the durability of wood-based materials and components. Fridley (2002) reported a list of critical research needs for wood structures that were identified following a related workshop. More durability studies were mentioned in 4 of the 12 identified needs. Reasons for limited durability studies include, but are not limited to, the relative costs of the materials and the importance of wood versus concrete structures. Wood-based materials and components have been primarily used in lower-performance building applications (e.g., residential) than structural concrete. In addition to fire, wood is primarily at risk to biological deterioration, which is driven by one culprit—moisture. Concrete, on the other hand, is prone to physical, chemical, biological, and mechanical damage caused by complex mechanisms that have received much scientific interest over the last century. In fact, much of the previous research attention has been positioned on the mechanical degradation of fasteners used in wood structures, as opposed to being focused on the degradation of the wood itself (see Carroll et al. 2010; Zelinka and Rammer 2009). The recent surge in international interest in tall wood structures will likely contribute to renewed attention to wood-based materials and structures and the development of models to predict in situ behavior.

Despite the limited number of service-life prediction tools, researchers have previously proposed models to predict mold growth, termite attack, and fungal attack in wood-based materials and components. Viitanen and Ojanen (2007) developed a numerical model to simulate mold growth on the surface of building materials, such as gypsum board, cement screed on concrete, porous wood fiberboard, and spruce plywood. In addition, the study aimed to better predict the effect of humidity on mold growth. Moon and Augenbroe (2003) identified the need for a performance indicator that best predicts mold growth via analysis of the physical state of materials over time such that the risk of mold growth occurrence can be better predicted. Clarke et al. (1997) developed a simulation tool to predict mold growth in buildings and proposed a technique to predict the conditions that lead to mold growth. Additional models related to mold formation, growth, and decay have been proposed by Krus et al. (2001), Ojanen et al. (2011), and Vereecken and Roels (2012). Leicester and Foliente (1999) developed a simple model for termite attack that was calibrated by expert opinion due to lack of quantitative information. Subsequently, Leicester et al. (2003, 2004) developed a risk-based model and a probabilistic model, respectively, for termite attack.

Fungal attack causes damage in wood by fungal degradation of the wood polymers—cellulose, hemicellulose, and lignin. Fungi grow in wood under certain environmental conditions. Most often, moisture contents must be in excess of 20%. As a result, the wood softens and decays. The mechanisms of fungal attack are described in detail by Hammel (1997). Yang et al. (1980) identified factors that influence fungal degradation of lignin, a primary biopolymer found in wood-based materials. Aside from a model to predict mold fungi formation developed by Sedlbauer and Krus (2002), very little additional work has been completed to develop models that predict damage due to fungi in wood-based materials.

TimberLife: Service-Life Prediction Software

TimberLife[5] is an Australia-based software that estimates timber's service-life performance. It considers decay due to ground contact and weather, as well as fastener corrosion within building envelopes and termites/marine borer attack. While simple and useful, the software is limited to Australian applications. An example result from *TimberLife* is shown in Fig. 4.5. In this example, the service life of 20 mm (width) × 5 mm (thickness) square roofing tiles is estimated when no sealant is used and the tiles lie flat in contact with wood roof support system. Similar to the concrete mixture design example shown in Fig. 4.1, these results can be used to either screen for

---

[5]TimberLife Software: https://www.woodsolutions.com.au/Articles/Resources/TimberLife-Educational-Software-Program

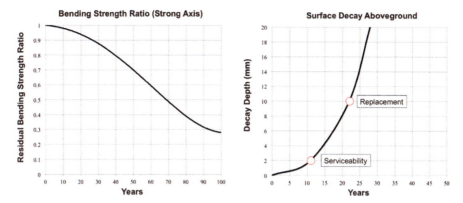

Fig. 4.5 Illustrative results (adapted) from *TimberLife* software, showing time-dependent reductions in flexural bending strength, expected surface decay, time to service or maintenance, and time to total replacement of an exposed structural wood building component

appropriate candidate materials or develop a long-term maintenance and replacement strategy.

Quantifying the EC of Buildings Under Acute Hazards

Overview

In addition to chronic deterioration, structures are vulnerable to acute hazard events or disasters. Disasters cause damage to critical assets such as buildings and contents. Damage and subsequent repair, demolition, and rebuilding increase the life cycle EC of the building, mapping to the B3 and B4 stages in the EN 15978 standard (CSN EN 15978 2011). As mentioned in section "Introduction" and discussed herein, in-service damage is typically ignored in WBLCA. It is critical, however, for the changes in EC caused by in-service damage and repair to be captured in the use phase of LCA to ensure that hazard mitigation and design decisions are considered in the sustainable development paradigm (Mileti and Gailus 2005).

To account for acute (disaster) events occurring during the service life of a building and their influence on the environmental impacts, such as EC, existing performance-based loss-estimation methodologies can be extended. These loss-estimation methodologies are intended to promote risk-informed decision-making for design and mitigation by providing estimates of the potential losses associated with future disasters or hazards at either the building or community scale. Linking hazard-related loss-estimation methodologies and LCA methodologies has the potential to connect design and assessment strategies to resilience and sustainability.

This section provides two examples to demonstrate how hazard loss estimation can be combined with an EC estimation for WBLCA, including repairs for hazard-induced damage. The examples demonstrate the decrease in losses and life cycle embodied carbon estimated following a seismic event when a building is constructed with initial hazard resistance strategies. The first example uses of a reinforced concrete-frame building as an archetypical building and uses *SimaPro* for the EC estimate. The second example uses a soft-story wood-frame building as the archetype and ATHENA for the EC estimate. The framework is provided in a general sense, for ease of adoption and extension, and the examples demonstrate the multiple approaches that can be taken in application of the framework.

A Review of Loss-Estimation Methodologies

Perhaps the most widely used loss-estimation model in the United States is HAZUS (DHS 2003), which was originally developed in 1997 by the Federal Emergency Management Agency (FEMA). HAZUS is applicable to earthquakes, floods, and hurricanes and uses simplified damage functions to compute aggregated loss estimates, including direct economic loss, casualties, and persons needing temporary shelter, at the building or community scale. HAZUS estimates integrate graphical information system (GIS) software, which is linked to databases of the US building stock and demography. More recently, the Mid-America Earthquake (MAE) Center and the National Center for Supercomputing Applications (NCSA) developed the seismic risk assessment software, MAEViz (Elnashai et al. 2008). MAEViz loss estimations are provided for business content loss, business interruption loss, business inventory loss, household and population dislocation, shelter requirements, and short-term shelter needs.

These loss models developed simultaneously with concepts of performance-based engineering (PBE). PBE is a methodology that links design and other decisions about built infrastructure to long-term performance, through probabilistic analysis to metrics that are relevant to building owners and other stakeholders, e.g., dollars, deaths, and downtime (Krawinkler and Miranda 2004; Moehle and Deierlein 2004). Although PBE can be carried out for individual buildings, groups of buildings, or regions, its unit of analysis has most often been a single building. One of the crucial metrics that results from a PBE assessment is the quantification of disaster-related economic losses in US dollars. This loss assessment provides a probabilistic accounting of the damage that may be incurred by the structure over its life cycle and the costs associated repair or replacement of critical components (see Mitrani-Reiser 2007; Ramirez et al. 2012). The assessment considers uncertainty in the damage and repair costs associated with each component in the building, as well as the severity of the hazard to which the building is subjected. Typically, these calculations are carried out using a Monte Carlo approach to consider correlations in damage in different components. This methodology is most mature in the context of earthquake engineering, but has also seen applications to wind (Huang et al. 2001), flood (Deniz et al. 2017; Taggart and van de Lindt 2009), and other extreme loads that may impact a structure's service life.

For loss estimation related to earthquake engineering, the FEMA P-58 project (ATC 2012) has compiled libraries of component fragilities and databases for repair costs and repair actions which are implemented in tools such as the Performance Assessment and Calculation Tool (PACT) software and now the SP3 software, developed by the Haselton Baker Risk Group. These tools provide a method for tracking building inventory details and to perform the intensive probabilistic computations for accumulation of losses based on inputs that include building system and component information and structural analysis results. The user may select which component fragilities—representing damage to components from demands on the building—to use from a database.

More recently, Sutley et al. (2016) developed a community-level seismic loss-estimation framework to incorporate social vulnerability. The methodology extends previous work by including an additional morbidity (injury, fatality, and PTSD diagnosis), based on the community's socioeconomic characteristics and demographics. The morbidity rates were determined as a function of the building damage levels and adjusted based on the social vulnerability of the population using empirically based odds ratios. These loss estimations focused specifically on acute hazard-induced losses yet excluded environmental impacts.

Quantifying EC by Extending Loss-Estimation Methodologies

Performance-based loss-estimation procedures provide a logical starting point to quantify the service-life EC impacts of extreme loads because they provide an accounting of the types of damage that may occur to each component of a structure, as well as the actions needed to repair that damage. Welsh-Huggins and Liel (2017a) extended the earthquake engineering loss methodology for reinforced concrete frames by cataloging each nonstructural and structural repair action recommended in FEMA P-58 (ATC 2012) by material needs and quantities, following typical construction practices described in Ching (2014). For example, seismic damage to curtain walls classified in damage state 1 is light enough to warrant only replacement of the glass, while the second, more severe, damage state requires new glass, as well as plywood to cover the area during replacement. Similar repair descriptions can be translated into material quantities to repair each damage state for each component. These quantities can be combined with environmental impacts for each material to quantify the overall EC impacts of hazard events of various intensities. This approach essentially extends life cycle EC calculations, which have been performed for different types of structures, as in, for example, the seminal study by Junnila and Horvath (2003), to include the additional repairs and construction associated with acute hazards.

Previous studies have quantified EC and economic losses from building construction and post-earthquake repairs to examine the influence of specific building design decisions for reinforced concrete buildings. For example, Hossain and Gencturk (2014) conducted Pareto optimization to minimize member size and reinforcement ratios, given desired design constraints, for two RC buildings, one with low initial cost and greater design story drifts and the other with higher upfront cost and lower drifts. They found that the low-cost building incurred larger seismic and, therefore, much higher environmental impacts than the more expensive building. However, the study also suggested that the overall life cycle environmental impact of the low-cost building was 40% lower than that of the high-cost building due to lower material volumes used in construction and removed during end-of-life disposal. Wei et al. (2016) evaluated trade-offs between the environmental, social, and economic costs associated with seismic retrofits to an existing RC building. The study computed the combined present value of losses by monetizing post-hazard carbon emissions and fatalities. Results showed that the retrofit design with lowest upfront cost and lowest hazard resistance offered the highest present value benefit.

At present, the authors are aware of only one article that provides a combined hazard and environmental loss assessment for wood buildings (Dong and Li 2016). The loss assessment considered hurricane events and computed financial loss as a percent of the building replacement value and injuries and deaths resulting from the hurricane event(s). The environment loss assessment estimated the greenhouse gas emissions (modeled as equivalent carbon dioxide emissions) using a carbon tax approach, which valuated the societal cost of CO_2 emissions.

It is worth noting that in all estimation methodologies, different studies use different parameters, factors, assumptions, datasets, scales, and boundaries. Thus, results must be carefully interpreted to account for what was and what was not included in the goal, scope, and impact assessment prior to drawing conclusions.

Exemplifying the Calculation of EC

This section exemplifies the calculation of embodied carbon during the service life of reinforced concrete-frame and wood-frame buildings exposed to and potentially damaged by earthquakes. Both examples focus solely on the potential significance of seismic repairs in relation to the environmental impacts of the upfront construction and neglect life cycle impacts of maintenance activities, operations, etc. All original and replacement materials and components are assumed in these examples to have a 50-year life to isolate the contribution of component damage to total life cycle EC. Furthermore, both examples provide a CO_2 equivalent emission estimation, namely, global warming potential (GWP) for EC. These estimations were determined through the respective software tools that use the IPCC (2013) GWP 100, a harmonized calculation method.

Reinforced Concrete Buildings

The process of quantifying embodied carbon for acute hazards is illustrated through two reinforced concrete-frame buildings in California at risk from earthquakes, adopted from Welsh-Huggins and Liel (2017b). The first structure is a modern, code-conforming four-story moment-frame concrete building designed for southern California with "special" seismic details (Building A). The second structure is designed for double the seismic forces but is otherwise identical in terms of design and detailing provisions (Building B). As a result of differences in design lateral loads, Building B has 40% greater ultimate strength (as determined from static pushover analysis); the difference in ultimate strength is less than the difference in the design force because of other factors (e.g., gravity loads, drift limits) that provide overstrength in design. The structure's deformation capacities, again as determined from static pushover analysis, and representing the deformation capacity from effective yield to a 20% post-peak drop in base shear strength, are virtually identical, corresponding to a ductility capacity of approximately 10.5. This similarity in deformation capacity is a product of the other design and detailing rules for special reinforced concrete moment-resisting frames (ACI 2011), which ensure that code-designed buildings have similar ductility at both the component and system levels.

To compare life cycle seismic and environmental performance of the two buildings, we first examine the impacts associated with material manufacturing and production needed to construct the buildings. Member sizes and reinforcement requirements for Building B are larger than Building A (see design details in Haselton 2006; Welsh-Huggins and Liel 2017b) in order to satisfy the larger force demands. The difference in member sizes requires larger material volumes for Building B, which correlates with higher environmental impacts. The architectural (nonstructural) components of the buildings (i.e., partition walls, ceiling tiles, piping) are identical in both buildings. To assess the upfront environmental impacts of the two different buildings, we use the *SimaPro* software to organize life cycle inventory quantities from the ecoinvent database (Goedkoop et al. 2013). The Tool for the Reduction and Assessment of Chemical and other Environmental Impacts (TRACI) is used to quantify resulting impacts from raw emissions associated with each inventory process. Here, impacts are quantified in terms of EC. The difference in upfront EC for the two buildings was determined to be approximately 1%. The relatively small difference in EC is due to the use of more reinforcement for Building B, which somewhat mitigates the need to increase member sizes, and the identical inventories of nonstructural components in the two buildings, which also contribute significantly to the EC.

We then examine the impacts of earthquakes over the lifespan of the buildings. This analysis considers the seismic hazard at a particular Los Angeles site (defined in Welsh-Huggins and Liel 2017b) and is based on dynamic analysis of nonlinear building simulation models. This analysis predicts the distribution of story drifts and floor accelerations on the building as a function of ground motion intensity. The seismic loss analysis uses the SP3 software to compute the distribution of losses as a function of ground motion intensity. This process takes the simulated dynamic

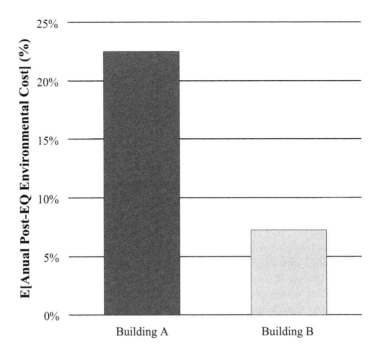

Fig. 4.6 EC associated with repairing seismic damage in four-story reinforced concrete-frame buildings, annualized over the lifetime of the structure and represented as a percentage of the upfront EC. This plot considers earthquakes producing various shaking intensities, as well as likelihood of occurrence for that shaking intensity

responses as inputs and estimates the damage to individual components in the building. The total loss is the sum of the repair and replacement costs for all components in the building, including structural beams, columns and joints and staircases, exterior-glazed curtain walls, exterior concrete cladding, interior wall partitions, suspended ceiling tiles, carpeted floor tiles, concrete roof tiles, HVAC ducts, hot and cold water pipes, sanitary waste pipes, and fire sprinkler systems.

These results show that Building B has lower seismic losses at the hazard levels considered due to less damage occurring in the stronger, stiffer building. When these results are summed over the lifetime of the building (considering the likelihood of different levels of ground shaking occurring), we find that Building A has an estimated life cycle earthquake-related EC that is 2.7 times more than the life cycle earthquake-related EC from Building B, as shown in Fig. 4.6.

Wood-Frame Buildings

Here, the process of quantifying embodied carbon for acute hazards is illustrated again through two light-frame wood buildings located in Los Angeles, California. We follow the same procedure as the previous example (see Table 4.4) for building

Table 4.4 Materials and quantities for wood-frame building archetypes

Material	Unit	Building A Total	Building B Total	Moderate damage Repaired components
#15 organic felt	m^2	2892.8	2892.8	1446.0
1/2″ regular gypsum board	m^2	3619.1	3619.1	1721.7
5/8″ regular gypsum board	m^2	1362.4	1362.4	670.78
Aluminum extrusion	Tonnes	0.5191	0.5191	0.5191
Ballast (aggregate stone)	Kg	23,168	23,168	11,581
Blown cellulose	m^2 (25 mm)	444.37	444.37	222.13
Concrete benchmark 3000 psi	m^3	88.689	88.689	88.689
FG batt R20	m^2 (25 mm)	3365.6	3365.6	1644.0
FG LF open blow R21–30	m^2 (25 mm)	6107.9	6107.9	3053.1
Galvanized sheet	Tonnes	0.8519	0.8519	0.4661
Glazing panel	Tonnes	1.014	1.014	1.014
Joint compound	Tonnes	4.9716	4.9716	2.3878
Large dimension softwood lumber, kiln-dried	m^3	65.812	65.812	31.202
Nails	Tonnes	1.2547	1.4037	0.6224
Oriented strand board	m^2 (9 mm)	1139.0	2056.8	556.37
Paper tape	Tonnes	0.0571	0.0571	0.0274
Roofing asphalt	Kg	15,519	15,519	7757.6
Screws, nuts, and bolts	Tonnes	0.3812	0.6814	0.1906
Small dimension softwood lumber, kiln-dried	m^3	21.987	31.745	11.016
Softwood plywood	m^2 (9 mm)	3133.0	3133.0	1460.5
Stucco over porous surface	m^2	1538.0	1538.0	751.27
Type III glass felt	m^2	5785.6	5785.6	2892.0
Vinyl clad wood window frame	Kg	128.78	128.78	128.78
Water-based latex paint	l	6987.5	6987.5	3353.7
Welded wire mesh/ladder wire	Tonnes	0.3819	0.3819	0.3819
Rebar, rod, light sections	Tonnes	0	0.733	0

details, material types, and quantities. The first building (Building A) is a three-story soft-story wood-frame building. The first story is used for tuck under parking and is "soft" meaning it has a significantly lower stiffness than the upper stories. The upper stories are apartment units, identical in plan with a total of ten units. Soft-story wood-frame buildings are at risk to pancake-like collapse in the soft story when exposed to ground shaking. Approximately 14,000 have been identified in Los Angeles, California, and need retrofitting. The second building (Building B) is identical in plan to the first building; however, it has been retrofitted so that the bottom story is no longer soft, and the entire building meets or exceeds current code requirements. The retrofit follows performance-based design methodology and uses oriented strand board (OSB) sheathing coupled with a steel anchor tie-down system as shear walls for the retrofit. As a result of the difference in design seismic hazard, the retrofitted building was designed for 5.7 times the base shear of the un-retrofitted building. The first mode periods of the buildings are 0.37 s and 0.23 s, respectively (see Sutley and van de Lindt 2016) for more details on the retrofit and seismic responses).

To compare life cycle seismic and environmental performance of the two buildings, we first examine the impacts associated with material manufacturing and production needed to construct the buildings. All structural and nonstructural members and components are assumed to be identical in the two buildings, except for the added shear walls and anchor tie-down systems in Building B. In total, 50 OSB shear walls were installed in Building B (18, 18, and 14 on the first, second, and third story, respectively) requiring over 16,000 additional nails, and each shear wall was accompanied by two steel rods (of varying diameter) for the anchor tie-down system. To assess the upfront environmental impacts of the two different buildings, we used ATHENA Impact Estimator for Buildings. Here, impacts are quantified in terms of CO_2 emissions or equivalents. Building B creates a 5.2% increase in upfront carbon emissions relative to Building A.

Now, we again examine the impacts of earthquakes over the lifespan of the building. This analysis considers the seismic hazard at a particular Los Angeles site and is based on nonlinear time history analysis performed in SAPWood (Pei and van de Lindt 2010) to compute the distribution of losses possible as a function of ground motion intensity. Peak inter-story drift is well-correlated with damage to wood-frame buildings (Folz and Filiatrault 2001) and is used to estimate the damage to individual components in the building. The total loss is the sum of the repair and replacement costs for all of the materials listed in Table 4.4, including interior walls, exterior walls, ceilings, windows, doors, and roofing materials in the building.

These results show that Building B has lower seismic losses at a range of hazard levels considered due to less damage occurring in the stiffened first story. Three seismic intensities were considered. A small earthquake (50% probability of occurring in 50 years) causes no damage to Building B and causes moderate damage to Building A. A moderate earthquake (10% probability of occurring in 50 years) similarly causes no damage to Building B, but complete damage to Building A. When exposed to a large earthquake (2% probability of occurring in 50 years), Building A again experiences complete damage, while Building B experiences moderate damage. The third column in Table 4.4 provides the material quantities for the replacement of damaged components to Building A caused by the small earthquake. For the moderate and large earthquakes, Building A would need to be completely replaced, and thus the first column in Table 4.4 can be assumed representative. Moderate damage is not enough to necessarily cause structural damage, and thus the retrofits themselves in Building B may not be damaged from the large earthquake. For brevity, this assumption is made, and the third column in Table 4.4 can provide a representation of the materials needing repair due to the moderate

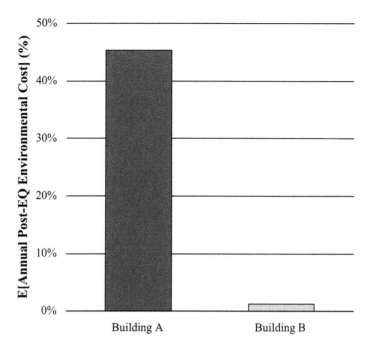

Fig. 4.7 EC associated with repairing seismic damage in three-story light-frame wood buildings, annualized over the lifetime of the structure and represented as a percentage of the upfront EC. This plot considers earthquakes producing three shaking intensities, as well as the likelihood of each intensity of shaking occurring

damage of Building B caused by the large earthquake. When these results are summed over the lifetime of the buildings, and considering the likelihood of the different levels of ground shaking occurring, we find that Building A has an estimated life cycle CO_2 emissions that are a factor of 1.4 times the life cycle CO_2 emissions from Building B considering the upfront and post-earthquake EC. Figure 4.7 provides a comparison of the life cycle earthquake-related EC as a function of the upfront EC for both buildings. Considering both of these, the retrofitted building, despite higher upfront environmental impacts from carbon emissions, is associated with significantly lower life cycle emissions due to its much superior seismic performance.

Conclusions

This chapter illustrated how maturing methods for predicting damage from chronic and acute hazard events can be incorporated into a holistic methodology for estimating total embodied carbon of concrete- and wood-based materials, components, and whole buildings. In addition to a review of LCA methodologies and challenges that are faced in quantifying embodied carbon during the use-face, simple mathematical predictions of in-service lifetime of wood and concrete materials exposed to chronic environmental stressors were demonstrated herein.

For acute stressors, such as earthquakes and floods, an extended loss-estimation methodology was used to predict anticipated repair and replacement actions for components or the whole building and, subsequently, to evaluate design alternatives based on total life cycle EC. The examples demonstrated the ability to include embodied carbon estimates in a hazard loss-estimation methodology by taking two different approaches, thus demonstrating the ease in adoptability. The results demonstrated the increase in life cycle embodied carbon that is gained when hazard damage and repair are considered (7–23% for the RC building example, considering all possible intensities of future earthquakes as well as their likelihood of occurrence). For the wood building example, the life cycle embodied carbon increases by 1–45% considering the three earthquake intensities and their likelihood of occurrence. These increases are substantial, indicating that including damage and repair due to acute hazards is crucial. Furthermore, the examples demonstrated that buildings designed with more hazard resistance provide lower losses and lower life cycle EC due to the lower levels of damage incurred during seismic events.

References

ACI (American Concrete Institute). (2011). *Building code requirements for structural concrete (ACI 318-11) and commentary (ACI 318R-11)*. Farmington Hills: American Concrete Institute.

Ahmad, S. (2003). Reinforcement corrosion in concrete structures, its monitoring and service life prediction-a review. *Cement and Concrete Composites, 25*(4–5), 459–471.

Ann, K. Y., Pack, S. W., Hwang, J. P., Song, H. W., & Kim, S. H. (2010). Service life prediction of a concrete bridge structure subjected to carbonation. *Construction and Building Materials, 24* (8), 1494–1501.

ASTM C666/C666M-15. (2015). *Standard test method for resistance of concrete to rapid freezing and thawing*. West Conshohocken: ASTM International.

ATC (Applied Technology Council). (2012). *Seismic performance assessment of buildings, FEMA P-58-1*. Redwood City: Applied Technology Council (ATC).

Attiogbe, E. K. (1996). Predicting freeze-thaw durability of concrete-a new approach. *ACI Materials Journal, 93*(5), 457–464.

Basbagill, J., Flager, F., Lepech, M., & Fischer, M. (2013). Application of life-cycle assessment to early stage building design for reduced embodied environmental impacts. *Building and Environment, 60*, 81–92.

Bazant, Z. P., Chern, J. C., Rosenberg, A. M., & Gaidis, J. M. (1988). Mathematical model for freeze-thaw durability of concrete. *Journal of the American Ceramic Society, 71*, 776–783.

Bribián, I., Aranda-Usón, A., & Scarpellini, S. (2009). Life cycle assessment in buildings: State-of-the-art and simplified LCA methodology as a complement for building certification. *Building and Environment, 44*(12), 2510–2520.

Brischke, C., Bayerbach, R., & Rapp, A. O. (2006). Decay-influencing factors: A basis for service life prediction of wood and wood-based products. *Wood Material Science and Engineering, 1* (3–4), 91–107.

Cabeza, L. F., Rincón, L., Vilariño, V., Pérez, G., & Castell, A. (2014). Life cycle assessment (LCA) and life cycle energy analysis (LCEA) of buildings and the building sector: A review. *Renewable and Sustainable Energy Reviews, 29*, 394–416.

Carroll, C., Leichti, R., & Clauson, M. (2010). Wood materials, nails, and sheathing connections from early 20th century residential buildings. *Journal of Materials in Civil Engineering, 22*(11), 1122–1128.

Chau, C. K., Leung, T. M., & Ng, W. Y. (2015). A review on life cycle assessment, life cycle energy assessment and life cycle carbon emissions assessment on buildings. *Applied Energy, 143*, 395–413.

Ching, E. (2014). *Building construction illustrated*. Hoboken: Wiley.

Cho, T. (2007). Prediction of cyclic freeze-thaw damage in concrete structures based on response surface method. *Construction and Building Materials, 21*(12), 2031–2040.

Clarke, J. A., Johnstone, C. M., Kelly, N. J., McLean, R. C., & Nakhi, A. E. (1997). Development of a simulation tool for mould growth prediction in buildings. *Proceedings of the Fifth International IBPSA Conference, 2*, 343–349.

Collins, F. (2010). Inclusion of carbonation during the life cycle of built and recycled concrete: Influence on their carbon footprint. *The International Journal of Life Cycle Assessment, 15*(6), 549–556.

CSN EN 15978. (2011). Sustainability of construction works – assessment of environmental performance of buildings – calculation method. European Standard 15978.

Deniz, D., Arneson, E. E., Liel, A. B., Dashti, S., & Javernick-Will, A. N. (2017). Flood loss models for residential buildings based on the 2013 Colorado floods. *Natural Hazards, 85*(2), 977–1003.

DHS (Department of Homeland Security). (2003). HAZUS-MH MRI technical manual. *DHS Emergency Preparedness and Response Directorate*. Washington, DC: FEMA Mitigation Division.

Dixit, M. K., Fernández-Solís, J. L., Lavy, S., & Culp, C. H. (2012). Need for an embodied energy measurement protocol for buildings: A review paper. *Renewable and Sustainable Energy Reviews, 16*(6), 3730–3743.

Dong, Y., & Li, Y. (2016). Risk-based assessment of wood residential construction subjected to hurricane events considering indirect and environmental loss. *Sustainable and Resilient Infrastructure, 1*(1–2), 46–62.

Elnashai, A. S., Elnashai, A., Hampton, S., Lee, J. S., McLaren, T., Myers, J. D., Navarro, C., Spencer, B., & Tolbert, N. (2008). Architectural overview of MAEviz-HAZTURK. *Journal of Earthquake Engineering, 12*, 92–99.

Erlandsson, M., & Borg, M. (2003). Generic LCA-methodology applicable for buildings, constructions and operation services, today practice and development needs. *Building and Environment, 38*(7), 919–938.

Folz, B., & Filiatrault, A. (2001). Cyclic analysis of wood shear walls. *Journal of Structural Engineering, 127*(4), 433–441.

Fridley, K. (2002). Wood and wood-based materials: Current status and future of a structural material. *Journal of Materials in Civil Engineering, 14*(2), 91–96.

García-Segura, T., Yepes, V., & Alcalá, J. (2014). Life cycle greenhouse gas emissions of blended cement concrete including carbonation and durability. *The International Journal of Life Cycle Assessment, 19*(1), 3–12.

Glasser, F. P., Marchand, J., & Samson, E. (2008). Durability of concrete—degradation phenomena involving detrimental chemical reactions. *Cement and Concrete Research, 38*(2), 226–246.

Goedkoop, M., Oele, M., Leijting, J., Ponsioen, T., & Meijer, E. (2013). *Introduction to LCA with SimaPro*. Amersfoort: PréSustainability.

Goel, R., Kumar, R., & Paul, D. K. (2007). Comparative study of various creep and shrinkage prediction models for concrete. *Journal of Materials in Civil Engineering, 19*(3), 249–260.

Grant, A., & Ries, R. (2013). Impact of building service life models on life cycle assessment. *Building Research and Information, 41*(2), 168–186.

Grant, A., Ries, R., & Kibert, C. (2014). Life cycle assessment and service life prediction. *Journal of Industrial Ecology, 18*(2), 187–200.

Hammel, K. E. (1997). Fungal degradation of lignin. In *Driven by nature: Plant litter quality and decomposition* (pp. 33–46). Wallingford: CAB International.

Haselton, C. B. (2006). *Assessing seismic collapse safety of modern reinforced concrete moment frame buildings* (Doctoral Dissertation). Stanford University, Palo Alto.

Hossain, K. A., & Gencturk, B. (2014). Life-cycle environmental impact assessment of reinforced concrete buildings subjected to natural hazards. *Journal of Architectural Engineering, 22*(4), A4014001-1-12.

Huang, Z., Rosowsky, D. V., & Sparks, P. R. (2001). Long-term hurricane risk assessment and expected damage to residential structures. *Reliability Engineering & System Safety, 74*(3), 239–249.

IPCC. (2013). Climate change 2013: The physical science basis. In T. F. Stoker, D. Qin, G. K. Plattner, M. Tignor, S. K. Allen, J. Boschung, A. Nauels, Y. Xia, V. Bex, & P. M. Midgley (Eds.), *Contribution of working group I to the fifth assessment report of the intergovernmental panel on climate change*. Cambridge, UK: Cambridge University Press.

Jiang, L., Lin, B., & Cai, Y. (2000). A model for predicting carbonation of high-volume fly ash concrete. *Cement and Concrete Research, 30*(5), 699–702.

Junnila, S., & Horvath, A. (2003). Life-cycle environmental effects of an office building. *Journal of Infrastructure Systems, 9*(4), 157–166.

Kashef-Haghighi, S., Shao, Y., & Ghoshal, S. (2015). Mathematical modeling of CO2 uptake by concrete during accelerated carbonation curing. *Cement and Concrete Research, 67*, 1–10.

Khasreen, M. M., Banfill, P. F., & Menzies, G. F. (2009). Life-cycle assessment and the environmental impact of buildings: A review. *Sustainability, 1*(3), 674–701.

Kosmatka, S. H., Panarese, W. C., & Kerkhoff, B. (2002). *Design and control of concrete mixtures*. Skokie: Portland Cement Association.

Krawinkler, H., & Miranda, E. (2004). Performance-based earthquake engineering. In *Earthquake engineering: From engineering seismology to performance-based engineering*. Boca Raton, Florida: CRC Press.

Krus, M., Sedlbauer, K., Zillig, W., & Künzel, H.M. (2001). A new model for Mould prediction and its application on a test roof. In *The second internal scientific conference on the current problems of building physics in the rural building*, Cracow.

Kurtis, K. E., Monteiro, P. J., & Madanat, S. M. (2000). Empirical models to predict concrete expansion caused by sulfate attack. *ACI Materials Journal, 97*(2), 156–161.

Lee, S., Park, W., & Lee, H. (2013). Life cycle CO2 assessment method for concrete using CO2 balance and suggestion to decrease LCCO2 of concrete in South-Korean apartment. *Energy and Buildings, 58*, 93–102.

Leicester, R. H. (2001). Engineered durability for timber construction. *Progress in Structural Engineering and Materials, 3*, 216–227.

Leicester, R. H., & Foliente, G. C. (1999). Models for timber decay and termite attack. In M. A. Lacasse & D. J. Vanier (Eds.), *Durability of building materials and components* (pp. 756–765). Ottawa: Institute for Research in Construction.

Leicester, R. H., Wang, C. -H., & Cookson, L. J. (2003). A risk model for termite attack in Australia. In *Proceedings of the 34th IRGWP annual meeting*, Brisbane.

Leicester, R.H., Wang, C-H., Cookson, L. (2004). A probabilistic model for termite attack. *Proceedings of the 8th World Conference on Timber Engineering*. Lahti.

Mileti, D., & Gailus, J. (2005). Sustainable development and hazards mitigation in the United States: Disasters by design revisited. *Mitigation and Adaptation Strategies for Global Change, 10*, 491–504.

Mitrani-Reiser, J. (2007). *An ounce of prevention: Probabilistic loss estimation for performance-based earthquake engineering* (Doctoral dissertation). California Institute of Technology.

Moehle, J., & Deierlein, G.G. (2004). A framework methodology for performance-based earthquake Engineering. In *13th world conference on earthquake engineering* (pp. 3812–3814).

Monteiro, P. J. M., & Kurtis, K. E. (2003). Time to failure for concrete exposed to severe sulfate attack. *Cement and Concrete Research, 33*(7), 987–993.

Monteiro, I., Branco, I. F., De Brito, J., & Neves, R. (2012). Statistical analysis of the carbonation coefficient in open air concrete structures. *Construction and Building Materials, 29*, 263–269.

Moon, H. J., & Augenbroe, G. (2003). Evaluation of hygrothermal models for mold growth avoidance prediction. In *Proceedings of the eighth international IBPSA conference*, Netherlands.

Mounanga, P., Khelidj, A., Loukili, A., & Baroghel-Bouny, V. (2004). Predicting Ca(OH)2 content and chemical shrinkage of hydrating cement pastes using analytical approach. *Cement and Concrete Research, 34*(2), 255–265.

Ojanen, T., Peuhkuri, R., Viitanen, H., Lähdesmäki, K., Vinha, J., & Salminen, K. (2011). Classification of material sensitivity–New approach for Mould growth modeling. *9th Nordic Symposium on Building Physics, 2*, 867–874.

Otieno, M. B., Beushausen, H. D., & Alexander, M. G. (2011). Modelling corrosion propagation in reinforced concrete structures–a critical review. *Cement and Concrete Composites, 33*(2), 240–245.

Pade, C., & Guimaraes, M. (2007). The CO2 uptake of concrete in a 100-year perspective. *Cement and Concrete Research, 37*(9), 1348–1356.

Papadakis, V. G., Vayenas, C. G., & Fardis, M. N. (1991). Experimental investigation and mathematical modeling of the concrete carbonation problem. *Chemical Engineering Science, 46*(5–6), 1333–1338.

Pei, S., & van de Lindt, J. (2010). User's manual for SAPWood for Windows Seismic Analysis Package for Woodframe Structures. NEEShub (http://www.nees.org).

Penttala, V. (2006). Surface and internal deterioration of concrete due to saline and non-saline freeze–thaw loads. *Cement and Concrete Research, 36*(5), 921–928.

Peuportier, B. L. P. (2001). Life cycle assessment applied to the comparative evaluation of single family houses in the French context. *Energy and Buildings, 33*(5), 443–450.

Ramirez, C. M., Liel, A. B., Mitrani-Reiser, J., Haselton, C. B., Spear, A. D., Steiner, J., Deierlein, G. G., & Miranda, E. (2012). Expected earthquake damage and repair costs in reinforced concrete frame buildings. *Earthquake Engineering & Structural Dynamics, 41*(11), 1455–1475.

Riding, K. A., Thomas, M. D., & Folliard, K. J. (2014). Apparent diffusivity model for concrete containing supplementary cementitious materials. *ACI Materials Journal, 110*(6), 705–714.

Roziare, E., Loukili, A., El Hachem, R., & Grondin, F. (2009). Durability of concrete exposed to leaching and external Sulphate attacks. *Cement and Concrete Research, 39*(12), 1188–1198.

Saetta, A. V., & Vitaliani, R. V. (2004). Experimental investigation and numerical modeling of carbonation process in reinforced concrete structures: Part I: Theoretical formulation. *Cement and Concrete Research, 34*(4), 571–579.

Sedlbauer, K., & Krus, M. (2002). Method for predicting the formation of Mould fungi. U.S. Patent Application 10/398,046.

Sfikas, I. P., & Ingham, J. P. (2016). Service life design of concrete structures using probabilistic modelling tools: Statistical analysis of input parameters. In M. Grantham, I. Papayianni, K. Sideris (Eds.), *Concrete solutions* (pp. 437–446). Taylor & Francis Group, Boca Raton, Florida: CRC Press.

Shah, S. P., Ouyang, C., Marikunte, S., Yang, W., & Becq-Giraudon, E. (1998). A method to predict shrinkage cracking of concrete. *Materials Journal, 95*(4), 339–346.

Sharma, A., Saxena, A., Sethi, M., & Shree, V. (2011). Life cycle assessment of buildings: A review. *Renewable and Sustainable Energy Reviews, 15*(1), 871–875.

Simonen, K. (2014). Life cycle assessment. Pocket Architecture: Technical Design Series.

Souto-Martinez, A., Delesky, E. A., Foster, K. E. O., & Srubar, W. V., III. (2017). A mathematical model for carbon sequestration potential of ordinary Portland cement (OPC) concrete. *Construction and Building Materials, 147*, 417–427.

Srubar, W. V., III. (2015a). Stochastic service-life modeling of chloride-induced corrosion in recycled-aggregate concrete. *Cement and Concrete Composites, 55*, 103–111.

Srubar, III, W.V. (2015b). The future of LCAs and EPDs: Incorporating service-life in the environmental impact assessments of green building materials. In *Proceedings of the 2015 Architectural Engineering Institute Conference* (pp. 606–615). Milwaukee.

Stambaugh, N. D. (2017). *Numerical service-life model of chloride-induced corrosion in recycled aggregate concrete and design optimization of sustainable and durable concrete mixtures* (MS Thesis). University of Colorado Boulder, Boulder.

Stambaugh, N. D., Bergman, T. L., & Srubar, W. V., III. (2018). Numerical service-life modeling of chloride-induced corrosion in recycled-aggregate concrete. *Construction and Building Materials, 161*, 236–245.

Steffens, A., Dinkler, D., & Ahrens, H. (2002). Modeling carbonation for corrosion risk prediction of concrete structures. *Cement and Concrete Research, 32*(6), 935–941.

Strand, S. M., & Hovde, P. J. (1999). Use of service life data in LCA of building materials. In M. A. Lacasse & D. J. Vanier (Eds.), *Durability of building materials and components* (pp. 1948–1958). Ottawa: Institute for Research in Construction.

Sutley, E. J., & van de Lindt, J. W. (2016). Evolution of predicted seismic performance for wood-frame buildings. *Journal of Architectural Engineering, 22*(3), B4016004.

Sutley, E. J., van de Lindt, J. W., & Peek, L. (2016). Community-level framework for seismic resiliency. I: Coupling social vulnerability and Engineering building systems. *Natural Hazards Review*: 04016014.

Taggart, M., & van de Lindt, J. W. (2009). Performance-based design of residential wood-frame buildings for flood based on manageable losses. *Journal of Performance of Constructed Facilities, 23*(2), 56–64.

Takewaka, K., Yamaguchi, T., & Maeda, S. (2003). Simulation model for deterioration of concrete structures due to chloride attack. *Journal of Advanced Concrete Technology, 1*(2), 139–146.

Tanaka, Y., Kawano, H., Watanabe, H., & Nakajo, T. (2006). Study on cover depth for Prestressed concrete bridges in airborne-chloride environments. *Prestressed Concrete Institute Journal, 51* (2), 42–53.

Tikalsky, P. J., Pospisil, J., & MacDonald, W. (2004). A method for assessment of the freeze–thaw resistance of preformed foam cellular concrete. *Cement and Concrete Research, 34*(5), 889–893.

Tixier, R., & Mobasher, B. (2003). Modeling of damage in cement-based materials subjected to external sulfate attack. I: Formulation. *Journal of Materials in Civil Engineering, 15*(4), 305–313.

Tuutti, K. (1982). *Corrosion of steel in concrete.* Stockholm: Swedish Cement and Concrete Research Institute.

Vereecken, E., & Roels, S. (2012). Review of Mould prediction models and their influence on Mould risk evaluation. *Building and Environment, 51*, 296–310.

Viitanen, H., & Ojanen, T. (2007). Improved model to predict mold growth in building materials. In *Proceedings of the Thermal Performance of the Exterior Envelopes of Whole Building X.* Florida.

Wang, X. Y., & Lee, H. S. (2009). A model for predicting the carbonation depth of concrete containing low-calcium fly ash. *Construction and Building Materials, 23*(2), 725–733.

Wei, H., Shohet, I. M., Skibniewski, J., & Shapira, S. (2016). Assessing the lifecycle sustainability costs and benefits of seismic mitigation designs for buildings. *Journal of Architectural Engineering, 22*(1), 1–13.

Welsh-Huggins, S. J., & Liel, A. B. (2017a). A life-cycle framework for integrating green building and hazard-resistant design: Examining the seismic impacts of buildings with green roofs. *Structure and Infrastructure Engineering, 13*(1), 19–33.

Welsh-Huggins, S. J., & Liel, A. B. (2017b). Is hazard resilience sustainable? Evaluating multi-objective outcomes from enhanced seismic design decisions for buildings. *Journal of Structural Engineering*, In press.

Wenting, L., Pour-Ghaz, M., Castro, J., & Weiss, J. (2012). Water absorption and critical degree of saturation relating to freeze-thaw damage in concrete pavement joints. *Journal of Materials in Civil Engineering, 24*(3), 299–307.

Williams, R. S., Jourdain, C., Daisey, G. I., & Springate, R. W. (2000). Wood properties affecting finish service life. *Journal of Coatings Technology, 72*(902), 35–42.

Yang, H. H., Effland, M. J., & Kirk, T. K. (1980). Factors influencing fungal degradation of lignin in a representative lignocellulosic, thermomechanical pulp. *Biotechnology and Bioengineering, 22*, 65–77.

Yang, K.-H., Seo, E.-A., & Tae, S.-H. (2014). Carbonation and CO2 uptake of concrete. *Environmental Impact Assessment Review, 46*(2014), 43–52.

Zelinka, S., & Rammer, D. (2009). Corrosion rates of fasteners in treated wood exposed to 100% relative humidity. *Journal of Materials in Civil Engineering, 21*(12), 758–763.

Chapter 5
Embodied Carbon of Surfaces: Inclusion of Surface Albedo Accounting in Life-Cycle Assessment

Tiziana Susca

Introduction

Earth is continuously irradiated by the Sun energy. A part of such energy is directly absorbed by the atmosphere, a part by the surface of the Earth, and a part is reflected back to space. Furthermore, Earth re-irradiates part of the absorbed energy to space, giving rise to energy balance (Fig. 5.1).

Any kind of perturbation, such as the modification of surface or atmospheric properties, can alter the energy balance. This phenomenon happens continuously exerting a radiative forcing (RF).[1] The difference between incoming and outgoing radiation compared to the same difference measured in the preindustrial era – usually 1750 is considered as reference value – gives a measure of the RF. Positive RFs provoke the warming of the system, while, negative RFs tend to cool it (IPCC – Intergovernmental Panel on Climate Change 2007).

RFs are usually divided into direct, indirect, and non-RF. Direct RFs affect the Earth's radiative budget. An example is carbon dioxide (CO_2) because it can absorb and emit infrared radiations. Indirect RFs are all the perturbations of the usual energy balance that can indirectly alter the radiation balance, such as precipitation efficiency of clouds. A non-RF does not directly involve the radiation, but its modification gives rise to an energy imbalance. An example of non-RF is the evapotranspiration flux deriving by agricultural irrigation that can alter the

---

[1]According to definition provided by the Intergovernmental Panel on Climate Change (IPCC) (2001): "The RF of the surface-troposphere system due to the perturbation in or the introduction of an agent is the change in net irradiance at the tropopause after allowing for stratospheric temperatures to readjust to radiative equilibrium, but with the surface and tropospheric temperatures and state held fixed at the unperturbed values."

T. Susca (✉)
Bundesanstalt für Materialforschung und –prüfung (BAM), Berlin, Germany
e-mail: tiziana.susca@gmail.com

© Springer International Publishing AG 2018
F. Pomponi et al. (eds.), *Embodied Carbon in Buildings*,
https://doi.org/10.1007/978-3-319-72796-7_5

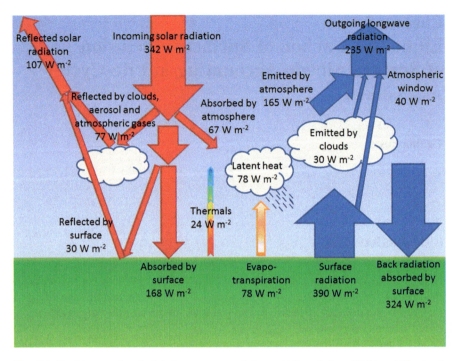

Fig. 5.1 Earth's annual and global mean energy balance (Adapted from Kiehl and Trenberth 1997)

radiation balance (e.g., IPCC 2001; Davin and de Noblet-Ducoudré 2010; Davin et al. 2007).

A RF is also defined as the instantaneous change in energy flux at the tropopause. However, frequently, in climate science, RFs are "adjusted" to the top of the atmosphere (TOA) – which is the layer to which forcings are commonly reported (W. G. I. IPCC – Intergovernmental Panel on Climate Change 2013) – throughout the stratosphere. These two metrics can be used indifferently in short-term analyses, since the RFs calculated at the tropopause are, considering the short-term climate response, a valid approximation to the adjusted RFs (Hansen et al. 1981).

RFs can be also distinguished, considering their origin, into natural and anthropogenic. The natural RFs are volcanos and solar changes, while the other RFs have an anthropogenic origin (IPCC – Intergovernmental Panel on Climate Change 2007).

The largest RFs are connected to the variation in the concentration of greenhouse gases (GHGs) in the atmosphere. Nevertheless, also changes in land use can consistently alter the energy balance. In particular, comparing the data related to 2011 with those related to 1750, the modification of the land use determined, during time, an increase in albedo that has provoked a decrease in the net radiance on the Earth's surface of 0.15 ($\pm$0.10) Wm^{-2} (W. G. I. IPCC – Intergovernmental Panel on Climate Change 2013). Every alteration of the landscape, such as urbanization, deforestation, desertification, and reforestation, produces RFs (Kabat et al. 2004; Köhler et al. 2010; Pitman 2003).

Land-cover and land-use changes can alter emissivity and surface albedo.[2] In turn, variations in emissivity and albedo can affect both absorbed and reflected short-waves, as well as long-waves altering the Earth's radiative balance. Moreover, the change in land use determines a modification in the amount of evaporated water, sensible heat, and transpiration. Any modification in sensible heat flux alters evapotranspiration and, consequently, also the surface temperature and the emitted long-waves. Such variation in net irradiance on a surface gives rise to a modification of the global mean surface response temperature (W. G. I. IPCC – Intergovernmental Panel on Climate Change 2013; Menon et al. 2010). Depending on the amount of surface on which the surface albedo is modified and depending on the latitude, the variation in temperature can be relevant at the regional or global scale and its extent can vary (Bright 2015; Otterman 1977; Pongratz et al. 2010; Cess 1978; Betts 2001). The importance of the amount of the incoming solar radiation, depending on the latitude (e.g., Kiehl and Trenberth 1997), explains the importance of using site-dependent climatological models for the assessment of the effect of the variation in albedo on RF.

IPCC (2013) uses global warming potential (GWP) – kilograms of carbon dioxide equivalents (CO_2 eq) per square meter – as a normalized metric to express the effect of RF at TOA. Therefore, also the effect of the variation in surface albedo is expressed by means of such metric. However, some authors argue that albedo effect and CO_2 effect involve different spatial and temporal scales (Davin et al. 2007). Moreover, the effect of GHGs on climate is independent of the site where they are emitted since they are well-mixed in the troposphere. Therefore, they have the same impact all over the globe (W. G. I. IPCC – Intergovernmental Panel on Climate Change 2013). Contrarily, the effect of the variation in surface albedo on RF is highly site-dependent because it depends on the amount of solar energy reaching the surface – that in turn depends on latitude – and the effects on temperature are predominantly local (e.g., Bright 2015; Otterman 1977). Other metrics have been explored to express the effect of the variation of surface albedo on RF; however, to date, not a unanimous consensus about a better metric has been reached.

Aim of the Research and Methodology

Over time, humankind has intervened on natural landscapes transforming Earth's land surface that, consequently, has exerted RF. Such interventions have become particularly significant starting from the industrial revolution (e.g., Goldewijk

---

[2]Albedo is defined as "The ratio of the light reflected by a body to the light received by it. Albedo values range from 0 (pitch black) to 1 (perfect reflector)" (National Aeronautics and Space Administration n.d.).

2001). Furthermore, the growing urbanization has led to an increase in the substitution of natural materials with artificial ones such as asphalt. The use of asphalt or dark materials in the building envelope can impact the urban energy budget.

Traditionally, life-cycle assessment (LCA) – a valuable tool for assessing the impact related to products, goods, and services – omits the effect of the modification of surface albedo due to land-use and land-cover changes (LULCC), including urbanization, on RFs (Røyne et al. 2016). Just in recent years, following IPCC, CO_2 eq has been increasingly used in LCA studies as a metric to include albedo (Bright 2015). However, if the published body of literature about the effect of LULCC on RF – referred mainly to forest management – is growing (Pongratz et al. 2010), the number of scientific publications about the effect of the variation of the albedo of urban surfaces on RF is still exiguous.

The present chapter aims, first, at examining the body of literature that investigated the inclusion of the direct effect of surface albedo on RF in the LCA methodology. Subsequently, the observation is narrowed to the case studies that investigated the effect of the variation of surface albedo in the building sector and that implemented such effect in the LCA methodology.

The studies presented in this review chapter have been selected among the international peer-reviewed literature already published at the time of this research. The selected studies have been found in Web of Science among those published from 2000 to 2016 and responding contemporarily to the keywords: "albedo" and "life-cycle assessment."[3]

Land-Use and Land-Cover Changes and Surface Albedo

Deforestation, afforestation, and substitution of native vegetation with croplands can affect climate. Two kinds of effect related to vegetation can be distinguished: biogeophysical and biogeochemical. Variation in surface albedo exerts biogeophysical mechanisms affecting RF.

Over the last years, the importance of biogeophysical effects related to the anthropogenic land cover change grew contemporarily due to the increasing attention to deforestation, afforestation, and change in land use for food production and for biofuel, bioenergy, and biomass production (e.g., Cherubini and Strømman 2013; McKechnie et al. 2014) widely promoted by the European Union (European Commission 2014).

In particular, the investigation about the potential environmental convenience of the production of biofuel compared to fossil fuels has triggered the exploration of the effect of the variation in surface albedo, due to the substitution of the natural

---

[3]Of the articles responding to such criteria, a number have been excluded from the review because they are not fitting to the topic, because they contain literature review, or because they were not published in scientific international journals.

biomes in favor of croplands and their management, on climate. The results of such exploration have been just recently included in LCA studies since, traditionally, bio-based products – such as biomass – have been considered climate neutral[4] (Singh et al. 2014; Holtsmark 2015). The carbon neutrality of bio-based products has been criticized because there is a time lag between the point in time when carbon is taken up and the time of carbon release (e.g., Holtsmark 2015; Zhang et al. 2010; Bright et al. 2011; Haberl 2013; Friedland and Gillingham 2010). Research has shown that not only bio-based product cannot be considered carbon neutral by default but also that the accounting of the effect of surface albedo on climate is fundamental since its contribution to the environmental impacts of biofuel is dominating (e.g., Caiazzo et al. 2014; Cherubini et al. 2012; Bright et al. 2012, 2015).

In the last years also, the evaluation of the effect of the variation of albedo due to forest management has gained momentum (e.g., Helin et al. 2013; Lohila et al. 2010; Schwaiger and Bird 2010). Some studies revealed that the increase in albedo due to temporarily deforestation has a more important cooling effect on the climate than the warming effect deriving from the release of CO_2 in the atmosphere (e.g., Cherubini et al. 2012; Randerson et al. 2006; Bala et al. 2007).

The attention toward deforestation, afforestation, and forest management for the production of biofuel entailed many progresses in the field of the accounting of the effect of the variation in surface albedo on RF. For instance, the models used for such evaluations became, over time, more sophisticated overcoming the limits of the first pioneering studies in this field.

Muñoz et al. (2010) evaluated the effect of the increase in albedo related to the greenhouse agriculture. For such evaluation, the authors assessed the RF at the TOA. Therefore, for the evaluation of the effect of the surface albedo, they introduced a parameter (i.e., T_a) accounting absorption and reflection of solar radiation through the atmosphere:

$$RF_{TOA} = -R_S T_a \Delta \alpha_S$$

where RF_{TOA} is the radiative forcing at the TOA, R_S is the downward solar radiation at the surface, $T_a{}^5$ is the atmospheric transmittance factor that expresses the fraction of solar radiation reaching the TOA, and $\Delta \alpha_S$ is the variation in surface albedo.

The equivalence between the variation in surface albedo and the correspondent amount of CO_2 eq that gives rise to the same RF is expressed by the following equation:

---

[4]Meaning that its production has a zero net impact on climate change considering the carbon uptakes and carbon release related to biomass growth and combustion, respectively.

[5]It has been evaluated that the global value for T_a is 0.854 (Lenton and Vaughan 2009).

$$CO_2 \ eq = \frac{A \ RF_{TOA} \ln 2 \ p_{CO_2,ref} M_{CO_2} m_{air}}{A_{Earth} \Delta F_{2X} \ M_{air} AF}$$

where A is the surface affected by the change in albedo [m^2], RF_{TOA} [W m^{-2}], $p_{CO2,ref}$ is the partial pressure of CO_2 in the atmosphere (i.e., 383 ppmv), MCO_2 is the molecular weight of CO_2 (i.e., 44.01 g mol^{-1}), m_{air} is equal to 5.148×10^{15} [Mg], A_{Earth} is the surface of Earth (i.e., 5.1×10^{14} m^2), ΔF_{2X} is the RF that results from doubling the current CO_2 concentration in the atmosphere (i.e., 3.7 W m^{-2}), M_{air} is the molecular weight of dry air (i.e., 28.95 g mol^{-1}), and AF is the CO_2 decay over time according to the Bern carbon cycle:

$$AF = \frac{\int_1^n f(t) dt}{n}$$
$$f(t) = 0.217 + 0.259 e^{\frac{-\Delta t}{172.9}} + 0.338 e^{\frac{-\Delta t}{18.51}} + 0.186 e^{\frac{-\Delta t}{1.186}}$$

Such equivalence has been used to evaluate the effect of tomato production covered by plastic greenhouse in Almeria, Spain, on climate. Specifically, the authors accounted the effect of the change of surface albedo due to the plastic greenhouse. The results reveal that the accounting of surface albedo has an important role in the evaluation of the LCA impacts ranging from 9% to 75% of the total impact on GWP depending on the time horizon used for the evaluation. Such evaluation is affected by uncertainties related to model used. For instance, the authors evaluated that the error related to T_a is $\pm30\%$ and the overall error related to the CO_2 eq is $\pm35\%$ excluding the uncertainties implicit in the life-cycle inventory.[6]

When the assessment of the effect of the variation in surface albedo is applied to the forest management at high latitude, the effect of the temporal variability of surface albedo due to snow cover should be included. Cherubini et al. (2012) evaluated the effect on RF due to temporary land-use change – such as forest rotation or plantations – for bioenergy systems including the effect of surface albedo. The variation in surface albedo is due to harvesting of biomass. Such variation can be particularly relevant in regions interested by snow cover. Therefore, the authors evaluated the effect on RF including the monthly (i.e., m) variation in surface albedo:

$$RF_\alpha(t) = \frac{\sum_{m=1}^{m=12} -\bar{R}(m) \bar{f}(m) \Delta \bar{\alpha}(m) A_{aff}}{M} y_\alpha(t) A_{Earth}^{-1}$$

As a consequence, they used the mean monthly incoming radiation at the TOA (i.e., $\bar{R}(m)$). M is the number of months involved in the forest rotation or plantation and it depends on the trees planted. $\bar{f}(m)$ is a two-way atmospheric transmittance parameter that accounts for the monthly mean reflection (i.e., T_a equal to 0.854

---

[6]In general, another source of uncertainties is the climate sensitivity (e.g., Holtsmark 2015; Roe and Baker 2007; Caldeira et al. 2003).

(Lenton and Vaughan 2009)) and absorption of solar radiation (i.e., the clearness index K(m)) throughout the atmosphere. $y_\alpha(t)$ is a function describing the interannual time evolution of the annually averaged local instantaneous forcing. Besides, A_{aff} is the surface interested by the change in land use and A_{Earth} is the surface of Earth. The authors multiplied the RF deriving from the variation of surface albedo by 1.94,[7] a coefficient that accounts for the climate efficacy (E) of surface albedo variation. Eventually, the authors calculated the characterization factor for surface albedo change as follows:

$$\text{GWP}_{\text{Albedo}} = \frac{\gamma^{-1} \int_0^{\text{TH}} E_{\text{albedo}} \Delta RF_\alpha(t)\text{dt}}{\int_0^{\text{TH}} E_{\text{CO}_2} \Delta RF_{\text{CO}_2}(t)\text{dt}}$$

Where γ is the carbon yield expressed in kg-bioCO_2 m^{-2} of the affected area. The assessment of the effect of the variation in surface albedo on GWP has been applied to different types of forests in different sites. The results display that the inclusion of the evaluation of the effect of the variation in surface albedo is relevant and can reduce significantly the effect of the production of bioenergy on climate change.

Jørgenson et al. (2014) pointed out the importance of translating the effect of surface albedo on the instantaneous change in RF into the variation in global surface temperature. The authors replaced the instantaneous RF with the adjusted RF multiplying the instantaneous RF by the albedo climate efficacy that is between 1.5 and 5 times higher than CO_2 (see footnote 13). Therefore, the authors used the effective forcing rather than the RF as a metric for assessing the effect of the variation of surface albedo in the case of the plantation of the *Miscanthus* instead of forest or fallow land in Wisconsin in the United States. *Miscanthus* was chosen for such study because it is a second-generation biomass plantation. The study shows that the highest difference in surface albedo between the *Miscanthus* plantation and forest or fallow land can be found in winter. Indeed, crop, which is harvested before winter, allows the deposition of snow that has a higher albedo than fallow land or forest. Contrarily, in summer, such a difference is minimized. According to the assumptions considered for such study, *Miscanthus* plantation leads to a net global cooling, with the cooling effect related to the albedo variation being predominant compared to other effects, such as CO_2 sequestration.

Similarly, Giuntoli et al. (2015) assessed the environmental impacts related to bioenergy for domestic heat production from logging residues and compared the results with the impacts related to natural gas. The results reveal that bioenergy from logging residues does not impact the climate per se, but the energy for its

---

[7]Hansen et al. (2005) define the efficacy of a climate forcing as "[. . .] the global mean temperature change per unit forcing produced by the forcing agent relative to the response produced by a standard CO_2 forcing from the same initial climate state." The climate efficacy (E) of surface albedo can range from 1.5 to 5 depending on the model considered (Hansen et al. 2005; Hansen and Nazarenko 2004; Bellouin and Boucher 2010).

processing and the change in carbon feedstock has an impact on climate. The authors calculated the contribution of surface albedo on climate assuming that albedo remains constant over time. Indeed, they postulate that the same amount of logging residues are removed every year, maintaining constant the balance – in terms of surface albedo – between vegetation regrowth and biomass removed. In this evaluation, the site-specific parameters, such as the specific amount of solar radiation reaching the soil have been included. The results of such investigation exhibit that the increase in albedo has a limited impact on climate compared to the impact related to the well-mixed GHG emissions.

Caiazzo et al. (2014) evaluated the effect of the change in land use in four geographical areas for the cultivation of biomass for biofuel production, including the effect of the variation of albedo. The data about the 8-h surface albedo related to a whole year and retrieved from Moderate Resolution Imaging Spectroradiometer (MODIS) were linearly interpolated to obtain the daily albedo evaluations. Eleven scenarios were investigated revealing that in two cases, the land-use change produces a warming effect, whereas in the remaining cases, the change in land use entails a cooling effect. Regardless of the specific results, it is of utmost importance that in all cases, the effect of the variation in surface albedo is dominant among the other causes of effects on climate related to biofuel production.

Notwithstanding the equivalences between change in surface albedo and CO_2 eq developed in the last years are quite advanced, the initial assumptions for the evaluation of the effect of the variation in surface albedo can be still problematic. In the literature, some disagreement can be found about the assumption at the base of the modeling in the management of forests. Such assumptions are fundamental for assessing whether biofuel is more sustainable than other fuels. For instance, Holtsmark (2015) states that the choice of the baseline scenarios is crucial to the conclusions of the LCA studies. The author evaluated the impact of wood fuels comparing the harvest scenario with a no-harvest one considered as a baseline. In such evaluation, the baseline scenario includes the accumulation of dead and living biomass that is not harvested and the related effect of the variation in surface albedo. Holtsmark found that the inclusion of albedo effect has an important contribution in the evaluation of the effect of forest management on climate, especially when a short time horizon is considered in the evaluations. Furthermore, differently from other studies, the author found that accounting also the effect of the modification in surface albedo, bioenergy from slow-growth forests has more impact on climate in a timeframe of 100 years than fossil fuels and gas.

Surface Albedo and the Building Sector

Embodied carbon, or embodied CO_2 eq, is defined as the amount of GHG of all the activities related to the life cycle – excluding the operational energy consumption – of a building, a building component, or a building material (Pomponi and Moncaster 2016; De Wolf et al. 2017). The TC350 standards delineate which are the phases to include for assessing the environmental impacts from cradle to grave[8] related to a building. In a building use phase, the impacts concerning components, maintenance, repair, replacement, and refurbishment are included in the embodied CO_2 eq assessment. Contrarily, the operational energy use and the operational water use are excluded from the embodied CO_2 eq assessment (De Wolf et al. 2017).

Surface albedo can have an effect on both the energy use for cooling and heating and RF. According to the aforementioned distinction, in this chapter, the inclusion of the CO_2 eq related to the variation in surface albedo in the embodied CO_2 eq assessment has been investigated omitting the discussion about the effect of the variation of surface albedo on operational energy.

Among the scientific literature, a plethora of studies about the influence of building surface albedo on energy use can be found (Belussi and Barozzi 2015; Cubi et al. 2016). Contrariwise, just an exiguous number of studies focuses on the effect of building surfaces' and built environment's albedo on RFs (Belussi and Barozzi 2015; AzariJafari et al. 2016; Santero et al. 2011) and include it in LCA studies. However, due to the growing urbanization and urban sprawl that the increase in urban populations entails (United Nations n.d.), this topic is gaining importance. In fact, the physical and optical features of the building materials used for replacing natural surfaces can affect microclimate contributing to exacerbate the urban heat island (UHI) effect (e.g., Taha 1997; Taha et al. 1988; Akbari and Konopacki 2005). In particular, urban albedo is usually lower than the albedo in non-urban areas, for instance, croplands (Jin et al. 2005). The progressive urban sprawl and the related decrease in urban albedo can contribute in increasing local temperature that, in turn, can affect human health (e.g., Tan et al. 2010; Lo and Quattrocchi 2003; Rosenzweig et al. 2005). Therefore, a detailed environmental information about the effect of the use of different materials in the urban areas is crucial for taking environmentally wise decisions. By the same token, the use of building materials can be of utmost importance and the inclusion of the effect of surface albedo in LCA can lead to the use of building materials which differ from those that currently are the most common.

The remainder of this section scrutinizes the literature published to date that investigates the effects of surface albedo of urban surfaces, including also the surfaces constituting the building envelopes, on climate. In particular, the radiative effect related to the variation of surface albedo is discussed.

When a high-albedo surface is substituted to a low-albedo one, it exerts a decrease in RF that can be translated into a negative amount of CO_2 eq. In this case, a negative amount of CO_2 eq corresponds to a sequestration of CO_2 eq. Contrarily, a decrease in surface albedo corresponds to an increase in RF and to an equivalent release of CO_2 eq. Some materials with different albedos can be characterized by the same manufacture carbon footprint if the manufacture phase does not differ for the two materials. However, for some other materials or

---

[8]The cradle-to-grave approach includes the product stage, the construction product stage, the use phase, and the end-of-life stage (BS EN 15978:2011 n.d.).

Table 5.1 LCA studies related to the built environment investigating the effect of surface albedo

Authors	Reference	Building envelope	Pavements/roads
Susca et al.	Susca et al. (2011)	x	
Santero and Horvarth	Santero and Horvath (2009)		x
Susca	Susca (2012a)	x	
Yu and Lu	Yu and Lu (2014)		x

The articles listed in Table 5.1 are those responding to the criteria used for the selection in Web of Science and related to the built environment

components, surface albedo can give rise to a different embodied carbon for the manufacturing phase. In this last case, it is important to include the effect of the surface albedo during the service life in order to prefer the materials with the lowest embodied carbon.

Although such theme is important, just few studies can be found in the literature reporting such evaluation (see Table 5.1).

Among the literature reviewed, two articles investigated the accounting of surface albedo in the life cycle of roads and pavements; besides, two articles introduced the accounting of surface albedo in a building component: the building rooftop (See Table 5.1).

Susca et al. (2011) evaluated the impact on climate change damage category related to the accounting of surface albedo in LCA. The authors used the surface albedo of an asphalt membrane rooftop with a surface albedo of 0.05 as a baseline, since its use is a widespread practice. Then, they evaluated the effect on RF considering the surface albedo of a green and a white roof with surface albedos of 0.6 and 0.2, respectively. The assessment of the effect of the variation in albedo on RF was conducted using the equivalence between the variation in surface albedo and RF provided by Akbari et al. (2009). The equivalence provided by Akbari et al. between the increase in albedo and the potential effect on the adjusted RF at TOA includes the combined effect of atmospheric and cloud scattering and absorption of the solar radiation. Such effect has been considered as average, and therefore, it is valid for the entire globe. However, the atmospheric and cloud effect on land and ocean, separately, should bring to different values. The equivalence developed by Akbari et al. comprises the short-term – 20–25 years – physical effects deriving by the variation in surface albedo and excludes the long-term effects. Based on the IPCC fourth assessment report estimates of RF (IPCC – Intergovernmental Panel on Climate Change 2007), Akbari et al. found that an increase of 0.01 in surface albedo corresponds to an offset of -2.55 kg CO_2 eq m^{-2}. Susca et al. evaluated the effect of the inclusion of surface albedo in carbon accounting for the green and the white roof in comparison with a black roof in a time horizon of 50 years. The results disclosed that, for both the green and the white roof, the accounting of the effect deriving by the inclusion of the surface albedo on RF is a source of important information because relevant in terms of percentage of avoided impacts. Such a result is relevant since can provide an orientation for the design of urban policies.

Santero[9] and Horvath (2009) evaluated the effect of the surface albedo of pavements on both RF and UHI mitigation. In order to account for the effect of surface albedo on RF, the authors used the equivalences provided by Akbari et al. (2009), which assign to an increase of 0.01 in surface albedo a correspondent sequestration of 2.55 or 4.90 kg CO_2 eq.[10] The results show that the inclusion of albedo in the life-cycle impact assessment of a road may significantly reduce the final impact. However, such evaluation is affected by uncertainties. The main source of uncertainty is the variation of albedo due to aging or weathering of the surfaces (Sen and Roesler 2016).

Susca (2012a) noted that the climatological model proposed by Akbari et al. (2009) for accounting the effect of the variation of urban surfaces on RF, when applied to the LCA methodology, has some limits. Indeed, Akbari et al. assigned to an increase of 0.01 in surface albedo in a 25–50 years evaluation a decrease in CO_2 eq equal to 2.55 kg/m². Besides, in the "decadal to centennial timescale," the authors assigned a decrease of 4.9 kg CO_2/m². The equivalences provided by Akbari et al. suffer from the use of climatological models featured by different initial assumptions. Furthermore, the equivalences between the increase in surface albedo and the variation in CO_2 eq refer to overlapping time ranges creating uncertainties about the value to use. Finally, even within the same range of time, the assessments performed at different points in time provide the same results, nullifying the importance of the time horizons in LCA studies (Susca 2012a). In order to overcome such limits, Susca developed the following equivalence – derived by Bird et al. (2008) – for translating an increase of 0.01 in surface albedo (α) into kg CO_2 eq:

$$+0.01\alpha = \frac{1.087\,\mathrm{RF}\,t}{0.217t - 44.78e^{-t/172.9} - 6.26e^{-t/18.51} - 0.22e^{-t/1.186} + 51.26}$$

where t is time.

Since the aim of the study conducted by Susca (2012a) was to overcome the limits related to the application of the equivalence developed by Akbari et al. (2009) to LCA studies, the enhanced equivalence respects the same assumptions considered in Akbari et al. (2009). Therefore, the author substituted to RF – 1.27 W/m².[11] Then, it resulted as follows:

---

[9]In a further research, Santero et al. (2013) evaluated, among other strategies to decrease the GWP related to pavements, the increase in surface albedo. In particular, the authors included the effect of the increase in pavements' albedo from 0.33 to 0.41 on the energy use for lighting. The analyses showed that the increase in albedo for local roads and collectors entails a decrease in the energy for lighting that corresponds to a decrease in GWP by 20%.

[10]The latter value has been assessed by Akbari et al. using an alternative methodology based on (Matthews and Caldeira 2008) which calculates the RF from CO_2 on a decadal to century timescale.

[11]This value has been used by Akbari et al. (2009).

$$+0.01\alpha = \frac{-1.38\,t}{0.217t - 44.78e^{-t/172.9} - 6.26e^{-t/18.51} - 0.22e^{-t/1.186} + 51.26}$$

Finally, Susca (2012a) applied the equivalence to the case study of a white and a black roof. In particular, the author calculated the GWP of the white and black roof comparing the surface albedos of the two rooftops. The results display that, in this case study, the inclusion of the evaluation of the effect of the increase in albedo on embodied carbon (i.e., CO_2 eq) provides an important information. Indeed, it results that considering the surface albedo of the black roof as a baseline, the accounting of the effect of surface albedo of the white roof on RF decreases the embodied carbon of the rooftop by about 50%.[12]

Susca compared the results deriving from the application of the equivalence provided by Akbari et al. to the case study of a cool roof, highlighting the importance of the inclusion of the temporal dimension in such studies. In order to make the equivalence developed comparable with that developed by Akbari et al., Susca applied the average values of the solar irradiation, omitting specific data related to punctual local irradiation. Nevertheless, the specific irradiation values can be easily substituted in future applications to case studies.

In this study (i.e., Susca 2012a), uncertainties are intrinsic to the climatological models used. Therefore, any enhancement in the knowledge related to the biogeophysical effects deriving from the variation in surface albedo on climate can be used for providing more precise equivalence between change in albedo and CO_2 eq.

Yu and Lu (2014) applied the same time-dependent methodology developed by Susca (2012a) for the evaluation of the effect of surface albedo of pavements on RF. However, differently from Susca, the authors attributed to an increase in surface albedo of 0.01 a mean RF equal to -1.54 W/m^2.[13] Then, the authors integrated the effect of surface albedo in a research study previously developed by Zhang et al. (2010). In this study, the authors considered a gray Portland cement concrete pavement that after 20 years is rehabilitated through two solutions: a Portland cement concrete overlay and a hot mixture asphalt overlay. The authors calculated the contribution of albedo on climate change impact category compared to the impacts deriving from the other life-cycle phases. Accounting for the albedo effect, it resulted that the CO_2 eq for the Portland cement concrete pavement decreases by 9.2% but increases by 19.1% for the hot mixture asphalt pavement. Also in this study uncertainties are implicitly related to the quantification of the mean RF.[14]

---

[12]The incidence of the decrease in embodied carbon related to the accounting of the effect of surface albedo on RF on the total embodied carbon (i.e., 50%) has been calculated considering the manufacture and the replacement phase of the materials and omitting the embodied carbon related to the end of life of the roofing system.

[13]The value was calculated as mean value of previous research: -1.27 W/m^2, -1.63 W/m^2 (Menon et al. 2010), -1.12 W/m^2 (Barnes et al. 2013) -2.14 W/m^2 (Hatzianastassiou et al. 2004).

[14]Considering the variability of the RF in the range -1.12---2.14 W/m^2, a Monte Carlo analysis shows that the corresponding offset of CO_2 eq has a standard deviation of about 30 and 160% for hot mixture asphalt and gray Portland cement concrete, respectively.

Both the studies by Yu and Lu (2014) and Susca (2012a) have used 1 as characterization factor while, other studies evaluated the specific characterization factor for albedo (e.g., Cherubini et al. 2012).

Discussion

This review chapter displays the importance of including the evaluation of the variation in surface albedo when an LCA study is carried out. In particular, the assessment of the effect of surface albedo on RF for building materials or components is of utmost importance to determine the embodied carbon and, thus, to guide decision makers to establish which are the most environmentally friendly materials and components to decrease the embodied carbon of buildings.

The exiguous number of scientific articles which investigated the impact of surface albedo of building material and components on climate revealed that the inclusion of surface albedo in LCA study is an innovative topic.

In the last years, progresses have been done to improve the methodology to account for the variation of surface albedo in LCA although the precision of the accounting greatly relies on the progress in climate science. To apply the equivalences between variation in surface albedo and CO_2 eq to an LCA case study, two requirements are fundamental: first, the equivalence has to be applied to a comparative LCA study; second, a baseline value for the surface albedo has to be considered. Such baseline value can be the surface albedo of a building material that is traditionally used or the surface albedo of the material to which a new material has to be substituted. In fact, the effect of albedo on climate cannot be evaluated per se. But the effect of the variation of surface albedo can be evaluated. Therefore, two surface albedos are necessary for the assessment.

Conclusions

The variation in surface albedo alters the Earth's energy budget that, in turn, affects climate change. Notwithstanding this effect is of great importance, it is typically omitted in LCA studies.

Just in recent years, some studies have translated the variation in surface albedo into CO_2 eq and applied such equivalence to LCA case studies. Most of such case studies explore the effect of the variation in surface albedo applied to land-use change. However, the effect of the variation in surface albedo is not just confined to land-use change. Indeed, nowadays we are facing an increase in urban population that in the next years will be also more conspicuous (United Nations n.d.). The increase in urban population is associated to a new urbanization process or to an urban sprawl. The substitution of natural material with man-made ones with a different albedo may exert an effect on the environment and in particular on climate. Albeit such phenomenon is of growing importance, just an exiguous number of studies including such effect for building materials can be found to date in the published literature.

The inclusion of the effect of the variation in surface albedo on RF would provide important pieces of information that might support urban decision makers in the choice of the most environmentally building materials or components. In fact, the reviewed scientific literature – that accounts for the effect of the variation in surface albedo in LCA studies related to building materials and components – shows that such contribution can be relevant depending on the specific application. Furthermore, the literature reviewed has shown that high-albedo materials and components may have a lower embodied carbon than the same materials with a lower albedo. However, to avoid providing misleading information, the carbon footprint of the operational phase of such materials should be carefully evaluated to avoid shifting the carbon burdens. For instance, depending on the specific characteristics of the climate, high-albedo rooftops or high-albedo building envelopes might entail a higher use of heating energy.[15] Contrariwise, some studies (e.g., Susca et al. 2011; Susca 2012a) showed that the installation of high-albedo rooftops may decrease the summer heat loads, decreasing the summer cooling energy demand and, consequently, the overall carbon footprint of the rooftops. For this reason and for the capacity to contribute in reducing UHI, high-albedo roofing systems have already been suggested in urban regulations of some cities such as New York (*Local laws of the City of New York for the year* n.d.).

Variation in albedo can also give rise to non-radiative effects that might influence the local climate and that might be of some importance for defining urban policies oriented to the safeguard of public health and energy saving, although, as previous studies demonstrate, the potential effect should be evaluated case by case (Yang et al. 2015). Some attempts to evaluate the effect of the increase in surface albedo, not only on RF but also on urban population, have already been conducted (e.g., Susca 2012b), though further research is necessary.

References

Akbari, H., & Konopacki, S. (2005). Calculating energy-saving potentials of heat-island reduction strategies. *Energy Policy, 33*(6), 721–756.
Akbari, H., Menon, S., & Rosenfeld, A. (2009). Global cooling: Increasing world-wide urban albedos to offset CO2. *Climatic Change, 94*(3–4), 275–286.
AzariJafari, H., Yahia, A., & Ben Amor, M. (2016). Life cycle assessment of pavements: Reviewing research challenges and opportunities. *Journal of Cleaner Production, 112*(Part 4), 2187–2197.
Bala, G., et al. (2007). Combined climate and carbon-cycle effects of large-scale deforestation. *Proceedings of the National Academy of Sciences, 104*(16), 6550–6555.

---

[15]Such possibility is however remote, since in winter the amount of solar radiation reaching the Earth is much lower than in summer.

Barnes, C. A., Roy, D. P., & Loveland, T. R. (2013). Projected surface radiative forcing due to 2000–2050 land-cover land-use albedo change over the eastern United States. *Journal of Land Use Science, 8*(4), 369–382.

Bellouin, N., & Boucher, O. (2010). *Climate response and efficacy of snow albedo forcing in the HadGEM2-AML climate model.* UK: Met Office Hadley Centre.

Belussi, L., & Barozzi, B. (2015). Mitigation measures to contain the environmental impact of urban areas: A bibliographic review moving from the life cycle approach. *Environmental Monitoring and Assessment, 187*(12), 745.

Betts, R. A. (2001). Biogeophysical impacts of land use on present-day climate: Near-surface temperature change and radiative forcing. *Atmospheric Science Letters, 2*(1–4), 39–51.

Bird, D. N., Kunda, M., Mayer, A., Schlamadinger, B., Canella, L., & Johnston, M. (2008). Incorporating changes in albedo in estimating the climate mitigation benefits of land use change projects. *Biogeosciences Discussions, 2008*, 1511–1543.

Bright, R. M. (2015). Metrics for biogeophysical climate forcings from land use and land cover changes and their inclusion in life cycle assessment: A critical review. *Environmental Science & Technology, 49*(6), 3291–3303.

Bright, R. M., Strømman, A. H., & Peters, G. P. (2011). Radiative forcing impacts of boreal forest biofuels: A scenario study for Norway in light of albedo. *Environmental Science & Technology, 45*(17), 7570–7580.

Bright, R. M., Cherubini, F., & Strømman, A. H. (2012). Climate impacts of bioenergy: Inclusion of carbon cycle and albedo dynamics in life cycle impact assessment. *Environmental Impact Assessment Review, 37*, 2–11.

Bright, R. M., Zhao, K., Jackson, R. B., & Cherubini, F. (2015). Quantifying surface albedo and other direct biogeophysical climate forcings of forestry activities. *Global Change Biology, 21* (9), 3246–3266.

BS EN 15978:2011. (n.d.). Sustainability of construction works. Assessment of environmental performance of buildings. Calculation method. [Online]. Available: http://shop.bsigroup.com/ProductDetail/?pid=000000000030256638. Accessed 04 Nov 2015.

Caiazzo, F., Malina, R., Staples, M. D., Wolfe, P. J., Yim, S. H. L., & Barrett, S. R. H. (2014). Quantifying the climate impacts of albedo changes due to biofuel production: A comparison with biogeochemical effects. *Environmental Research Letters, 9*(2), 024015.

Caldeira, K., Jain, A. K., & Hoffert, M. I. (2003). Climate sensitivity uncertainty and the need for energy without CO2 emission. *Science, 299*(5615), 2052–2054.

Cess, R. D. (1978). Biosphere-albedo feedback and climate modeling. *American Meteorological Society, 35*(9), 1765–1768.

Cherubini, F., & Strømman, A. H. (2013). Bioenergy vs. natural gas for production of district heat in Norway: Climate implications. *Energy Procedia, 40*, 137–145.

Cherubini, F., Bright, R. M., & Strømman, A. H. (2012). Site-specific global warming potentials of biogenic CO 2 for bioenergy: Contributions from carbon fluxes and albedo dynamics. *Environmental Research Letters, 7*(4), 045902.

Cubi, E., Zibin, N. F., Thompson, S. J., & Bergerson, J. (2016). Sustainability of rooftop technologies in cold climates: Comparative life cycle assessment of white roofs, green roofs, and photovoltaic panels. *Journal of Industrial Ecology, 20*(2), 249–262.

Davin, E. L., & de Noblet-Ducoudré, N. (2010). Climatic impact of global-scale deforestation: Radiative versus nonradiative processes. *Journal of Climate, 23*(1), 97–112.

Davin, E. L., de Noblet-Ducoudré, N., & Friedlingstein, P. (2007). Impact of land cover change on surface climate: Relevance of the radiative forcing concept. *Geophysical Research Letters, 34* (13), L13702.

De Wolf, C., Pomponi, F., & Moncaster, A. (2017). Measuring embodied carbon dioxide equivalent of buildings: A review and critique of current industry practice. *Energy and Buildings, 140*, 68–80.

European Commission. (2014). *A policy framework for climate and energy in the period from 2020 up to 2030*. Brussels: European commission. http://eur-lex.europa.eu/legal-content/EN/TXT/PDF/?uri=CELEX:52014DC0015&from=EN.

Friedland, A. J., & Gillingham, K. T. (2010). Carbon accounting a tricky business. *Science, 327* (5964), 411–412.

Giuntoli, J., Caserini, S., Marelli, L., Baxter, D., & Agostini, A. (2015). Domestic heating from forest logging residues: Environmental risks and benefits. *Journal of Cleaner Production, 99*, 206–216.

Goldewijk, K. K. (2001). Estimating global land use change over the past 300 years: The HYDE database. *Global Biogeochemical Cycles, 15*(2), 417–433.

Haberl, H. (2013). Net land-atmosphere flows of biogenic carbon related to bioenergy: Towards an understanding of systemic feedbacks. *GCB Bioenergy, 5*(4), 351–357.

Hansen, J., & Nazarenko, L. (2004). Soot climate forcing via snow and ice albedos. *Proceedings of the National Academy of Sciences of the United States of America, 101*(2), 423–428.

Hansen, J., et al. (1981). Climate impact of increasing atmospheric carbon dioxide. *Science, 213* (4511), 957–966.

Hansen, J., et al. (2005). Efficacy of climate forcings. *Journal of Geophysical Research-Atmospheres, 110*(D18), D18104.

Hatzianastassiou, N., Katsoulis, B., & Vardavas, I. (2004). Global distribution of aerosol direct radiative forcing in the ultraviolet and visible arising under clear skies. *Tellus B, 56*(1), 51–71.

Helin, T., Sokka, L., Soimakallio, S., Pingoud, K., & Pajula, T. (2013). Approaches for inclusion of forest carbon cycle in life cycle assessment – A review. *GCB Bioenergy, 5*(5), 475–486.

Holtsmark, B. (2015). A comparison of the global warming effects of wood fuels and fossil fuels taking albedo into account. *GCB Bioenergy, 7*(5), 984–997.

IPCC. (2001). *Climate change 2001: The scientific basis*. Cambridge, UK: Cambridge University Press.

IPCC – Intergovernmental Panel on Climate Change. (2007). *IPCC fourth assessment report: Climate change* 2007. Working group I: The physical science basis.

IPCC. (2013). In T. F. Stocker, D. Qin, G.-K. Plattner, M. Tignor, S. K. Allen, J. Boschung, A. Nauels, Y. Xia, V. Bex, & P. M. Midgley (Eds.), *Climate Change 2013: The Physical Science Basis. Contribution of Working Group I to the Fifth Assessment Report of the Intergovernmental Panel on Climate Change* (p. 1535). Cambridge\New York: Cambridge University Press. https://doi.org/10.1017/CBO9781107415324.

Jin, M., Dickinson, R. E., & Zhang, D. (2005). The footprint of urban areas on global climate as characterized by MODIS. *Journal of Climate, 18*(10), 1551–1565.

Jørgensen, S. V., Cherubini, F., & Michelsen, O. (2014). Biogenic CO2 fluxes, changes in surface albedo and biodiversity impacts from establishment of a miscanthus plantation. *Journal of Environmental Management, 146*, 346–354.

Kabat, P., et al. (2004). *Vegetation, water, humans and the climate: A new perspective on an interactive system*. Springer Science & Business Media, Berlin.

Kiehl, J. T., & Trenberth, K. E. (1997). Earth's annual global mean energy budget. *Bulletin of the American Meteorological Society, 78*(2), 197–208.

Köhler, P., et al. (2010). What caused Earth's temperature variations during the last 800,000 years? Data-based evidence on radiative forcing and constraints on climate sensitivity. *Quaternary Science Reviews, 29*(1), 129–145.

Lenton, T. M., & Vaughan, N. E. (2009). The radiative forcing potential of different climate geoengineering options. *Atmospheric Chemistry and Physics, 9*(15), 5539–5561.

Lo, C. P., & Quattrocchi, D. A. (2003). Land-use and land-cover change, urban heat island phenomenon, and health implications: A remote sensing approach. *Photogrammetric Engineering and Remote Sensing, 69*(9), 1053–1063.

Local laws of the City of New York for the year 2011 No. 21 (n.d.).

Lohila, A., et al. (2010). Forestation of boreal peatlands: Impacts of changing albedo and greenhouse gas fluxes on radiative forcing. *Journal of Geophysical Research – Biogeosciences, 115*(G4), G04011.

Matthews, H. D., & Caldeira, K. (2008). Stabilizing climate requires near-zero emissions. *Geophysical Research Letters, 35*(4), L04705.

McKechnie, J., Colombo, S., & MacLean, H. L. (2014). Forest carbon accounting methods and the consequences of forest bioenergy for national greenhouse gas emissions inventories. *Environmental Science & Policy, 44*, 164–173.

Menon, S., Akbari, H., Mahanama, S., Sednev, I., & Levinson, R. (2010). Radiative forcing and temperature response to changes in urban albedos and associated CO offsets. *Environmental Research Letters, 5*(1), 014005.

Muñoz, I., Campra, P., & Fernández-Alba, A. R. (2010). Including CO2-emission equivalence of changes in land surface albedo in life cycle assessment. Methodology and case study on greenhouse agriculture. *International Journal of Life Cycle Assessment, 15*(7), 672–681.

National Aeronautics and Space Administration. (n.d.). Albedo. [Online]. Available: http://neo.jpl.nasa.gov/glossary/albedo.html

Otterman, J. (1977). Anthropogenic impact on the albedo of the earth. *Climatic Change, 1*(2), 137–155.

Pitman, A. J. (2003). The evolution of, and revolution in, land surface schemes designed for climate models. *International Journal of Climatology, 23*(5), 479–510.

Pomponi, F., & Moncaster, A. (2016). Embodied carbon mitigation and reduction in the built environment – What does the evidence say? *Journal of Environmental Management, 181*, 687–700.

Pongratz, J., Reick, C. H., Raddatz, T., & Claussen, M. (2010). Biogeophysical versus biogeochemical climate response to historical anthropogenic land cover change. *Geophysical Research Letters, 37*(8), L08702.

Randerson, J. T., et al. (2006). The impact of boreal forest fire on climate warming. *Science, 314* (5802), 1130–1132.

Roe, G. H., & Baker, M. B. (2007). Why is climate sensitivity so unpredictable? *Science, 318* (5850), 629–632.

Rosenzweig, C., Solecki, W. D., Parshall, L., Chopping, M., Pope, G., & Goldberg, R. (2005). Characterizing the urban heat island in current and future climates in New Jersey. *Global Environmental Change Part B Environmental Hazards, 6*(1), 51–62.

Røyne, F., Peñaloza, D., Sandin, G., Berlin, J., & Svanström, M. (2016). Climate impact assessment in life cycle assessments of forest products: Implications of method choice for results and decision-making. *Journal of Cleaner Production, 116*, 90–99.

Santero, N. J., & Horvath, A. (2009). Global warming potential of pavements. *Environmental Research Letters, 4*(3), 034011.

Santero, N. J., Masanet, E., & Horvath, A. (2011). Life-cycle assessment of pavements part II: Filling the research gaps. *Resources, Conservation and Recycling, 55*(9–10), 810–818.

Santero, N., Loijos, A., & Ochsendorf, J. (2013). Greenhouse gas emissions reduction opportunities for concrete pavements. *Journal of Industrial Ecology, 17*(6), 859–868.

Schwaiger, H. P., & Bird, D. N. (2010). Integration of albedo effects caused by land use change into the climate balance: Should we still account in greenhouse gas units? *Forest Ecology and Management, 260*(3), 278–286.

Sen, S., & Roesler, J. (2016). Aging albedo model for asphalt pavement surfaces. *Journal of Cleaner Production, 117*, 169–175.

Singh, B., Guest, G., Bright, R. M., & Strømman, A. H. (2014). Life cycle assessment of electric and fuel cell vehicle transport based on forest biomass. *Journal of Industrial Ecology, 18*(2), 176–186.

Susca, T. (2012). Multiscale approach to life cycle assessment. *Journal of Industrial Ecology, 16* (6), 951–962.

Susca, T. (2012a). Enhancement of life cycle assessment (LCA) methodology to include the effect of surface albedo on climate change: Comparing black and white roofs. *Environmental Pollution, 163*, 48–54.

Susca, T., Gaffin, S. R., & Dell'Osso, G. R. (2011). Positive effects of vegetation: Urban heat island and green roofs. *Environmental Pollution, 159*(8–9), 2119–2126.

Taha, H. (1997). Urban climates and heat islands; albedo, evapotranpiration, and anthropogenic heat. *Energy and Buildings, 25*, 99–103.

Taha, H., Akbari, H., Rosenfeld, A., & Huang, J. (1988). Residential cooling loads and the urban heat island—the effects of albedo. *Building and Environment, 23*(4), 271–283.

Tan, J., et al. (2010). The urban heat island and its impact on heat waves and human health in Shanghai. *International Journal of Biometeorology, 54*(1), 75–84.

United Nations, department of economic and social affairs, population division. (n.d.). *The world's cities in 2016 – Data booklet.* http://www.un.org/en/development/desa/population/publications/pdf/urbanization/the_worlds_cities_in_2016_data_booklet.pdf

Yang, J., Wang, Z.-H., & Kaloush, K. E. (2015). Environmental impacts of reflective materials: Is high albedo a "silver bullet" for mitigating urban heat island? *Renewable and Sustainable Energy Reviews, 47*, 830–843.

Yu, B., & Lu, Q. (2014). Estimation of albedo effect in pavement life cycle assessment. *Journal of Cleaner Production, 64*, 306–309.

Zhang, Y., et al. (2010). Life cycle emissions and cost of producing electricity from coal, natural gas, and wood pellets in Ontario, Canada. *Environmental Science & Technology, 44*(1), 538–544.

Chapter 6
Quantifying Environmental Impacts of Structural Material Choices Using Life Cycle Assessment: A Case Study

D. Davies, L. Johnson, B. Doepker, and M. Hedlund

Life Cycle Assessment: History and Limitations

Environmentally focused LCA took a major step forward in the 1990s, when the International Organization for Standardization (ISO) defined its ISO 14040 series, providing a general LCA approach. ISO 14044 defines LCA as the "compilation and evaluation of the inputs, outputs and the potential environmental impacts of a product system throughout its life cycle." LCA methods have also been formalized by the British Standards Institution's PAS 2050, the World Resources Institute protocols, and ISO 14067 (Skone 2013).

In the United States, the American Society for Testing and Materials (ASTM), the International Building Code (IBC), the American Society of Civil Engineers/ Structural Engineering Institute (ASCE/SEI), and the Council on Tall Buildings and Urban Habitat (CTBUH) have all initiated environmental LCA working groups. Most recently, the Athena Sustainable Materials Institute defined LCA as "a scientific method for measuring the environmental footprint of materials, products, and services over their entire lifetime" (Athena SMI 2016).

Green rating systems, including USGBC LEED v4, the Living Building Challenge (LBC), IgCC, calGreen, and others have all recently included whole-building LCA elements within their ratings. Architecture 2030's Challenge for Products and the Carbon Leadership Forum are organizations spearheading the application of LCA to the US building industry.

While there is considerable activity around the topic—most of it well intentioned and with the ultimate hope of reducing the speed of global warming—there is no consensus on how best to use LCA principles and data. The green rating systems also do not yet provide consistent guidance for how

D. Davies · L. Johnson (✉) · B. Doepker · M. Hedlund
Magnusson Klemencic Associates, Seattle, WA, USA
e-mail: ddavies@mka.com; ljohnson@mka.com; bdoepker@mka.com; mhedlund@mka.com

© Springer International Publishing AG 2018   123
F. Pomponi et al. (eds.), *Embodied Carbon in Buildings*,
https://doi.org/10.1007/978-3-319-72796-7_6

LCA should be fully executed within their standards. Many practitioners who attempt to track carbon emissions for upcoming or completed projects also find the data sometimes inaccessible or inconsistent or the processes to evaluate the data difficult to implement.

LCA is an evolving field, and it would be misleading to imply that all involved parties have the same objectives. However, there are more common themes and activities than differences. To mitigate the impacts of building design on the environment and global warming, LCA has the potential to transform many of our design and decision-making processes. As the impacts from global warming increase, LCA evaluations become an effective method to answer public comments with credible scientific and quantified data in ways not previously possible.

There are several LCA terminologies that are useful to understand:

- LCA provides quantified data of the environmental impacts of a building design within a defined study boundary. The data includes environmental impacts such as global warming, ozone depletion, land/water acidification, eutrophication, tropospheric ozone, and nonrenewable energy use (US Green Building Council 2013).
- Product Category Rules (PCRs) are guidelines that define industry-specific measurements for the purpose of producing an Environmental Product Declaration (EPD). PCRs provide the structure needed to report the results of EPDs and are typically developed with the input of the industry trade organizations which the PCR covers (Carbon Leadership Forum 2015).
- An EPD declares the environmental impacts of a product over its expected life, similar to a food nutrition label. An EPD should be third-party verified and made public upon completion. An EPD and the respective PCR should, at a minimum, be compliant with ISO 14025 and 21930 and be posted in their entirety (see Fig. 6.1).

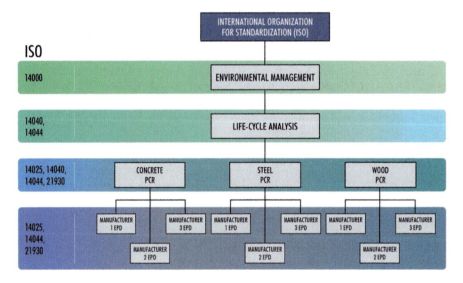

Fig. 6.1 Relationships between ISO Standards, LCA, PCRs, and EPDs

Herein resides one of today's fundamental challenges with LCAs. The industry- and material-specific PCRs are often not cross-compatible, as they do not always use the same boundary conditions or consider variables in the same way. Even LCA experts knowledgeable about the differences cannot always make practical comparisons between differing materials.

ASTM recently published a standard, ASTM E2921-16, to try to tackle this issue. It presents the minimum criteria for comparing whole-building life cycle assessments. However, it only applies to comparative assessments of a baseline building with a modified one for the purposes of an LCA study. As an example of the challenge, it is difficult to directly evaluate timber land use issues against steel and concrete environmental impacts. This is a significant issue that is yet to be addressed with industry consensus but is fundamental to any credible whole-building LCA comparative studies between material types.

Attempting a whole-building LCA with today's data sets can be approached through a strategy of reducing the number of variables considered and focusing on the most significant environmental impacts as a means of getting closer to more comparable results. While not perfect, by looking at statistical correlations and percentage reductions of single largest impact variables from one option to another, reliable data can often be achieved to allow for more informed decision-making (Pomponi and Moncaster 2017). Such focus begins to bring clarity to data comparisons within the design and decision-making processes. While LCA often requests and reports data from multiple impact categories, analyzing the Global Warming Potential (GWP), often defined as the embodied carbon of a design, is perhaps the most understood and meaningful of these variables and has thus been the focus of this study.

LCA studies that focus on GWP and consider the relative impact of system choices and optimization studies within the same collection of materials are some of the most meaningful LCA studies to date. Within the building industry, prior LCA studies by Magnusson Klemencic Associates (MKA) and others have also shown that the structure is typically the single largest component of the embodied carbon footprint for new construction projects, provided the building itself is the only consideration and ignoring building use (Mahler and Schneider 2016; Davies and Klemencic 2014). Given the limited number of materials and industries that produce the structural frame for buildings, and with LCA focusing on optimizing those materials and reducing their production carbon footprints, the structure becomes a high-value investment of time and effort for current LCA discussions and comparisons.

The life cycle stages included in the scope of an LCA are important to define (Pomponi and Moncaster 2016). Two common examples are referred to as "Cradle to Gate" and "Cradle to Grave." Cradle to Gate encompasses the material extraction and product manufacturing up to the factory gate. Meanwhile, Cradle to Grave includes the Cradle to Gate environmental impact while also including the environmental impact of the material during construction and through to its end of life. While more globally conclusive, a Cradle to Grave LCA typically involves speculation on the future, so its accuracy is often more debated than the Cradle to Gate study. It is an attempt, though, for closed-loop considerations of all impacts. Whether the Cradle to Grave impacts of recycling, depositing in a landfill, or end-of-life decomposition/incineration are captured in an LCA depends on the governing PCR for the materials considered and is a point of inconsistency between many current PCRs. Standardization around this topic is an area for future work.

LCA Software Used for Study

Numerous software tools are currently available for completing LCAs. Of these, Tally is used for the purpose of this case study. Tally is a recently released LCA module for Autodesk Revit®, and its goal is to help to streamline the process of quantifying the environmental impact of building materials for whole-building analysis as well as comparative analyses of design options. While working on a Revit model, the user can define relationships between Building Information Modeling (BIM) elements and construction materials from the Tally database, which relies on the GaBi databases (2013 LCI: http://www.gabi-software.com/international/databases/gabi-database/) from Thinkstep.

Tally does not provide the ability to directly model LCA data for anything other than products specified within its database. This limitation, however, controls the quality and reliability of the LCA data reported. While the database is extensive and growing, there exists a limitation when attempting to identify project-specific and unique material traits, such as concrete mix designs. Tally can, however, output a bill of quantities to an Excel spreadsheet, allowing for data manipulation outside of the program at later project stages.

Case Study Parameters

The building used for this case study includes seven above-grade floors, with one floor of below-grade parking. The height of the building is 85 ft above grade and has approximately 290,000 ft^2 of floor space. The occupancy is mostly office with retail at the ground level. The building itself is located in a low seismic region where wind controls the lateral design, with a basic wind speed of 90 mph per ASCE7-05. The foundation used for all four gravity systems is an array of spread footings with conventional slab on grade, which was based on a bearing pressure of 10,000 psf. Note that, while the foundation was included for each building configuration in this study, they were not uniquely designed for each gravity system. Given the relatively high bearing pressure of the site and minimum foundation system dimensions that would not change between gravity systems (such as basement wall height or building footprint area), this was not seen as an area of significant difference between the options studied. This may not be the case, though, at sites with poorer soil conditions.

The lateral system is a concrete core for each of the designs considered. This choice allowed for a consistency of variable control between designs, including impacts from architectural programing of the building core program.

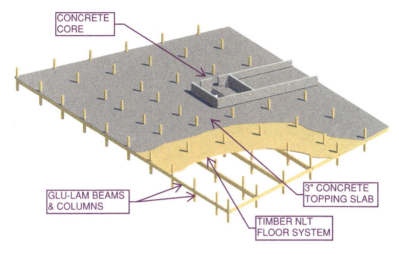

Fig. 6.2 Portion of floor plate: design A

The following structural gravity systems were selected for comparisons of their embodied carbon and are illustrated in Figs. 6.2, 6.3, 6.4, and 6.5:

- **Design A**: Glulam beams and columns with nail-laminated timber (NLT) flooring,(including a concrete topping slab for acceptable vibration and acoustical performance) above grade, with a mildly reinforced concrete slab at levels 1 and 2 with concrete columns below level 2
- **Design B**: Steel beams and columns with 3″ concrete slab on 3″ corrugated metal deck above grade and 10″ mild concrete slab at level 1 with concrete columns below grade
- **Design C**: 8″ post-tensioned concrete slabs above grade and 10″ mild concrete slab at level 1 with concrete columns
- **Design D**: 10″ mild concrete slabs with concrete columns

All three designs were fully designed and modeled in separate Revit models. Using Tally, specific materials and reinforcing quantities were defined for each of the structural elements, which allowed the software to capture the embodied environmental impacts in the concrete, steel, and timber elements. Concrete mixes were industry standard (as defined in the Tally database) and not carbon optimized for the design of this building. Appendix A summarizes material quantities for each design. The embodied global warming potential of non-primary structure, such as structure to support mechanical systems, exterior enclosure, architectural components, and secondary finishes (i.e., exposed wood ceiling versus dropped ceilings), as well as any secondary structure required to support these elements, were not considered within the scope of this study.

The building was analyzed for a 60-year life cycle. Tally includes Stage D (reuse, recovery, and recycling) in its Cradle to Grave analysis, which is sometimes only included in other LCA's Cradle to Cradle scopes. To be consistent with Tally's

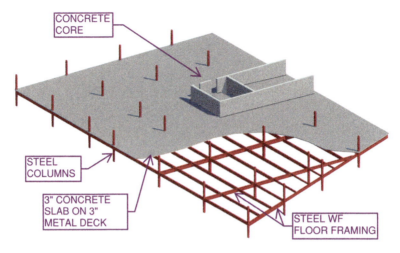

Fig. 6.3 Portion of floor plate: design B

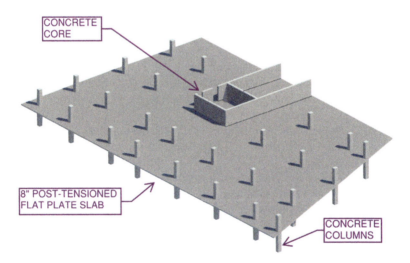

Fig. 6.4 Portion of floor plate: design C

definitions, "end of life" includes Stage D and is included in the Cradle to Grave analysis.

Tally includes the following life cycle phases for its data per ISO 14040 and 14044:

Product manufacturing:

- A1: Raw material supply
- A2: Transport (raw materials to the location where the product is manufactured)
- A3: Manufacturing

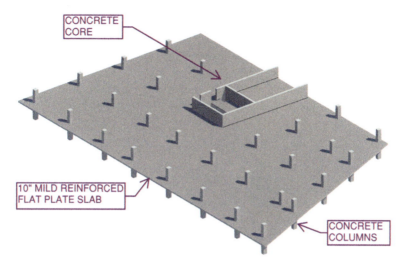

Fig. 6.5 Portion of floor plate: design D

Construction:

- A4: Transportation (between manufacturer and building site)
- A5: On-site construction

Building use:

- B2: Maintenance
- B3: Repair
- B4: Replacement

Note: Life cycle phases B1 (use), B5 (refurbishment), and B6 (operational energy) are outside the scope of Tally and were not included in this study. At the time of the study, life cycle phase B6 (operational energy) was not built into Tally and thus was also outside the scope of this study.

End of life:

- C2: Transport
- C3: Waste processing
- C4: Disposal
- Note: Life cycle phase C1 (demolition) is not included in Tally's database and was outside the scope of this study.
- D: Reuse, recovery, and recycling potential

"Cradle to Gate" comparisons were also investigated. This includes life cycle phases A1–A3 as defined above.

The following figures summarize the different structural systems investigated.

Tally Results and Observations

The following figures show the results of the Tally LCA studies for both Cradle to Gate and Cradle to Grave boundary conditions (Figs. 6.6 and 6.7).

In reviewing these findings, the cleanest data set comparison is between Design C (PT slab system) and Design D (mild reinforced slab system), as both structural

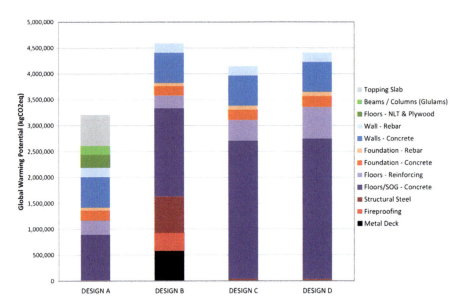

Fig. 6.6 Tally results – Cradle to Gate

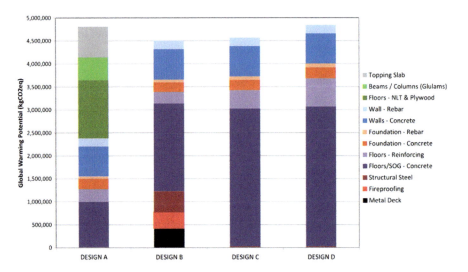

Fig. 6.7 Tally results – Cradle to Grave

systems are governed by the same PCRs. The LCA results of this study, with its noted qualifications and industry average concrete mix data, indicate that the PT slab system (Design C) has approximately 5% less GWP than the mild slab system (Design D). PT slabs can achieve the same spans with reduced thickness, thus resulting in less concrete volume. However, due to high early strength requirements of PT slabs, the benefits of lower overall concrete quantities are partially offset by the use of mixes with higher cement requirements. As a result, unless time to PT stressing is adjusted from current industry practice, the largest improvement between the respective GWPs between Designs C and D is the lower reinforcement quantities within the PT slabs.

When comparing designs in the Cradle to Gate scope, the composite steel slab scheme (Design B) exhibits a lower GWP than the mild concrete slab scheme (Design D) and similar GWP to the concrete PT slab scheme (Design C). It is worth noting that while spray-applied fireproofing for the steel framing is included in this study, the added GWP associated with architecturally finishing around steel columns and added dropped ceilings is not included. These components tend to be more prevalent in steel buildings than concrete and heavy timber buildings that frequently leave columns and slab soffits exposed. Another issue outside the scope of this study is that many mild and PT slab systems allow for MEP system routing directly under the slab in lieu of below or through beam framing, allowing floor to floors to often be 1–2 ft shorter when compared to steel or wood system alternatives. Larger floor-to-floor dimensions create greater cladding surface area and interior partition wall heights, which can impact the overall project GWP. These secondary systems were not considered as part of this study but should be for any whole-building LCA efforts that extend beyond the structural frame.

Perhaps most interesting to note is that based upon the embodied carbon data sets used by Tally, the timber floor scheme (Design A) contains a higher global warming potential than might have been anticipated, especially when compared with the steel system in the Cradle to Grave study. This can be attributed to two significant factors. First is the significant increase in GWP for the timber components when factoring in the end-of-life phase. The timber PCR assumes that the wood components will decompose or be incinerated at end of life. This issue is discussed in greater length in subsequent pages.

Another unique aspect contributing to the GWP of the heavy timber building is that a 3-in. concrete topping was required for comparable vibration and acoustic performance to the other systems, in order to meet Class-A office standards. If the mass timber building design included different alternatives for vibration and acoustical control, there may be a way to decrease in the overall global warming potential for the mass timber design, provided the concrete topping could be reduced or eliminated. However, it is common that many architects and building owners do not want to hide the mass timber structure frame for aesthetic reasons; therefore this paper assumes the concrete topping slab as a more accurate representation of the timber solution as it would be designed and constructed.

Figure 6.8 was derived from the LCA results and illustrates Tally's assumption of the percent of embodied carbon, for the most common materials found in the four building designs, attributed to each life cycle relative to the total Cradle to Grave

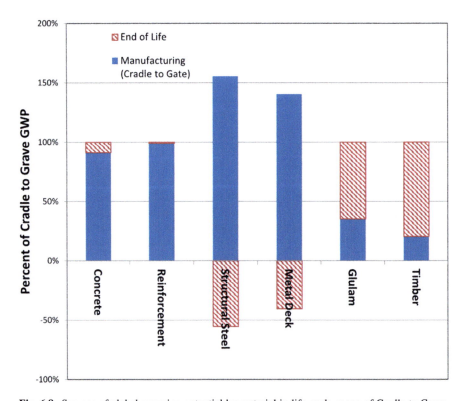

Fig. 6.8 Sources of global warming potential by material in life cycle scope of Cradle to Grave

GWP. Note that the summation of "manufacturing (Cradle to Gate)" and "end-of-life" phases represents 100% of the total Cradle to Grave environmental impact.

Unlike building components such as interior finishes and MEP equipment, which are more prone to updating and replacing, the primary structure (barring low recurrence, high-impact events such as earthquakes, hurricanes, or fires) sees little in the form of maintenance, repair, and replacement over a 60-year building life. For this reason, life cycle stages B2–B4, while included in the Cradle to Grave study, result in a negligible delta when comparing Cradle to Gate with Cradle to Grave. The end-of-life phases C2–C4 and D, by contrast, can result in a significant delta between Cradle to Gate and Cradle to Grave boundary conditions. Due to inconsistencies in the varying material's PCRs, the magnitude and even the sign of this delta can vary between the materials of steel, concrete, and timber.

The global warming potential of wood (glulam and softwood timber) and steel (structural steel and metal deck) are both highly dependent on the scope of the LCA, with inconsistencies and in some conditions unique industry slanted data sets coming from the PCRs and their resulting EPDs of both materials. When considering a Cradle to Gate scope within Tally, wood has a very low GWP relative to concrete and steel structures, but the North American wood database referenced by Tally does not consider FSC-certified wood as the referenced EPDs are industry averages and specifically ignores land impact issues associated with the harvesting of the materials, such as road building and site disruption, or lost forest carbon sequestration by the trees not continuing to grow on the site. The database for wood also assumes that at the end of the building life, the timber is not reused and the carbon sequestered in the wood is eventually released into the atmosphere (e.g., through burning or decaying), which is the reason for the significant increase in GWP in the mass timber structure in the end of use phase.

LCA studies by others (e.g., see Mahler and Schneider 2016) have also shown considerable data scatter on how embodied carbon footprints for timber should be accounted for. LCA data sets other than Tally sometimes report timber as a negative carbon footprint when consumed in a project. The conclusion of Mahler and Schneider is that the databases account for CO2 using different methods (Mahler and Schneider 2016). The differences led to "[surprising results] and deviations of this order of magnitude were not expected" (Mahler and Schneider 2016). Despite these differences, one cannot today conclude that the quality of one database exceeds that of another, from a data comparison perspective. This is an area in need of further independent research for how to accurately handle timber LCA findings.

Conversely, when compared with the timber PCR, the PCR for rolled structural steel provides a negative end-of-life global warming potential, which the US industry argues is indicative of the high rates of recycled steel content that goes into these specific steel building products. Tally uses a database containing a mixture of domestic and international steel, so it is difficult to quantify how much of the steel is recycled and how much is virgin. This partially picks up on the difference in energy consumed when steel comes from virgin ore feed stocks, or from recycled products that are remelted and used again, and highlights the value of using higher recycled material contents in the steel. This is a relevant point for industry average LCA data for steel, but it should be specifically understood what production processes are used in the making of the steel product for more detailed LCA investigations. One finding is that the majority of sheet and plate steel (metal decking and metal studs being one example), where dimensional tolerances are more exacting, typically comes from virgin iron ore feed stock and a blast furnace at the mill. Rebar and rolled beam shapes more frequently come from a largely recycled material feed stock and an electric arc furnace at the mill. The carbon footprints between these two mill processes are too significant to ignore but are currently not reported separately by industry.

For this paper, we have not attempted to debate the validity of the above assumptions by these industries; we have simply reported and used what is currently stated within their PCRs and EPDs for industry average information, as reported within Tally. For long-term credibility, these noted areas need future validation, most likely from independent third parties and not the institutions with conflicts of interest who are promoting their particular material.

It is important to maintain consistency between PCR scopes when making LCA comparisons, but what happens if the results of Cradle to Gate are compared to Cradle to Grave? Figure 6.9 shows the global warming potential in kilograms of

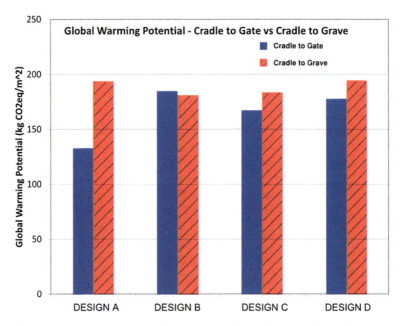

Fig. 6.9 Case study results – relative difference in global warming potential for Cradle to Gate vs Grave

carbon dioxide equivalent per square meter for the various schemes. The Cradle to Gate and Cradle to Grave results are reported as well as the relative difference of these two values. The relative magnitude of this difference is partly the result of the differing assumptions made in the respective materials' PCRs.

The range shown in Fig. 6.9 and the above discussions illustrate that making a decision on a sustainability "winner" between the different alternative building systems with today's available LCA information is not appropriate. There are inherent assumptions and approximations that could change the results even within a given life cycle boundary.

Limitations of This Study

Data Set Limitations and Findings

As previously explained, with today's data it is difficult to compare the GWP of timber, concrete, and steel because of the differences in EPDs and PCRs developed by the material industries. Differences in the applicable PCRs for each material are some of the causes of the differences in the GWP of the "end of life" between timber, concrete, and steel.

While GaBi was one of the data sets used in the previously discussed Mahler and Schneider study and all data sets showed a similar "lowest carbon" solution regardless of the database considered, variations on how wood is considered within the LCA industry vary widely. The source of much of this variation comes from how timber is handled, including ignoring upstream impacts within their PCR boundary conditions and the handling of negative carbon credits due to the use of a bio-based material.

It is common in the LCA community to disregard any attempt at comparing the embodied carbon between materials, taking the stance that LCA studies should only compare variations within the same material. For data sets as they exist today and from findings presented in this paper, this would seem to be appropriate. It would be useful to address this topic within building industry PCRs for the further advancement of the science of Life Cycle Analysis.

Software and Database Limitations

Results of this study are limited by the selection of Tally's industry average data sets, which are based upon the GaBi database, along with the quality of the Revit model detail provided for the study. While this Tally-based study is based upon design findings and not as-built findings, this is consistent with what can be achieved today with early design LCA efforts. More exacting LCAs by the research community can and should be accomplished around limitations noted within this study, to allow such future efforts to occur with greater confidence and surety.

Parameter Limitations

Several limitations resulted from the parameter selection for the building. The low seismicity location of the building site indicates that the demands and lateral resisting system did not significantly differ from design to design.

In addition, the building height is the same for these comparisons despite the fact that the structural depths differ. The impact to the overall building height was therefore not considered.

Structural Design Limitations

The seismic and wind hazards of a project site also impact the relevance of this study, as different sites yield different load demands, which would thus result in different structural proportioning. This case study project was not located in a high seismic zone, with the lateral design being controlled by wind. If this building were situated on a seismically active site on the West Coast of the United States, the lateral system and foundations would be likely to be controlled by seismic forces which will penalize buildings with a higher mass more. Note that Design A (mass timber floor) and Design B (composite structural steel) have comparable floor system weights and are much lighter than the concrete designs. The floor diaphragm of Design A, though, would likely require more reinforcement, to meet the lateral system requirements of high seismic loading. The floor diaphragms of Designs B, C, and D would externally look the same in wind or high seismic locations, but the internal rebar to each system would increase. A future effort will be to run this study again but based upon a high seismic location to directly quantify those impacts.

Conclusions

Given these limitations, this paper does not show that a decisive GWP winner can be chosen between the four different building frame options studied, based upon a material system choice alone and the data sets considered for this study. This was not the anticipated conclusion at the start of this investigation, but we look forward to future efforts which help clarify the shortcomings and inconsistencies of LCA data sets. This paper affirms that designers should choose materials that are most materially efficient for the intended building use and then optimize and economize the design to save on quantities while also finding ways to decrease the embodied carbon of that material choice.

Moving forward, it is important that efforts work to better align the data from the PCRs and EPDs between timber, concrete, and steel, especially in how each looks at the Cradle to Gate and Cradle to Grave approaches of their data set boundaries. It is also critical to attain material sourcing data specific to the manufacturing stage in order to attain meaningful embodied carbon footprint information. For structural steel and reinforcing steel, this needs to include mill-specific information; for concrete, this needs to include plant production information; and for timber, this needs to include impacts associated with the logging process such as building roads, and sequestered carbon losses had the trees remained growing or whether the forest is sustainably managed.

The results of this study should not be taken as hard evidence of the "carbon winner" between concrete, timber, and steel whole-building systems. From the LCA information considered from this study, such conclusions were not possible to make. Instead, it should be seen as a comparative look at where we are today with at least one thoughtful LCA tool that is available to the design community and where the limitations are that need to continue to be worked on. Tally and similarly related software provide a major step forward in being able to systematically quantify within the design process the baseline of a building's embodied carbon footprint and to begin to quantify the sustainable value of design optimization and the use of fewer materials. This allows for identifying trends in embodied carbon impacts between various building system combinations to be explored and helps designers and owners make more informed decisions. The Life Cycle Analysis system and PCR and EPD processes in place today in the United States, though do not yet give statistically compatible data for directly comparing the embodied carbon footprint of one material type against another, and claims made today to the contrary should be very critically evaluated before believing them to be true.

Appendix A: Material Quantity Tables for Four Design Schemes

Tables 6.1a, 6.1b, 6.1c, and 6.1d: Quantities Based on Element Tributary Area
Table 6.2a, 6.2b, 6.2c, and 6.2d: Quantities Based on Gross Project Floor Area (279,532 ft^2)

Table 6.1a Design A: NLT floor system

		Area/volume quantities		Steel quantities normalized by tributary floor area/ volume	
				Average quantities [(rebar, PT, steel tonnage) ÷ (trib floor area or volume)]	
		Floor area	Volume		
	Material properties	(ft^2)	(ft^3 wood, yd^3 concrete)	(lb/ft^2)	(lb/yd^3)
Glulam columns, spruce-pine-fir	f'b = 2400 PSI	–	7981	–	–
Glulam beam, spruce-pine-fir	f'b = 2400 PSI	–	22,900	–	–
Floors – NLT and plywood, spruce-pine-fir	f'b = 2400 PSI	–	120,600	–	–
Miscellaneous steel	Fy = 46 KSI	279,532	–	0.02	–
Metal deck	Fy = 50 KSI	2166	–	2.3	–
Concrete basement walls	f'c = 4 KSI	–	380	–	150
Concrete shear walls	f'c = 5 KSI	–	1368	–	180
Concrete columns	f'c = 5 KSI	–	166	–	190
Topping slab	f'c = 4 KSI	178,440	–	0.7	–
Slab on grade	f'c = 4 KSI	29,740	–	2	–
Concrete slabs (mild reinforced)	f'c = 5 KSI	69,186	–	4	–
Mat foundation	f'c = 4 KSI	–	8942	–	160
Spread footings	f'c = 4 KSI	–	13,578	–	80
	Gross project floor area	279,532			

Table 6.1b Design B: steel framing and slab on metal deck floor system

| | | Total quantities | | Steel quantities normalized by tributary floor area | |
	Material properties	Floor area (ft^2)	Volume (yd^3)	Average quantities [(rebar, PT, steel tonnage) ÷ (floor area or volume)] (lb/ft^2)	(lb/yd^3)
Steel beams and columns	Fy = 50 KSI	180,606	–	12	–
Miscellaneous steel	Fy = 46 KSI	279,532	–	0.02	–
Steel spray applied fireproofing	–	–	638	–	15
Metal deck (roof + composite deck)	Fy = 50 KSI	180,606	–	2.3	–
Concrete basement walls	f'c = 4 KSI	–	380	–	150
Concrete shear walls	f'c = 5 KSI	–	1368	–	180
Concrete columns	f'c = 5 KSI	–	166	–	190
Concrete (slab on metal deck)	f'c = 4 KSI	178,440	–	1	–
Slab on grade	f'c = 4 KSI	29,740	–	2	–
Concrete slabs (mild reinforced)	f'c = 5 KSI	69,186	–	4	–
Mat foundation	f'c = 4 KSI	–	8942	–	160
Spread footings	f'c = 4 KSI	–	13,578	–	90
	Gross project floor area	279,532			

Table 6.1c Design C: post-tensioned slab floor system

| | | Total quantities | | Steel quantities normalized by tributary floor area | |
	Material properties	Floor area (ft^2)	Volume (yd^3)	Average quantities [(rebar, PT, steel tonnage) ÷ (floor area or volume)] (lb/ft^2)	(lb/yd^3)
Miscellaneous steel	Fy = 46 KSI	279,532	–	0.02	–
Metal roof deck	Fy = 50 KSI	2166	–	2.3	–
Concrete basement walls	f'c = 4 KSI	–	380	–	150
Concrete shear walls	f'c = 5 KSI	–	1368	–	180
Concrete columns	f'c = 5 KSI	–	636	–	190
Slab on grade	f'c = 4 KSI	29,740	–	2	–
Post-tensioned concrete slabs	f'c = 5 KSI	208,180	–	2.5 (rebar), 0.9 (PT)	–
Mild reinforced concrete slabs	f'c = 5 KSI	39,446	–	4	–
Mat foundation	f'c = 4 KSI	–	8942	–	170
Spread footings	f'c = 4 KSI	–	13,578	–	100
	Gross project floor area	279,532			

Table 6.1d Design D: mild slab floor system

		Total quantities		Steel quantities normalized by tributary floor area	
				Average quantities [(rebar, PT, steel tonnage) ÷ (floor area or volume)]	
		Floor area	Volume		
	Material properties	(ft^2)	(yd^3)	(lb/ft^2)	(lb/yd^3)
Miscellaneous steel	Fy = 46 KSI	279,532	–	0.02	–
Metal roof deck	Fy = 50 KSI	2166	–	2.3	–
Concrete basement walls	f'c = 4 KSI	–	380	–	150
Concrete shear walls	f'c = 5 KSI	–	1368	–	180
Concrete columns	f'c = 5 KSI	–	636	–	190
Slab on grade	f'c = 4 KSI	29,740	–	2	–
Mild reinforced concrete slabs	f'c = 5 KSI	247,626	–	4	–
Mat foundation	f'c = 4 KSI	–	8942	–	180
Spread footings	f'c = 4 KSI	–	16,293	–	110
	Gross project floor area	279,532			

Table 6.2a Design A: NLT floor system

| | | Steel quantities normalized by gross project floor area | Concrete and timber quantities normalized by gross project floor area |
| | Material properties | Average quantities [(rebar, PT, steel tonnage) ÷ (gross proj. floor area or volume)] | Average quantities [(concrete or wood volume) ÷ (gross proj. floor area)] |
		(lb/ft^2)	(ft^3/ft^2)
Glulam columns, spruce-pine-fir	f'b = 2400 PSI	–	0.029
Glulam beam, spruce-pine-fir	f'b = 2400 PSI	–	0.082
Floors – NLT and plywood, spruce-pine-fir	f'b = 2400 PSI	–	0.431
Miscellaneous steel	Fy = 46 KSI	0.02	–
Metal deck	Fy = 50 KSI	0.02	–
Concrete basement walls	f'c = 4 KSI	–	0.037
	f'c = 5 KSI	–	0.132

(continued)

Table 6.2a (continued)

	Material properties	Steel quantities normalized by gross project floor area	Concrete and timber quantities normalized by gross project floor area
		Average quantities [(rebar, PT, steel tonnage) ÷ (gross proj. floor area or volume)] (lb/ft^2)	Average quantities [(concrete or wood volume) ÷ (gross proj. floor area)] (ft^3/ft^2)
Concrete shear walls			
Concrete columns	f'c = 5 KSI	–	0.016
Topping slab	f'c = 4 KSI	–	0.160
Slab on grade	f'c = 4 KSI	–	0.035
Concrete slabs (mild reinforced)	f'c = 5 KSI	–	0.206
Mat foundation	f'c = 4 KSI	–	0.864
Spread footings	f'c = 4 KSI	–	1.311

Table 6.2b Design B; steel framing and slab on metal deck floor system

	Material properties	Steel quantities normalized by gross project floor area	Concrete/timber quantities normalized by gross project floor area
		Average quantities [(rebar, PT, steel tonnage) ÷ (gross proj. floor area or volume)] (lb/ft^2)	Average quantities [(concrete or wood volume) ÷ (gross proj. floor area)] (ft^3/ft^2)
Steel beams and columns	Fy = 50 KSI	3.5	–
Miscellaneous steel	Fy = 46 KSI	0.02	–
Steel spray applied fireproofing	–	–	0.002
Metal deck (roof + composite deck)	Fy = 50 KSI	1.9	–
Concrete basement walls	f'c = 4 KSI	–	0.037
Concrete shear walls	f'c = 5 KSI	–	0.132
Concrete columns	f'c = 5 KSI	–	0.016

(continued)

Table 6.2b (continued)

	Material properties	Steel quantities normalized by gross project floor area	Concrete/timber quantities normalized by gross project floor area
		Average quantities [(rebar, PT, steel tonnage) ÷ (gross proj. floor area or volume)] (lb/ft^2)	Average quantities [(concrete or wood volume) ÷ (gross proj. floor area)] (ft^3/ft^2)
Concrete (slab on metal deck)	f'c = 4 KSI	–	0.040
Slab on grade	f'c = 4 KSI	–	0.035
Concrete slabs (mild reinforced)	f'c = 5 KSI	–	0.206
Mat foundation	f'c = 4 KSI	–	0.864
Spread footings	f'c = 4 KSI	–	1.311

Table 6.2c Design C: post-tensioned slab floor system

	Material properties	Steel quantities normalized by gross project floor area	Concrete/timber quantities normalized by gross project floor area
		Average quantities [(rebar, PT, steel tonnage) ÷ (gross proj. floor area or volume)] (lb/ft^2)	Average quantities [(concrete or wood volume) ÷ (gross proj. floor area)] (ft^3/ft^2)
Miscellaneous steel	Fy = 46 KSI	0.02	–
Metal roof deck	Fy = 50 KSI	0.02	–
Concrete basement walls	f'c = 4 KSI	–	0.037
Concrete shear walls	f'c = 5 KSI	–	0.132
Concrete columns	f'c = 5 KSI	–	0.061
Slab on grade	f'c = 4 KSI	–	0.035
Post-tensioned concrete slabs	f'c = 5 KSI	–	0.496
Mild reinforced concrete slabs	f'c = 5 KSI	–	0.118
Mat foundation	f'c = 4 KSI	–	0.864
Spread footings	f'c = 4 KSI	–	1.311

Table 6.2d Design D: mild slab floor system

	Material properties	Steel quantities normalized by gross project floor area	Concrete/timber quantities normalized by gross project floor area
		Average quantities [(rebar, PT, steel tonnage) ÷ (gross proj. floor area or volume)] (lb/ft^2)	Average quantities [(concrete or wood volume) ÷ (gross proj. floor area)] (ft^3/ft^2)
Miscellaneous steel	Fy = 46 KSI	0.02	–
Metal roof deck	Fy = 50 KSI	0.02	–
Concrete basement walls	f'c = 4 KSI	–	0.037
Concrete shear walls	f'c = 5 KSI	–	0.132
Concrete columns	f'c = 5 KSI	–	0.061
Slab on grade	f'c = 4 KSI	–	0.035
Mild reinforced concrete slabs	f'c = 5 KSI	–	0.738
Mat foundation	f'c = 4 KSI	–	0.864
Spread footings	f'c = 4 KSI	–	1.574

References

Carbon Leadership Forum. (2015). *EPDs and PCRs*. Web. http://carbonleadershipforum.org/epds-and-pcrs.html

Davies, D., & Klemencic, R. (2014). *"Life cycle analysis: Are we there yet?" Future cities: Towards sustainable vertical urbanism: A collection of state-of-the-art, multi-disciplinary papers on tall buildings and sustainable cities, Proceedings of the CTBUH 2015 Shanghai Conference, China, 16–19 September 2014*. Chicago: Council on Tall Buildings and Urban Habitat.

Mahler, P., & Schneider, P. (2016). The influence of databases on the Life Cycle Assessment (LCA) of building components – a comparison of databases using three different wall constructions. In F. Bakker & v. Breugel (Eds.), *Life-cycle of engineeirng systems: Emphasis on sustainable civil infrastructure* (pp. 967–974). Munich: Technische Universitat Munchen.

Pomponi, F., & Moncaster, A. (2016). Embodied carbon mitigation and reduction in the built environment – what does the evidence say? *Journal of Environmental Management, 181*, 687–700.

Pomponi, F., & Moncaster, A. (2017). *Renewable and sustainable energy reviews*. https://doi.org/10.1016/j.rser.2017.06.049.

Skone, T. J. (2013). *LCA at the Department of Energy (DOE), National Energy Technology Laboratory (NETL)*. Presentation accessed via Web. http://lcacenter.org/lcaxiii/final-presentations/964.pdf.

U.S. Green Building Council. (2013). *Materials and resources credit: Building life-cycle impact reduction*. LEED Reference Guide for Building Design and Construction.

Chapter 7
Analysis of Embodied Carbon in Buildings Supported by a Data Validation System

Nuri Cihan Kayaçetin and Ali Murat Tanyer

Introduction

Life cycle assessment (LCA) has been considered as a reliable and effective environmental impact analysis method developed within ISO 14040 standards (ISO 1997). According to further development in EN 15978 (BS EN 2011), the method considers environmental impacts of a particular building component or system or the whole building, starting from material extraction and construction phase to end-of-life. In literature, LCA has been utilized as a means of finding environmentally friendly solutions in an increasing number of studies (e.g. Tsai et al. 2011; Peuportier et al. 2013; Ron et al. 2014).

LCA provides a holistic approach to sustainability studies. A comprehensive LCA study yields a better understanding of total impact of building. Hence, it may be misleading to brand a building as "sustainable" without considering a broad set of environmental impacts during its expected life cycle. In the construction industry, LCA studies have led to the recognition of the increasingly fundamental role that embodied carbon in buildings plays (Heinonen et al. 2011). This is because, through an LCA, manufacture and construction phases can be measured and assessed, as well as end-of-life activities. Newer evidence have shown that such phases may have an overall embodied carbon that is close to the carbon emissions of the operational phase (Pomponi and Moncaster 2017).

LCA studies are data-intensive procedures that demand time and effort (Pomponi et al. 2017). The quality of a study depends on the availability of different types of data such as environmental data on evaluated processes; system data on flow of raw materials, energy or products; and performance data to compare different product systems. In most cases, the scope of the study varies according

N. C. Kayaçetin · A. M. Tanyer (✉)
Middle East Technical University, Ankara, Turkey
e-mail: nckayacetin@gmail.com; tanyer@metu.edu.tr

© Springer International Publishing AG 2018
F. Pomponi et al. (eds.), *Embodied Carbon in Buildings*,
https://doi.org/10.1007/978-3-319-72796-7_7

to the boundaries and requirements that are set by the researcher. These aspects consequently cause subjectivity and uncertainty in LCA studies, which imply lack of standardized works that are hard to compare.

Moreover, data on the material production and environmental impact of building products are also often missing. Available data on construction materials are fragmented, hard-to-reach and even harder to confirm in terms of quality and credibility. In such a research environment, handling data gaps becomes one of the crucial aspects of conducting an LCA. In literature, there are several ways of treating data gaps, but, regardless of approach, all of them consequently increase uncertainty (Heijungs and Huijbregts 2004).

A general approach to improve the data quality is to utilize a national database. In literature, national databases are considered the most important source for an LCA study (Trusty and Horst 2003). On the other hand, the use of national databases may not be possible for developing countries where industry does not invest on research and development adequately. Furthermore, concern about sharing data on material production is a solid obstacle against research efforts. In the absence of national databases, the quality of data in LCA studies must be clearly reported, and a certain level of validation is required for healthy results.

Lloyd and Ries (2007) defined understanding uncertainty in LCA results as a requirement for a correct analysis. In order to deal with uncertainty, a standardized method is required to explicitly display the problems regarding data quality. This chapter introduces a hybrid life cycle assessment method which is based on data quality. Different environmental impact assessment methods are utilized depending on the quality score that is determined by the pedigree matrix. The aim is to provide a quality check for the available data for LCA studies. The proposed system is explained through a case study in which carbon footprint of an office building has been calculated.

Literature Review

In this section, the use of the LCA methodology for the assessment of embodied carbon in buildings is reviewed. The standard EN 15978 provides the general idea on how to conduct a building-related LCA.

Results of an LCA study can be displayed in different categories according to the impact assessment methods. Regardless of the type of assessment, there are several categories which evaluate issues such as resource extraction, acidification, ozone depletion, etc. One of the most common impact categories for LCA in the built environment is the global warming potential (GWP) defined by the Intergovernmental Panel on Climate Change (2014).

In the sections below, different LCA approaches are reviewed with a particular focus on issues such as lack of data, data quality and methodology-specific errors.

LCA Approaches

In the literature, LCA can be grouped in two conceptually different approaches: process-based LCA (SETAC-EPA approach) and economic input-output LCA (EIO-LCA) (Hendrickson et al. 1998, 2006). The major difference between the two approaches is that while process-based LCA focuses on the individual phases that are used to make a product or generate a service, the latter uses a macroeconomic framework that includes all the monetary and sectoral exchanges of a country's economy.

While a growing body of literature discusses the importance of embodied carbon, most studies rely on process analysis for the quantification. On the other hand, an increasing number of studies utilize a hybrid method which is the combination of both techniques. It consists of using process data where available and filling the gaps with input-output data in order to assess the entirety of the supply chain of a product (Treloar 1997).

It is observed that studies which utilize a single LCA method, either process-based LCA or EIO-LCA, display technical errors due to the nature of the method itself. Studies that rely on one LCA technique suffer from technique-related problems, such as truncation error in process-based LCA and aggregation error in EIO-LCA. As stated by Stephan et al. (2013), quantification of embodied impacts is one of the controversial issues in the literature. It is claimed that process-based LCAs may omit up to 87% of the embodied energy due to truncation error. As a result, the number of studies that benefits from both methods tends to increase.

The two approaches above are integrated to provide a more accurate or cost-effective LCA or to provide alternative estimates for comparison purposes. In particular, the EIO-LCA can be applied for the materials extraction and manufacturing stage assessments to use an economy-wide boundary and to take advantage of its focus on specific processes; the SETAC-EPA LCA approach can be utilized in product use and end-of-life phase assessments (Hendrickson et al. 1998). Treloar (1997) proposed a hybrid LCA method that integrates traditional process LCA and IO LCA data within the IO model. The proposed model enables reliable comparisons of construction products with less work and costs compared to a fully fledged process LCA.

LCA Studies in Buildings

In literature, there are some distinct approaches to research studies regarding the LCA methodology. A group of studies put emphasis on a few components of buildings with a detailed life cycle analysis, whereas other studies attempted to perform a whole building LCA with limited impact categories. From another perspective, the scale of the research studies differs from building specific to urban scale.

The key element of the LCA studies on buildings is the consideration of the embodied carbon of the manufacture and demolition phases, as well as the ratio between such impacts and those occurring during the operational phase. In general, the impact of the construction phase is often considered minor (Scheuer et al. 2003; Junnila and Horvath 2003). However, some more recent studies show that the construction phase of an energy-efficient passive house may account for more than 50% of the building's total life cycle primary energy use (Saynajoki et al. 2012).

Junnila et al. (2006) conducted two detailed LCAs by quantifying the environmental impacts of two new high-end office buildings in Europe and United States over 50 years of service life. Scheuer et al. (2003) conducted an LCA on a six-story building of 7500 m^2 at the University of Michigan campus for a projected 75-year life span. It was concluded that the optimization of operations phase performance should still be the primary emphasis for design (83% of the overall life cycle impacts), until it is evident that there is a significant shift in distribution of life cycle burdens. On the other hand, Stephan et al. (2012) presented a comprehensive life cycle analysis framework for residential buildings. Comparing two case studies in Brussels and Melbourne, they confirmed that embodied, operational and transport requirements are of equal importance.

LCA Studies on Embodied Carbon

As the importance of the construction phase increases, there is a good amount of research effort on LCA studies on embodied carbon. Some studies focus on embodied carbon of office buildings, which are also utilized in this study for comparison.

Airaksinen and Matilainen (2011) investigated the carbon footprint of different design alternatives for an office building in Helsinki, Finland. The authors evaluated the design alternatives with regards to the lowest CO_2 equivalent emissions. It was found out that when the energy use is minimized, emissions originated from materials and products become crucial.

Basbagill et al. (2013) introduced a method for applying LCA to early stage decision-making in order to inform designers of the relative environmental impact importance of building component material and dimensioning choices. A case study was performed to identify the building components with concentrated embodied impacts and to realize which design decisions achieve the greatest reductions in embodied impact.

Ibn-Mohammed et al. (2013) evaluated the impact of embodied carbon against operational carbon emissions in buildings. The authors examined existing studies in the literature to understand the trends and their potential reasons in different countries and regions. It was concluded that embodied carbon is responsible for an increasing ratio of the life cycle emissions associated to new buildings and is subject to rise due to improvement of building regulations for better operational performance.

Pomponi and Moncaster (2016) applied a systematic critical review on academic knowledge regarding the existing approaches to deal with embodied carbon. Seventeen mitigation strategies were identified and concluded that these strategies were not able to tackle the problem alone. Also, an analysis on 77 studies displayed that most of the studies include only the construction phase and exclude the following life cycle phases.

De Wolf et al. (2016) analysed data from over 200 buildings to evaluate the embodied carbon through the development of database of building structures. The results were given, for example, according to building typologies, size, etc. which includes offices, commercial, residential and other building types. It was found out that all the buildings ranged between 150 and 600 $kgCO_2$-eq/m^2.

Initiatives such as Embodied Carbon Benchmark Study (Simonen et al. 2017), Waste Reduction Action Program (WRAP 2017), the methodology of Royal Institution of Chartered Surveyors (2014) and the database of the embodied quantity outputs (deQo 2017) aimed to develop a common methodology on the calculation of embodied carbon and also aimed to facilitate collective databases on embodied carbon in buildings. In common, these initiatives conducted research on numerous buildings in different regions in order to come up with statistical values on buildings. The Embodied Carbon Benchmark Study included more than 1000 entities and demonstrated a range between 266 and 515 $kgCO_2$-eq/m^2 among 291 office buildings.

Data Quality

Data quality is one of the key aspects of an LCA which may facilitate or hinder the validity of the study. Several characteristics of data in LCA determine the quality of the assessment, including data types, uncertainty of data, variety, etc.

According to the acquisition method, LCA data can be grouped into two basic categories: (i) process data which is provided by the producer or directly derived from the production line and (ii) generic data which includes input-output values that are based on national economic frameworks (Dahlstrom et al. 2012). In literature, process data implies high-quality data from the source, whereas generic data is the average value of similar products which may represent the target. Uncertainty plays an important role for LCA studies, especially when it is used for decision-making. It is natural that LCA practitioners seek for quality and credibility in their works. Uncertainty may originate from several sources in an LCA (Lloyd and Ries 2007), and it can be referred to as lack of knowledge: no data is available, or the data available is wrong or ambiguous. The available methods for tackling data gaps aim to either reduce the uncertainty level or explicitly incorporate it (Heijungs and Huijbregts 2004).

While there are several issues which may hinder the credibility of an LCA study, most of the studies refrain from declaring data quality properly. In their study, Junnila et al. (2006) declared that out of 30 components assessed, only four in the

European Union and seven in the US case study were considered to have average or lower-than-average data quality. This lowers the possibility of comparing or adopting the outputs. As Heijungs and Huijbregts (2004) put forward, the interest in data quality has not been common practice since the development of LCA and the rise of its use. Even though the quality of LCA results should be considered at an early stage of LCA development, assessment of this quality is still not a standard step, and a holistic method has not been introduced within the LCA literature.

One of the most common methods on determining and reporting on the data quality of an LCA has been proposed by Weidema and Wesnaes (1996). In this pedigree matrix-based method, a set of predetermined characteristics of data are evaluated in a semi-quantitative 1–5 scale. The characteristics are categorized as "data quality indicators" according to acquisition, independence, representativeness and temporal, geographical and technological correlation. Indicator scores of 1–5, where 1 implies the best and 5 implies the worst condition, are assigned to a predetermined qualitative description as seen in Fig. 7.1.

The pedigree matrix has been utilized in previous LCA research (Weidema and Wesnaes 1996; Junnila and Horvath 2003; Heijungs and Huijbregts 2004; Ciroth 2009; Koffler et al. 2016). In these studies, pedigree matrix is used for determining the level of uncertainty. Depending on this level of uncertainty, a coefficient of variation (CV) is generated which implies the amount of variation as percentage. To ensure credibility, LCA data are modified by multiplying the results with an overall CV (combination of CVs from all quality indicators). In this sense, it is useful for reviewing data quality, pinpointing areas to improve data collection method, but it does not have an effect on the selection of impact assessment method.

In order to fulfil these disadvantages, two main improvements are introduced in the scope of this paper while using the pedigree matrix as a data evaluation tool:

- It is crucial that the impact of LCA data quality has a significant effect on how the LCA is conducted. Therefore, level of data quality is introduced as a means of determining the method of life cycle impact assessment.
- The importance of data quality indicators may differ according to the scope of the study. A weighting coefficient (w) is introduced for all quality indicators for enabling flexibility while determining data quality.

By implementing these improvements, the aim is to integrate data quality as a crucial and transparent part of an LCA model. In the sections below, the proposed method is explained with a focus on data quality and display.

Methodology

Following on the concerns briefly explained above, there is a need to have a standardized framework to assess data quality in LCA studies. This framework should:

Indicator Category	Indicator Score				
	1	2	3	4	5
Acquisition Method	Measured Data	Calculated data based on measurement	Calculated data partly based on assumptions	Qualified estimate by an expert	Non-qualified estimate
Independence of data supplier	Verified data from public or other sources	Verified information from enterprise interested in the study	Inpependent source, but based on non verified information from industry	Non-verified information from the industry	Non-verified information from enterprise interested int he study
Representitiveness of the study	Representitive data from sufficient sample of sites over an adequate period	Representitive data from a smaller number of sites but for adequate periods	Representitive data from adequate number of sites, for shorter periods	Representitive data from a smaller number of sites for shorter periods	Representitiveness unknown or incomplete data from a smaller number of sites and/or from shorter periods
Temporal Correlation	Less than 3 years of different to year of study	Less than 5 years difference	Less than 10 years difference	Less than 15 years difference	Age of data unknown or more than 15 years difference
Geographical Correlation	Data from area under study	Average data from larger area including the area under study	Data from area with similar production conditions	Data from area with slightly similar production conditions	Data from unknown area or area with very different production
Technological Correlation	Data from producer, process and materials under study	Data from different producers, process and materials under study	Data from processes and materials under study but from different	Data on related processes or materials but same technology	Data on related processes or materials but different technology

Fig. 7.1 Pedigree matrix (Adapted from Weidema and Wesnaes 1996 and Junnila and Horvath 2003)

- Explicitly display data sources
- Evaluate the quality of data
- Explain the methods of tackling data gaps and uncertainty, if there are any

Currently, it is not mandatory to include such a framework in academic studies. Regardless of this situation, data quality display should be considered as a great asset for achieving studies that are more credible and also capable of being compared and adopted.

In the following sections, the details of the model are given in a step-by-step manner which follows the standard LCA phases.

Data Validation System in LCA Studies

In this section, data validation model is going to be explained in accordance with the LCA steps given in ISO 14040. It is aimed to enhance the LCA procedure step by step to achieve a higher transparency and better data quality/impact assessment methodology. A general framework of an LCA can be seen in Fig. 7.2. In literature, frameworks do not generally include data evaluation as a separate phase. For the sake of clarity, data evaluation is shown separately from life cycle modules, but within life cycle inventory (LCI) phase.

The proposed model can be seen in Figs. 7.2 and 7.3. The model focuses on the data requirements, data evaluation and data quality-based impact assessment. It utilizes a hybrid impact assessment methodology. The following subsection explains how the model provides additions to the LCA steps.

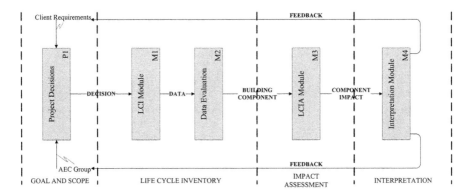

Fig. 7.2 LCA framework

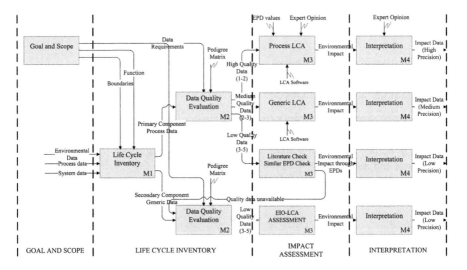

Fig. 7.3 Data validation framework

Goal and Scope

Goal and scope is considered as a crucial step of any LCA, as the concept and boundaries of the study are determined here. Among the necessary steps like function and system boundaries, the critical sub-step for this model is the data requirements. In ISO standards, there are already predefined data requirements of a component. However, the interpretation of these requirements should be made by the LCA practitioner according to the goal and scope of a study. In this model, data requirements are going to determine the method to be used in life cycle impact assessment phase.

The model utilizes the average values derived from six different data quality indicators. Data requirements may vary according to goal and scope of a specific LCA. Hence, for enabling flexibility of the model, weighting of the data quality indicators is introduced so that certain indicators may imply higher importance than others in different cases.

At this point, the necessary action is to determine the level of data quality at which the study should differentiate high, average and low. This critical level of data quality is called quality benchmark. For certain benchmarks, different life assessment methods are assigned which are explained in detail in the following sections.

Another data requirement is related with the building components which build up the life cycle inventory. It is assumed that components which build up the inventory can be categorized and treated differently according to two parameters: (i) embodied carbon of functional unit and (ii) total amount of the component. In this regard, the components can be considered as primary or secondary. Components with higher embodied carbon and amount are defined as primary component, whereas the other components are considered as secondary components. This may provide a higher ratio of time/effort vs. efficiency.

Life Cycle Inventory

In the process of building life cycle inventory, it is important to define which components have larger or smaller impact for any LCA study. Specifically, for simplified LCAs, it is crucial to be able to define priorities among evaluated components in order to increase the efficiency of assessment.

Components are considered as primary components if they have higher embodied carbon per functional unit and if they are higher in quantity when compared to total building components. In this approach, secondary components can be omitted and system boundaries can be redefined for the sake of simplicity and efficiency of the study. Primary components can also be utilized for sensitivity analysis.

Carbon footprint of commonly used materials can be seen in Fig. 7.4. For demonstrative purposes, components with higher than 5% of total amount and embodied carbon with more than 1 $kgCO_2$-eq/kg are considered as primary components in this study.

Table 7.1 shows the template in which necessary information regarding environmental impacts and the data quality of a component is displayed. Information under system data, output and environmental data sections are utilized for modelling the life cycle of the component. Then, the calculated value is multiplied with the corresponding quantity for the total impact.

Evaluation regarding data quality should be filled in accordance with the predetermined scores provided by pedigree matrix as shown in Fig. 7.1.

When the information is provided, data quality should be calculated as given below:

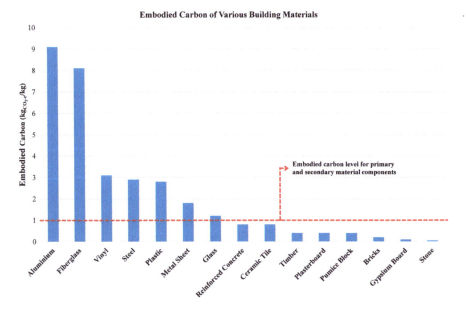

Fig. 7.4 Embodied carbon of materials (Adapted from Hammond and Jones 2011)

Data quality

$$= \frac{(v_{acq} * w_{acq}) + (v_{ind} * w_{ind}) + (v_{rep} * w_{rep}) + (v_{temp} * w_{temp}) + (v_{geo} * w_{geo}) + (v_{tech} * w_{tech})}{\text{Sum of } w \text{ coefficients}}$$

$$(7.1)$$

where:

- w = Weighting coefficient
- $v_{(i)}$ Where $1 \leq v \leq 5$ and i is the related indicator as given below
- acq = Acquisition method
- ind = Independence of data supplier
- rep = Representativeness of the study
- temp = Temporal correlation
- geo = Geographical correlation
- tech = Technological correlation

By assigning weighting coefficients (w) and evaluating each indicator with a value ($v_{(i)}$) between 1 and 5, data quality values of all components are defined, and then components are categorized as primary and secondary.

Life Cycle Impact Assessment (LCIA)

For LCIA of building components, a hybrid LCA methodology is used in which available data are utilized in a process-based approach and the lack of data is

Table 7.1 Necessary data for quality assessment

SYSTEM DATA		
Data Type	**Quantity**	**Units**
Raw Material Composition		
X		
Y		
Z		
Energy Input		
Electricity		
Heat		
Other		

OUTPUT		
Data Type	**Quantity**	**Units**
End Product		
Waste		

ENVIRONMENTAL DATA		
Impact Assessment Method: CML 2001 - Nov. 2010	**Quantity**	**Units**
Acidification Potential		$kg_{SO2\text{-}eq.}$
Eutrophication Potential		$kg_{PO4\text{-}eq.}$
Global Warming Potential		$kg_{CO2\text{-}eq.}$
Ozone Depletion Potential		$kg_{Sb\text{-}eq.}$
Abiotic Depletion, Fossil		MJ

PROJECT VALUES		
Data Type	**Quantity**	**Units**
Quantity in the Project (tons)		
Notes:		

Building Component	Raw Material Inputs	Production Inputs	Life Cycle Model
Acquisition Method			
Independence of Data Supplier (Reliability)			
Representativeness of the Study (Completeness)			
Temporal Correlation (Data Age)			
Geographical Correlation			
Technological Correlation			

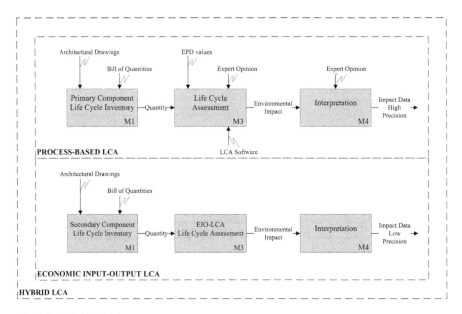

Fig. 7.5 Hybrid LCA impact assessment

compensated by generic data from multiregional input-output tables. After the LCA calculations, the total environmental impact of the building is displayed in a single unit of $kgCO_2$-eq, which is suitable for the study.

Primary components with data quality between data quality scores of 1–2 are assessed with process-based LCA, 2–3 are assessed with generic LCA and 3–5 are assessed with EIO-LCA after a literature check. Secondary components are directly assessed with EIO-LCA (Fig. 7.5).

In process-based LCA, the utilization of software for assessment is recommended. For primary components, it is advised to collect process data from the manufacturers. It is important to focus on these components and to have outputs with high precision. As these components are supposed to have higher ratio of environmental impacts, it will directly increase the accuracy of the whole study. For primary components with partially complete data, it is advised to make use of generic databases to fill in the gaps. While making use of generic data, the data quality should be clearly displayed and evaluated accordingly.

For secondary components, EIO-LCA is advised where generic economy-based data are utilized with low precision. The assessment with this method should take minimum effort, whereas the main focus is on the primary components.

The proposed model in this paper is evaluated through an LCA study of a sample project, a university research centre building (MATPUM) which is used for office purposes.

Evaluation of the Model with a Case Study

In this section, the proposed data validation model is evaluated through a case study, in which the embodied carbon of an office building is assessed. The case under consideration is an office building which is used as a research centre at the campus of Middle East Technical University (METU). The building, called MATPUM (Research Centre for the Built Environment), was commissioned in 2006 and has a rectangular layout with long facades facing north and south, measuring approximately 18×41 m. The building is shaped around a linear atrium around which offices, studios and other work units are arranged on three floors.

Designed with energy efficiency concerns, northern facade of the building introduces minimum openings, while southern facade was designed to be transparent with adequate sun breakers in order to benefit from light and solar energy. The roof of the building is considered as a double-layered and insulated wall that will form continuity with the northern facade and is intended to protect the building against thermal effects.

The building has a combination of structural system where the load-bearing system is reinforced concrete and the roof system is supported by steel beams. The roof system comprises double-skin sandwich panels and is a shell structure that also covers the northern façade of the building. Walls are constructed with either concrete or aerated blocks (Fig. 7.6).

Goal and Scope

The scope of the study is to evaluate the building with regards to the embodied carbon.

Function of the system was defined as the total of the CO_2 or equivalent greenhouse gases emitted to the atmosphere as a result of resource extraction, manufacture and transportation of product stages, which is referred as an A1-A4 assessment according to EN 15978 (BS 2011).

Fig. 7.6 MATPUM building at the university campus

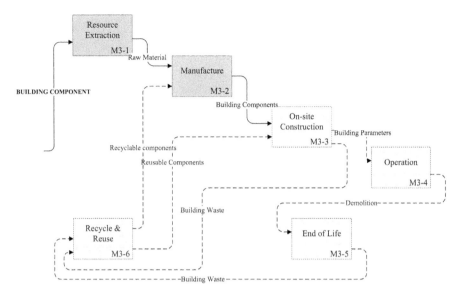

Fig. 7.7 System boundaries

Functional unit was 1 m^2 of useful area. The environmental impacts were calculated according to CML 2001 assessment methodology, and the results were given in kgCO$_2$-eq/m^2 for embodied carbon which corresponds to global warming potential impact category.

In the system boundaries, factors which were not directly related to building LCA were omitted. These omissions were common practice for such a study and were unable to be modelled. As can be seen in Fig. 7.7, on-site construction emissions, maintenance and end-of-life processes were excluded due to lack of data.

Most of the building inventory data relied on the architectural drawings and building model. Pedigree matrix was applied for assessing the quality of data regarding its source, qualitative or quantitative aspect, date and geographical context.

Data requirements on how the data on components were used are as follows:

- Components with:

  (i) Embodied carbon higher than 1 kgCO$_2$-eq/kg and (ii) ratio of total quantity in the building above %5 are considered as primary components.
  (i) Embodied carbon lower than 1 kgCO$_2$-eq/kg and (ii) ratio of total quantity in the building above %5 are considered as secondary components.
  (i) Embodied carbon lower than 1 kgCO$_2$-eq/kg and (ii) ratio of total quantity in the building below %5 are not considered in LCA.

- Primary components with data quality indicator score of:

  1–2 are assessed with process-based LCA.
  2–3 are assessed with generic LCA.
  3–5 are assessed with EIO-LCA after a literature check.

- Secondary components were assessed with EIO-LCA.

Life Cycle Inventory

For structural breakdown of the building, Uniformat II (ASTM 2009) categories were preferred. In the data collection process, respective manufacturers were contacted and provided with the data template. For each component, information on resource extraction, production and transportation to construction site were retrieved.

As the inventory of the study was completed, it was evaluated by pedigree matrix with weighted indicator categories as mentioned earlier (Fig. 7.1). According to the requirements of the study, the components that were evaluated in the LCA process can be seen in Table 7.2. Components were displayed according to the building clusters, and necessary information on amount, unit and indicator scores were given. Data quality was calculated depending on the indicator category weights by using formula (7.1) mentioned earlier.

According to weighted quality indicator score, data retrieved from the manufacturers can be grouped into three:

The first group of manufacturers had provided a detailed description and input/output values that were necessary to model the production as a whole. The data quality of these individual building components has been labelled with a score of 1 or 2, which indicates "high-quality" environmental data. The composition of raw material, amount emissions during processes and waste in the end were given as well as the location of production. The following components have high-quality process data:

- Structural steel
- Aluminium curtain wall profile
- Double-skin sandwich panel
- Vinyl floor tile

The second group of manufacturers had provided partial information on the production processes. The data quality of these individual building components has been labelled with a score of 2 or 3, which indicates "medium-quality" environmental data. The information was not complete in order to build a whole model. These components were (i) plasterboard walls, (ii–iii) pumice wall blocks, (iv) curtain wall glass and (v) concrete.

The third group of manufacturers had not provided any information regarding these components, or the researchers could not communicate with them. The data

Table 7.2 Building inventory of MATPUM building

Building Group Elements	Building Individual Elements	Quantity	Unit	PEDIGREE MATRIX						Average Rate (Weighted)
				Indicator Category Weights						
				Acquisition Method	Independence of Data Supplier (Reliability)	Representativeness of the Study (Completeness)	Temporal Correlation (Data Age)	Geographical Correlation	Technological Correlation	
				2	1	2	1	2	2	
Roof	Double-Skin Sandwich Panel	1,234.23	m²	1	1	2	1	1	1	1.2
	Structural Steel	9.51	tons	1	1	2	1	1	1	1.2
Exterior Walls	Pumice Block - 19cm	539.68	m²	2	2	3	3	2	2	2.3
Interior Walls	Pumice Block - 9cm	1,007.57	m²	2	2	3	3	2	2	2.3
	Plasterboard Walls	329.03	m²	2	2	3	3	2	2	2.3
Windows & Doors	Aluminium Curtain Wall Profile	0.97	tons	1	1	2	1	1	1	1.2
	Aluminium Curtain Wall Glass	360.16	m²	2	2	3	3	2	2	2.3
Floors & Ceilings	Ceramic Tile	877.10	m²	3	2	3	2	4	3	3.0
	Vinyl Floor Tile	518.70	m²	1	1	2	1	1	2	1.4
	Timber Floor Covering	30.00	m²	3	2	3	2	3	4	3.0
	Gypsium Board Suspended Ceiling	379.80	m²	3	2	3	2	3	4	3.0
Basement	Concrete	2,670.19	tons	2	2	3	3	2	2	2.3

quality of these individual building components has been labelled with a score below 3, which indicates "low-quality" environmental data. These components were (i) ceramic tile, (ii) timber floor covering and (iii) gypsum board suspended ceiling.

Life Cycle Assessment

According to the amount and embodied carbon per functional unit of the component, the primary components were determined as:

- Reinforced concrete
- Structural steel
- Pumice block walls
- Double-skin sandwich panel
- Aluminium curtain wall profile

Among the primary components, structural steel, double-skin sandwich panel, aluminium curtain wall profile and vinyl floor tile were to be assessed with process LCA, whereas reinforced concrete, pumice block walls and plasterboard walls were to be assessed with generic LCA (Thinkstep 2017). Rest of the components was considered as secondary and to be assessed with EIO-LCA (Carnegie Mellon University 2008).

For the impact assessment of components, CML 2001 method was used, and environmental impacts are calculated for global warming potential (GWP) category. GWP category is one of the common impact categories in which total equivalent of carbon dioxide emissions is evaluated. The result of the analysis is given in Fig. 7.8.

Interpretation

According to the results of the analysis, MATPUM building embodies an amount of 586,000 $kgCO_2$-eq. The building has a floor area of 2093 m^2 and that concludes to 280 $kgCO_2$-eq/m^2.

For the validation of the study, there is a need to compare the results with available benchmarks and other studies in the literature. Yet, regulations in Turkey introduce overall reduction in carbon emissions in building sector but do not provide a benchmark for carbon emissions of specific building typologies. In literature, despite the fact that there are ongoing developments on embodied carbon databases (deQo 2017; WRAP 2017), several studies put forward the lack of benchmarking studies in this field (RICS 2014; De Wolf et al. 2017). In order to develop benchmarks for the construction sector, more comprehensive and peer-reviewed datasets are needed.

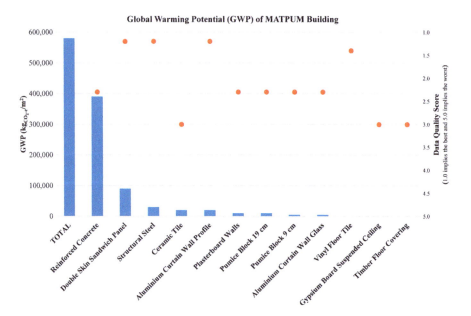

Fig. 7.8 Global warming potential (GWP) of the selected components of MATPUM building

For the validation of this research, MATPUM building was compared with the impacts of a number of office buildings in the literature. At the end of the LCA study, it was also recognized how some of the building components have higher impact than others. This situation is observed to comply with the differentiation of components as primary and secondary.

In a study conducted by De Wolf et al. (2016), 200 buildings were evaluated and 22 of them were office buildings with different characteristic. The range for the office buildings were observed as between 130 and 340 $kgCO_2$-eq/m^2. According to a more recent analysis (deQo 2017), the embodied carbon outputs of 115 office buildings (35 from Europe) are between 227 and 418 $kgCO_2$-eq/m^2. The study of Simonen et al. (2017) on embodied carbon benchmark demonstrated a range between 266 and 515 $kgCO_2$-eq/m^2 among 291 office buildings. Also, a study on an office building of larger scale (Airaksinen and Matilainen 2011) presented similar value which is around 308 $kgCO_2$-eq/m^2 (Fig. 7.9).

Evaluation of the model with the presented case study was considered limited due to the following reasons. The case study on MATPUM building introduced a single value of 280 $kgCO_2$-eq/m^2 in a deterministic manner, whereas compared studies displayed a range of values. It was not possible to identify a range for embodied carbon of MATPUM building due to the limitation on the scope of the study and data acquisition. This condition prevented to see whether case study results statistically overlap with other studies.

Nevertheless, when the GWP value of MATPUM building was compared with other studies, it was observed that the measured value was lower than the median

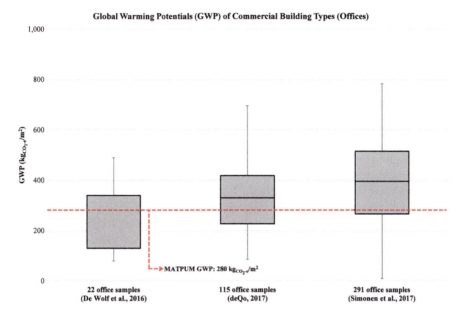

Fig. 7.9 Comparison of embodied carbon in commercial buildings (Adapted from De Wolf et al. 2016; deQo 2017; Simonen et al. 2017)

value obtained by deQo (2017) and Simonen et al. (2017), and the value remained between the two quartiles Q1 and Q2. However, according to the study of De Wolf et al. (2016), it was seen that the value of MATPUM was still within the limits close to the median and between the two quartiles Q2 and Q3. It was seen that the GWP value of MATPUM drifted further away from median when the sample group enlarged (Simonen et al. 2017); and, yet, it still remained within the first quartile.

Conclusion

The proposed model aims to increase the precision of LCA studies according to quality of data inventory. It also emphasizes the necessity to set a standard for data quality display. Without providing the data source explicitly, the level of uncertainty in LCA studies is going to prevent comparison of LCA studies and adoption of LCA results, decrease the credibility of outputs and hinder the ability to build up databases.

Comparison of LCA studies is of crucial importance to evaluate the different methods of impact assessment and system boundaries in order to develop best practices. For investigating best practices, the outputs must have high quality, and credibility of results should be justified. Only when the LCA studies become standardized and credible enough to share and adopt inputs/outputs, then common databases on LCA results can be achieved.

Currently, ISO standards do not consider data quality display as a mandatory step of any LCA. But it is increasingly argued that credibility of LCA studies is questionable. Lack of common practice and varying results according to subjective system boundaries are among the main reasons. The researchers also have right to differ in system boundaries due to unavailability of data.

The proposed framework suggests the differentiation between building components with high environmental impact and low environmental impact. For this categorization, the average values in available databases are utilized. As the precision and quality of these databases increase, the definition of primary and secondary components can be enhanced. It provides practitioners with flexibility on data quality assessment. By using weighting on data quality indicators, the practitioners are capable of putting forward data quality with better precision.

The study also introduces guidance on selection of assessment methodology according to the quality of life cycle inventory. The framework suggests a hybrid impact assessment methodology that directs users to use process-based LCA with generic data or EIO-LCA depending on (i) the environmental impact level of component and (ii) quality of collected data.

The purpose is to make the most of available data and increase the efficiency of research efforts. It is also aimed to introduce a standardized framework for explicitly displaying data sources and evaluation.

It is also shown that transparency of data sources may be a great asset and tool for strengthening the credibility of LCA studies and it can facilitate a better comparison and adoption medium for input and output data. In case of a mandatory data quality step in LCA standards, the credibility and precision of studies will at least be explicitly put forward. This will enable other researchers to take necessary actions while making use of other sources.

References

Airaksinen, M., & Matilainen, P. (2011). A carbon footprint of an office building. *Energies, 4*, 1197–1210.

ASTM. (2009). *E1557-09, standard classification for building elements and related Sitework-UNIFORMAT II*. West Conshohocken: ASTM International.

Basbagill, J., Flager, F., Lepech, M., & Fischer, M. (2013). Application of life cycle assessment to early stage building design for reduced embodied environmental impacts. *Building and Environment, 60*, 81–92.

BS EN 15978. (2011). Sustainability of construction works. Assessment of environmental performance of buildings. Calculation method.

Carnegie Mellon University Green Design Institute. (2008). Economic input-output life cycle assessment (EIO-LCA), US 1997 Industry Benchmark model [Online], available at: http://www.eiolca.net. Accessed 1 Jan 2008.

Ciroth, A. (2009). Cost data quality considerations for eco-efficiency measures. *Ecological Economics, 68*(6), 1583–1590.

Dahlstrom, O., Sornes, K., Eriksen, S. T., & Hertwich, E. G. (2012). Life cycle assessment of a single-family residence built to either conventional- or passive house standard. *Energy and Buildings, 54*, 470–479.

De Wolf, C., Yang, F., Cox, D., Charlson, A., Hattan, A. S., & Ochsendorf, J. (2016). Material quantities and embodied carbon dioxide in structures. *Proceedings of the Institution of Civil Engineers – Engineering Sustainability, 169*(4), 150–161.

De Wolf, C., Pomponi, F., & Moncaster, A. (2017). Measuring embodied carbon dioxide equivalent of buildings: A review and critique of current industry practice. *Energy and Buildings, 140*, 68–80.

deQo. (2017). Analyze embodied carbon data. [Online]. Available from: https://www.carbondeqo.com/database/graph. Accessed 20 Oct 2017.

Hammond, G., & Jones, C. (2011). Embodied carbon. In F. Lowrie & P. Tse (Eds.), *The inventory of carbon and energy (ICE)*. Bath: University of Bath and Bracknell: BSRIA.

Heijungs, R., & Huijbregts, M. A. J. (2004). A review of approaches to treat uncertainty in LCA, Proceedings of the IEMSS Conference, Osnabruck, Austria.

Heinonen, J., Saynajoki, A., & Junnila, S. (2011). A longitudinal study on the carbon emissions of a new residential development. *Sustainability, 3*, 1170–1189.

Hendrickson, C. A., Horvath, S. J., & Lave, L. B. (1998). Economic input-output models for environmental life cycle analysis. *Environmental Science & Technology, 32*(7), 184–191.

Hendrickson, C. T., Lave, L. B., & Matthews, H. S. (2006). *Environmental life cycle assessment of goods and services: An input-output approach*. Routledge: Resources for the Future Press.

Ibn-Mohammed, T., Greenough, R., Taylor, S., Ozawa-Meida, L., & Acquaye, A. (2013). Operational vs. embodied emissions in buildings – a review of current trends. *Energy and Buildings, 66*, 232–245.

IPCC. (2014). *Intergovernmental panel on climate change – climate change 2014: Impacts, adaptation, and vulnerability*. Cambridge, UK: Cambridge University Press.

ISO International Standard 14040. (1997). *Environmental management – Life cycle assessment – Principles and framework*. Geneva: International Organization for Standardization (ISO).

Junnila, S., & Horvath, A. (2003). Life-cycle environmental effects of an office building. *Journal of Infrastructure Systems, 9*(4), 157–166.

Junnila, S., Horvath, A., & Guggemos, A. (2006). Life-cycle assessment of office buildings in Europe and the United States. *Journal of Infrastructure Systems, 12*(1), 10–17.

Koffler, C., Shonfield, P., & Vickers, J. (2016). Beyond pedigree – Optimizing and measuring representativeness in large-scale LCAs. *International Journal of Life Cycle Assessment., 22*(7), 1065–1077.

Lloyd, S. M., & Ries, R. (2007). Characterizing, propagating, and analyzing uncertainty in life-cycle assessment: A survey of quantitative approaches. *Journal of Industrial Ecology, 11*, 161–179.

Peuportier, B., Thiers, S., & Guiavarch, A. (2013). Eco-design of buildings using thermal simulation and life cycle assessment. *Journal of Cleaner Production, 39*, 73–78.

Pomponi, F., & Moncaster, A. M. (2016). Embodied carbon mitigation and reduction in the built environment – What does the evidence say? *Journal of Environmental Management, 181*, 687–700.

Pomponi, F., & Moncaster, A. (2018). Scrutinising embodied carbon in buildings: The next performance gap made manifest. *Renewable and Sustainable Energy Reviews, 81*(2), 2431–2442.

Pomponi, F., D'Amico, B., & Moncaster, A. M. (2017). A method to facilitate uncertainty analysis in LCAs of buildings. *Energies, 10*, 524.

Ron, S., Sungho, T., Sungwoo, S., & Jeehwan, W. (2014). Development of an optimum design program (SUSB-OPTIMUM) for the life cycle CO assessment of an apartment house in Korea. *Building and Environment, 73*, 40–54.

Royal Institute of Chartered Surveyors (RICS). (2014). *Methodology to calculate embodied carbon, RICS guidance note, global* (1st ed.). Parliament Square, London, SW1P 3AD, UK.

Saynajoki, A., Heinonen, J., & Junnila, S. (2012). A scenario analysis of the life cycle greenhouse gas emissions of a new residential area. *Environmental Research Letters, 7*(3), 034037.

Scheuer, C., Keoleian, G. A., & Reppe, P. (2003). Life cycle energy and environmental performance of a new university building: Modelling challenges and design implications. *Energy and Buildings, 35*(10), 1049–1064.

Simonen, K., Rodriguez, B., Barrera, S., Huang, M., McDade, E., & Strain, L. (2017). Embodied carbon benchmark study: LCA for low carbon construction. Available at http://hdl.handle.net/1773/38017

Stephan, A., Crawford, R. H., & Myttenaere, K. (2012). Towards a comprehensive life cycle energy analysis framework for residential buildings. *Energy and Buildings, 55*, 592–600.

Stephan, A., Crawford, R. H., & Myttenaere, K. (2013). A comprehensive assessment of the life cycle energy demand of passive houses. *Applied Energy, 112*, 23–34.

Thinkstep. (2017). Professional database. [Online] Available at: http://www.gabi-software.com/international/databases/gabi-databases/professional/. Accessed 12 Jul 2017.

Treloar, G. J. (1997). Extracting embodied energy paths from input–output tables: Towards an input–output-based hybrid energy analysis method. *Economic Systems Research, 9*(4), 375–391.

Trusty, W. B., & Horst, S. (2003). *Integrating LCA tools in green building rating systems.* Merrickville, Ontario: The Athena Sustainable Materials Institute.

Tsai, W.-H., Lin, S.-J., Liu, J.-Y., Lin, W.-R., & Lee, K.-C. (2011). Incorporating life cycle assessments into building project decision-making: An energy consumption and CO emission perspective. *Energy, 36*(5), 3022–3029.

Weidema, B. P., & Wesnaes, M. S. (1996). Data quality management for life cycle inventories – An example of using data quality indicators. *Journal of Cleaner Production, 4*(3–4), 167–174.

WRAP (2017) Embodied Carbon Database: Share and compare embodied carbon data. [Online] Available from: ecdb.wrap.org.uk. Accessed 20 Oct 2017.

Part II
Management

Chapter 8
Embodied Carbon Tools for Architects and Clients Early in the Design Process

R. Marsh, F. Nygaard Rasmussen, and H. Birgisdottir

Introduction

Following decades of regulatory attention, contemporary architectural practice is skilled in considering the operational impacts of new buildings, mainly in terms of operational energy consumption and the related carbon emissions. Design experience within the field of embodied emissions is still limited, although wider societal changes push towards the integration of environmental concerns in the earliest phases of the design process.

Most environmental calculation tools are used for technical specification and require access to large amounts of precise data. However, this precise data is typically not available in the early design process where many design variables are still not set. The aim of this research is to explore how complex environmental knowledge, stemming from performance-based standards and used in the environmental assessment of buildings, can be simplified for strategic use early in the design process. The subject of analysis is how embodied carbon data for building elements can be made more accessible for non-technical clients and construction professionals when designing low embodied carbon buildings, within the context of the Danish construction sector.

R. Marsh · F. Nygaard Rasmussen (✉) · H. Birgisdottir
Danish Building Research Institute, Aalborg University, Copenhagen, Denmark
e-mail: fnr@sbi.aau.dk

© Springer International Publishing AG 2018
F. Pomponi et al. (eds.), *Embodied Carbon in Buildings*,
https://doi.org/10.1007/978-3-319-72796-7_8

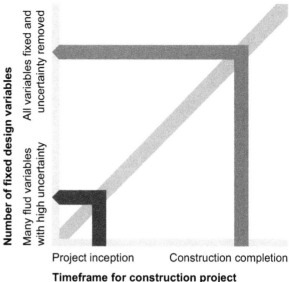

Fig. 8.1 The design process can be characterised by a progression over the time span of a construction project, where uncertainty is progressively eliminated and more design variables become fixed (Marsh 2016)

Environmental Assessment Early in the Design Process

In both the theoretical and empirical literature, the design process in construction projects is characterised as a series of iterative decision processes over time, each of which progresses the design to a greater level of detail, so that uncertainty is progressively reduced, as illustrated in Fig. 8.1. Through the passage of time, the design matures from the initial conceptual stages, where many variables are fluid and the design team explores various strategic and parametric variations, through to the project completion, where all uncertainty is removed in the finished building (Lawson 2005; Mitchell et al. 2011; Tunstall 2006). Construction professionals have also codified this approach in their contractual organisation of the procurement process (Phillips and Lupton 2000).

Reflecting wider societal changes, there has in recent years been a growing demand for the improved energy and environmental performance of buildings. This has been implemented through the broad development of performance-based standards (European Committee for Standardization 2008, 2011a), the regulation of building construction (European Parliament and the Council of the European Union 2010) and the promotion of voluntary initiatives such as the DGNB certification for sustainable building (DGNB 2013). In Denmark, policy initiatives have been formulated to ensure the integration of low-energy strategies at the beginning of the procurement process (National Agency for Enterprise and Construction 2011). Within this rapidly changing framework, construction professionals are now required to integrate environmental concerns in the earliest phases of the design process.

Fig. 8.2 Typical
relationship between
precision of environmental
approach, timeframe for
construction project and
number of fixed design
variables

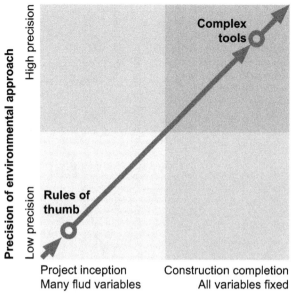

Project inception   Construction completion
Many flud variables   All variables fixed

**Timeframe fore construction project &
Number of fixed design variables**

However, integrating environmental concerns early in the design process can give rise to a contradiction, since most design variables are fluid at this stage and only few strategic parameters are fixed. In contrast, most energy and environmental assessments demand large amounts of data and use complex calculation methods for technical specification and code compliance and cannot be used in the early design stages (Gervásio et al. 2014).

Since few parameters are fixed early in the design process, it can be argued that it is not possible to base simplified methods on the exact same data and precision that is used in complex calculation and specification tools. This reflects the traditional relationship between the precision of the approach and its time of use in the design process, where rules of thumb with low precision have typically been used in the early design process, whilst complex calculation tools for specification and documentation have been used at the end of procurement, as shown in Fig. 8.2.

Hence, the assessment of energy and environmental performance in the early design stages can be problematic. A survey of 390 North American energy assessment solutions shows that 90% were evaluative tools aimed at engineers and that only 1% was aimed at the early design stages (Attia et al. 2012). A survey of common building performance simulation tools shows they were too complex and detailed to fit the workflow of architects, providing excessive amounts of information without communicating the results visually and making dialogue with clients difficult (Weytjens et al. 2011). Furthermore, studies of environmental assessment methods show assessment often happens after design and specification are complete (Schlueter and Thesseling 2009; Schweber and Haroglu 2014). Recently, there has

Fig. 8.3 Use of simplified
tools and proposed new
relationship between
precision of environmental
approach, timeframe for
construction project and
number of fixed design
variables

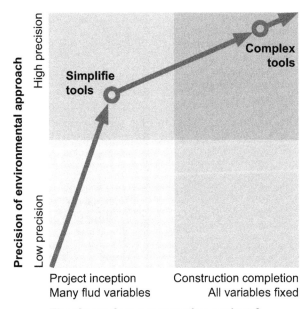

been a rise of interest in developing tools that are able to provide feedback and optimisation strategies at the level of Building Information Models (BIM). However, BIM can entail high complexity and is therefore not used on small projects nor at the very early stages of building design (Hollberg and Ruth 2016). Decisions about building shape and building frame primarily take place in the conceptual stages of the building design (Häkkinen et al. 2015). Hence, if environmental concerns are to be integrated more effectively in the early design process, there is a need for simplified tools which can deliver greater precision whilst using few generalised parameters in the early design process, as shown in Fig. 8.3.

The approach presented in this research takes a twofold approach:

- Instead of quantifying the consumption of building elements after a chosen building design has been completed, it is proposed that construction professionals could use a simplified geometric model, which estimates areas of building elements on the basis of limited geometric input data. Such a model could be used in the early design process to quantify the impact of built form in relation to embodied carbon. This approach would create a parallel between the precision of the geometric model and its time of use.
- Instead of specifying, detailing and calculating the embodied carbon of building elements after a chosen building design has been completed, it is proposed that construction professionals could select suitable building elements from a database containing embodied carbon results for a very large number of prespecified and precalculated building elements. This approach would create a greater

precision, since the precalculated elements would have a high degree of accuracy and data quality, than would be normally expected in the early (or late) design phases.

In combining these two approaches, the aim of this research is to develop a simple tool, where embodied carbon data can be made more accessible and useable for non-technical clients and construction professionals early in the design process, when designing low embodied carbon buildings. This research is presented as follows:

1. Building Geometry Calculation: A simplified geometric building model is developed, which estimates areas of construction elements on the basis of limited geometric input data.
2. Building and Material Lifespan: Since the embodied carbon of building elements is impacted by the lifespan of both the building and the individual materials, the method for defining these has to be detailed first.
3. Parametric Variation of Building Elements: Based on multiple workshops with construction professionals and reflecting contemporary construction practice, the choice of 1095 predefined building elements covering the multiplicity of design solutions for all primary internal and external elements is explained, and the specification of typological variations detailed.
4. Embodied Carbon Calculation: An embodied carbon life cycle assessment is carried out for each typological variation. The results are collated into a library of predefined construction elements and attendant embodied carbon data.
5. Embodied Carbon Design Tool: The proposed design tool is presented, where the geometric data is paired with the predefined construction elements, so that the embodied carbon of buildings can be quickly estimated.
6. Simplified vs. Detailed Building LCA: The results of using the simplified tool are compared with the results from using a detailed building LCA, and the results are discussed.

Building Geometry Calculation

Despite the widespread use of CAD for detail design work, it is still common for architects and clients to explore design options in the early design phase by using simplified built forms and volumes that are represented in scale models by the use of foam or timber, as shown in Fig. 8.4. It is at this level of design that this research is addressing, with low levels of precision early in the design process.

To limit design complexity and simplify the proposed calculation tool, all buildings are to be rectilinear with a rectangular plan and are to have the same floor area on each floor. These parameters reflect the reality of current economical constraints in the construction sector and as such represent the overwhelming majority of new buildings constructed today. The vast majority of new buildings are between one and five storeys in height, with loadbearing external walls and

Fig. 8.4 Architects and clients explore design options in the early design phase by using simplified built forms and volumes that are represented in scale models by the use of foam

where necessary loadbearing internal party or diaphragm walls. It is assumed that multistorey buildings, whether residential or commercial, have basements.

The primary building elements from the major building types found in contemporary Danish construction practice have been analysed, based on a survey of relevant publications (V&S Byggedata 2013; Hansen et al. 2012; Danish Architectural Press 2012) and workshops with construction professionals representing Denmark's major consulting architects and engineers. For the purpose of this research, 17 primary building elements have been selected, representing external, internal and subsurface constructions, as shown in Table 8.1.

Geometric Input Parameters

The aim of this research is to use a limited number of geometric parameters to describe the building and from that basis generate the areas of the different building elements that are used. The following inputs are used to describe the overall geometry of the external building fabric:

L Length of building (m)
D Depth of building (m)
H Floor-to-floor height (m)
S_n Number of storeys (-)
W Window area as a percentage of heated floor area (%)

Table 8.1 Primary building components found in contemporary Danish construction practice, which are covered by this research

Component category	Primary building components
External building components	External wall
	Window/glazing system
	Roof
	Ground floor
Internal building components	Loadbearing internal wall
	Non-loadbearing internal wall
	Columns
	Beams
	Internal floor
Subsurface components	Foundations
	Basement external wall
	Basement floor
	Loadbearing basement internal wall
	Non-loadbearing basement internal wall
	Basement columns
	Basement beams
	Basement internal floor

If the proposed building design has an irregular form in either plan or section, then it is seen as acceptable to make simplified generalisations that can equate the building to a rectangular form, as shown in Fig. 8.5.

The following inputs are used to define the overall geometry of the internal building fabric:

L_1 Loadbearing grid along the length of building, spacing of gridlines (m)
L_d Loadbearing grid along the depth of building, number of gridlines (-)
N_1 Non-loadbearing grid along the length of building, spacing of gridlines (m)
N_d Non-loadbearing grid along the depth of building, number of gridlines (-)
L_c Columns/beams to internal loadbearing structure (1 = *yes*; 0 = *no*)

It is assumed that internal walls follow a relatively regular grid structure and that this can be defined by the number and spacing of the grids, as shown in Fig. 8.6.

The following inputs are used to describe the overall geometry of the subsurface building fabric:

S_b Number of basement storeys (-)
B_c Columns/beams to basement internal loadbearing structure (1 = *yes*; 0 = *no*)
B_p Percentage of basement floor area included in total floor area (%)

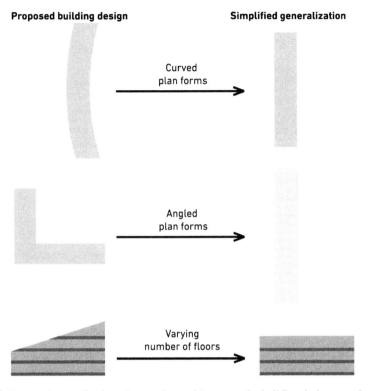

Fig. 8.5 Proposed generalisations that can be used to equate the building design to a simplified rectangular form (Marsh 2014)

Extent of Primary Building Elements

On the basis of the geometric input parameters, the areas of the external building fabric elements can be calculated as follows:

Window area (m^2):

$$W_a = LDS_nW \tag{8.1}$$

External wall area (m^2):

$$E_a = 2HS_n(L+D) - W_a \tag{8.2}$$

Roof area (m^2):

$$R_a = LD \tag{8.3}$$

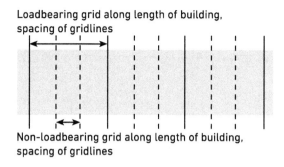

Loadbearing grid along length of building, spacing of gridlines

Non-loadbearing grid along length of building, spacing of gridlines

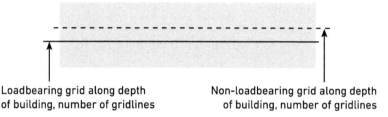

Loadbearing grid along depth of building, number of gridlines

Non-loadbearing grid along depth of building, number of gridlines

Fig. 8.6 Proposed generalisations that are used to simplify the layout of internal walls, following a relatively regular grid structure (Marsh 2014)

Ground floor area (m^2):

$$G_a = LD \text{ if } S_b = 0; \text{otherwise } G_a = 0 \tag{8.4}$$

The areas and lengths of the internal building fabric elements are calculated as follows:

Loadbearing internal wall area (m^2):

$$I_{la} = HS_n\left(LL_d + D(\lfloor L/L_l \rfloor - 1)\right) \text{ if } L_c = 0; \text{otherwise } I_{la} = 0 \tag{8.5}$$

Non-loadbearing internal wall area (m^2):

$$I_{na} = HS_n\left(LN_d + D(\lfloor L/L_d \rfloor - 1)\right) \tag{8.6}$$

Column total height (m):

$$C_h = L_c HS_n(\lfloor L/L_l \rfloor - 1 + C_f)(L_d + C_f) \tag{8.7}$$

where C_f is the column/facade constant, indicating whether there are columns on the facade grid line, with $C_f = 2$ for steel- or timber-framed external walls and $C_f = 0$ for concrete- and brickwork-based external walls.

Beam total length (m):

$$B_l = L_c S_n \left(D(\lfloor L/L_1 \rfloor - 1 + C_f) + L(L_d + C_f) \right) \qquad (8.8)$$

Internal floor area (m^2):

$$F_a = LD(S_n - 1) \text{ if } S_b = 0; \text{ otherwise } F_a = LDS_n \qquad (8.9)$$

The areas and lengths of the subsurface building fabric elements are calculated as follows:

Foundation length (m):

$$F_l = 2(L + D) \text{ if } S_b = 0; \text{ otherwise } F_l = 0 \qquad (8.10)$$

Basement external wall area (m^2):

$$B_{ea} = 2HS_b(L + D) \qquad (8.11)$$

Basement floor area (m^2):

$$B_{fa} = 0 \text{ if } S_b = 0; \text{ otherwise } B_{fa} = LD \qquad (8.12)$$

Loadbearing basement internal wall area (m^2):

$$B_{la} = HS_b \left(LL_d + D(\lfloor L/L_1 \rfloor - 1) \right) \text{ if } B_c = 0; \text{ otherwise } B_{la} = 0 \qquad (8.13)$$

Non-loadbearing basement internal wall area (m^2):

$$B_{na} = HS_b \left(LN_d + D(\lfloor L/L_d \rfloor - 1) \right) \qquad (8.14)$$

Basement column total height (m):

$$B_{ch} = B_c HS_b L_d (\lfloor L/7.5 \rfloor - 1) \text{ if } L_1 = 0; \text{ otherwise } B_c HS_b L_d (\lfloor L/L_1 \rfloor - 1) \qquad (8.15)$$

Basement beam total length (m):

$$B_{bl} = B_c S_b \left(D(\lfloor L/L_1 \rfloor - 1) + LL_d \right) \qquad (8.16)$$

Basement internal floor area (m^2):

$$B_{ia} = LD(S_b - 1) \text{ if } S_b > 0; \text{ otherwise } B_{ia} = 0 \qquad (8.17)$$

Building and Material Lifespan

To be able to select suitable building elements from a database containing embodied carbon results for a very large number of precalculated building elements, it is first necessary to take account of issues related to building and material lifespan.

An individual building typically consists of hundreds of separate materials, and they all have unique production, replacement and disposal processes and also different service lives (Haapio and Viitaniemi 2008). The ISO standard 15686-1 (ISO 2011) defines the service life of the materials according to their accessibility. As such, inaccessible, irreplaceable or structural materials should have the same intended service life as the building, whilst other materials with progressively shorter lifespans should be layered around these, such that materials with a shorter lifespan can be replaced without needing to remove those with a longer lifespan. This gives rise to functionally layered building elements, which have become dominant in the Nordic countries since the 1970s (Marsh and Lauring 2011). A consequence of this is that building elements do not as such have a lifespan. Rather, it is the individual materials used in that element that have a lifespan, and this is determined by technical, economic, functional and design factors (Grant et al. 2014). The same material can of course be used in functionally different layers with different lifespans, such that timber cladding has a shorter lifespan than loadbearing timber framing. It is therefore important to distinguish between the lifespan of the building and the lifespan of the materials that are used in each element of the building.

Lifespan of Materials

For each material used in every layer of a building component, it is necessary to assign a lifespan. This is determined by using the lifespan tables for construction materials based on the functional layering of contemporary building elements, as developed by Aagaard et al. (2013) as part of the Danish Energy Agency's policy initiative for sustainable construction. These tables are based on national data and draw from relevant theoretical research and North European empirical experiences within the framework of the ISO standard 15686-1 (ISO 2011). The range of lifespans contained in the tables is between 10 and 120 years. An example of the lifespan data for an external wall is shown later in Table 8.4.

Lifespan of Buildings

The lifespan of a building reflects whether it still fulfils the many functions for which it was established. Traditionally, a lifespan of between 35 and 60 years has been used in the calculation of a building's life cycle cost, mainly because it is seen as economically irrelevant to calculate with a longer time span for the depreciation of construction investments (Aagaard et al. 2013; Goh and Sun 2016), and it is for this reason that the Danish DGNB certification system uses 50 years as one of two reference study periods for a building LCA (DK-GBC 2012). The other reference study period applied in the Danish DGNB system reflects factual data regarding current building lifespans, coupled with annual rates of new construction, renovation and demolition. This approach indicates that an average building lifespan of 100 years or more would be more accurate in sustainability assessments, reflecting changing social, technical and economic factors in a wider European context (Brown et al. 2011; Ravetz 2008).

An analysis of the Danish building mass shows that different building types have differing expected average lifespans, thus reflecting the balance between construction economics and changes in building usage, with housing having a longer lifespan and buildings for commerce and production having a shorter lifespan (Aagaard et al. 2013), as shown in Table 8.2. For the purposes of this research, building lifespans of 50, 80, 100 and 120 years are selected. Fifty years is chosen to reflect the DGNB system, as used in Denmark. The other three lifespans are chosen to represent all major Danish building types, comprising over 65% of the total completed floor area of all new construction over the last 10 years (Statistics Denmark 2016).

Table 8.2 Average lifespan for various building types in Denmark

Building type	Average lifespan
Agricultural buildings	40
Storage buildings	60
Production facilities	80
Office buildings	
Retail buildings	
Nursery/education	100
University/research	
Medical buildings	
Sports facilities	
Detached housing	120
Terrace housing	
Multistorey housing	
Cultural buildings	

Source: Aagaard et al. (2013)

Parametric Variation of Building Elements

To generate embodied carbon data for a large number of building elements, it is necessary to define many parametric variations of the 17 primary building elements.

Constructive Variations

As described previously, modern building elements are typically functionally layered, allowing each element to be divided into three layers: an internal loadbearing layer, an insulating layer and an external cladding. Typical materials reflecting contemporary Danish construction practice, and covering both traditional materials and newer solutions with supposedly lower environmental impacts, have therefore been selected for each layer of all the building elements shown in Table 8.1. Each variation of the external building elements is specified on the basis of three different U-values, representing the levels used in the current and expected future building regulations. Each variation of the internal building elements is specified on the basis of three different thicknesses, representing varying structural or fire-related demands.

These variations can therefore be seen as covering most functional and aesthetical demands with a variety of in situ and prefabricated constructional solutions, reflecting the diversity of factors that construction professionals may need to address early in the procurement process. The selected variations for one of the elements are shown in Table 8.3. Danish construction practice is broadly similar to that found in other North European countries.

On the basis of the 11 primary building elements with differing layer variations, U-values and thicknesses, it is possible to generate 1095 constructive variations. For each variation an inventory is prepared, describing the material consumption required for a functional unit of one square metre of each building element, except for the foundation typology, which uses a functional unit of one linear metre. All materials are included, and for each variation there are typically between 10 and 30 discrete materials used, including fixings, sealants and finishes.

Lifespan Variations

For each variation, it is necessary to produce further parametric variants based on the lifespan of the building that the element is used in. Therefore, for each of the four building lifespans of 50, 80, 100 and 120 years, the total material consumption is calculated with a consumption multiple, which includes both the originally installed materials and the materials which are replaced due to having a lifespan

Table 8.3 Parametric variations of materials used for each layer of the external wall with heavyweight loadbearing and lightweight cladding

Layers of external wall with heavyweight loadbearing and lightweight cladding	
	Number of variations
Internal layer: heavyweight loadbearing	5
100 mm lightweight concrete elements with steel reinforcement and paint finish	
150 mm lightweight concrete elements with steel reinforcement and paint finish	
120 mm concrete with steel reinforcement and paint finish	
180 mm concrete with steel reinforcement and paint finish	
250 mm concrete with steel reinforcement and paint finish	
Insulating layer: insulation and framing (U-value, 0.16/0.14/0.12 W/m^{2}K)	9
230/260/300 mm mineral wool insulation with C-post galvanised steel framing	
230/260/300 mm mineral wool insulation with I-post timber framing	
250/290/340 mm cellulose insulation with I-post timber framing	
External layer: lightweight finish	8
25 mm pine timber cladding on timber battens on 8 mm fibre cement sheet	
25 mm larch timber cladding on timber battens on 8 mm fibre cement sheet	
10 mm fibre cement panel with finish on timber battens on 8 mm fibre cement sheet	
30 mm natural stone on galvanised steel profiles on 8 mm fibre cement sheet	
35 mm clay tile panel on galvanised steel profiles on 8 mm fibre cement sheet	
1.5 mm aluminium sheet on galvanised steel profiles on 8 mm fibre cement sheet	
1.0 mm zinc sheet on galvanised steel profiles on 8 mm fibre cement sheet	
1.0 mm copper sheet on galvanised steel profiles on 8 mm fibre cement sheet	
Total number of parametric variations (5 × 9 × 8)	360

shorter than the building's. This replacement is determined by using the material's average lifespan in relation to the building lifespan at a decimalised annual replacement rate using data from Aagaard et al. (2013), as described in the previous section. This ensures a consistent methodology, allowing for the calculation of all

Table 8.4 Resource consumption, lifespan and consumption multiple for all materials used in each layer of a functional unit of one square metre of a typical external wall with a U-value of 0.14 W/m²K, based on building lifespans of 50, 80, 100 and 120 years

Internal layer units/m² façade area	Thickness mm	Weight kg	Material lifespan Years	Consumption multiple based on building lifespan of			
				50 years	80 years	100 years	120 years
Paint finish	–	0.13	15	3.33	5.33	6.67	8.00
Concrete, C30–37	180	414.00	120	1.00	1.00	1.00	1.00
Steel reinforcement	–	10.00	120	1.00	1.00	1.00	1.00
Insulating layer							
Polyethylene vapour barrier	–	0.13	120	1.00	1.00	1.00	1.00
Galvanised steel framing	–	4.40	120	1.00	1.00	1.00	1.00
Mineral wool insulation	250	7.50	120	1.00	1.00	1.00	1.00
Fibre cement sheet	8	13.60	120	1.00	1.00	1.00	1.00
Galvanised steel fixings	–	0.10	120	1.00	1.00	1.00	1.00
External layer							
Timber battens	22	1.05	120	1.00	1.00	1.00	1.00
Galvanised steel fixings	–	0.10	120	1.00	1.00	1.00	1.00
Timber cladding with finish	25	13.23	50	1.00	1.60	2.00	2.40
Galvanised steel fixings	–	0.10	50	1.00	1.60	2.00	2.40

environmental impacts from production, use and end of life. With 1095 constructive variations, this gives 4380 discrete parametric variations in total. An example of one constructive variation including the construction multiples for the four lifespan variations is shown in Table 8.4.

Embodied Carbon Calculation

The embodied carbon of all the variations of building elements was calculated by carrying out a life cycle assessment (LCA) using the calculation tool *DK LCA-Calc*, which was developed for the Danish implementation of the DGNB system. This tool draws LCA inventories for materials from the *ESUCO* and *Ökobau* 2011 databases (Birgisdottir et al. 2013; DK-GBC 2012). These databases use the CML 2002 life cycle impact assessment methodology, using a problem-oriented approach with midpoint impact category indicators that translate impacts into environmental themes (Joint Research Centre 2010). In line with the CML 2002 methodology and the DGNB system, the optional steps of normalisation and weighting are not part of the LCA.

The system boundaries for the LCA include the production stage processes, the replacement of materials, the final waste processing and disposal and the benefits and loads beyond the system boundary as these stages are defined by the CEN standard for the sustainability of construction works EN 15978 (European Committee for Standardization 2011b) and the DGNB system (DK-GBC 2012).

Inventory Data

Uncertainty regarding the accuracy of environmental data is an important factor, with the use of environmental product declarations (EPDs) for product-specific materials being the primary recommendation within the CEN standard for the sustainability of construction works EN 15978 (European Committee for Standardization 2011b). However, there are valid arguments as to why product-specific EPDs cannot be used and why the use of generic data is an accepted alternative. Firstly, whilst a growing number of products from international businesses are covered by EPDs, very few are currently available for specific Danish construction materials, and this is especially true for the great number of generic materials used, such as concrete or gravel. Secondly, issues relating to public procurement policies within the EU often mean that buildings are constructed with different materials than those specified earlier in the procurement process, thus reducing the accuracy of a product-specific LCA approach early in the procurement process. Finally, it has been shown that the use of generic data in a national context has a methodological validity and relevance (Silvestre et al. 2015).

Table 8.5 Embodied carbon results for a functional unit of one square metre for two external wall typology variations with a U-value of 0.16 W/m^{2}K and building lifespan of 120 years

Global warming potential CO$_2$ equivalent kg/m^2 year	External wall variation 1	External wall variation 2
Internal layer	120 mm reinforced concrete	25 mm plasterboard
Insulating layer	Mineral wool insulation with ties	Steel framing with mineral wool
External layer	100 mm brickwork	25 mm pine timber cladding
Building lifespan:		
50 years	2.03	0.96
80 years	1.28	0.65
100 years	1.03	0.54
120 years	0.86	0.47

Both the *ESUCO* and *Ökobau* databases contain specific and generic data, which is representative of typical European or German construction materials, and their suitability for use within the Danish context has been determined (Schmidt 2012). The databases contain LCA inventories covering upstream extraction and manufacturing and downstream end-of-life processes.

Calculation Procedure

An LCA includes various categories of potential environmental impact and resource consumption that have been scientifically validated. Embodied carbon is represented by the global warming potential impact category from the LCA. For the purposes of comparison, the total primary energy consumption and the non-renewable primary energy consumption categories are also included.

The embodied carbon and other categories are calculated for each of the 4380 typological variations, and all results are calculated for the functional unit of each variation on an annualised basis. Typical embodied carbon results for two external wall typologies are shown in Table 8.5. These clearly show the impact of building lifespan, with short lifespans giving high levels of annualised embodied carbon and long lifespans giving lower levels.

The parametric results from the LCA are then stored in a spreadsheet database, tagged in relation to the primary building element type, building lifespan, internal loadbearing layer, insulating layer and external cladding layer.

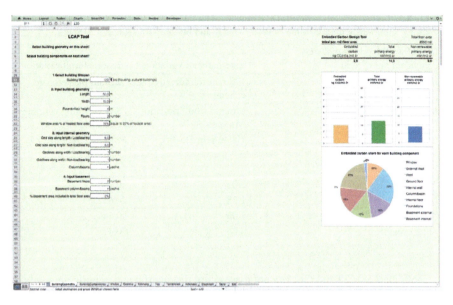

Fig. 8.7 Interface of the building geometry sheet of the proposed embodied carbon design LCAP tool

LCA Profile Tool

The proposed design tool, the *LCA profile tool* (*LCAP* tool), has been developed as a spreadsheet with the use of Microsoft Excel. The *LCAP* tool consists of three elements:

1. Building geometry: a visible spreadsheet, where data for the geometric input parameters is entered and where the overall results for the building are shown.
2. Building elements: a visible spreadsheet, where suitable building elements are selected for the 17 primary elements, including variations for U-value or size. The overall results for the building are also shown.
3. Spreadsheet database: a series of hidden worksheets containing the results of the embodied carbon calculations for all the parametric variations for all building elements.

Building Geometry

In this visible spreadsheet, data for the geometric input parameters is entered, as shown in Fig. 8.7. Based on these input parameters, areas and lengths for the 17 primary construction elements can be calculated using the previously presented Eqs. 8.1, 8.2, 8.3, 8.4, 8.5, 8.6, 8.7, 8.8, 8.9, 8.10, 8.11, 8.12, 8.13, 8.14, 8.15, 8.16,

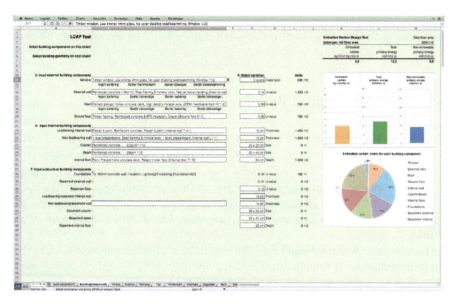

Fig. 8.8 Interface of the building component sheet of the proposed embodied carbon design LCAP tool

and 8.17. The lifespan of the building is also selected from a drop-down menu here, either 50, 80, 100 or 120 years. The overall results for the building are shown, expressed as units per square metre floor area per year:

A pie chart shows the percentage of the total embodied carbon that each of the different building elements is responsible for. For the sake of clarity, some of the elements have been collated, such as loadbearing and non-loadbearing walls.

Column charts show the embodied carbon, total primary energy and non-renewable primary energy consumption. These results are also shown numerically.

Building Elements

In this visible spreadsheet, suitable building elements are selected from drop-down menus for the 17 primary elements, and variations for U-value or size are also selected, as shown in Fig. 8.8. It is also possible to sort the primary external building elements, where there are most parametric variations, after the size of the variation's embodied carbon or alphabetically after the material type used in each of the three layers.

The calculated area and length of all building elements are also shown. Based on the selections, the relevant embodied carbon data is pulled from the spreadsheet database and multiplied by the calculated areas and lengths for the 17 primary construction elements to give the total results, which are then divided by the total floor area of the building. The overall results for the building are also shown again.

Simplified Vs. Detailed Building LCA

The complexity of regular life cycle calculations on buildings has prompted many attempts to simplify the process at different levels. One such simplified approach is to limit the amount of life cycle stages included in the calculations. This means omitting some single processes (e.g. replacements) or omitting whole life cycle stages (e.g. end of life). In practice, variations of these types of simplifications are carried out more often than not in building scale LCAs (Soust-Verdaguer et al. 2016).

Limiting the inventory input can also constitute a simplified approach, either by omitting materials of limited quantities in a building or by omitting building parts that often contribute negligibly to the total results of a building.

Both these simplification approaches are convenient measures at the early design stage, where few parameters are set, and the simplifications are thus a way to get started on the environmental assessment of a building. However, the accuracy of results becomes naturally impaired to an extent that may prove difficult to judge. On the other hand, other factors than just simplifications at the assessment level may challenge the accuracy of results, for example, the accuracy of environmental data used for the calculations. The question of accuracy can be seen as being at an expert level and thus out of the scope for building designers, who are developing skills in environmental design.

Precision of the Simplified Embodied Carbon Design Tool

A different perspective on LCA simplifications can be defined from the tool user's point of view. The *LCAP* tool can be characterised as a simplified tool because the inputs are limited to 14 data entries and up to 25 choices from drop-down menus. In contrast, *DK LCA-Calc*, the LCA tool used for the Danish DGNB certification, requires entries of all volumes of materials and areas of building surfaces. This type of extensive, detailed data is typically only known at the later stages of the building design.

As a simplified tool, the precision of *the LCAP* tool's results can be questioned, when being compared to the results of a more detailed assessment later in the design process. However, the *DK LCA-Calc* tool is consistent with the *LCAP* tool in terms of background database and system boundaries for the building life cycle. The major difference is thus the input for the two tools. A comparison has therefore been carried out between the results obtained from modelling four buildings in the

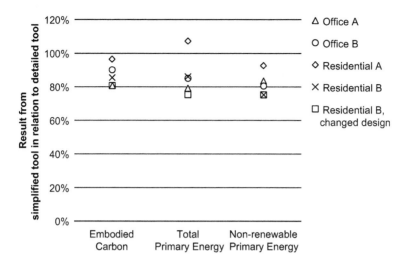

Fig. 8.9 Difference in results of five case buildings modelled in embodied carbon LCAP tool compared with results from the same five buildings modelled in a detailed LCA tool

detailed *DK LCA-Calc* tool and the *LCAP* tool, as shown in Fig. 8.9. The comparison has been made horizontally and vertically:

- Horizontally, where four different building cases have been analysed to determine how the results correspond in both tools for the same building.
- Vertically, where residential building B has been compared in both tools on the basis of a changed design from heavyweight to lightweight building elements for the wall and roof.

When comparing the results from the two tools, the following tendencies can be seen:

- The embodied carbon, representing the global warming potential, is between 4% and 19% lower in the simplified tool when compared to the detailed tool. The mean value is 13% lower, with a standard deviation of 7% from the mean.
- The embodied total primary energy is between 7% higher and 25% lower in the simplified tool. The mean value is 14% lower, with a standard deviation of 12% from the mean.
- The embodied non-renewable energy is between 7% and 25% lower in the simplified tool. The mean is 19% lower, with a standard deviation of 7% from the mean.

The precision of embodied carbon, total primary energy and non-renewable primary energy results in the *LCAP* tool is both consistent and explainable, where it can be argued that errors of between 10% and 15% can be seen as acceptable in the early design process. When comparing the two tools, the simplified *LCAP* tool does not contain data for building services and secondary building elements, such as internal doors, whilst the detailed *DK LCA-Calc* tool does. It has, for example, been shown that the building services can be responsible for between 5% and 10% of the total environmental impact in a Danish context (Rasmussen and Birgisdottir 2015). This would indicate that the actual margin of error between the simplified and detailed tools is very low, at about only 5–10%, when accounting for the missing building elements.

Conclusion

The results from this research show that it is possible to develop simplified embodied carbon tools, such as the *LCAP* tool, which deliver greater precision on the background of using few generalised parameters in the early design process:

- The tool is shown to be easy to use, allowing quick, parametric modelling within minutes. This compares to the many hours typically needed to input data into traditional detailed tools.
- The results indicate a margin of error of only 5–10% compared to traditional detailed tools.
- The use of simplified geometric models, which estimate areas of building elements on the basis of limited geometric input data, is shown to be a valid approach.
- The selection of suitable building elements from a database containing embodied carbon results for a very large number of prespecified and precalculated building elements is shown to be a valid approach.

The results show that the further development of the tool would benefit from applying a similar approach to the secondary building elements and building services, which are not currently included in the model. This would include making estimates of material consumption for doors, staircases, heating/ventilation systems, etc. per square metre floor area for different building types. This would allow the tool to be extended with these elements and thus increase the accuracy of the tool even further.

References

Aagaard, N.-J., Brandt, E., Aggerholm, S., & Haugbølle, K. (2013). *SBi 2013:30 – Levetider af bygningsdele ved vurdering af bæredygtighed og totaløkonomi. [lifespan of building elements for assessing sustainability and life-cycle costs]*. Copenhagen: Danish Building Research Institute, Aalborg University.
Attia, S., Gratia, E., De Herde, A., & Hensen, J. (2012). Simulation-based decision support tool for early stages of zero-energy building design. *Energy and Buildings, 49*, 2–15. https://doi.org/10.1016/j.enbuild.2012.01.028.

Birgisdottir, H., Mortensen, L., Hansen, K., & Aggerholm, S. (2013). *SBi 2013:09 – Kortlægning af bæredygtigt byggeri. [Review of sustainable construction]*. Copenhagen: Danish Building Research Institute, Aalborg University.

Brown, N., Malmqvist, T., Peuportier, B., Zabalza, I., Krigsvoll, G., Wetzel, C., et al. (2011). *Low resource consumption buildings and constructions by use of LCA in design and decision making: Report on scenarios in constructions, LoRe-LCA-WP4-D4.2*. Stockholm: KTH Royal Institute of Technology.

Danish Architectural Press. (2012). *Arkitektur DK. [Architecture Denmark]*. Copenhagen: Danish Architectural Press.

DGNB. (2013). *Excellence defined: Sustainable building with a systems approach*. Stuttgart: DGNB GmbH.

DK-GBC. (2012). *DGNB system Denmark 2012 new office and administrative buildings – Environmental quality*. Copenhagen: Green Building Council Denmark.

European Committee for Standardization. (2008). *EN ISO 13790:2008 – Energy performance of buildings – Calculation of energy use for space heating and cooling*. Brussels: European Committee for Standardization.

European Committee for Standardization. (2011a). *EN 15643-2:2011 – Sustainability of construction works – Sustainability assessment of buildings – Part 2: Framework for the assessment of environmental performance*. Brussels: European Committee for Standardization.

European Committee for Standardization. (2011b). *EN 15978: 2011 – Sustainability of construction works – Assessment of environmental performance of buildings – Calculation method*. Brussels: European Committee for Standardization.

European Parliament and the Council of the European Union. (2010). The directive 2010/31/EU of the European parliament and of the council of 19 May 2010 on the energy performance of buildings (recast). *Official Journal of the European Union, 153*, 13–35.

Gervásio, H., Santos, P., Martins, R., & de Silva, S. (2014). A macro-component approach for the assessment of building sustainability in early stages of design. *Building and Environment, 73*, 256–270. https://doi.org/10.1016/j.buildenv.2013.12.015.

Goh, B. H., & Sun, Y. (2016). The development of life-cycle costing for buildings. *Building Research and Information, 44*, 319–333. https://doi.org/10.1080/09613218.2014.993566.

Grant, A., Ries, R., & Kibert, C. (2014). Life cycle assessment and service life prediction: A case study of building envelope materials. *Journal of Industrial Ecology, 18*, 187–200. https://doi.org/10.1111/jiec.12089.

Haapio, A., & Viitaniemi, P. (2008). Environmental effect of structural solutions and building materials to a building. *Environmental Impact Assessment Review, 28*, 587–600. https://doi.org/10.1016/j.eiar.2008.02.002.

Hansen, E. J., Stang, B. D., Ginnerup, S., Kirkeby, I. M., Buch-Hansen, T., Aagaard, N. J., et al. (2012). *SBi guidelines 230 – Guidelines on building regulations 2010* (2nd ed.). Copenhagen: Danish Building Research Institute, Aalborg University.

Hollberg, A., & Ruth, J. (2016). LCA in architectural design – A parametric approach. *The International Journal of Life Cycle Assessment, 7*, 943. https://doi.org/10.1007/s11367-016-1065-1.

Häkkinen, T., Kuittinen, M., Ruuska, A., & Jung, N. (2015). Reducing embodied carbon during the design process of buildings. *Journal of Building Engineering, 4*, 1–13. https://doi.org/10.1016/j.jobe.2015.06.005.

ISO. (2011). *ISO 15686-1:2011 buildings and constructed assets – Service life planning – Part 1: General principles and framework*. Geneva: International Standards Organisation.

Joint Research Centre. (2010). *ILCD handbook: Analysis of existing environmental impact assessment methodologies for use in life cycle assessment* (1st ed.). Ispra: European Commission Joint Research Centre.

Lawson, B. (2005). *How designers think: The design process demystified* (4th ed.). Oxford: Architectural Press/Elselvier.

Marsh, R. (2016). LCA profiles for building elements: Strategies for the early design process. *Building Research and Information, 44*(4), 358–375. https://doi.org/10.1080/09613218.2016. 1102013.

Marsh, R. (2014). *LCA profiler for bygninger og bygningsdele – vejledning til værktøj til brug tidligt i designprocessen. [LCA Profiles for buildings and building elements – Guide for tool for use early in the design process]*. Copenhagen: Danish Building Research Institute, Aalborg University.

Marsh, R., & Lauring, M. (2011). Architecture and energy: Questioning regulative and architectural paradigms for Danish low-energy housing. *Architectural Research Quarterly, 15*, 165–175. https://doi.org/10.1017/S1359135511000583.

Mitchell, A., Frame, I., Coday, A., & Hoxley, M. (2011). A conceptual framework of the interface between the design and construction processes. *Engineering, Construction and Architectural Management, 18*, 297–311. https://doi.org/10.1108/09699981111126197.

National Agency for Enterprise and Construction. (2011). *Bygningsklasse 2020 [Building Class 2020]*. Copenhagen: National Agency for Enterprise and Construction.

Phillips, R., & Lupton, L. (2000). *Architect's plan of work*. London: RIBA Publications.

Rasmussen, F. N., & Birgisdottir, H. (2015). *SBi 2015:09 – Bygningens livscyklus [the lifecycle of the building]*. Copenhagen: Danish Building Research Institute, Aalborg University.

Ravetz, J. (2008). State of the stock – What do we know about existing buildings and their future prospects? *Energy Policy, 36*, 4462–4470. https://doi.org/10.1016/j.enpol.2008.09.026.

Schlueter, A., & Thesseling, F. (2009). Building information model based energy/exergy performance assessment in early design stages. *Automation in Construction, 18*, 153–163. https://doi. org/10.1016/j.autcon.2008.07.003.

Schmidt, A. (2012). *Analysis of five approaches to environmental assessment of building components in a whole building context*. FORCE Technology: Lyngby.

Schweber, L., & Haroglu, H. (2014). Comparing the fit between BREEAM assessment and design processes. *Building Research and Information, 42*, 300–317. https://doi.org/10.1080/09613218. 2014.889490.

Silvestre, J. D., Lasvaux, S., Hodková, J., de Brito, J., & Pinheiro, M. D. (2015). NativeLCA – A systematic approach for the selection of environmental datasets as generic data: Application to construction products in a national context. *International Journal of Life Cycle Assessment, 20*, 731–750. https://doi.org/10.1007/s11367-015-0885-8.

Soust-Verdaguer, B., Llatas, C., & García-Martínez, A. (2016). Simplification in life cycle assessment of single-family houses: A review of recent developments. *Building and Environment, 103*, 215–227. https://doi.org/10.1016/j.buildenv.2016.04.014.

Statistics Denmark. (2016). *BYGV80: Total floor area (adjusted for delays) by phase of construction and use*. Retrieved from http://www.statistikbanken.dk/statbank5a/default.asp?w=1440. Statistics Denmark.

Tunstall, G. (2006). *Managing the building design process* (2nd ed.). Oxford: Butterworth-Heinemann/Elselvier.

V&S Byggedata. (2013). *V&S Prisdata: Nybyggeri – Bygningsdele*. [V&S Price data: New build - Building elements]. V&S Byggedata. Ballerup.

Weytjens, L., Attia, S., Verbeeck, G., & De Herde, A. (2011). The 'architect-friendliness' of six building performance simulation tools: A comparative study. *International Journal of Sustainable Building Technology and Urban Development, 2*, 237–244. https://doi.org/10.5390/ SUSB.2011.2.3.237.

Chapter 9
Embodied Carbon Research and Practice: Different Ends and Means or a Third Way

B. Cousins-Jenvey

Introduction

The theory of embodied carbon and life cycle assessment is still relatively new, and there is a limited understanding of whether there is – or should be – a 'gap' between embodied carbon and life cycle assessment research and assessment practice. However, there is plenty of evidence that the needs or 'criteria' of researchers and practitioners differ, and it is therefore a strong possibility that different types of embodied carbon and life cycle assessments of buildings are necessary to meet the different needs of researchers and practitioners.

Understanding and bridging research-practice gaps is an interest in many sectors including construction (Seymour and Rooke 1995), and the research methods used to study construction have been scrutinised for their ability to address the needs of practitioners (Zou et al. 2014). The challenge Guba (1981) suggests is that there are two competing criteria to consider. One criterion is it 'doesn't matter what you do so long as you do it well', while the other is '[a]nything not worth doing at all is certainly not worth doing well!' Both are reasonable perspectives, but the difficulty is the trade-off: the more controlled the research, the less the relevance of the research in the real world. Clemson (2012) addresses a similar idea but with the criteria of truth and utility – terminology that this chapter adopts to capture the idea that the needs or 'criteria' of researchers and practitioners differ.

Advancing the theory of embodied carbon and life cycle assessment will undoubtedly rely on contributions from both researchers and practitioners, but the issue of truth and utility seems to require careful management. Too much emphasis

B. Cousins-Jenvey (✉)
Expedition Engineering/Useful Projects Ltd, London, UK

Department of Architecture and Civil Engineering, University of Bath, Bath, UK

Systems Centre, University of Bristol, Bristol, UK
e-mail: bengt.c@expedition.uk.com

© Springer International Publishing AG 2018
F. Pomponi et al. (eds.), *Embodied Carbon in Buildings*,
https://doi.org/10.1007/978-3-319-72796-7_9

on truth, and an assessment might be limited to only what can be accurately measured. Too much emphasis on utility, and an assessment might be untrustworthy or – worse – completely misleading. There are multiple motives. Firstly, careful management of the issue could reduce the risk that assessments produced by researchers will be peer reviewed unfairly and that the results or conclusions will be interpreted unreasonably. Secondly, if practitioners can find the right balance, they can ensure they use the time, fee and other available resources intelligently. Thirdly, there is a risk that collaborations between researchers and practitioners will be highly challenging, inefficient or ineffective without some way to find the right balance of truth and utility for those involved.

The concept explored by this chapter is that a 'taxonomy of assessments' has the potential to help manage the production of different assessments to meet the different needs of researchers and practitioners as well as to explore the possibility of a middle ground or Third Way. The concept of a taxonomy arose from observations that it is possible to translate the assessment process into sets of questions each with alternative answers. If these questions and alternative answers are suitably engaging for both researchers and practitioners, they can make certain that assessments are considered suitably holistically and pluralistically by dealing with the ends (purpose or 'what') and the means (process or 'how') as well as other influences on an assessment such as the available resources. Providing the taxonomy is well designed and correctly used according to any usage guidance, nothing important will be unintentionally overlooked when describing and categorising assessments or planning, interpreting and evaluating them. A further benefit of a taxonomy is its potential to provide assessors with a way to comprehensively but concisely communicate a large amount of detail to help with the delivery of assessments on time and on budget.

Aim

The aim of this chapter is to capture the questions and alternative answers that those producing embodied carbon or life cycle assessments of buildings should consider in a 'taxonomy of assessments'. It seeks to show how a taxonomy can help to manage the production of different types of assessment to meet the different needs of researchers and practitioners.

Background

Observations and Informal Correspondence

The author has spent half a decade between academia and industry, studying assessments published in journals and producing assessments as an EPSRC[1] funded researcher-practitioner. Observations and informal correspondence with many researchers and practitioners during this period will be mentioned at various points in this chapter to acknowledge their influence. In academia, correspondence with researchers highlighted the narrow boundaries of many assessments and the possibility that assessing a building could be a much more complex proposition than assessing a product. In industry, correspondence with practitioners continually re-emphasised the limited time, fees and other available resources (such as data in the public domain and assessment skills in the construction industry) that cannot be ignored if assessments of every project are to be part of a solution to the problem of climate change. Aspirations that assessments of buildings will become more complete and more complex are clearly at odds with concerns about time, fees and other available resources.

Embodied Carbon and Life Cycle Assessments

In the experience of the author, embodied carbon assessments are widely thought of as assessments of cradle to gate emissions of carbon dioxide, which is essentially a limited scope life cycle assessment of the global warming potential of the product phase of a building's life. All mentions of assessment(s) in this chapter refer to not just embodied carbon assessments, but more ambitious assessments of buildings that include other life phases from 'factory gate to grave' and other gases with global warming potential. In the future, it might be expected that more holistic assessments will replace assessments of just production phase carbon dioxide emissions. However, the barriers associated with time, fees and other available resources will clearly need to be addressed.

Building and Product Assessments

Buildings may need to be considered and assessed differently from products because of their large bills of quantities, different materials, hard to define functions (or functional units), long expected lives, many stakeholders, many actors, high variety of design and continually changing supply chains. Others have discussed the complexity of buildings, citing many of these characteristics (Dixit et al. 2012).

---

[1]Engineering and Physical Sciences Research Council.

Therefore, the proposed taxonomy of assessments considers buildings as complex systems and the possibility that it might not be realistic to predict their contributions to climate change during their entire life cycles with high certainty.

Literature

The most relevant sections of life cycle assessment standards or guides, assessment meta-research and closely connected taxonomies are the focus of this review. The review is cross-disciplinary because while the reviewed standards, guides and meta-research deal specifically with assessments of buildings, the taxonomies relate to understanding research methods and software quality.

Parts of the Life Cycle Assessment Standards and Guides

All steps of the assessment process described by the ISO 14000 series of standards and the building-specific EN 15978 are clearly relevant (goal and scope definition, life cycle inventory analysis, impact assessment and interpretation). As a result of the broad applicability of the standards, assessors have considerable freedom to tailor assessments to a particular situation, and there are concerns that even the building-specific EN15978 is insufficiently prescriptive. The supplementary guides produced in response to these concerns include the Building Research Establishment (BRE) Guidance Note 8 (BREEAM 2013), EeBGuide Part B (2012) and the closely related Institute for Environment and Sustainability ILCD Handbook, ENSLIC Building (ENSLIC) and Lore-LCA (LoRe-LCA) projects. Of particular relevance to the concept of a taxonomy of assessments are BRE's Guidance Note 8 and EeBGuide's Part B. BRE's Guidance Note 8 outlines minimum requirements relating to scope that assessors must comply with the exemplary level requirements of the Mat01 BREEAM credit. EeBGuide's Part B proposes that practitioners consider three types of assessment: screening, simplified and complete. Part B suggests that when choosing between the different types, considerations could include the assessor's experience, data available and maturity of the design of the building. Choice of a type has implications for completeness, data representativeness, documentation and communication, and these are outlined in detail by the document.

Assessment Meta-research: Reviews and Meta-studies

A variety of meta-research undertaken in the last decade offers analysis of published cases of assessment. Six are particularly relevant, two are literature reviews (Khasreen et al. 2009; Dixit et al. 2010) and four can be classified as meta-studies with varying interests in particular aspects of assessments (Sartori and Hestnes 2007; Ramesh et al. 2010; Sharma et al. 2011; Yung et al. 2013). Among the meta-studies, there is a clear focus on assessment results, and there are clearly different ideas about the details of the assessment processes that provide important context. The coding of assessments by Yung et al. (2013) provides the most detail, but even so, they concentrate on capturing the building assessed, building area described (usable, gross, habitable, heated) and scope assessed. An advantage of their approach is their ability to study and report the results for 38 cases of assessment which is not possible with a more holistic study. More recently, Pomponi and Moncaster have produced a larger meta-analysis of 77 cases of assessment but with a focus on four aspects: the included life cycle stages, suggested mitigation strategies, geographical breadth and scale of the study (Pomponi and Moncaster 2016, 2017). Importantly, none of these sources make a distinction between cases produced by researchers and cases produced by practitioners. The means (process) of assessment is a clear focus with the ends (purpose) of assessment receiving limited or no attention at all.

Taxonomies

Taxonomies are a subject of ongoing research in many different disciplines. Natural scientists, educationalists, knowledge managers and information system designers develop taxonomies. Particular taxonomies identified for their potential relationship with a taxonomy of assessments are the work of Ferris (2009) for its use of Varro's taxonomy of philosophies to establish 'a fundamental link between the knowledge sought and research methodology used to generate that knowledge'. The software quality literature offers a taxonomy too – the SERP test (Engström et al. 2016) – with the specific ambition to improve collaboration between researchers and practitioners. The designers of the SERP test used a particular method (Bayona-Oré et al. 2014) to develop their taxonomy.

Method

The work of Ferris (2009) was the biggest single influence on the overall design of this chapter and the design of the proposed taxonomy, but the method adopted acknowledges established taxonomy development steps too, including the work of Bayona-Oré et al. (2014) referenced by Engström et al. (2016).

Articulating the Questions Posed by Assessment

A set of questions about the assessment process were articulated based on observations and informal correspondence:

What goal or why assess?
What subject and what approach to its selection?
What scope and what impacts, phases and parts?
What inputs and what data?
What model(s) and theories of cause and effect?
What outputs and what units, tables and charts?
What detail, discussion and conclusion to report?

These questions are clearly closely related to the previously mentioned life cycle assessment steps of goal and scope definition, life cycle inventory analysis, impact assessment and interpretation. The questions are, however, intended to be more descriptive for those unfamiliar with the ISO 14000 series and EN 15978 standard while placing extra emphasis on questioning the included parts, inputs and data as well as what to report. How thoroughly assessors explore alternative answers to these questions can be expected to vary depending on the resources available and whether an assessor uses an off-the-shelf software tool or creates their own.

A second set of questions about the assessment purpose were articulated, again based on observations and informal correspondence:

How practical?

- Are there available resources, the time and the budget?

How representative?

- Is there an interest in a specific case or a general understanding?

How faithful?

- Are theories and data based on different or similar subject(s)?

How conventional?

- Is the approach partly or wholly original or even radical?

How objective?

- Are judgements by the assessor(s) required as well as acceptable?

These questions and the language reflect correspondence with practitioners but have been worded deliberately to make a connection with the criteria of Guba (1981) and Lincoln and Guba (1986). Questions about assessment purpose may have received less attention to date than the questions about process, but they are logically linked, and when evaluating assessments it is important to consider whether the answers to questions about process are coherent with answers to the questions about purpose.

Table 9.1 Questions 1–6 and part one of the proposed taxonomy: assessment ends

No.	How	'C' answers	'B' answers	'A' answers
1	Ambitious?[a]	Only retrospective, describing	Proactive, short-term decision supporting	Proactive, long-term decision supporting
2	Practical?	Low resource-intensity, experience and complexity	Medium resource-intensity, experience and complexity	High resource-intensity, experience and complexity
3	Representative?	One building, context and scenario	Mid-range: multiple buildings, contexts and scenarios	General: all buildings, contexts and scenarios
4	Faithful?[b]	Generic and global data or theory	Appropriate and regional data or theory	Specific and local data or theory
5	Conventional?[c]	Ignores assessment conventions or deliberately explores radical new ideas	Based on well-established, prescriptive guidance	Compliant with well-established, prescriptive guidance
6	Objective?	Assessor role unclear or high significance	Assessor role clear and low significance	Assessor passive or role minimised

[a]Note a much wider variety of goals is clearly possible – see the ISO/EN standards
[b]Relates to the idea of verisimilitude or low/high fidelity – see Guba (1981)
[c]Relates to guidance such as the EeBGuide

Developing the Taxonomy

Having established questions about assessment purpose as well as process, it was necessary to establish alternative answers to each of the questions. Using the two sets of questions as a starting point, an iterative process of design began with documented theories and terminology where appropriate, before piloting the design on a few cases of assessment and then redesigning as necessary. Tables 9.1 and 9.2 are the product of this process. Several goals emerging during the design are worthwhile stating explicitly:

- Capture assessments comprehensively but concisely in a 21 × 3 schema.
- Apply to any assessment produced by researchers or practitioners.
- Be consistent with established theories and terminology.
- Minimise user judgements by adopting quantitative alternative answers.

(Note that a 21 × 3 constraint emerged because it seemed to create a manageable amount of data to handle with radar charts and alphanumeric code (see the section 'Communicating the Taxonomy' later).)

The influence of the criteria of Guba (1981) on these objectives should be apparent. The 'aspects' of the EeBGuide and the influence of the work of Ferris (2009) should also be evident among the alternative answers.

Table 9.2 Questions 7–21 and part two of the proposed taxonomy: assessment means

No.	What	'C' answers	'B' answers	'A' answers
7	Subject type?	Wholly hypothetical or early-stage design	Partly as-built or late-stage design	Wholly as-built with supporting evidence
8	Subject approach?	Convenient, local or instrumental case(s)	Representative, typical or unusual case(s)	Randomly sampled subject(s)
9	Subject number?	One	Two	Three or more
10	Scope of impacts?	One impact or resources only	Multiple impacts, but not all or excluded indicators	All ISO/EN standard impacts and indicators
11	Scope of processes?	One phase or module	Multiple phases, but not all or excluded modules	All ISO/EN standard phases, modules and processes
12	Scope of elements?[a]	One element or sub-element	Multiple elements, but not all or excluded sub-elements	All NRM elements and sub-elements
13	Input type?	Estimates and new, unpublished theory	Secondary data and peer-reviewed theory	Primary data and established theory
14	Input approach?[b]	One value, no uncertainty[c]	Ranges	Distributions
15	Input number?	One source or only one data point	Benchmarked – comparison sources or data points	Triangulated – several sources and data points
16	Model type?	Phase-level detail	Module-level detail	Process-level detail
17	Model approach?	Deterministic, static and linear	Partly probabilistic, dynamic, non-linear	Fully probabilistic, dynamic, non-linear
18	Model no.?[d]	Single (e.g. either P-B or I-O)	Paired (e.g. a P-B augmented with I-O)	Multiple – nested or overlapped (e.g. a tiered I-O or I-O revisited with P-B to understand cause and effect)
19	Outputs type?[e]	Insufficient description for confident transfer of findings and details method/data	Provides or cites descriptions for confident transfer of findings and details method/data	Provides all description for confident transfer of findings and details method/data
20	Outputs approach?	Contributes only to decisions with no publication or dissemination	Contributes theory about impacts of buildings or production of assessments	Contributes theory about impacts of buildings and production of assessments
21	Outputs no.?	Unadulterated indicators with no characterisation or normalisation	ISO/EN standard impact categories based on characterisation only	One output produced by characterisation and normalisation

[a]Relates to RICS New Rules of Measurement (NRM)
[b]Note a wider variety of ranges (full, interquartile, etc.) and distributions is possible
[c]Refers to aleatory and epistemic uncertainty – see Walker et al. (2003)
[d]Refers to process-based (P-B) and input-output (I-O) modelling
[e]Note reasons to detail methods include reproducing, reviewing and reusing data

An important structuring principle was the organisation of alternative answers into columns of A, B and C answers so that assessments might be categorised as:

- A-oriented or truth-type – high effort and complexity
- C-oriented or utility-type – low effort and complexity
- B-oriented or Third Way – a potential middle ground

The connotation that A answers are better than C answers is deliberate and reflects the production of this chapter from a research perspective, but clearly the practicality and 'utility' of assessments are important considerations if assessments are to inform the decisions made by practitioners in industry. Incorporating both the perspectives of truth and utility as a spectrum is potentially but deliberately provocative to spark discussion and debate. The taxonomy should not be considered an endorsement of every possible permutation however.

Communicating the Taxonomy

A further requirement arising repeatedly during correspondence with practitioners is the use of accessible language, avoiding potentially off-putting terminology while acknowledging the need to make a link with important academic concepts. Careful consideration of the communication of the taxonomy extended to thinking about two different ways to represent the taxonomy other than in a table: one being a radar chart (Fig. 9.1) and the other being an alphanumeric code, i.e. (1A, 2B...21C). Ways of formatting such a code by using dots/ellipses to indicate the exclusion of parts of the taxonomy were found helpful to highlight the omissions at a glance. In order to illustrate the options, both ways of representing the taxonomy were eventually adopted in this chapter. To assist with the interpretation of the radar charts, Fig. 9.1 shows a radar chart for an A-oriented assessment, C-oriented assessment and the possibility of the Third Way – an 'all B answers' assessment, which will be discussed in more detail later in the chapter.

Using the Taxonomy

With any new taxonomy, its initial uses are a continuation of the development process. Trialling the proposed taxonomy on a few pilot cases of assessment had provided an initial opportunity to iterate the 21 questions and alternative A, B and C answers. Once confidence in the proposed scheme had been established, attention turned to identifying candidate cases of assessment produced by both researchers and practitioners.

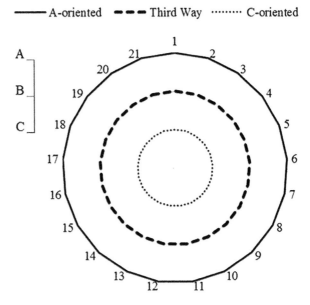

Fig. 9.1 How 'all A', 'all C' and Third Way assessments would be represented with a radar chart

Identifying Candidate Cases of Assessment

Candidate cases logically came from academic and industry contexts, either having been produced in academia by researchers for journal articles or produced in industry by the author as a practitioner in industry for project reports. An attempt to obtain all of the assessments produced by researchers included in the Ramesh et al. (2010) and Yung et al. (2013) meta-studies returned 30 assessments that are available digitally and do not require subscriptions to individual publications. These 30 assessments are a large sample of the assessments published in academic journals since 1995.

Assessments produced by the author as a practitioner offered multiple convenient cases of assessments produced to meet design team needs. All of these assessments were produced for a UK-based organisation offering structural, civil and sustainability consultancy between 2011 and 2016. Clearly the author's 'dual role' as essentially a researcher-practitioner is an important context for these cases that should not be ignored. A close involvement with researchers and the EPSRC support incentivised closing the gap between research and practice industry, and therefore some of the cases presented unusual opportunities and unusual available resources. They should not therefore be assumed representative of all assessments being produced in industry generally. The advantage presented by these cases is the author's access to the documentation of these cases and knowledge of them.

Identifying Extreme Cases of Assessment

Only by focussing on extreme cases would detailed discussion be possible within the word and page constraints of a single book chapter. The most extreme cases were quickly identified for the full, formal classification that appears in the results (a step that the results could not possibly elaborate in detail). Despite the decision to focus on extreme cases, it was decided that these sufficed to illustrate the concept of a taxonomy and to indicate how assessments produced by researchers compared with assessments produced by the author as a practitioner. Whether or not there was an overlap was determined as sufficient to inform discussion about a possible Third Way.

Established Guiding Principles

Despite efforts to refine the design of the proposed taxonomy so that it is as intuitive as possible, several guiding principles (or instructions) were established when using the taxonomy to ensure its consistent use:

- When in any doubt, classify B instead of A, and C instead of B.
- Record no answer at all when there is no relevant description.
- Avoid inferring the ends/purpose from just the means/process.
- Note any reliance on companion articles and reference articles.

One further specific point relevant to the decision taken to focus on the extreme cases of assessment: it was difficult to establish the most extreme, most A-oriented and C-oriented cases without a highly subjective weighting of answers to particular questions. Two contenders for the extreme cases produced in academia by researchers are consequently included in the following results.

Results

Figures 9.2, 9.3, 9.4 and 9.5 are radar charts summarising the six extreme cases of assessment found to be the most A-oriented or C-oriented. In addition to this version of the results, descriptions of the six cases of assessment are provided with their classification according to the proposed taxonomy subsequently communicated using the alphanumeric code. Note that the shapes of Figs. 9.2 and 9.3 reflect questions 1–6 not being discernible for the cases of assessments produced by researchers.

Fig. 9.2 Classification of the most C-oriented assessments produced by researchers

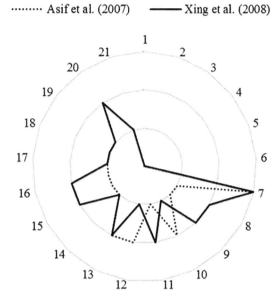

Fig. 9.3 Classification of the most A-oriented assessments produced by researchers

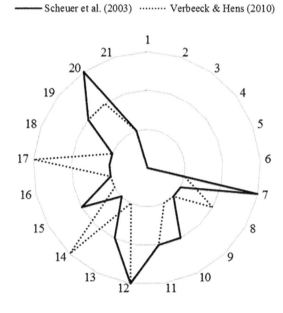

Fig. 9.4 Classification of
the most C-oriented
assessments produced by
the author as practitioner

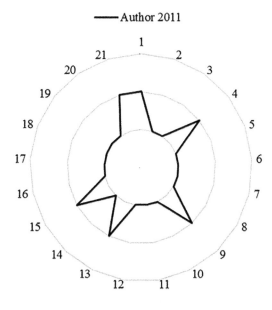

Fig. 9.5 Classification of
the most A-oriented
assessments produced by
the author as practitioner

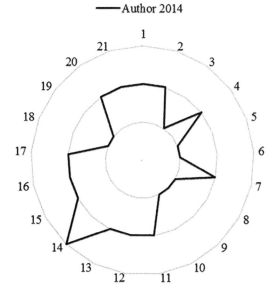

C-Oriented Assessments Produced by Researchers

Xing et al. (2008) is an assessment produced in 2007 and published in Energy and Buildings with three named authors from Tongji University, Shanghai. The subjects are two, apparently real office buildings in China that are 'typical' for Shanghai. One building has a steel superstructure, while the other has a reinforced concrete superstructure.

The scope is to assess the global warming, acidification, eutrophication and toxicity potential of the steel and aerated concrete bricks during all phases. However, the scope does not include all modules and sub-elements. The inputs for material quantity data are from unclear sources, while the environmental impacts are from the BESLCI programme and the China Iron and Steel Industry Annual published in 1998 (some time before 2007). No uncertainty is considered even for the prediction 50 years into the future. Description of the model is limited, but by consulting other sources to find out about the BESLCI programme, it appears to be process-based and mainly process-level from the data reported. The outputs for different emissions are presented separately for the 50-year period.

Consequently, the assessment classifies as a:

(. . .7A, 8B, 9B, 10C, 11B, 12C, 13B, 14C, 15B, 16B, 17C, 18C, 19C, 20B, 21C).

Asif et al. (2007) is an assessment produced in 2004 and published in Building and Environment with three named authors from Napier University, Edinburgh. The subject is a real, semi-detached residential building with three bedrooms in Scotland. Concrete, timber and ceramic tiles are the main materials by mass.

The scope is the CO_2, SOx and NOx of eight main construction materials (three are unreported) during the product phase. The inputs for material quantity data are of high quality from reports, observation and interviews with contractors or local housing personnel, while environmental impact data are all from *The Ecology of Building Materials* (Berge 2009) apart from the data for aluminium (International Aluminium Institute). Notably, *The Ecology of Building Materials* is a Norwegian synthesis of sources and of questionable transferability to the Scottish context. The model is not explicitly described, but appears to be deterministic and process-based and only offers a phase-level understanding. The outputs are unadulterated emissions.

Consequently, the assessment classifies as a:

(. . .7A, 8C, 9C, 10B, 11C, 12B, 13B, 14C, 15C, 16C, 17C, 18C, 19C, 20B, 21C)

A-Oriented Assessments Produced by Researchers

Scheuer et al. (2003) is an assessment produced in 2002 and published in Energy and Buildings with three named authors all affiliated with the University of Michigan. The subject is Sam Wyly Hall: a real, mixed-use building in the USA that is located on the Ann Arbor Campus of the University of Michigan. The building comprises three storeys of classrooms or offices and three storeys of hotel. Concrete, steel and bricks are the main materials by mass (all more than 100 tonnes). Description in the form of a detailed bill of materials provides some detail about these materials such as fly ash being used as a cement replacement and steel being produced in an electric arc furnace.

The scope is the global warming potential, ozone depletion potential, nitrification potential and acidification potential of all elements during all phases. However, the scope does not extend to all modules, and from the bill of quantities alone, it is difficult to confirm the assessment of all sub-elements of the building. The inputs for material quantities are of high credibility being established in a variety of ways including on-site take-offs. Environmental impact data are from multiple sources of questionable transferability to the US context (TEAM™ or DEAM™, SimaPro, Swiss Agency of the Environment, Forests and Landscape, a report by Franklin Associates). There is no consideration of uncertainty even for the future phases (75 years away), and accordingly scenario analysis is a modelling challenge suggested in the discussion of future assessments provided by the authors. The single, solely process-based model appears to be process-level, linear, deterministic and static and seems to provide a module-level understanding of some modules, not their constituent processes. However, description of the model is limited, but these are further 'modelling challenges' highlighted by the assessors. The outputs are unadulterated emissions.

Consequently, the assessment classifies as a:

(...7A, 8C, 9C, 10B, 11B, 12A, 13B, 14C, 15B, 16C, 17C, 18B, 19B, 20A, 21B)

Verbeeck and Hens (2010a, b) is an assessment produced in 2009 and published as a pair of articles in Building and Environment with two named authors. The subjects are four 'typical' designs for residential buildings in Belgium, but the articles only describe the assessment of the terraced design. The designs have 'massive', 'cavity' and lightweight (wood frame) variations.

The scope is global warming potential plus emissions of NOx, SOx and (unreported) non-methane VOCs of seemingly all elements during all phases. However, the scope does not extend to all modules and sub-elements. The inputs appear to be randomly sampled from a distribution for material quantities that reflect changing the geometry of representative designs, while impact data are from one source of questionable accuracy in the Belgian context (the ecoinvent database V2.0). The model description is limited but can be traced as the output of the EL^2EP research project (Verbeeck 2007) and appears to be a solely process-based model that provides some module-level and even process-level understanding. A Monte Carlo analysis of multiple variables associated with production and construction phases is clearly evident from the description provided. The outputs are unadulterated emissions.

Consequently, the assessment classifies as a:

(...7C, 8B, 9C, 10C, 11B, 12A, 13C, 14A, 15C, 16C, 17A, 18C, 19B, 20B, 21C)

C-Oriented Assessments Produced by the Author as Practitioner

The following is a short description of an assessment produced by the author in 2011 for a project at RIBA Stage 2. The subject was a mixed-use building in the UK (London) for which the design team considered two alternative designs: one in situ, reinforced concrete design and one cross-laminated timber (CLT) design.

The scope was the global warming potential of the superstructure (frame) and any formwork during the product phase. The model was process-based, deterministic and only at phase-level. The inputs were based on material quantity data from drawings and correspondence with the design team, while environmental impact data were from the University of Bath Inventory of Carbon and Energy (ICE) V2.0, ignoring the stated uncertainties. The output was the global warming potential with characterisation according to the ICE V2.0 factors.

The involvement of the author in the case means it is possible to offer a classification of the purpose. The assessment responded to a brief to offer short-term decision support in the form of a choice between the two alternative designs, there was no ambition to draw conclusions about buildings generally. Both budget and programme only allowed for approximately two days of work in total, and part of the briefing was a need to report the details of the assessment on a single slide of a presentation. The ICE V2.0 was an obvious choice of regionally appropriate data at the time. There was no stipulation about minimising the role of the author as the assessor.

Consequently, the assessment classifies as a:

(1B, 2C, 3C, 4B, 5C, 6C
...7C, 8C, 9B, 10C, 11C, 12C, 13B, 14C, 15B, 16C, 17C, 18C, 19C, 20C, 21B)

A-Oriented Assessments Produced by the Author as Practitioner

The following is a short description of an assessment produced by the author in 2014 for a project at RIBA Stage 1 and later revisited at Stage 2. The subject was a reference design for a now built factory building in India (the state of Maharashtra). Steel and concrete were the main materials by mass and were the only materials assessed.

The scope was the global warming potential of the substructure, superstructure and envelope during the product, construction and use phases, though reporting focussed on the product and construction phases. Inputs were three-point estimates of a distribution based on material quantity data supplied by the client for previous factories supplemented by estimates by a structural engineer and the assessor (the author), while environmental impact data were initially from the synthesis of sources offered by the (ICE) V2.0, and these were then corroborated or adapted based on literature offering data for concrete and steel specified and produced in India (Tiwari 2001; Venkatarama Reddy and Jagadish 2003; Venkatarama Reddy 2009). The model was process-based, permitted scenario analysis (within the scope) and offered some module-level understanding. The output was energy (with no distinction between renewable and non-renewable) and greenhouse gases as characterised by the sources of environmental impact data.

The involvement of the author in this case means it is again possible to offer a classification of the purpose of the assessment. A reference design offered a proxy for the emerging design at the early stages of the design process. The assessment was not intended to provide design option decision support, but was instead produced with the intention of pinpointing high-impact parts to inform the design team prioritising their efforts (particularly the structural engineers). There was a strong interest in opportunities to reduce energy use and greenhouse gas emissions by using less material and minimally processed materials or influencing specifiers and suppliers. Undertaking the assessment showed that by focussing attention on the superstructure, the structural engineers could achieve larger savings than by optimising the superstructure and its roof trusses. Conducted after the C-oriented assessment produced by the author, this case was part of an agenda to demonstrate ideas from several research articles. The budget and programme allowed for this exploration with the author able to spend approximately two weeks working on the assessment over a two-year period.

Consequently, the assessment classifies as a:

(1B, 2B, 3C, 4B, 5C, 6C

...7B, 8C, 9C, 10C, 11B, 12B, 13B, 14A, 15B, 16B, 17B, 18C, 19C, 20B, 21B)

Discussion

The results suggest a research-practice overlap rather than a gap. A research-practice gap would have been indicated if the assessments produced by researchers had classified with predominantly A answers, while the cases of assessment produced by the author as practitioner had classified with predominantly C answers. Instead however, the results show that cases of assessments produced by researchers frequently classify with many more C answers than might be expected. At the same time, the A-oriented assessments produced by researchers are notably more A-oriented than the assessments produced by the author as practitioner in some areas (see Fig. 9.6).

The most C-oriented cases of assessment produced by researchers might have been published in academic journals, but they actually classify similarly to the most C-oriented cases of assessment produced by the author under the influence of commercial pressures in industry (see Fig. 9.7). Interestingly, all four assessments produced by researchers appear in two journals (*Energy and Buildings* and *Building and Environment*), and each journal publishes an A-oriented and C-oriented case as opposed to one journal publishing the A-oriented cases and the other journal

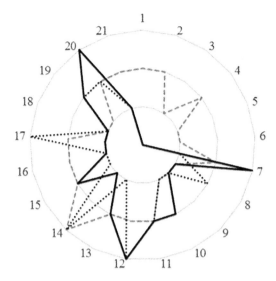

Fig. 9.6 Comparison of the most A-oriented assessments

publishing the C-oriented cases. The journals do not appear to have had a clear orientation based on the cases of assessments considered.

Range Produced by Researchers

Clearly distinguishing the most A-oriented assessments produced by Scheuer et al. (2003) and Verbeeck and Hens (2010) from the most C-oriented assessments produced by Asif et al. (2007) and Xing et al. (2008) are questions about scope. The Scheuer et al. (2003) assessment adopts the most comprehensive scope in terms of impacts, phases and parts (see questions 10–12). Additionally, Scheuer et al. (2003) provide more description (see question 18) and extensive discussion of implications (see question 20) than all of the other cases of assessment. The Verbeeck and Hens (2010) is clearly distinguishable for running Monte Carlo simulations, which is the reason for notable spikes at the input approach and model approach (see questions 14 and 17). One question where Xing et al. (2008) are more A-oriented than Scheuer et al. (2003) and Verbeeck and Hens (2010) is the model-type with the BESLCI tool offering module-level detail (see question 16).

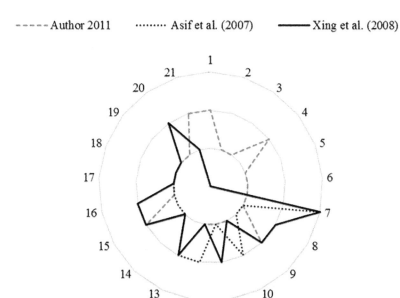

- - - - - Author 2011   ⋯⋯⋯ Asif et al. (2007)   ——— Xing et al. (2008)

Fig. 9.7 Comparison of the most C-oriented assessments

Range Produced by the Author as Practitioner

Modelling and input approach distinguish the two cases of assessment produced by the author as practitioner. The purpose of these cases is known to differ (as a result of the author's involvement), and this is reflected in the extra resources invested and an interest in specific, local data (questions 2 and 4). There are process differences too because of a more comprehensive scope and using estimated distributions for the modelling of scenarios (see questions 11, 12, 14 and 16).

A Third Way?

With the results suggesting that the range of assessments produced by researchers overlap with the range of assessments produced by practitioners, the idea of a middle ground or the Third Way suggests promise. The cases of assessment produced by Xing et al. (2008) and the most A-oriented assessment produced by the author as practitioner reinforce this promise as they classify with the most B answers. They are therefore the closest cases to a Third Way-type assessment between an all A truth-type assessment and an all C utility-type assessment. The obvious motive for adopting B answers associated with a Third Way assessment is the apparent balance of truth and utility that it offers. However, there is a risk that it will be too much of a compromise, attracting criticism from both researchers and practitioners. A Third Way type of assessment may not meet researcher expectations for universal rules, measurements or other high-quality data, certainty, precedent and objectivity. At the same time, busy practitioners will still have to commit or secure more fees, more time and other resources. Perhaps only by exploring the Third Way will it be possible to identify which B answers to change to As and Cs to increase its uptake by both researchers and practitioners.

Perhaps it is important to note that the case of assessment produced by the author as practitioner of the factory building in India classifies with the largest number of 'B' answers. The author benefitted from reduced commercial pressures thanks to research funding, and it is possible that similar funding in the form of Knowledge Transfer Partnerships, industrial doctorates and industry-sponsored PhDs might be crucial to enable exploration of the Third Way. Additionally, the importance of evidencing the need for adopting B answers rather than C answers should not be underestimated, and the author undoubtedly benefitted from periodic immersion in an academic context to gather or create such evidence.

Although the Third Way suggests promise, it is equally conceivable that a gap between assessments produced by researchers and those produced by practitioners will be created if academic journals introduced more stringent, minimum requirements of assessments produced in an academic context. Assessments accepted by journals in the past might be rejected in the future. The potential issue with publications introducing minimum requirements is that there could be a complete disconnect between how researchers and practitioners assess buildings without the creation of easy-to-follow guidance, up-to-date 'open' data sets and timesaving software that enable practitioners to produce truth-type assessments quickly and cheaply.

Other Recommendations

Co-assessors, editors and reviewers or developer clients, future building owners, consultants and those writing standards, guides and rating systems could all potentially benefit from using a taxonomy of assessments to support the categorisation of assessment types, specification of assessments precisely and communication of assessments concisely.

A taxonomy can support the description and communication of assessments concisely by acting simply as a 'checklist' to encourage more targeted reporting of important written detail. A radar chart or alphanumeric code can ensure that a report clearly communicates exactly what is assessed, why it is assessed and how it is assessed. While using the proposed taxonomy to classify the cases of assessment studied in this chapter, there were occasions when it was necessary to cautiously classify possible A answers as B answers or possible B answers as C answers because of a lack of clarity about particular details. A radar chart or alphanumeric code produced by the assessor would provide this clarity. A radar chart or alphanumeric code could offer assessors the opportunity to reduce the words and page space that they would currently devote to providing details of the ends and means of their assessment. They could then use these words for other important details and discussion of the contributions to theory about both the impact of buildings and production of assessments. A radar chart or alphanumeric code can feasibly provide the basis for rapid interpretations and evaluations of assessments if the interpreter or evaluator is highly familiar with the taxonomy of assessments represented. When looking for an assessment that adopts particular answers to particular questions, this could be a dramatically more efficient than an equivalent search at present.

A taxonomy can support the categorising of assessments by establishing types just like the truth-oriented type (all A answers), utility-oriented type (all C answers) and Third Way (all B answers) suggested here. There may of course be other 'taxon' of all the possible 'taxa' represented by the proposed taxonomy of assessment that are a combination of A, B and C answers that works well in a particular situation. Once well-established and widely understood, types of assessment could become a highly efficient way to imply or infer a lot of detail about a past or future assessment, and this might further help assessors to specify assessments as well as improve the description they provide.

A taxonomy can support the specification of assessments precisely by either identifying particular target answers (e.g. target 1A) or setting minimum expectations (e.g. no C answers). Considering the proposed taxonomy in this way, there are specific alternative answers based on the results that it would be interesting for assessors to explore to contribute to assessment production theory. Specific A answers to target for highly A-oriented assessments could include those related to assessing a large sample of real buildings selected randomly (targeting 8A and 9A), gathering more primary data and developing well-established theories about future phases (targeting 13A and 15A) and process-level, fully probabilistic modelling (targeting 14A and 17A). There are assessments not studied in this chapter that have explored some individual A answers, e.g. larger samples (Suzuki et al. 1995) and multi-model P-B and I-O comparisons (Holmberg et al. 2007). Targeting 'A's in combination rather than individually though would be both interesting and challenging. Another interesting proposition specifically for researchers would be an informal policy not to adopt C answers to questions about scope and therefore targeting 10B, 11B and 12B as a minimum.

Limitations

The proposed taxonomy does not address all questions that arise during production of an assessment, with notable examples such as: what evidence to provide to support the material quantities used? What ranges or distributions to use for particular inputs? What calculation rules to adopt for calculating material demands during the use phase of a building? When is a model fully dynamic and non-linear? Perhaps though, it is a positive thing that the taxonomy highlights such questions for more attention. The development of the proposed taxonomy should clearly be considered ongoing, and any future use should acknowledge that although it is holistic, there are details that it does not capture.

Although its use on different cases of assessment and different types of assessment has been demonstrated, use of the taxonomy to classify an even wider variety of cases of assessment would reinforce its promise. More cases of assessment produced by practitioners other than those produced by the author as practitioner would provide greater insight into practice. As a consequence of the single location of the cases of assessment produced by the author as researcher, these cases are not necessarily representative of assessment by practitioners in general as has already been emphasised. The A-oriented assessment could be an unusual case considering the subsidised cost of the author, the incentives for knowledge transfer, organisations involved and opportunities to work on particular projects.

Other researchers and practitioners should use the proposed taxonomy, interrogate any discrepancies and consider how to eliminate these by adapting the design of the taxonomy or developing the usage guidelines provided in the method. Most simply, such testing could include other users of the proposed taxonomy reproducing the results of this chapter. Redesigns of the proposed taxonomy are strongly encouraged, but this will obviously necessitate some form of 'version control' to record the changes and ensure that it is always clear to assessors what iteration is being used (for both 'encoding' and 'decoding'). Such an approach would allow the taxonomy to evolve while always avoiding the confusion caused by references to different versions. If a taxonomy of assessments has been through this process and becomes stable, this will clearly reinforce its credibility.

Further Work

There are two obvious areas of further work that would advance the concept of a taxonomy of assessments and reinforce the results of using the proposed taxonomy in its current form. Others using and refining as necessary the proposed taxonomy with the ultimate aim of creating a stable taxonomy is one area of work. To support this work, it would be helpful to study the sensitivity of assessment results to alternative answers to particular questions. These studies should consider the potential risk that seemingly minor assessment details might affect the results of major decision altering consequences, and this could include details not dealt with sufficiently by the proposed taxonomy. Studying other effects on the efficiency and effectiveness of assessments will be important too, and consequently surveys and interviews to obtain researchers and practitioners will be necessary. One benefit of a taxonomy is that it can provide some structure to these surveys and interviews as well as less formal discussions between many different actors involved in an assessment of a building.

Use of the taxonomy – or some new version of it – to describe many more diverse cases of assessments produced by different practitioners is a second area of work. One way for this to happen would be for researchers or practitioners to review and classify other cases of assessments as demonstrated in this chapter. Another way would be for the taxonomy to inform embodied carbon database submission forms or 'schema' used already when reporting assessments to organisations such as WRAP in the UK, Building Research Establishment and US Green Building Council.

Conclusion

A proposed taxonomy of assessments seeks to capture the questions and alternative answers that those producing embodied carbon or life cycle assessments of buildings should consider. The underlying motive is to help manage the production of different types of assessment to meet the different needs of researchers and practitioners as well as to explore the possibility of a middle ground or Third Way. The proposed taxonomy is based on 21 questions each with three alternative answers that are frequently reinforced by references to important theories from the literature. Importantly, the questions relate to both the ends (purpose) and means (process) of assessment, and the organisation of the alternative answers reflects three notional types of assessment: truth type, Third Way and utility type. Based on the theory that the needs or 'criteria' of researchers and practitioners differ, it might be expected that assessments produced by researchers are truth oriented, while assessments produced by practitioners are utility oriented and few – if any – assessments are produced that suggest the promise of the Third Way.

Using the proposed taxonomy to describe and categorise cases of assessment produced by researchers and cases of assessments produced by the author as practitioner does not indicate a gap between embodied carbon and life cycle assessment research and assessment practice as might be expected. The range of assessments produced by researchers is broad and appears to include assessments similar to those produced in industry by the author as practitioner. The use of the proposed taxonomy to produce the results and aid discussion of them indicates how a proposed taxonomy can support describing, categorising, planning, interpreting and evaluating. In addition, the proposed taxonomy makes it possible to outline a precise agenda for future assessments and supporting studies. Others are encouraged to explore the concept of a taxonomy of assessments for these different purposes as well as to use and refine the proposed taxonomy.

Acknowledgements   The author gratefully acknowledges the support of the EPSRC funded Industrial Doctorate Centre in Systems (Grant EP/G037353/1), Useful Projects Ltd. and Expedition Engineering Ltd.

References

Asif, M., Muneer, T., & Kelley, R. (2007). Life cycle assessment: A case study of a dwelling home in Scotland. *Building and Environment, 42*, 1391–1394. https://doi.org/10.1016/j.buildenv. 2005.11.023.

Bayona-Oré, S., Calvo-Manzano, J. A., Cuevas, G., & San-Feliu, T. (2014). Critical success factors taxonomy for software process deployment. *Software Quality Journal, 22*, 21–48. https://doi.org/10.1007/s11219-012-9190-y.

Berge, B. (2009). *The ecology of building materials*. Oxford: Architectural Press.

BREEAM. (2013). Assessor guidance note GN08.

Clemson, B. (2012). What is systems thinking ? A personal perspective.

Dixit, M. K., Fernández-Solís, J. L., Lavy, S., & Culp, C. H. (2010). Identification of parameters for embodied energy measurement: A literature review. *Energy and Buildings, 42*, 1238–1247. https://doi.org/10.1016/j.enbuild.2010.02.016.

Dixit, M. K., Fernández-Solís, J. L., Lavy, S., & Culp, C. H. (2012). Need for an embodied energy measurement protocol for buildings: A review paper. *Renewable and Sustainable Energy Reviews, 16*, 3730–3743. https://doi.org/10.1016/j.rser.2012.03.021.

EeBGuide. (2012). EeBGuide guidance document part B: Buildings.

Engström, E., Petersen, K., Bin, A. N., & Bjarnason, E. (2016). SERP-test: A taxonomy for supporting industry-academia communication. *Software Quality Journal*, 1–37. https://doi.org/10.1007/s11219-016-9322-x.

ENSLIC Energy Saving through Promotion of Life Cycle Analysis in Buildings (ENSLIC BUILDING). (n.d.). https://ec.europa.eu/energy/intelligent/projects/en/projects/enslic-building

Ferris, T. L. J. (2009). On the methods of research for systems engineering. 7th Annu Conf Syst Eng Res 2009 (CSER 2009) 2009:7.

Guba, E. (1981). Criteria for assessing the trustworthiness of naturalistic inquiries. *ECTJ, 29*, 75–91.

Holmberg, J., Wadeskog, A., & Nyman, M. (2007). Direct and indirect energy use and carbon emissions in the production phase of buildings: An input – Output analysis. *Energy*. https://doi.org/10.1016/j.energy.2007.01.002.

Khasreen, M. M., Banfill, P. F. G., & Menzies, G. F. (2009). Life-cycle assessment and the environmental impact of buildings: A review. *Sustainability, 1*, 674–701.

Lincoln, Y. S., & Guba, E. G. (1986). But is it rigorous? Trustworthiness and naturalistic evaluation. *New Directions Program Evaluation, 1986*, 73–84. https://doi.org/10.1002/ev. 1427.

LoRe-LCA. (n.d.). Low resource consumption buildings and constructions by use of LCA in design and decision making. http://www.sintef.no/Projectweb/LoRe-LCA/

Pomponi, F., & Moncaster, A. (2016). Embodied carbon mitigation and reduction in the built environment – What does the evidence say? *Journal of Environmental Management, 181*, 687–700. https://doi.org/10.1016/j.jenvman.2016.08.036.

Pomponi, F., & Moncaster, A. (2017). Scrutinising embodied carbon in buildings: The next performance gap made manifest. *Renewable and Sustainable Energy Reviews*. https://doi.org/10.1016/j.rser.2017.06.049.

Ramesh, T., Prakash, R., & Shukla, K. K. (2010). Life cycle energy analysis of buildings: An overview. *Energy and Buildings, 42*, 1592–1600.

Sartori, I., & Hestnes, A. G. (2007). Energy use in the life cycle of conventional and low-energy buildings: A review article. *Energy and Buildings, 39*, 249–257. https://doi.org/10.1016/j. enbuild.2006.07.001.

Scheuer, C., Keoleian, G. A., & Reppe, P. (2003). Life cycle energy and environmental performance of a new university building: Modeling challenges and design implications. *Energy and Buildings, 35*, 1049–1064.

Seymour, D., & Rooke, J. (1995). The culture of the industry and the culture of research. *Construction Management and Economics, 13*, 511–523. https://doi.org/10.1080/01446199500000059.

Sharma, A., Saxena, A., Sethi, M., & Shree, V. (2011). Life cycle assessment of buildings: A review. *Renewable and Sustainable Energy Reviews, 15*, 871–875. https://doi.org/10.1016/j.rser.2010.09.008.

Suzuki, M., Oka, T., & Okada, K. (1995). The estimation of energy consumption and CO 2 emission due to housing construction in Japan. *Energy and Buildings, 22*, 165–169.

Tiwari, P. (2001). Energy efficiency and building construction in India. *Building and Environment, 36*, 1127–1135.

Venkatarama Reddy, B. V. V. (2009). Sustainable materials for low carbon buildings. *International Journal of Low Carbon Technologies, 4*, 175–181. https://doi.org/10.1093/ijlct/ctp025.

Venkatarama Reddy, B. V., & Jagadish, K. S. (2003). Embodied energy of common and alternative building materials and technologies. *Energy and Buildings, 35*, 129–137. https://doi.org/10.1016/S0378-7788(01)00141-4.

Verbeeck, G. (2007). Optimisation of extremely low energy residential buildings.

Verbeeck, G., & Hens, H. (2010a). Life cycle inventory of buildings: A calculation method. *Building and Environment, 45*, 1037–1041. https://doi.org/10.1016/j.buildenv.2009.10.012.

Verbeeck, G., & Hens, H. (2010b). Life cycle inventory of buildings: A contribution analysis. *Building and Environment, 45*, 964–967. https://doi.org/10.1016/j.buildenv.2009.10.003.

Walker, W. E., Harremoës, P., Rotmans, J., et al. (2003). Defining uncertainty: A conceptual basis for uncertainty management in model-based decision support. *Integrated Assessment, 4*, 5–17.

Xing, S., Xu, Z., & Jun, G. (2008). Inventory analysis of LCA on steel- and concrete-construction office buildings. *Energy and Buildings, 40*, 1188–1193. https://doi.org/10.1016/j.enbuild.2007.10.016.

Yung, P., Lam, K. C., & Yu, C. (2013). An audit of life cycle energy analyses of buildings. *Habitat International, 39*, 43–54.

Zou, P., Sunindijo, R., & Dainty, A. (2014). A mixed methods research design for bridging the gap between research and practice in construction safety. *Safety Science, 70*, 316–326.

Chapter 10
Embodied Carbon in Construction, Maintenance and Demolition in Buildings

G. K. C. Ding

Introduction

With the rapid growth of the construction industry worldwide during the last decades due to population growth, construction, maintenance and demolition waste (hereafter referred to as waste for the chapter) has become a major source of solid waste, in particular in developing and rapidly growing countries such as China. Hence, while construction plays an important role in national growth and wealth worldwide, it also adversely affects the environment. Waste is often neglected as it is regarded as less important compared to the waste generated during the operating activities.

Globally, the construction industry is responsible for 20–60% of the solid waste stream (Crowther 2015). Of all solid waste produced, construction-related waste constitutes approximately 20–30% in the European Union (EU), 30% in Canada, 29% in the USA, 26% in Hong Kong and 30–40% in Australia (Arslan et al. 2012; Malia et al. 2013; Yeheyis et al. 2013; Yuan et al. 2013; Randell et al. 2014). Since waste has become a major environmental problem in the European countries, it has been categorised by the EU as a priority waste for reduction among the member states (Manfredi and Pant 2011).

The common practices of dealing with wastes are landfilling and incineration which have caused enormous damage to the society both economically and environmentally in terms of waste accumulation and escalation of the problem of global warming through the generation of carbon dioxide (CO_2) in the processes (Banias et al. 2010; Marshall and Farahbakhsh 2013). Many previous studies have indicated

G. K. C. Ding (✉)
School of the Built Environment, Faculty of Design, Architecture & Building, University of Technology Sydney, Broadway, Ultimo, NSW, Australia
e-mail: grace.ding@uts.edu.au

© Springer International Publishing AG 2018
F. Pomponi et al. (eds.), *Embodied Carbon in Buildings*,
https://doi.org/10.1007/978-3-319-72796-7_10

217

that the construction sector can play an important part in properly controlling and reducing carbon emission through construction activities (Hu et al. 2017).

This chapter aims to establish an understanding of the nature of waste in a building's lifecycle and the effects of its associated embodied carbon. This chapter begins with a discussion of the building lifecycle and the origin of waste, followed by an examination of policies, initiatives and international regulations in dealing with the problem of waste. This chapter also reviews the calculation methods in the assessment of embodied carbon of waste in various stages of a building's life. The chapter ends with a discussion on reducing waste and the associated embodied carbon using a case study.

Literature Review

Problems of Waste

Waste is a significant problem worldwide due to it being generated in large quantities throughout the life of a building. Waste is present in all stages from raw material extraction and construction to maintenance and renovation and finally through to demolition of the building. Waste can increase project cost and add to environmental pollution. The negative impact on the environment includes consuming landfills and natural unrecoverable resources. The environmental problem posed by waste is not only from its increasing quantities but also from its handling and treatment. The common type of treatments for wastes is landfilling and/or incineration (Banias et al. 2010; Marshall and Farahbakhsh 2013). The negative environmental effects of waste start by the dumping of untreated waste into uncontrolled or illegal open dumps, forests and waterways, contaminating surface or underground water and causing erosion (Arslan et al. 2012; Randell et al. 2014). The generation of waste is not just causing problems to the man-made and natural environment, but it also depletes valuable natural resources and generates CO_2 in the process.

Landfilling causes additional environmental problems, land depletion and deterioration; it is hazardous to the health of humans and the ecosystems, pressures limited landfill space and causes a deterioration of the landscape (Banias et al. 2010; Yeheyis et al. 2013). In addition, landfill leaching causes freshwater eutrophication and air pollution due to the generation of landfill gases which typically consist of CO_2 and methane (CH_4) with traces of other gases from anaerobic degradation of organic wastes. These may contaminate surface and groundwater, and the level of contamination may vary depending on the height of water table, permeability of the soil, waste types, landfill construction and management (Yuan et al. 2013; Randell et al. 2014; Butera et al. 2015).

In addition, soil may be contaminated with heavy metals, hydrocarbons and salts from waste (Yuan et al. 2013; Butera et al. 2015). According to Randell et al. (2014), the operation of landfills that involves the uses of machinery may also be a major source of greenhouse gas (GHG) emissions for the waste sector. Incineration is another common way of dealing with the increasing waste. However, incineration may cause toxicity to the environment as emissions may contain heavy metals, dioxins, particulates and furans, smoke and ash. According to Dajadian and Koch (2014), the waste produced by burning trash furnaces includes more than 200 different dioxin compounds and generates a large amount of CO_2 emissions.

The environmental problem with waste is not just about its volume but also its close link to carbon embodied in this waste. CO_2 is one of the GHGs and has the most impact on the greenhouse effect contributing to global warming as there is a lot of it (Tingley and Davison 2011). The common types of gases that are related to the construction industry are carbon dioxide (CO_2), methane (CH_4) and nitrous oxide (N_2O) (Cole 1999; Mao et al. 2013) and account for more than 97% of the total global warming potential (GWP) (Nadoushani and Akbarnezhad 2015).

CO_2 equivalent (CO_2-e) is usually used as a reference standard for greenhouse effect (Cole 1999). All other GHGs are expressed as a ratio of the GWP caused by an equivalent mass of CO_2 and is termed CO_2-e. Table 10.1 presents the GWP values of CH_4 and N_2O relative to CO_2 over a 100-year timeframe. From the table, CH_4 has a CO_2-e of 28. Therefore, the calculation of GWP of the energy-related GHG is regulated over a 100-year timeframe, and the GWPs for each of these emissions are added to give total CO_2-e intensities for the construction sector (Acquaye and Duffy 2010; Brander 2012; IPCC 2013). This chapter only considers CO_2 when referring to "embodied carbon".

Carbon is embodied in each stage in the entire lifecycle of a building including the extraction and manufacturing process of building materials; construction, operating and demolition activities; and the fuel-related activities that occur throughout the lifespan of a building. All the waste generated in the activities related to material, construction, maintenance and renovation during operation stage and demolition, as well as the transport caused by these activities, results in embodied carbon emissions (RICS 2014).

Previous studies on embodied carbon have concentrated on the emissions during material extraction and manufacturing stage, but more recent studies on embodied carbon have been expanded to include a whole of life approach. Therefore, the focus of this chapter is waste generated during:

- Onsite activities during the construction of a building
- Maintenance, renovation and refurbishment of a building during operation but not including operation waste
- End-of-life demolition

Waste Definition and Composition

With regard to the definition of waste, there is no consensus in exactly what waste means and what constitutes a waste. Manfredi and Pant (2011) state that there are

Table 10.1 GWP values relative to CO_2

GHG	Chemical formula	GWP for 100 years
Carbon dioxide	CO_2	1
Methane	CH_4	28
Nitrous oxide	N_2O	265

Source: IPCC (2013)

different definitions throughout the EU so it makes cross-country comparison difficult. Some EU countries include land levelling materials as part of the waste but some do not. However, excavated materials are largely excluded in most studies of waste. According to the EU Directive, waste is defined as an object the holder discards, intends to discard or is required to discard (Randell et al. 2014). The definition is somehow vague and broad. The US Environmental Protection Agency (EPA) has a clearer indication of the origin of waste and has defined waste that is produced in the process of construction, renovation or demolition of structures that include buildings of all types as well as roads and bridges (US EPA 2017).

Lu and Yuan (2011) conduct a study examining the definitions of waste. They reviewed 147 articles in the literature and found 3 common ways of defining waste. The most common way was an activity-based definition that defines waste according to points of origin, such as waste generated from construction activities onsite. This is closely related to the definition used by the US EPA. Some others define waste as a composition-based definition in accordance with the category of waste as, for instance, the European Waste Categories. Other studies define waste based on property and define waste as construction by-products that may be reused and recycled. This definition is commonly used in Japan.

Waste generally consists of a mixture of inert and non-inert materials arising from construction-related activities. Malia et al. (2013) conducted research to develop indicators for waste and found that the average composition of waste generated mostly onsite consists of around 80% of inert materials such as concrete and ceramic.

Waste often contains bulky and heavy materials that include a wide range of mostly inert materials that are not organic and not readily degradable materials including concrete, brick, glass, plastic, tiles, etc. Composition varies for different types of activities and structures, and some are suitable for land reclamation and site formation (Bakshan et al. 2015). They generally do little damage to the environment apart from occupying land space, but the treatment processes and the loss of natural resources can be significant (Wu et al. 2014). This type of waste has a great potential for recycling according to the EU Waste Strategy (Banias et al. 2010), but only a small fraction is actually recovered (Malia et al. 2013). However, the inert materials may be contaminated by hazardous components if not properly sorted and stored. Hazardous materials are wastes from the industrial sector as well as contaminated soil, asbestos and particulate matters (Randell et al. 2014). These contaminated inert materials may need to be managed and disposed of in ways that minimise negative impacts on the environment and human health as well as being economically feasible.

Waste also contains non-inert materials that are largely organically based and chemically active substances that will decompose over time. These include timber, paper, vegetation and other organic materials (Yuan et al. 2013; Bakshan et al. 2015). The inert waste can be contaminated if mixed with non-inert materials. Onsite sorting is important to improve the reusability and recyclability of the inert waste.

International Policies and Regulations

Reflecting the growing concern with the effects of waste on the environment, important policies and regulations have been incorporated into national and international laws, with the aim to reduce waste and promote reuse and recycling (Laquatra and Pierce 2011; Wonschik et al. 2014). International conventions and agreements relevant to waste have been developed at different levels, and some of them have long histories. Table 10.2 summarises policies and regulations for waste.

The Basel Convention on the Control of Transboundary Movements of Hazardous Wastes and their Disposal is an international agreement initiated in 1989 to establish a waste management framework and to control the movement of hazardous waste across boundaries of signatory countries. It aims to protect human health and the environment against the adverse effects of the generation, management, transboundary movements and disposal of hazardous and other wastes. The Convention became effective in 1992 with 184 countries (Wonschik et al. 2014; DEE 2013).

The Waigani Convention is a multi-lateral agreement under Article 11 of the Basel Convention. It is an international convention to ban the importation into Forum Island countries of hazardous and radioactive wastes within the South Pacific region. It aims to control and minimise the movement of hazardous and radiative wastes into countries such as Fiji, New Zealand, Niue, etc. It is also to ensure that disposal of wastes is in an environmentally sound manner and as close to the source as possible (DEE 2013).

The Kyoto Protocol was passed in 1997 and is an additional protocol to the United Nations Framework Convention on Climate Change (UNFCCC) for the configuration of the climate convention framework. It came into force in 2005 and sets binding goal values for the emission of GHG within the developed countries. It aims to reduce the emissions of six main GHGs: CO_2, methane (CH_4), nitrous oxide (N_2O), hydrofluorocarbons (HFCs), perfluorocarbons (PFCs) and sulphur hexafluoride (SF6). In assessing the total GHG emissions, these gases are converted into tones of CO_2 equivalence (Wonschik et al. 2014; Anonymous 2012; DEE 2013).

According to Malia et al. (2013), EU has no specific legislation for construction, maintenance and demolition waste as it does for other types of waste. The first European law on waste was published in 1975 (Directive 75/442/CEE) to regulate the disposal of waste that is harmful to human health and the environment. The

Table 10.2 International waste-related policies and regulations

Policies/ regulations	Year	Countries	Aims
Basel Convention	1992	International (opened for signature in 1989 and has 184 member countries)	Protect human health and the environment against the adverse effects resulting from the generation, management, transboundary movements and disposal of hazardous and other wastes
Waigani Convention	2001	International (open for signature in 1995 and has 24 member countries)	Reduce or eliminate production and transboundary movements of hazardous and radioactive waste into and within the Pacific region and ensure that disposal of wastes is completed in an environmentally sound manner and as close to the source as possible
Kyoto Protocol (Kyoto Protocol to the UNFCCC)	2005	International	Reduce the emissions of six main greenhouse gases (CO_2, CH_4, N_2O, HFCs, PFCs and SF_6) over a commitment period from 2008 to 2012
			Amended in 2012 to incorporate a second commitment period from 2013 to 2020 and include a new gas, NF_3
EU Waste Framework Directive 2008/98/EC	2008	European Union	A legal framework for the treatment of waste to preserve and protect the environment, to diminish the harmful effects of waste has on the environment and human health
Resource Conservation and Recovery Act (RCRA)	1976	USA	Protect the environment, conserve resources and reduce waste

Sources: DEE (2013), Malia et al. (2013), Wonschik et al. (2014), US EPA (2017)

latest is the Directive 2008/98/EC of the European Parliament and of the Council in 2008 on waste and repealing certain Directives – also called the Waste Framework Directive (WFD). It establishes a legal framework for the treatment of waste within the EU countries. It contains important subsections regarding the hierarchy of waste, the waste management, permits and registration and plans and programmes for future waste treatments. It defines and determines key policies and concepts for waste, waste recovery and recycling (WFD 2017). The WFD sets target for the recycling of nonhazardous waste at a minimum of 70% of its weight by 2020 (Article 11.2) and confirms the waste hierarchy as the legally binding order of priority for waste management (Article 4) (Wonschik et al. 2014; Manfredi and Pant 2011; Dahlbo et al. 2015; WFD 2017).

The USA has a long history of dealing with solid waste. The solid waste disposal was covered in the Resource Conservation and Recovery Act (RCRA) of 1976 and aims at protecting the environment, conserving resources and reducing the amount of waste being generated (US EPA 2017). Construction, maintenance and demolition waste has not been explicitly regulated in the RCRA at the federal level as waste has been classified as neither hazardous nor part of the municipal solid waste (MSW) and will only be so when waste is included in the MSW (US EPA 2017). Nonetheless, construction, maintenance and demolition waste has been regulated at the state level but with a high degree of variations among states. The variations range from basic definition of waste through to disposal of waste and the operation and management of landfills (Laquatra and Pierce 2011).

Building Lifecycle and the Associated Waste

A construction project will have various impacts economically, socially and environmentally at different stages of the building's lifecycle. Waste is also generated at different stages and at different magnitudes with the bulk being produced during the renovation and demolition phases (Malia et al. 2013). Reducing waste will certainly reduce energy use, minimise degradation of the environment as well as reduce embodied carbon emissions.

According to EN 15978 Standard (CEN 2011), the building lifecycle is divided into stages of product, construction, use and end-of-life:

- Product (A1–3) – Raw material supply, transport and manufacturing
- Construction process (A4–5) – Transport, and construction and installation
- Use (B1–7) – Installed products in use, maintenance, repair, replacement and refurbishment
- End-of-life (C1–4) – Deconstruction/demolition, transport, waste processing and disposal

A building lifecycle therefore consists of initial raw material extraction and production, transportation, construction, operation and eventual demolition on site at the end of the lifecycle (Gangolells et al. 2009; Wu et al. 2012). In the past decades, many research studies have focused on waste at the end-of-life demolition only, but wastes are generated when the work starts onsite, and this continues through its use stage until the end of its useful life. Therefore, it is important to develop waste reduction strategies with a lifecycle perspective. According to US EPA (2009), the estimated amount of waste suggested that 9% of wastes were generated during initial construction, 42% from the operation – which includes replacements, renovations and refurbishments – and 49% from the demolition of buildings. The construction stage may require more materials to be installed, but it produces the smallest amount of waste. The end-of-life stage demolition generates the largest amount of waste. However, the operating stage produces the renovation waste which is a combination of both construction and demolition.

At the product stage (A1–3), wastes are generated during extraction of raw materials from the ground, the process of manufacturing raw materials into construction materials and components and eventually packaging materials for the delivery of products to suppliers or building sites. The waste generated at this stage includes waste due to cutting or reformation during production and losses due to non-standard or defective products. The wastes generated at the product stage are not normally included in the evaluation of waste of a building's lifecycle as these wastes have most likely been included in the municipal solid waste or waste for their respective industrial sectors.

According to Bakshan et al. (2015), construction waste constitutes more than 10% of the waste generated worldwide. Waste generated at the construction process stage (A4–5) is a major concern in the construction industry as more wastes mean more costs of material consumption plus the associated treatment cost (Yeheyis et al. 2013; Randell et al. 2014). These wastes can be reduced if managed properly. The level of construction waste caused depends on the construction system, nature of the project and the variety of construction materials and components used. Waste is caused by many factors, and these include unused material due to over order and incorrect material, poor workmanship and management onsite, errors in the design and lack of communication between designers and contractors, design changes, inaccurate cuttings and measurement, improper equipment use, adverse weather conditions, double handling, schedule delays, incorrect storage and manufacturing defects (Arslan et al. 2012; Dajadian and Koch 2014; RICS 2014). According to Panos and Danai (2012), approximately 33% of onsite waste is due to architects' failure to implement waste reduction measures during the design stage. According to Bossink and Brouwers (1996), construction waste is approximately 1–10% of the purchased construction materials. Hammond and Jones (2008) examine embodied carbon of construction waste for 14 residential dwellings in the UK and found that construction waste was approximately $0.12 \ m^3/m^2$ of constructed floor area and contributed to about $2 \ kgCO_2/m^2$ of embodied carbon, equalling 19% of the total embodied carbon in the material for the building. The embodied carbon-related waste at this stage includes material waste in relation to construction activities onsite due to the factors stated above and transport of these waste materials from construction site to waste disposal/reprocessing/recovery gate (RICS 2014).

During the use stage (B1–7), the building is maintained and refurbished throughout its lifecycle in order to sustain the functional capacity. Building elements or components are required to be maintained, repaired, replaced and refurbished, and waste is generated due to deterioration, corruption, changing needs, changing technology or fashion (RICS 2014). Therefore, many construction materials and components are changed, and the old or obsolete materials and components become waste. These processes sometimes occur with short intervals over this longest stage in a building's lifecycle. Therefore, significant levels of wastes are generated at this stage. Approximately 30–50% of overall construction waste results from renovation activities (Arslan et al. 2012), and approximately 5–10% ends up in a landfill (Terry and Moore 2008). A previous study in Turkey by Cosgun (2005) reveals an even more disturbing picture with very frequent renovations and about 74% of construction materials and components ending up as landfill (Cosgun 2005).

When a project approaches the end of its useful life (C1–4), demolition is almost inevitable. However, buildings may be demolished not because they are old or structurally unstable. Some buildings are demolished for economic, social or locational and stylistic obsolescence. Increasing building demolition activities are anticipated as there are numerous construction activities occurring all over the world every year (Pun et al. 2006). End-of-life demolition generates about as much waste as the combined effect of construction and operation. The demolition activity has become one of the major waste flows in the world. These wastes have a very negative impact on the environment and take up precious land space (Moffatt and Kohler 2008; Huang et al. 2013; Gao et al. 2015). According to Gong et al. (2012), this stage includes the relevant processes concerned with the recovery and utilisation of the dismantled building and abandoned building materials.

According to Scheuer et al. (2003), the demolition activities represent about 2–5% of resource consumption and 0.4% of total energy consumption. Liu and Pun (2006) state that the building demolition can be viewed as the reverse process of building construction as the issues involved in a demolition project are fundamentally the same or similar to those in the construction of a project. Moncaster and Symons (2013) examine the whole life embodied carbon and energy on a masonry residential building in the UK and found that the embodied carbon at the end-of-life demolition contributes to about 21% of the whole life embodied carbon. Different system boundaries and demolition methods used may have contributed to the different percentages of demolition embodied carbon in these studies.

Broadly speaking, incineration and disposal of demolition waste impact on the conservation of resources. By recycling and reuse, demolition waste can be reduced and hence so can the associated EC emissions. According to Gong et al. (2012), the demolition stage of a building lifecycle provides an opportunity for recovery and utilisation of dismantled and abandoned building materials. Nonrecycled waste causes the loss of raw materials and takes up land space for final disposal (Moffatt and Kohler 2008).

Estimation Methods

The wastes generated at the product stage are not normally included in the construction waste as they may be included in the municipal solid waste or waste of their respective industrial sector. The stream of waste, on a lifecycle perspective, therefore includes construction, renovation and demolition waste and the associated waste from the required transportation.

The quantification of lifecycle embodied carbon of waste is complicated, and it is important to understand the magnitude and material composition of the waste stream. The approach in quantifying embodied carbon on waste relates to applying the carbon intensity per unit of waste material generated. The embodied carbon intensity can be obtained from either the published sources or computer software. With regard to quantifying waste, there are two commonly used estimation methods: material waste rate (MWR) and an overall waste generation rate (WGR). The data collection for quantifying lifecycle waste includes direct field observation via site visits, measuring and recording waste onsite, questionnaire survey and interviews of key personnel of the project, data on purchased materials and the actual quantities used in the building provided by sub-contractors or suppliers (Li et al. 2013; Wu et al. 2014).

The MWR is commonly used by research studies to estimate the amount of waste generated in a project by material categories (Cochran et al. 2007; Tam et al. 2007; Lu et al. 2011; Li et al. 2013; Mercader-Moyano and Ramirez-de-Arellano-Agudo 2013; Nagapan et al. 2013). The MWR is the ratio of waste material to that purchased or required by the design. It indicates the waste generation level of a construction project and is expressed as a percentage. Key materials are identified from the project, and wastes are either measured directly via site observation and measurement or from purchase orders from sub-contractors or suppliers. The common key materials include concrete, brick, timber, tiles, metal and drywalls. Table 10.3 summarises the MWR of key materials at various stages of a building's lifecycle per gross floor area (GFA). The three wastes that are commonly measured in projects are concrete, brick and timber. The MWR varies a great deal due to different construction techniques used, size of sample, system boundary and data collection methods.

Waste can also be calculated as an overall WGR per GFA of total construction floor space using volume (m^3) or mass (e.g. kg or tonne) without distinguishing between materials (Cochran et al. 2007; US EPA 2009; Lu et al. 2011; Wu et al. 2014). Table 10.4 summarises the WGR in research studies. The waste is calculated by dividing the average WGR per GFA, e.g. 21 kg/m^2 (US EPA 2009) and 0.14 m^3/m^2 (Llatas 2011). Similar to MWR, WGR varies greatly due to different construction techniques, work procedures and common practices. The estimation of waste using WGR can be calculated at either project or regional level.

These estimation methods used in various research studies provide approximate data through various data collection methods, but they are invariably not sufficient to develop detailed waste indicators as there are too many variables such as building types and conditions, construction/demolition methods and so forth. Some target a particular type of buildings, and others depend on existing sources of data, e.g. regional databases, waste guides, industry surveys or site personnel's perception. Thus, the results may not be applicable in other regions with different typological characteristics and construction techniques. Most of the studies were undertaken in developed regions and may not be equally applicable to less developed regions (Bakshan et al. 2015).

The various equations used in the estimation of embodied carbon are discussed in the following sections by building lifecycle phases.

Table 10.3 Summary of material waste rate in research studies

Research studies	Type	Stage	Concrete kg/m² of GFA	Brick	Timber	Tiles	Gypsum	Metal
Cochran et al. (2007)	Residential – wood frame	Construction	0.26	0.51	12		5.2	0.3
	Residential – concrete frame	Construction	22.9		6.4		4.9	0.9
	Non-residential – wood frame	Construction			7		0.5	
	Non-residential – concrete frame	Construction	33		3.3		5.2	1.4
	Residential	Use	25		29	2.3	9.4	0.75–6.8
	Non-residential	Use			4.2		13	1
	Residential – wood frame	End-of-life	240–840	90	70–90		30–100	10
	Non-residential – concrete frame	End-of-life	690		1.5			44
Lu et al. (2011)	Lab and residential	Construction	0.36–2.39	0.037–0.821	1.68–1.91			0.014–0.073
Mercader-Moyano and Ramirez-de-Arellano-Agudo (2013)	Residential	Construction	68	8.4	0.44			0.63
Li et al. (2013)	Residential	Construction	18	3.4	7.6	0.5		4
Nagapan et al. (2013)	Residential	Construction	1.44–7.44	6.48–17.27	19.69–39.54			0.44–1.54
	Non-residential	Construction	9.14	16.5	241.43			13.49

(continued)

Table 10.3 (continued)

Research studies	Type	Stage	Concrete kg/m² of GFA	Brick	Timber	Tiles	Gypsum	Metal
Malia et al. (2013)	Residential – wood frame	Construction	0.3–1.9	0.5–0.8	5.6–17.9		2.4–7.2	0.1–0.9
	Residential – concrete frame	Construction	17.8–32.9	19.2–58.6	2.5–6.4	1.7–3.2	3.7–7.6	0.9–3.9
	Non-residential – wood frame	Construction			4.7–10.7		0.5–3.4	0.2–2.9
	Non-residential – concrete frame	Construction	18.3–40.1	15.6–54.3	1.7–5.4	0.4–3.2	2.6–6.3	1.0–7.2
	Residential	Use	18.9–45.9	63.3–319.5	2.0–37.9	1.1–12.6	2.4–23.5	0.4–6.8
	Non-residential	Use	18.9–191.2	11.2–62.0	2.3–42.6	0.2–16.9	2.3–22.9	0.2–16.4
	Residential – wood frame	End-of-life	137–300	84–90	70–275		10.9–105.4	4.8–22.5
	Residential – concrete frame	End-of-life	492–840	170–486	12–58	10.6–17.6	10.8–64.3	9.8–28.4
	Non-residential – concrete frame	End-of-life	401–768		20–159		10.8–75.7	28.4–53.0
Bakshan et al. (2015)	Mixed	Construction	8.7	17.44	4.35	2.0	0.31	1.25

Table 10.4 Summary of GWR in research studies

Research studies	Building type	Stages		
		Construction	Use	End-of-life
Poon et al. (2004)	Residential and non-residential	0.14–0.21 m^3/m^2		
Hammond and Jones (2008)	Residential	0.12 m^3/m^2		
Solis-Guzmun et al. (2009)	Residential	0.11 m^3/m^2		
US EPA (2009)	Residential Non-residential	21.43 kg/m^2	16–348 kg/m^2	244 kg/m^2
		21.19 kg/m^2	15–139 kg/m^2	771 kg/m^2
Katz and Baum (2011)	Residential	0.2 m^3/m^2		
Llatas (2011)	Residential	0.14 m^3/m^2		
Bakshan et al. (2015)	Mixed	38–43 kg/m^2		

Construction Process (A4–5)

The estimation of embodied carbon for construction waste will vary a great deal according to project size, building and structural type, construction technology, workmanship and management onsite as discussed earlier. The embodied carbon for construction waste can be estimated by summing the waste generated at this stage and multiplied by the carbon intensity per unit of waste material using the following equation (Li et al. 2013):

$$\text{EC}_{\text{construction}} = \sum_{i=1}^{n} M_i \times R_i \times \text{MEC}_i \qquad (10.1)$$

where:

$\text{EC}_{\text{construction}}$ = Embodied carbon for construction waste (kgCO$_2$)
M_i = Purchased or designed amount of material$_i$ (kg, m^2 or m^3)
R_i = Material waste rate for material$_i$ (%)
MEC_i = CO$_2$ intensity per unit of material$_i$ (kgCO$_2$ per kg, m^2 or m^3)
n = Number of material types

The generated wastes are also delivered by trucks either to landfills, incineration or recycling facilities. The emission in relation to transportation is directly related to the fuel type, number of truck loads, total distance travelled (outward and return trip), fuel consumption and CO$_2$ coefficient per fuel type using the following equation (Zhang et al. 2013; De Wolf et al. 2016):

$$EC_{transport} = \sum_{i=1}^{n} D_i \times T_i \times TF_i \times TEC_i \qquad (10.2)$$

where:

$EC_{transport}$ = Embodied carbon from fuel combustion in the transportation of waste from site to disposal facilities ($kgCO_2$)
D_i = Total distance travel for $waste_i$ (km)
T_i = No. of truck loads for transporting waste $material_i$ (No)
TF_i = Fuel consumption per truck load (litre per km)
TEC_i = CO_2 intensity per unit consumption of $fuel_i$ ($kgCO_2$ per litre)

Use Stage (B2–5)

The building use stage includes activities B1–7. However, waste generated from this stage will only cover activities from B2 to 5: maintenance, repair, replacement and refurbishment that include quantities of materials, transport of these materials from final manufacture to site and wastes to landfill or recycling facilities. Stage B1 is the installed products in use, Stage B6 is the operating energy and Stage B7 is the operating water use and is therefore not included in calculations for the embodied carbon. There are many different types of renovations and replacements during the use stage of a building. The need for scenario predictions makes it difficult to ascertain the amount of waste generated at this stage (RICS 2014).

Embodied carbon at this stage is referred to as recurrent embodied carbon when they relate to the use of material and components to renovate or refurbish the building over its lifetime (Zhang et al. 2013). Therefore, the service life and durability of materials and components are among the most important factors affecting the embodied carbon in this stage. Over a building's service life, one or more replacements of material or components may be required. The shorter the service life of a material or component, the greater the quantity of waste of ongoing maintenance and repair and the greater will be the embodied carbon associated with the waste generated throughout a building's life.

Table 10.5 summarises the lifespans of some common materials and components for buildings, and they vary from 5 to 75 years (Menzies and Wherrett 2005; Kubba 2010). This means that the lifetime of many systems or internal finishes is shorter than the lifespan of the building. For example, while in mechanical systems the lifespan of steel air ducts and liners is 75 years, that of carpets for floors is about 10 years. Therefore, there will be no replacement for steel air ducts and liners, but carpeted floors will require re-carpeting five times if a building lifetime of 50 years is considered. In addition, on occasions building services and internal finishes need to be repaired and generate waste which is extremely difficult to predict and estimate (Verbeeck and Hens 2010).

The embodied carbon at the use stage can be calculated using the following equation (Kua and Wong 2012; Zhang et al. 2013):

Table 10.5 Economic lifespan of building components

Building services		Building interior and finishes	
Components	Lifespan	Components	Lifespan
Duct liner, acoustic	75	Facing brick	80
Steel air ducts (sheet metal)	75	Stone	75
Pipe (metal or PVC)	50	Laminated timber frame	60
Metal sinks	50	Concrete columns, beams	60
Sprinkler system	50	Aerated lightweight block	60
Radiators (baseboard)	40	Ceramic tile	50–75
Cable tray	30	Glass block	50
HV cable, switchgear	30	Timber door frames	50
Lifts	25–30	Drywall	50
Storage tanks (plaster or PVCu)	25–30	Metal doors	50
Sanitary fixtures	20–50	Wood panelling	50
HVAC system	20–30	Aluminium windows	50
Phone and data wiring (copper)	25	Softwood windows	30
Sprinkler heads	25	Hardwood windows	40
Conduit	25	Aluminium roof decking	32
Insulation	20	Asphalt roof covering	30
Kitchen fittings	15	Galvanised steel roof decking	30
Extract fans	15	Softwood doors	30
Smoke detectors	15	Acoustical wall panels	20
Hot water system	12	Ceiling tiles	20
		Rubber or vinyl tile	10
		Carpet	10
		General painting	5–10

Source: AIQS (2002), RICS (2006)

$$EC_{maintenance} = \sum_{i=1}^{n} \left[Q_i \times \left(\frac{BSL}{ESL_i} - 1 \right) \right] \times OEC_i \qquad (10.3)$$

where:

$EC_{maintenance}$ = Embodied carbon for material used at the use stage for maintenance or refurbishment ($kgCO_2$)

Q_i = Quantity of waste generated from material$_i$ for the maintenance or refurbishment of a building (kg, m^2 or m^3)

BSL = Reference service life of the building

ESL_i = Estimated service life of material$_i$

OEC_i = Carbon intensity per unit of waste generated from material$_i$ used in maintenance or refurbishment ($kgCO_2$ per kg, m^2 or m^3)

The waste generated due to maintenance or refurbishment is also transported either to landfills, incineration or recycling facilities. The embodied carbon associated with transport of waste is established using Eq. 10.2. However, it is very difficult to predict the maintenance requirements or the exact year of the replacement in advance as it may be impacted by the user patterns, quality of materials, workmanship and so forth (RICS 2014).

End-of-Life Stage (C1–4)

Embodied carbon at this stage will include the onsite operation of demolition activities, sorting or processing and transporting these wastes to landfills or recovery facilities, disposing or processing at the recovering facilities (Moncaster and Symons 2013). Demolition waste is a major source of global waste flow, and the choice of demolition methods used is regarded as a solution to reduce waste and promote reuse and recycling (Wu et al. 2014).

All carbon emissions in relation to energy used onsite during the process of demolishing a building include erecting site infrastructure, operating the plant to carry out the demolition work and sorting materials arising from the demolition operations. The embodied carbon at this stage will include the estimated running time of the plants and equipment used in demolishing the building multiplied by the average consumption of electricity and/or fuel per unit of time and the associated carbon intensity per litre of fuel consumed using the following equation (Kua and Wong 2012; Zhang et al. 2013):

$$EC_{demolition\ (plant)} = \sum_{i=1}^{n} P_i \times PF_i \times PEC_i \qquad (10.4)$$

where:

$EC_{demolition\ (plant)}$ = Embodied carbon-associated plant operation to the end-of-life stage ($kgCO_2$)
P_i = Duration of plant/equipment$_i$ used for demolition (hours)
PF_i = Fuel consumption per plant/equipment$_i$ used (kWh or litre per hour)
PEC_i = Carbon intensity per unit consumption of fuel$_i$ ($kgCO_2$ per litre)

Embodied carbon will also be calculated from the wastes generated as a result of demolishing the building. It is assumed that demolition waste at the end of the building's life is equal to the mass of material in the constructed building, not including the waste factor, and has a similar breakdown of type using the following equation (Liu and Pun 2006; Kua and Wong 2012):

$$EC_{demolition\ (building)} = \sum_{i=1}^{n} M_i \times MEC_i \qquad (10.5)$$

where:

$EC_{demolition\ (building)}$ = Embodied carbon for construction waste ($kgCO_2$)
M_i = Quantity of waste by material$_i$ (kg, m^2 or m^3)
MEC_i = CO_2 intensity per unit of material$_i$ ($kgCO_2$ per kg, m^2 or m^3)

At this stage, transportation of demolition waste will also be required to either landfill or recycling facilities. The emission will, therefore, be associated with the truck loads, distance of travel and fuel consumption, and it will be estimated using Eq. 10.2.

Mitigation Methods

Waste is seen as a burden both economically and environmentally, and the recovery of waste is an appropriate approach to the protection of the environment (Banias et al. 2010; Lu and Yuan 2011; Yeheyis et al. 2013). Waste recovery mitigates climate change by saving energy, conserving natural resources through substitution and preventing illegal dumping. Strategies to minimise waste will need to start early in a building's lifecycle. Early planning, design and material selection play an important role in this respect. In order to reduce embodied carbon in waste, it is important to select materials that sequester carbon for long periods of time as it can be seen as a way to reduce the embodied carbon of a project. However, the embodied carbon may be released back to the environment if reuse becomes impractical and the waste is incinerated at the end of the building's life (Tingley and Davison 2011).

Strategies have been developed to minimise waste over the years. The best known is the waste minimisation hierarchy (Fig. 10.1).

The hierarchy is arranged in six levels and in descending order of their adverse impacts on the environment from low to high and from most to least preferable methods (Lu and Yuan 2011; Manfredi and Pant 2011; Randell et al. 2014). Prevention is at the top as the most preferred strategy as opposed to disposal which is the least acceptable way to deal with waste.

Prevention is the best possible strategy for the environment as resources are not lost and negative impacts to the environment do not occur and reduce the cost of disposing of waste (Malia et al. 2013). This approach also includes reducing the use of harmful substances in materials and products to reduce the adverse impacts on the environment and on human health. One such programme is the Waste and Resources Action Program (WRAP) that has been developed in the UK for waste prevention and other good management practices related to construction, maintenance and demolition activities (Manfredi and Pant 2011).

This strategy involves rethinking design and design-out-waste to generating less waste onsite and offsite, e.g. designing buildings and infrastructure that can be constructed efficiently and specifying work procedures and method to avoid waste and enable efficient use of material from using advanced technologies (RICS 2014). Lifecycle thinking also plays an important role in preventing waste well before the building is built. Planning and designing buildings to maximise potential material recovery at the end-of-life deconstruction stage is the most effective way to increase the amount of materials that can be reused with minimal refabricating (Tingley and Davison 2011; Arrigoni et al. 2016). Prevention of waste may also be

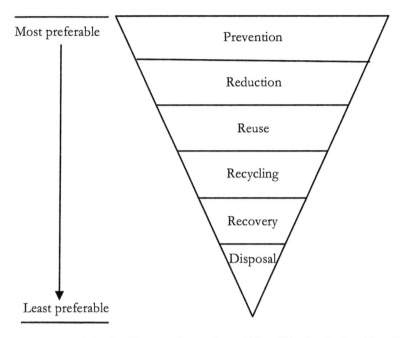

Most preferable

Least preferable

Prevention

Reduction

Reuse

Recycling

Recovery

Disposal

Fig. 10.1 Waste minimisation hierarchy (Source: Lu and Yuan 2011; Manfredi and Pant 2011; Randell et al. 2014)

enhanced through the use of prefabrication to improve the integrity of the building design and construction by avoiding waste caused by excessive customisation and fitting onsite (Yeheyis et al. 2013; Dajadian and Koch 2014).

The next level down is reduction which includes optimising resources and reducing packaging. This strategy does not only minimise waste but also reduces the cost for waste transporting, disposal and recycling (Lu and Yuan 2011). It can be implemented through government legislation in increasing landfill charges for disposing of waste to public landfills. This is an effective approach to drive towards higher reuse and recycling rates and act as an economic incentive for stakeholders to reduce landfill disposal (Yuan et al. 2013).

Reuse is different from recycling as it means that a product and its components are returning whole or partially for the same purpose without much input of additional materials and reprocessing. If neither prevention nor reduction is feasible to minimise waste, reuse would be a more favourable strategy than recycling. For products to be reused for the same purpose, it will require various activities that include a separate collection and return system, cleaning or reconditioning stage (Yeheyis et al. 2013). Modular design and deconstruction instead of demolition can enhance the potential for reuse, and this can only be achieved if planning and design are considered early so that buildings can be dissembled to recover components and materials for reuse at the end-of-life stage.

Recycling is a popular strategy to reduce waste when a reduction or reuse is less viable. This strategy helps reducing demand for new resources, cutting down transport and production energy and utilising waste which would otherwise be lost to landfill sites. However, additional resources and reprocessing are required to turn waste into new products (Lu and Yuan 2011; Yeheyis et al. 2013). Recycling of waste can be hampered by contamination during sorting and storing. Therefore, provision of separate bins for onsite collection and sorting plays an important role to improve the quality and recyclability of wastes, but this may be restricted by limited site space and budget (Malia et al. 2013; Dajadian and Koch 2014; Brennan et al. 2014). The reprocessing of waste into new products will require the input of energy which will release carbon into the environment. Therefore, the preferred strategy is to minimise reprocessing and locate recycling facilities as close as possible to the sites. Waste has high reuse and recycling potentials such as crushed concrete and brick as aggregates in roads, drainage and other construction projects. However, it will require that the technology for separation and recovery is readily accessible and inexpensive.

Waste can also be used for energy recovery from the energy embodied within waste materials, e.g. wood. This can lead to significant environmental benefits, particularly for materials with a high calorific content. This is usually done in special facilities using processes such as incinerations, pyrolysis or anaerobic digestion or by collection and combustion of methane gas collected from landfill sites (Manfredi and Pant 2011).

If none of these strategies is feasible, the waste will only be disposed at the landfills which are usually for low-grade inert or contaminated materials. Disposal may also be used if the recycling facilities are so far away that transportation for reprocessing is neither feasible nor economical. Disposal of waste will lead to the remanufacturing of building materials and products that involve use of energy and emission of CO_2. When waste is landfilled, the embodied carbon can be considered lost as new and virgin materials would be needed to replace them (Randell et al. 2014).

Challenges and Barriers

The waste minimisation hierarchy presents an ideal framework to reduce waste, but in reality, it is not always easily implemented (Lu and Yuan 2011). There are challenges and barriers inherited in the minimisation processes that will require a close collaboration of stakeholders from the construction industry. A waste minimisation strategy can only be successful if planning and management utilise a whole lifecycle approach. This means that the strategy will need to be developed as early as at the feasibility stage and managed throughout the entire lifespan.

There are many challenges and barriers for waste minimisation. The most common barrier to resource recovery is that the wastes may be contaminated with other materials such as soil or asbestos when landfilling may be the only option as decontamination may be too expensive (Monier et al. 2011; Randell et al. 2014). Secondly, the recycled materials are competing with virgin raw materials in the market. There are a high availability and low cost of many virgin raw materials, such as stones, sand or gravel, which decrease the demand for and interest in developing business from recycling (Monier et al. 2011; Dahlbo et al. 2015). This could be overcome by improving the competitiveness of recycled materials by either raising the price of primary raw materials through taxation or subsidising recycled materials if used in construction.

Thirdly, the cost of disposal can be a barrier too as the low levy on landfilling may discourage recovery. According to the Australian National Waste Report (DEE 2013), among the states, Queensland has the lowest landfill levies ($30/t), approximately 78% lower than the highest of New South Wales (NSW) ($133/t). As such 477,000 and 398,000 tonnes of waste for 2014 and 2015, respectively, travelling by road and rail from NSW to Queensland contribute to 15,000 heavy truck movements and the associated embodied carbon. In addition, recycling facilities in NSW lost the opportunity to recover materials from the stream.

Fourthly, the lack of infrastructure for source separating onsite may also hinder resource recovery. This may be due to limited budget and space onsite for collecting, sorting and recycling waste (Randell et al. 2014). The engagement of a waste contractor may be considered an economic and effective way of reducing and recovering waste (Dajadian and Koch 2014). Finally, the misconception that the quality of recycled products is not as good as that of virgin material in the structural application is an important barrier. Confidence in using secondary materials can be improved through proper quality certification and communicating the benefits of secondary materials to stakeholders (Monier et al. 2011). Further research on the application of recycled waste and particularly on long-term behaviour could contribute to reducing uncertainty linked to the use of recycled products.

The Case Study

Considering the complex nature of the research and the lack of available data to understand the impact of embodied carbon of waste at each stage of a building lifecycle, case study approach is deemed to be the preferable method. The use of a case study to investigate the nature and magnitude of waste will provide insight into its types, mass and final destination so that strategies for its management and minimisation can be developed. A lifecycle analysis approach is therefore used to analyse the embodied carbon in relation to waste at different stages of a building's lifecycle.

The case study project represents a typical house design in Australia. This type of house includes concrete floor, timber wall frames with brick veneer and timber roof structure with concrete tiles for a lifespan of approximately 50 years. The case study project characterises the conventional use of high-density and high-energy intensity materials. The project is located in the North of Sydney which is the main

Table 10.6 Construction materials used for the project

Building component	Materials
Structure	Subfloor to include sand blinding, concrete piers and 200 μm waterproof membrane
	Ground floor to be RC slab on ground and EPS foam/RC ribs
	First floor to be timber I-beam and chipboard flooring
Walls	External wall to be pine timber frame, vapour barrier and extruded clay brick cladding with 10 mm plasterboard lining
	Internal wall to be pine timber frame with 10 mm plasterboard lining on both sides
Roof	Pine timber frame with concrete tiles covering and sarking
Windows and doors	Windows to be aluminium glazed
	Doors to be timber and paint finish
Finishes	Floor finishes to be ceramic tile on ground floor and carpet to the first floor
	Wall finishes to be painted plasterboard and ceramic tiles to kitchen, bathroom and laundry
	Ceiling finishes to be painted plasterboard on pine timber frame

residential area of NSW. The project is a two-storey detached house with a total GFA of 143 m². Table 10.6 summarises the materials used in the project. The construction method for the project is a traditional brick veneer construction for external walls, timber framed with plasterboard for internal walls, concrete roof tile and reinforced concrete (RC) slab on ground.

The project construction started in May 2016, and the building was completed at the end of the year, after a construction period of approximately 7 months. Architectural and structural drawings were obtained from the architect. Inventory of material quantities measured from the drawings is summarised in Table 10.7. The construction of the building used high embodied carbon materials such as bricks, RC, aluminium, carpet and tiles as the main materials. Painting, carpets and ceramic tiles have required frequent repair and replacement during the operating stage of a building's lifecycle which will add more embodied carbon into the lifecycle of the building.

The embodied carbon for construction waste was calculated based on the data collected onsite at the construction stage. Construction waste was documented onsite from several site visits during the construction period. The construction waste was further verified from interviewing the site manager and labourers to examine the causes of waste and its quantities. From the study, the construction waste for the project ranged from approximately 1% for metals to 20% for concrete and timber products (Table 10.7). In assessing the quantity of construction waste, the report published by the Department of Environment and Climate Change (DECC) (2007) was also used to supplement the data collected onsite, and the results agreed with the outcomes from the report.

Table 10.7 Inventory of material quantities and waste for the project

Material	Quantity per unit	Waste per unit	Material	Quantity per unit	Waste per unit
Waffle pod (m³)	27	1.4	Bricks (m³)	781	62.5
Reinforcement (t)	1.1	0.1	Aluminium windows (m²)	21.2	–
Concrete (m³)	22.5	2.3	Timber doors (m³)	0.16	–
Timber (m³)	11.6	2.4	Glass (m²)	1.6	–
Chipboard (m²)	110	22	Plasterboard (m²)	569	29
Sarking (m²)	180	10	Painting (litre)	50	3
Concrete roof tile (m²)	180	14	Ceramic tiles (m²)	44	4
Insulation (m²)	233	8	Carpets (m²)	110	11

The embodied carbon for maintenance and replacement waste during the operating stage was calculated based on a building lifecycle of 50 years. Owners and operating waste were excluded in the calculation as these are part of the municipal solid waste. Maintenance and replacement cycles were calculated based on the lifespan of materials used in the project in accordance to the *Australian Management Manual* (AIQS 2002) and the *Life Expectancy of Building Components* (RICS 2006).

At the end of the building's lifecycle, the house is deemed to be demolished and sent to landfill as the usual practice in Australia. The demolition waste and associated embodied carbon at the end of the building's life are assumed to be equal to the mass of material in the constructed building with a similar breakdown structure (Liu and Pun 2006). The waste factor related to construction activities was excluded from the calculation and to have a similar breakdown by type. No allowance was made for the recycling of waste or for any further emissions relating to the operation of the landfill station.

The embodied carbon emissions from plant and equipment involved in the demolition are calculated based on the diesel fuel consumption and the running time of these plant and equipment used onsite. Total emissions were estimated by multiplying the predicted running time of the plant and equipment by the average consumption of diesel fuel per unit of time. The average fuel consumption for an excavator was sourced from demolition contractors and equipment manufacturers. The fuel consumption for a 15 tonne excavator for demolition was estimated to be approximately 15 l/h for approximately 20 h onsite for the demolition of the building of this size and type.

Embodied carbon also includes emissions from the fuel combustion for the transportation of construction, maintenance and demolition waste to landfill. Waste materials were transported offsite and deposited at the nearest public landfill, approximately 15 km away from the site on rigid trucks. The fuel consumption for a rigid truck with a capacity of 11 m³ is approximately 40 l/100 km. The embodied carbon model was created using the computer program GaBi 6 and supplemented

Table 10.8 Summary of embodied carbon of waste on a building's lifecycle

Building elements	Construction stage	Operating stage	End-of-life stage		
	Onsite construction waste ($kgCO_2$)	Maintenance and replacement waste ($kgCO_2$)	Demolition waste ($kgCO_2$)	Total ($kgCO_2$)	%
Structure	1069	–	11,159	12,228	22.6
Walls	845	–	10,694	11,539	21.3
Roof	420		3215	3635	6.7
Windows and doors	–	1140	1467	2607	4.8
Finishes	426	16,328	7307	24,061	44.6
Total	2760	17,468	33,842	54,070	100
kg CO_2/m^2 (GFA)	19.3	122.2	236.6	378.1	
%	5.1	32.3	62.6	100	

by data from BPIC LCI (BPIC 2017), literature, and government published reports from the DEH (2006) and DCCEE (2010).

Results

The collected data is analysed and summarised in Table 10.8. The table presents the embodied carbon waste from a lifecycle perspective. The embodied carbon associated with the fuel combustion of plant, equipment and trucks is included in the figures in the table. The embodied carbon of the waste has been assessed on the following parameters:

- Waste generated during the construction activities onsite
- Waste generated for the maintenance of the building during the operating stage on a 50-year lifecycle
- Waste generated at the end-of-life demolition
- Emissions of fuel for plant and equipment during demolition and trucks in the delivery of waste to landfills

The study reveals that the end-of-life stage generates the majority of the waste in the building's lifecycle. The demolition waste was the highest in the study, approximately 63%, and generated 237 $kgCO_2$/m^2 of embodied carbon from the demolition of the building. Included in the figure is approximately 17% of the embodied carbon associated with fuel consumption of the excavator used in demolition and trucks in delivering waste to the landfill. The table also reveals that the major part of the embodied carbon originated from the building structure (51%) that includes the main structure, walls and roofs. This was followed by the finishes (45%). Building structure has a long lifespan, approximately 50–100 years (AIQS 2002; RICS

2006). Therefore, approximately 51% (192 kgCO$_2$/m^2) of embodied carbon can be saved if the building is continued to be used.

As revealed in the literature, embodied carbon related to waste from the maintenance of buildings during the operating stage has rarely been considered in previous waste studies. Maintenance waste constitutes approximately 32% of the lifecycle waste and generates 122 kgCO$_2$/m^2 of embodied carbon. Some wastes were generated from replacing windows and doors, and finishes, but most of the maintenance wastes were generated from replacing finishes to floors such as tiles and carpets. There were no maintenance wastes generated from the building structure as structural defects are unlikely if the building has been erected properly. The embodied carbon-associated fuel combustion from transporting waste to landfill was only approximately 2% of the total embodied carbon at the operating stage.

Construction waste has been the main focus of studies in waste. However, in this case study, embodied carbon from construction waste was approximately 5% only over the building's lifecycle. The construction of the building structure generates most of the waste, approximately 54%, followed by the walls 31%. However, the embodied carbon in the transportation of construction waste to landfill constitutes approximately 12% of the total embodied carbon in this stage.

The findings from the case study confirm results from previously published studies. The structure of a building generates most of the embodied carbon in the construction phase (Cochran et al. 2007; Thomas 2015) as the traditional construction methods rely heavily on high-density and high-energy intensity materials such as RC, bricks, steel and aluminium. These materials have long lifespans which should be recovered and recycled to minimise energy consumption and the associated embodied carbon from the manufacturing of raw materials into new building material. More importantly, these high-density and high-energy intensity materials should have been replaced with more materials from renewable sources to minimise the embodied carbon.

Discussion

This chapter reviews the literature and research studies on waste and its associated embodied carbon of buildings on a lifecycle perspective. Firstly, the literature review found that construction industry is responsible for 20–60% of the solid waste stream globally. The origin and magnitude of waste in buildings are significant at each stage of a building's lifecycle with demolition stage generating most of the waste, followed by building maintenance and construction.

Secondly, the nature, types, mass and waste stream in various stages of a building's lifecycle include mainly inert materials such as concrete and ceramic that are not biodegradable but have reuse and recycle potential. These are valuable materials for the construction industry that can help to reduce demand for virgin materials and waste to landfills.

Thirdly, there are potentials to minimise waste production at each stage of a building's lifecycle, and the design-out-waste principles and lifecycle thinking are to be implemented at the feasibility stage and managed throughout the entire lifespan.

Fourthly, the investigation of policies, initiatives and collaborations in reducing waste and its associated embodied carbon is now a top priority at both national and international agenda. Finally, this chapter establishes an assessment framework for assessing embodied carbon of waste of a building on a lifecycle perspective.

Conclusion

The rapid population growth has intensified construction activities which have exacerbated the generation of waste and its associated embodied carbon in various stages of a building's lifecycle. This chapter reveals that demolition activities generate the majority of the waste and associated embodied carbon, while construction waste constitutes the lowest amount of waste in the total waste stream. However, even though construction stage generates the least amount of waste, it can be immediately and efficiently reduced by better site management and improved information flow and communication among design team members during the design and construction period. The maintenance waste has been largely ignored in previous studies, but it has a higher embodied carbon than construction waste, approximately six times more over a 50-year lifecycle as revealed in the case study. This impact will be intensified if the lifespan of a building is extended further. How frequently materials/products need to be maintained, replaced or repaired will depend on the quality and durability of materials/products. Therefore, rethinking design and adopting lifecycle thinking are fundamental ways to reduce waste and to maximise potential recovery.

The waste generation and the associated embodied carbon at the end-of-life stage are by far the highest. In some circumstances, demolition at the end of a building's lifecycle is inevitable due to various types of obsolescence or other reasons. However, if lifecycle thinking and design-out-waste principles are incorporated at the outset, materials or components can be reused or recycled without ending up in a landfill. In addition, building lifespan can be extended if buildings are properly constructed and maintained, protecting the majority of the embodied carbon that is trapped in the building structure.

The wastes generated at the various stages of a building's lifecycle are valuable resources for the construction industry if they are collected, treated and reprocessed properly. They can be important resources to replace raw materials and can help minimise embodied carbon and landfill. Transportation was shown to be important for the embodied carbon impacts related to waste. Therefore, small variations in transport distances might change the outcomes from bad to worse. This is the case when landfill sites are filling up fast and potential new landfill sites are further away from build-up areas and will require a longer distance of travel. The most important strategy is to prevent the generation of waste so that resources are neither lost nor impacting negatively on the environment. The success of this strategy is the need to consider minimising waste through design optimisation and collaboration with stakeholders at the early design stage.

References

Acquaye, A. A., & Duffy, A. P. (2010). Input-output analysis of Irish construction sector greenhouse gas emissions. *Building and Environment, 45*, 784–791.

AIQS. (2002). *Australian cost management manual volume 3 – Life cycle costing.* Canberra: Australian Institute of Quantity Surveyors.

Anonymous. (2012). *Review of Australia's international waste-related reporting obligations.* Queensland: Sinclair Knight Merz Pty Ltd.

Arrigoni, A., Collatina, D., Zucchinelli, M., & Dotelli, G. (2016). The environmental relevance of the construction and end-of-life phases of a building: A temporary structure LCA case study, Proceedings: Sustainable Built Environment (SBE) Regional Conference (pp. 436–441), 15–17, Zurich.

Arslan, H., Cosgun, N. & Salgin, B. (2012). Construction and demolition waste management in Turkey. In L. F. M. Rebellion (Ed.), *Chapter 14, Waste management – An integrated vision.* Intech.

Bakshan, A., Srour, I., Chehab, G., & El-Fadel, M. (2015). A field based methodology for estimating waste generation rates at various stages of construction projects. *Resources, Conservation and Recycling, 100*, 70–80.

Banias, G., Achillas, C., Vlachokostas, C., Moussiopoulos, N., & Tarsenis, S. (2010). Assessing multiple criteria for the optimal location of a construction and demolition waste management facility. *Building and Environment, 45*, 2317–2326.

Bossink, B. A. G., & Brouwers, H. J. H. (1996). Construction waste: Quantification and source evaluation. *Journal of Construction Engineering and Management, 122*(1), 55–60.

BPIC LCI project is available from the BPIC website. (2017). http://www.bpic.asn.au/bpic/the-building-products-innovation-council

Brander, M. (2012). Greenhouse gases, CO2, CO2e and carbon: What do all these terms mean? *Ecometrica*, 1–3.

Brennan, J., Ding, G., Wonschik, C. R., & Vessalas, K. (2014). A closed-loop system of construction and demolition waste recycling. In *Proceedings of the 31st international symposium on automation and robotics in construction and mining* (pp. 499–505), Sydney, Australia, 9–11 July.

Butera, S., Christensen, T. H., & Astrup, T. F. (2015). Life cycle assessment of construction and demolition waste management. *Waste Management, 44*, 196–205.

CEN. (2011). *Sustainability of construction works – Assessment of environmental performance of buildings – Calculation method.* UK: British Standard Institute.

Cochran, K., Townsend, T., Reinhart, D., & Heck, H. (2007). Estimation of regional building-related C&D debris generation and composition: Case study for Florida US. *Journal of Waste Management, 27*, 921–931.

Cole, R. J. (1999). Energy and greenhouse gas emissions associated with the construction of alternative structural systems. *Building and Environment, 34*(3), 335–348.

Cosgun, T. E. (2005). Ecological analysis of reusability and recyclability of modified building materials and components at use phase of residential buildings in Istanbul, UIA 2005 Istanbul XXII World Congress of Architecture-Cities: Grand Bazaar of Architectures, 310, July, Istanbul.

Crowther, P. (2015). Re-valuing construction materials and components through design to disassembly'. In: *Proceedings: Unmaking waste 2015 conference*, 22–24 May, Adelaide.

Dahlbo, H., Bacher, J., Lahtinen, K., Jouttijarvi, T., Suoheimo, P., Mattila, T., Sironen, S., Myllymaa, T., & Saramaki, K. (2015). Construction and demolition waste management- a holistic evaluation of environmental performance. *Journal of Cleaner Production, 107*, 333–341.

Dajadian, S. A., & Koch, D. C. (2014). Waste management models and their applications on construction site. *International Journal of Construction Engineering and Management, 3*(3), 91–98.

DCCEE. (2010). *National greenhouse accounts (NGA) factors*. Canberra: Department of Climate Change and Energy Efficiency.

De Wolf, C., Bird, K., & Ochsendorf, J. (2016). Material quantities and embodied carbon in exemplary low carbon case studies. In G. Habert & A. Schlueter (Ed.), *Proceedings: Sustainable Built Environment (SBE) regional conference, expand boundaries: system thinking for the built environment* (pp. 726–733). Zurich.

DECC. (2007). *Report into the construction and demolition waste stream audit 2000–2005*. NSW: Department of Environment and Climate Change.

DEE. (2013). *National waste reporting 2013*, Department of Environment and Energy, Australian Government. http://www.environment.gov.au/resource/national-waste-reporting-downloads. Date of access: 16/1/2017.

DEH. (2006). *AGO factors and methods workbook*. Canberra: Department of the Environment and Heritage, Australian Greenhouse Office.

Gangolells, M., Casals, M., Gassó, S., Forcada, N., Roca, X., & Fuertes, A. (2009). A methodology for predicting the severity of environmental impacts related to the construction process of residential buildings. *Building and Environment, 44*(3), 558–571.

Gao, T., Shen, L., Shen, M., Chen, F., Liu, L., & Gao, L. (2015). Analysis on differences of carbon dioxide emission from cement production and their major determinants. *Journal of Cleaner Production, 103*, 160–170.

Gong, X., Nie, Z., Wang, Z., Cui, S., Gao, F., & Zuo, T. (2012). Life cycle energy consumption and carbon dioxide emission of residential building designs in Beijing. *Journal of Industrial Ecology, 16*(4), 576–587.

Hammond, G. P., & Jones, C. I. (2008). Embodied energy and carbon in construction materials. *Proceedings of the Institution of Civil Engineers & Energy, 161*(2), 87–98.

Hu, X., Si, T., & Liu, C. (2017). Total factor carbon emission performance measurement and development. *Journal of Cleaner Production, 142*, 2804–2815.

Huang, T., Shi, F., Tanikawa, H., Fei, J., & Han, J. (2013). Materials demand and environmental impact of buildings construction and demolition in China based on dynamic material flow analysis. *Resources, Conservation and Recycling, 72*, 91–101.

IPCC. (2013). In T. F. Stocker, D. Qin, G. -K. Plattner, M. Tignor, S. K. Allen, J. Boschung, A. Nauels, Y. Xia, V. Bex, & P. M. Midgley (Eds.), *Climate Change 2013: The Physical Science Basis. Contribution of Working Group I to the Fifth Assessment Report of the Intergovernmental Panel on Climate Change*. Cambridge: Cambridge University Press.

Katz, A., & Baum, H. (2011). A novel methodology to estimate the evolution of construction waste in construction sites. *Waste Management, 31*, 353–358.

Kua, H. W., & Wong, C. L. (2012). Analysing the life cycle greenhouse gas emission and energy consumption of a multi-storied commercial building in Singapore from an extended system boundary perspective. *Energy and Buildings, 51*, 6–14.

Kubba, S. (2010). *Green construction project management and cost oversight*. Oxford: Architectural Press.

Laquatra, J., & Pierce, M. (2011). Waste management at the construction site. In S. Kumar (Ed.), *Integrated waste management volume 1* (pp. 281–300). Intech.

Li, J., Ding, Z. X., & Wang, J. (2013). A model for estimating construction waste generation index for building project in China. *Resources, Conservation and Recycling, 74*, 20–26.

Liu, C., & Pun, S. K. (2006). A framework for material management in the building demolition industry. *Architectural Science Review, 49*(4), 391–399.

Llatas, C. (2011). A model for quantifying construction waste in projects according to the European waste list. *Waste Management, 31*(6), 1261–1276.

Lu, W., & Yuan, H. (2011). A framework for understanding waste management studies in construction. *Waste Management, 31*, 1252–1260.

Lu, W., Yuan, H., Li, J., Hao, J. J. L., Mi, X., & Ding, Z. (2011). An empirical investigation of construction and demolition waste generation rates in Shenzhen city, South China. *Waste Management, 31*(4), 680–687.

Malia, M., de Brito, J., Pinheiro, M. D., & Bravo, M. (2013). Construction and demolition waste indicators. *Waste Management & Research, 31*(3), 241–255.

Manfredi, S., & Pant, R. (2011). Supporting environmentally sound decisions for construction and demolition (C&D) waste management – A practical guide to life cycle thinking (LCT) and life cycle assessment (LCA), Joint Research Centre, Institute for Environment and Sustainability, EUR 24918 EN.

Mao, C., Shen, Q., Shen, L., & Tang, L. (2013). Comparative study of greenhouse gas emissions between off-site prefabrication and conventional construction methods: Two case studies of residential projects. *Energy and Buildings, 66*, 165–176.

Marshall, R. E., & Farahbakhsh, K. (2013). Systems approaches to integrated solid waste management in developing countries. *Waste Management, 33*(4), 988–1003.

Menzies, G. F., & Wherrett, J. R. (2005). Multiglazed windows: Potential for savings in energy, emissions and cost. *Building Services Engineering Research and Technology, 26*(3), 248–259.

Mercader-Moyano, P., & Ramirez-de-Arellano-Agudo, A. (2013). Selective classification and quantification model of construction and demolition waste from material resources consumed in residential building construction. *Waste Management & Research, 31*(5), 458–474.

Moffatt, S., & Kohler, N. (2008). Conceptualizing the built environment as a social-ecological system. *Building Research and Information, 36*(3), 248–268.

Moncaster, A. M., & Symons, K. E. (2013). A method and tool for cradle to grave embodied carbon and energy impacts of UK buildings in compliance with the new TC350 standard. *Energy and Buildings, 66*, 514–523.

Monier, V., Mudgal, S., Hestin, M., Trarieux, M., & Mimid, S. (2011). Management of construction and demolition waste – SRI: Final report task 2. European Commission (DG ENV), February.

Nadoushani, Z. M., & Akbarnezhad, A. (2015). Computational method for estimation of life cycle carbon footprint of buildings. In C. Blackman (Ed.), *Chapter 2, carbon footprinting*. Hauppauge: Nova Science Publishers.

Nagapan, S., Rahman, I. A., Asmi, A., & Adnan, N. F. (2013). Study of site's construction waste in Batu Pahat, Johor. *Procedia Engineering, 53*, 99–103.

Panos, K., & Danai, G. I. (2012). Survey regarding control and reduction of construction waste, PLEA 2012-28th Conference, Opportunities Limits & Needs Towards an Environmentally Responsible Architecture, Lima, Peru, 7–9 Nov, 7–9.

Poon, C. S., Yu, A. T. W., Wong, S. W., & Cheung, E. (2004). Management of construction waste in public housing projects in Hong Kong. *Construction Management and Economics, 22*(7), 675–689.

Pun, S. K., Liu, C., & Langston, C. (2006). Case study of demolition costs of residential buildings. *Construction Management and Economics, 24*(9), 967–976.

Randell, P., Pickin, J., & Grant, B. (2014). *Waste generation and resource recovery in Australia reporting period 2010/11*. Canberra: Department of Sustainability, Environment, Water, Population and Communities.

RICS. (2006). *Life expectancy of building components*. London: BCIS.

RICS. (2014). *Methodology to calculate embodied carbon* (1st ed.). London: RICS.

Scheuer, C., Keoleian, G. A., & Reppe, P. (2003). Life cycle energy and environmental performance of a new university building: Modeling challenges and design implications. *Energy and Buildings, 35*(10), 1049–1064.

Solís-Guzmán, J., Marrero, M., Montes-Delgado, M. V., & Ramírez-de-Arellano, A. A. (2009). Spanish model for quantification and management of construction waste. *Waste Management, 29*(9), 2542–2548.

Tam, V. W. Y., Shen, L. Y., & Tam, C. M. (2007). Assessing the levels of material wastage affected by sub-contracting relationships and project types with their correlations. *Building and Environment, 42*, 1471–1477.

Terry, A., & Moore, T. (2008). Waste and sustainable commercial buildings, your building, property council of Australia. http://www.yourbuilding.org/Search/Search.aspx?p=87&q=Waste+and+sustainable+commercial+buildings. Date of access 30/2/2017.

Thomas, D. N. (2015). The increase of timber use in residential construction in Australia: Towards a sustainable residential development model, PhD thesis, University of Technology Sydney, Australia.

Tingley, D. D., & Davison, B. (2011). Design for deconstruction and material reuse. *Proceedings: The Institution of Civil Engineers, 164*(EN4), 195–204.

US EPA. (2009). *Estimating 2003 building-related construction and demolition (C&D) materials amounts*. United States Environmental Protection Agency. http://www.epa.gov/cd-materials/estimating-2003-building-related-construction-and-demolition-materials-amounts. Date of access 23/2/3017.

US EPA. (2017). *Solid waste – Laws and regulations*. http://www3-epa.gov/region9/waste/solid/laws.html. Date of access: 23/2/2017.

Verbeeck, G., & Hens, H. (2010). Life cycle inventory of buildings: A calculation method. *Building and Environment, 45*, 1037–1041.

WFD. (2017). Directive 2008/98/EC of the European parliament and of the council of 19 November 2008 on waste and repealing certain directives. http://eur-lex.europa.eu/legal-content/EN/TXT/PDF/?uri=CEIEX:32008L0098&from=EN. Date of access: 9/1/2017.

Wonschik, C. R., Brennan, J., Ding, G., Heilmann, A., & Vessalas, K. (2014). Implications of legal frameworks on construction and demolition waste recycling – A comparative study of the German and Australian systems. In: *Proceedings: The 31st international symposium on automation and robotics in construction and mining* (pp. 523–530), Sydney, Australia, 9–11 July.

Wu, Z., Yu, A. T. W., Shen, L., & Liu, G. (2014). Quantifying construction and demolition waste: An analytical review. *Journal of Waste Management, 34*, 1683–1692.

Wu, H., Yuan, Z., Zhang, L., & Bi, J. (2012). Life cycle energy consumption and CO2 emission of an office building in China. *The International Journal of Life Cycle Assessment, 17*(2), 105–118.

Yeheyis, M., Hewage, K., Alam, S., Eskicioglu, C., & Sadiq, R. (2013). An overview of construction and demolition waste management in Canada: A lifecycle analysis approach to sustainability. *Clean Technology Environmental Policy, 15*, 81–91.

Yuan, H., Lu, W., & Hao, J. J. (2013). The evolution of construction waste sorting on-site. *Renewable and Sustainable Energy Reviews, 20*, 483–490.

Zhang, X., Shen, L., & Zhang, L. (2013). Life cycle assessment of the air emissions during building construction process: A case study in Hong Kong. *Renewable and Sustainable Energy Reviews, 17*, 160–169.

Chapter 11
Carbon and Cost Hotspots: An Embodied Carbon Management Approach During Early Stages of Design

Michele Victoria and Srinath Perera

Introduction

The need for embodied carbon mitigation mechanisms is well recognised by research scholars and government bodies of the many developed nations. The potential for mitigation is in fact high during early stages of designs. However, lack of standardised measurement procedures and a shortage of comprehensive embodied carbon data make embodied carbon estimating challenging and less attractive to industry practitioners, and hence, mitigation options are not much explored during early stages of design (Lockie 2012; Dixit et al. 2012; Victoria et al. 2015a). However, it has been proposed and proved that focusing on the design of the carbon significant building elements can lead to significant emissions reduction. These building elements are referred to as the 'carbon hotspots' (RICS 2014) that contribute up to 80% of the embodied carbon of buildings. Even though there is a general understanding about the carbon hotspots in buildings, there is a need for empirical research in this area (see, Perera and Victoria 2017). Further, it is also evident that carbon hotspots will vary depending on the type or the function of the building but these aspects are under-explored. Therefore, this chapter explores issues related to embodied carbon measurement and utilising empirical data analyses typical carbon hotspots of office buildings in the UK. In addition, carbon is considered as a currency of construction projects giving birth to the dual currency concept. Especially, with increasing awareness towards the dual currency of

M. Victoria
Scott Sutherland School of Architecture and Built Environment, Robert Gordon University, Aberdeen, UK

S. Perera (✉)
School of Computing, Engineering and Mathematics, Western Sydney University, Penrith, Australia
e-mail: srinath.perera@westernsydney.edu.au

© Springer International Publishing AG 2018
F. Pomponi et al. (eds.), *Embodied Carbon in Buildings*,
https://doi.org/10.1007/978-3-319-72796-7_11

construction projects (cost and carbon), the need to estimate, control and manage carbon alongside construction cost becomes fundamental for construction professionals and businesses to be sustainable (Perera and Victoria 2017; Victoria et al. 2015b). Further, Victoria et al. (2016) reported that there is a positive correlation between embodied carbon and capital cost of buildings. Accordingly, a concurrent analysis of carbon and cost will be more sensible and will add a different dimension to the discussion. Therefore, a comparison between carbon and cost critical elements is presented in the chapter and insights are drawn from the findings.

A Review of Studies on Embodied Carbon

Hammond and Jones (2011), the developers of the Inventory of Energy and Carbon (ICE), define embodied carbon as 'the sum of fuel related carbon emissions and process related carbon emissions'. As such, the carbon emitted during the manufacturing process of building materials attributable to fuel consumption and chemical processes involved contribute towards embodied carbon of that material. In contrary, some materials sequester the carbon in the atmosphere (e.g. wood). Hence, embodied carbon of materials is better considered as the net emissions related to the fuel consumption and processes related to the manufacturing. Embodied carbon can be categorised into mainly two types: 'initial embodied carbon' and 'recurring embodied carbon' (Chen et al. 2001; Ramesh et al. 2010). Initial embodied carbon is the emissions associated with raw material extraction, manufacturing, transport and construction, while recurring embodied carbon includes emissions during the use of the building such as repair, maintenance and replacement of building materials and equipment. In addition to those, there are two more types, namely, 'end-of-life embodied carbon' that includes emissions associated with the demolition of buildings and 'benefits beyond system boundary' (or end of life) which includes emissions saved as a result of reuse, recovery and recycling potential (as identified in TC350, BS EN 15978 standard). A cradle-to-cradle system boundary includes the embodied impact of all of the four identified types.

Royal Institution of Chartered Surveyors (RICS) (2014) defines 'carbon hotspots' as the carbon significant aspects of projects, which are not only carbon intensive but also easily measurable and with high reduction potential. This carbon significant aspect could be defined in terms of building elements or processes in the supply chain. In compliance with the RICS (2014) definition of carbon hotspots that it should be easily measurable, building elements were chosen as the focus of the study. Hence, in this chapter, carbon hotspots referred to as the carbon critical or significant building elements. Generally, floors (ground and upper floors), frame, external wall and roof are identified as carbon hotspots in building case studies (Clark 2013; Davies et al. 2014; Halcrow Yolles 2010). However, it was noticed that the element definition or classification differs from one study to the other (e.g. New Rules of Measurements (NRM), Standard Methods of Measurements (SMM)/Building Cost Information Services (BCIS) – older version, British Council of Offices 2011, some studies did not follow any standards) which makes the findings to be incomparable. Further, carbon hotspots may vary from one building to the other depending on the type or function of the building (Ashworth and Perera 2015). Hence, knowledge in this area is limited and yet to be developed.

Monahan and Powell (2011) highlighted the importance of identifying hotspots in buildings by modelling a two-storied residential building (in the UK) in three different scenarios – 'timber frame and larch cladding', 'timber frame and brick cladding' and 'conventional masonry cavity wall'. Substructure, external walls and roof were identified as the carbon hotspots in the timber frame and larch cladding building (elements responsible for 81% of embodied carbon, however, not all the building elements were included in the accounting), while the substructure alone (including foundation and ground floor) accounted for 50% of the embodied carbon. Further, the same building (timber frame with larch cladding) substituted with timber frame and brick cladding and conventional masonry resulted in additional embodied carbon of 32% and 51%, respectively. The major difference in embodied carbon was attributed to the difference in foundations and external walls. Hence, the findings of Monahan and Powell (2011) reveal substructure and external walls as carbon hotspots in this particular residential building and highlighted the potential embodied carbon reduction that could be achieved through the building design and selection of materials.

Similarly, Shafiq et al. (2015) studied a two-storied office building in Malaysia by modelling six different scenarios for structural composition using Building Information Model (BIM). However, Shafiq et al. (2015) used UK databases to estimate embodied carbon due to the absence of embodied carbon databases pertinent to Malaysia. Different grades or classes of concrete and steel were combined to generate different compositions, which resulted in different material quantities producing varying embodied carbon impacts. Only a few elements were studied including foundation, beams, slabs, columns and staircases, which can be related to the substructure, frame, upper floors and stairs as per NRM element definition. Shafiq et al. (2015) found that it was possible to reduce up to 31% of embodied carbon by using different classes of concrete and steel to meet the given design criteria. However, it should be noted that only the elements that constitute concrete and steel are considered in this study of Shafiq et al. (2015). Particularly, upper floors were identified as the key carbon hotspot followed by substructure, frame and stairs.

It is clear that embodied carbon studies in different types of buildings (Monahan and Powell 2011; Shafiq et al. 2015) have different focus and hence, limit the analyses to few elements. However, analysis of the whole building provides a holistic picture for the embodied carbon contribution of each element. This, in turn, will highlight the potential areas of embodied carbon reduction. Even though services account for 10–25% of total embodied carbon emissions, measuring services is complex and can be daunting due to absence of detail design information and relevant embodied carbon data. Hence, services are not widely recognised as a carbon hotspot (Hitchin 2013; RICS 2014). However, Cole and Kernan (1996) reported that services are to be one of the biggest components of recurring embodied carbon emissions considering the case of an office building in Canada. In fact, it was highlighted that in a 50-year life cycle recurring embodied carbon will be almost same as initial embodied carbon and for a longer life cycle, it would be greater than the initial embodied carbon. Hence, services cannot be disregarded in the carbon hotspot analysis.

Further, Hitchin (2013) investigated the building services element with respect to its embodied carbon contribution for a typical office building in London where space heating and air treatment and electrical installations were found to be contributing a significant proportion compared to other building services. However, only fundamental services have been considered in this analysis, and embodied carbon data on more sophisticated services (such as communication installations and building management systems) are lacking in the literature. Hence, reported studies on carbon hotspots of different types of buildings are scarce and at most incomplete resulting in a gap in the knowledge that needs to be fulfilled.

Identifying Carbon and Cost Hotspots

Identifying carbon and cost intensive elements or hotspots of buildings depends on the definition of hotspots itself. The Pareto principle was adopted to define the carbon (or cost) hotspots due to its popularity and applicability, especially in economics, business and management-related disciplines. Economist Vilfredo Pareto (1848–1943) found that 80% of the wealth of his country (Italy) was owned by 20% of the people. Then, Pareto applied the same theory to other states like Russia, France and Switzerland and found the same results. In 1940s, Joseph Juran (1904–2008), an American engineer, recognised the 80:20 theory and named it after Vilfredo Pareto. The Pareto principle defines that 80% of the results (or consequences) are attributable to 20% of the causes which implies an unequal relationship between the inputs and the outputs (Koch 2011; Delers 2015).

Munns and Al-Haimus (2000) highlighted that the seminal texts in the cost management literature (Ashworth and Perera 2015; Seeley 1996; Ashworth and Skitmore 1983) recognise the applicability of Pareto principle to identify cost significant items. The works of Munns and Al-Haimus (2000) and Tas and Yaman (2005) are examples of embracing the 80:20 Pareto principle to identify the cost significant items in a bill of quantities (BQ) and, eventually, developing prediction models using this cost significant modelling technique. Hence, it is evident that 80:20 Pareto principle is widely accepted as a popular method of capturing cost significant items in BQs. However, to identify cost significant items, the BQ items have to be grouped (to minimise complexity by reducing the number of items) according to the work packages (trades) or functional elements as exemplified in previous studies (Munns and Al-Haimus 2000; Tas and Yaman 2005). Of these two types, elemental grouping was adopted because BCIS data adheres to elemental grouping as prescribed in and New Rules of Measurements 1 (NRM1). (BCIS is an online cost database maintained by RICS and NRM1 is one

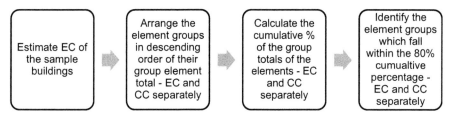

Fig. 11.1 The process of identifying embodied carbon (EC) and capital cost (CC) hotspots of the sample buildings

of the documents in the NRM suite of documents which is developed by RICS to provide guidance on cost management of construction projects.)

Accordingly, it can be hypothesised that 80% of the embodied carbon is emitted from 20% of the building elements. The building elements responsible for 80% of embodied carbon emissions are referred to as the carbon hotspots in this chapter. Even though 80:20 is accepted as the universal ratio, the Pareto principle dictates neither that the 80:20 ratio is applied to all situations nor should the two figures add up to 100 (say, it could be 80:30). Therefore, this 80:20 ratio was tested in the case of the relationship between embodied carbon of building elements. Therefore, to perform the carbon hotspot analysis, historical project data were obtained from BCIS online cost database (RICS 2016) and a sample of 41 office buildings was composed. Figure 11.1 illustrates the steps involved in identifying the hotspots of the sample. Accordingly, embodied carbon was estimated using the UK Building Blackbook (Franklin and Andrews 2011), ICE (Hammond and Jones 2011) and manufacturers' data for the sample buildings. Then, the sum total of embodied carbon and of each element group was estimated and the element groups were arranged in a descending order of their group totals. Cumulative percentage of the element group totals were calculated to identify the elements contributing up to 80% of the total embodied carbon and total capital cost separately, which are referred to as the carbon or cost significant elements or the hotspots of office buildings.

Analysis of Data

Building elements that are responsible for the 80% of the embodied carbon were identified for each building using the 80:20 Pareto principle as discussed in the Methodology section and presented in Table 11.1. Columns in the table present the building elements as per the NRM definition. All the buildings are listed in rows. A cell marked with an 'x' implies that the respective element in that building was identified as a 'carbon hotspot'. The bottom row of the table presents the probability of occurrence which indicates the probability of each element being identified as a carbon hotspot in the selected sample. For instance, the probability of frame was

Table 11.1 Carbon hotspot analysis

Building ID	1 Substructure	2A Frame	2B Upper Floors	2C Roof	2D Stairs	2E External Walls	2F External Windows and Doors	2G Internal Walls and Partitions	2H Internal Doors	3A Wall Finishes	3B Floor Finishes	3C Ceiling Finishes	4 Fittings and Furnishings	5 Services
#1	x		x	x		x		x						x
#2	x	x		x		x		x						x
#3	x	x		x		x								x
#4	x	x	x			x								x
#5	x	x	x			x								x
#6	x	x	x	x										x
#7	x	x	x	x		x								x
#8	x	x	x											x
#9	x	x	x	x		x								x
#10	x	x	x			x								x
#11	x		x	x		x					x			x
#12	x		x	x		x					x			x
#13	x		x	x		x					x			x
#14	x	x	x											x
#15	x	x	x	x		x								x
#16	x	x	x			x								x
#17	x		x	x		x						x		x
#18	x	x	x	x		x								x
#19	x	x	x					x						x
#20	x	x	x					x						x
#21	x	x	x					x						x
#22	x	x		x		x								x
#23	x	x	x	x										x
#24	x	x	x	x		x								x
#25	x	x	x									x		x
#26	x	x	x			x								x
#27	x	x	x			x								x
#28	x	x	x			x								x
#29	x	x	x			x								x
#30	x	x	x			x								x
#31	x	x	x			x								x
#32	x	x	x			x								x

(continued)

Table 11.1 (continued)

Building ID	1 Substructure	2A Frame	2B Upper Floors	2C Roof	2D Stairs	2E External Walls	2F External Windows and Doors	2G Internal Walls and Partitions	2H Internal Doors	3A Wall Finishes	3B Floor Finishes	3C Ceiling Finishes	4 Fittings and Furnishings	5 Services
#33	x	x	x			x								x
#34	x	x		x		x	x							x
#35	x	x	x	x										x
#36	x	x	x	x										x
#37	x	x	x	x										x
#38	x	x	x	x							x			x
#39	x	x	x			x								x
#40	x	x	x	x		x								x
#41	x	x		x		x				x				x
Probability of occurrence	1.0	0.9	0.9	0.5	0	0.7	0.02	0.1	0	0.02	0.1	0.05	0	1.0
Lead Positions														
Special Positions														
Remainder Positions														

0.9, which means, frame was identified as a carbon hotspot in 90% of the buildings in the sample (which is 36 out of 41). Based on the probability, building elements were classified into three categories such as:

1. Lead positions: elements that are found to be a carbon hotspot most of the buildings (probability of occurrence > 0.8)
2. Special positions: elements that are found to be a carbon hotspot in some of the buildings (0 < probability of occurrence < 0.8)
3. Remainder positions: elements that are not found to be a carbon hotspot in any of the building (probability of occurrence = 0)

Accordingly, building elements were grouped into one of the three 'hotspot category' based on the probability statistics and presented in Table 11.2.

In addition to the individual analysis of buildings, carbon hotspots for the whole sample were analysed and presented in Table 11.3 and Fig. 11.2. The embodied

Table 11.2 Mapping building elements against the carbon hotspot category

Carbon hotspot category	Building elements
Lead positions	Substructure, frame, upper floors, external walls and building services
Special positions	Roof, windows and external doors, internal walls and partitions, wall finishes, floor finishes and ceiling finishes
Remainder positions	Stairs, internal doors, fittings, furnishings and equipment

Table 11.3 Carbon hotspot analysis of the sample and the respective cost contribution

Elements (NRM Classification)	Avg. EC per GIFA ($kgCO_2/m^2$)	Average CC per GIFA ($£/m^2$)	Cumulative EC %	Cumulative CC %
1 Substructure	161	89	23.6	7.0
5 Services	145	419	44.9	39.6
2A Frame	100	102	59.6	47.6
2B Upper Floors	69	57	69.7	52.0
2E External Walls	60	159	78.5	64.5
2C Roof	43	91	84.8	71.6
3B Floor Finishes	26	75	88.6	77.4
2G Internal Walls and Partitions	23	39	92.0	80.5
3C Ceiling Finishes	19	36	94.8	83.3
2F External Windows and Doors	16	94	97.2	90.7
3A Wall Finishes	9	34	98.6	93.3
2D Stairs	8	27	99.7	95.4
2H Internal Doors	1	31	99.9	97.8
4 Fittings and Furnishings	1	28	100.0	100.0

carbon and capital cost values presented in Table 11.3 are the average values of building elements of the sample. Accordingly, substructure, services, frame, upper floors, external walls and roof were identified as the carbon hotspots of the sample. Further, it is interesting to see that all the building elements categorised as 'lead positions' have been identified as carbon hotspots (see, Tables 11.2 and 11.3). On the other hand, it was also noticed that the identified carbon hotspots accounted for 72% of the capital cost.

A similar analysis was repeated for cost hotspots. Table 11.4 presents the building elements grouped according to the hotspot category. Four elements including substructure, frame, external walls and services capture 'lead positions'

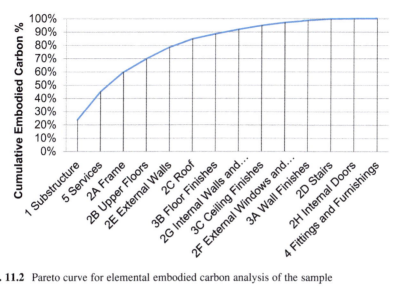

Fig. 11.2 Pareto curve for elemental embodied carbon analysis of the sample

Table 11.4 Mapping building elements against the cost hotspot category

Cost hotspot category	Building elements
Lead positions	Substructure, frame, roof, external walls, windows and external doors and services
Special positions	Upper floors, stairs, internal walls and partitions, internal doors, wall finishes, floor finishes, ceiling finishes, fittings, furnishings and equipment
Remainder positions	None

in both cost and carbon hotspot category. The rest of the elements were identified as 'special positions' leaving none of the elements under the 'remainder position' category. This showcases the hotspot sensitivity to building designs. Services, external walls, frame, external windows and doors, roof, substructure and floor finishes were identified as the cost hotspots of the sample (see, Table 11.5). Similar to the findings of carbon hotspots, all the identified 'lead positions' were found to be cost hotspots. Interestingly, these cost hotspots are also accounted for 80% of the embodied carbon.

Based on the hotspot analysis, cost and carbon hotspots were mapped as shown in Table 11.6 to gain a better understanding and infer relationships between capital cost and embodied carbon. Accordingly, substructure, frame, external walls and building services were found to be leading carbon and cost hotspots. Roof and windows and external doors were identified as leading cost hotspots and special position carbon hotspots. Similarly, upper floor element was identified as a leading carbon hotspot and a special position cost hotspot. Internal walls and partitions, wall finishes, floor finishes and ceiling finishes were found to be special position

Table 11.5 Cost hotspot analysis of the sample

Elements (NRM Classification)	Avg. CC (£/m² GIFA)	Avg. EC (kgCO₂/m² GIFA)	Cumulative CC %	Cumulative EC %
5 Services	419	145	32.7	21.3
2E External Walls	159	60	45.1	30.1
2A Frame	102	100	53.1	44.8
2F External Windows and Doors	94	16	60.4	47.1
2C Roof	91	43	67.5	53.4
1 Substructure	89	161	74.5	77.0
3B Floor Finishes	75	26	80.3	80.8
2B Upper Floors	57	69	84.7	90.9
2G Internal Walls and Partitions	39	23	87.8	94.3
3C Ceiling Finishes	36	19	90.7	97.2
3A Wall Finishes	34	9	93.3	98.6
2H Internal Doors	31	1	95.7	98.8
4 Fittings and Furnishings	28	1	97.9	98.9
2D Stairs	27	8	100.0	100.0

Table 11.6 Mapping carbon hotspots and cost hotspots against the hotspot category

Lead Positions	Special Positions	Remainder Positions
Upper Floors Substructure	Roof Windows and External Doors	Stairs Internal Doors
Frame	Internal Walls and Partitions	Fittings, Furnishing and Equipment
External Walls Services	Wall Finishes Floor Finishes Ceiling Finishes	
Roof Windows and External Doors	Upper Floors Stairs	**CARBON HOTSPOTS**
	Internal Doors Fittings, Furnishing and Equipment	**COST HOTSPOTS**

Table 11.7 Carbon and cost hotspots of the sample office building

Level of significance	Carbon hotspots	Cost hotspots
1	1 substructure	5 services
2	5 services	2E external walls
3	2A frame	2A frame
4	2B upper floors	2F external windows and doors
5	2E external walls	2C roof
6	2C roof	1 substructure
7		3B floor finishes

carbon and cost hotspots. Further, stairs, internal doors and fittings, furnishing and equipment were identified as the remainder position carbon hotspots and special position cost hotspots.

The summary of carbon and cost hotspot analysis of the sample office buildings is presented in Table 11.7. Accordingly, substructure was identified as the most significant carbon hotspot, while it is less significant as a cost hotspot. Services were identified as the most cost significant hotspot and the second most carbon significant hotspot. Frame was identified as the third most significant carbon and cost hotspot. Interestingly, upper floors were not found as cost significant, while it was identified as the fourth carbon significant hotspot. External walls were identified as the second most cost significant building element, while carbon significance of external walls was found to be low. Roof was identified as the least carbon significant hotspot, while it was found to be more cost significant than the substructure. Floor finishes were identified as the least cost significant hotspot, though, it was not found as a carbon hotspot.

Further, the concept of cost and carbon hotspot emerged from the Pareto principle, which suggests that 80% of the embodied carbon (or cost) is attributable to 20% of the elements. However, 80:20 ratio is not supported in this case. The findings suggest that 80% of the embodied carbon emissions are caused by 43% of the elements (6 of the 14 elements) and 80% of the building cost is spent on 50% of the elements (7 of the 14 elements) on average.

It can also be noticed that the level of significance of each element in terms of carbon and cost varies even though most of the elements identified as hotspots were found to be both carbon and cost hotspots (see, Table 11.7). Therefore, achieving optimisation between carbon and cost is not simple as it first appeared. For instance, efforts to minimise EC in the substructure might lead to increases in the frame cost as a result of the need to add wind bracings to the frame which might offset the cost savings achieved in the substructure. Hence, the findings highlight the need for in-depth case studies of buildings, exploring different design options and studying the change in EC and CC for alternative design options. It would require a platform for quick modelling of the impact of EC and CC on building design. This will inform the impact of different specifications on the EC and CC of the building and when an optimum point can be achieved. For instance, Fig. 11.3 presents average EC and CC values for three different types of foundation in sample buildings.

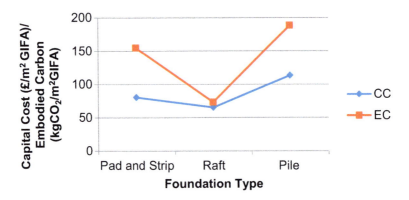

Fig. 11.3 EC and CC for different foundation types

Accordingly, both EC and CC values (mean values) are found to be the lowest in raft foundations and the highest in pile foundations. This means that the choice of raft foundations will reduce both EC and CC. Similar graphs can be produced for other building elements and that can lead to informed decisions made by the designers. Especially, with this kind of knowledge, cost and carbon reconciliation can be exercised by compromising the design of elements which do not produce significant savings in cost or carbon and focusing on the design of the most cost and carbon hotspots.

Implications

It is clear from the above analysis that there is an urgent need to understand the relationship between carbon and costs in building elements as the optimum balance of these two can be fine-tuned through design variations. Therefore, a greater focus on elements falling under the category of 'special position' can play an important role in influencing carbon and cost accountability of the building. Some building elements such as substructure and frame had minimal design options in the sample, while elements such as roof and internal finishes were found to have many alternative specification options. In fact, it was noticed that the elements classified as special position had a range of specification options in the sample compared to elements classified as lead position, highlighting the significance of choices of the specification for the elements classified as special position. Therefore, further studies and detailed analysis of the impacts of different choices of design in each element will open new avenues for achieving cost and carbon reduction through building designs.

For instance, Tables 11.8 and 11.9 present two design options for floor finishes in a particular building. Design option A proposes a combination of vinyl sheet, ceramic tiles (to toilet area) and raised access floors with carpet tiles on top; design

Table 11.8 Floor finishes – design option A

Floor finishes	Qty	Unit	CC	EC	Total cost	Total carbon
Vinyl sheet	797	m^2	28.71	7.69	22,896.65	6130.41
Ceramic tiles	399	m^2	84.14	15.37	33,547.42	6127.22
Carpet tiles	2791	m^2	26.69	10.45	74,484.77	29,159.32
Raised access floor	2791	m^2	28.30	25.03	78,982.47	69,847.85
					209,911.31	111,264.81
	3987	m^2		Per m^2	52.65	27.91

Table 11.9 Floor finishes – design option B

Floor finishes	Qty	Unit	CC	EC	Total cost	Total carbon
Ceramic tiles	399	m^2	84.14	15.37	33,547.42	6127.22
Carpet tiles	3588	m^2	26.69	10.45	95,766.13	37,490.56
Raised access floor	3588	m^2	28.30	25.03	101,862.44	89,804.38
					230,862.44	133,422.16
	3987	m^2		Per m^2	57.90	33.46

option B replaces the area covered by vinyl sheet with raised access floor with carpet tiles. Design option B is certainly expensive and has higher embodied carbon than option A. Hence, replacing vinyl sheet with access floor finished with carpet tiles has increased the rates of CC and EC by approximately 100% and 400%, respectively, which eventually resulted in an increase of 10% of CC per m2 EUQ (element unit quantity – this varies for elements, and measurement rules are defined in NRM compliant BCIS and 20% of EC per m2EUQ (which implies 10% and 20% increase in total CC and EC of the building, respectively). Therefore, what-if analysis (simulations) can be run during the detailed design stages and designers can choose the most efficient design option if this type of analysis is entertained by construction professionals and practices.

Further, the analysis of the whole sample gives a different insight. It was found that substructure, services, frame, upper floors, external walls and roof were the most carbon significant building elements (in descending order) which also contribute up to 72% of the CC of the buildings. On the other hand, services, external walls, frame, external windows and doors, roof, substructure, and floor finishes (in descending order) were found to be the most cost significant elements which contribute up to 81% of the EC of the building. This finding implies that tackling carbon hotspots also means tackling the building elements that are responsible for 72% of the cost in general. Similarly, tackling the identified cost hotspots implies tackling the elements accountable for 81% of the EC of the building. In comparison, treating cost hotspots seems to be a better option than treating carbon hotspot as it includes both the elements responsible for 80% of the CC and EC. Given that the list of cost hotspots includes all of the identified carbon hotspots except upper floors.

Conclusions

It cannot be denied that carbon is becoming an important and significant metric of sustainability in the built environment. Apart from that, carbon is also being recognised as another currency of construction projects; thus, carbon and cost evaluations in construction projects in the commercial sector are becoming commonplace and referred to as dual currency evaluations. This necessitates the management of carbon in the built environment to subdue the impacts of climate change. Accordingly, embodied carbon analysis of buildings is gaining momentum since the introduction and implementation of policy towards zero (operational) carbon buildings. Embodied carbon is highly affected by the processes involved in the building material and component manufacturing. Developing benchmarks for the embodied carbon of materials is highly challenging. The ICE inventory of carbon developed by Hammond and Jones (2011) is considered as the fundamental building block of most databases currently in use. In fact, it can be said that most of the research on embodied carbon in the UK are reliant on the ICE and this research is no exception. However, the main data source used in this research is the UK Building Blackbook which was developed using the ICE and other industry data. The UK Building Blackbook was used mainly because it presents embodied carbon data in compliance with BQ format which simplifies the estimating process.

The literature revealed that the building elements that contribute up to 80% of the embodied carbon are regarded as the hotspots of a particular type of building. These hotspots vary for different types of buildings due to different element intensities. The findings of the research suggest that substructure, frame, upper floors, external walls, roof and services (elements as per NRM classification) were identified as the carbon hotspots of office buildings that are responsible for 80% of the embodied carbon. Generally, accepted norm is that 20% of the causes are attributable to 80% of the effects though in these cases the general norm of the Pareto principle was not supported. The Pareto ratio in this context was found to be 80:43 where 43% of the elements are responsible for 80% of the embodied emissions.

Further, substructure, frame, upper floors, external walls and services were always or mostly found to be carbon hotspots which are named as 'lead positions'. On the other hand, roof, windows and external doors, internal walls and partitions, wall finishes, floor finishes and ceiling finishes were identified as carbon hotspots in some of the buildings and identified as 'special positions'. The remaining elements (stairs, internal doors, fittings, furnishings and equipment) were not found as hotspots in any of the buildings evaluated, hence, named as 'remainder positions'. This suggests that the element categories that need more focus on design are 'lead positions' and 'special positions' to achieve maximum reduction in embodied carbon. On the other hand, elements falling under the 'remainder position' category can be disregarded when exploring low carbon options as the influence of these elements on total embodied carbon is almost negligible.

Further, it was also found that frame, external walls and services are leading position carbon and cost hotspots, while wall, floor and ceiling finishes are special position carbon and cost hotspots. This implies that it is possible to reduce capital cost and embodied carbon at the same time by focusing on these elements. Moreover, tackling the carbon hotspots also means dealing with the elements responsible for 72% of the capital cost and tackling cost hotspots means dealing with the elements that are responsible for 81% of embodied carbon. This implies that both cost and carbon, which are the dual currency of construction projects, can be managed by focusing on the carbon or cost critical elements or the hotspots. These findings inform designers of the cost and carbon-intensive sources of buildings in terms of building elements and showcase the elements in which the most capital cost and embodied carbon are locked in, which can be omitted by wisely choosing materials and designing material efficiently during the detailed design stages.

References

Ashworth, A., & Perera, S. (2015). *Cost studies of buildings*. Oxon: Routledge.

Ashworth, A., & Skitmore, R. M. (1983). *Accuracy in estimating, CIOB Occasional Paper No. 27*. London: CIOB.

Chen, T., Burnett, J., & Chau, C. (2001). Analysis of embodied energy use in the residential building of Hong Kong. *Energy, 26*, 323–340.

Clark, D. H. (2013). *What colour is your building?: Measuring and reducing the energy and carbon footprint of buildings*. UK: RIBA Enterprises Limited.

Cole, R. J., & Kernan, P. C. (1996). Life-cycle energy use in office buildings. *Building and Environment, 31*, 307–317.

Davies, P. J., Emmitt, S., & Firth, S. K. (2014). Challenges for capturing and assessing initial embodied energy: A contractor's perspective. *Construction Management and Economics, 32*, 290–308.

Delers, A. (2015). Pareto's principle: Expand your business!, 50Minutes.com

Dixit, M. K., Fernández-Solís, J. L., Lavy, S., & Culp, C. H. (2012). Need for an embodied energy measurement protocol for buildings: A review paper. *Renewable and Sustainable Energy Reviews, 16*, 3730–3743.

Franklin, & Andrews. (2011). *Hutchins UK building blackbook: The cost and carbon guide: Hutchins' 2011: Small and major works*. Croydon: Franklin & Andrews.

Halcrow Yolles. (2010). *Sustainable offices – Embodied carbon*. UK: South West Regional Development Agency.

Hammond, G., & Jones, C. (2011). *A BSRIA guide embodied carbon the inventory of carbon and energy (ICE)*. UK: BSRIA.

Hitchin, R. (2013). *CIBSE research report 9: Embodied carbon and building services*. UK: CIBSE.

Koch, R. (2011). *The 80/20 principle*. London: Nicholas Brealey Publishing.

Lockie, S. (2012). Embodied carbon – New guidelines from the RICS [Online]. Available: http://www.fgould.com/uk-europe/articles/embodied-carbon-guidelines/. Accessed 6 Jan 2014.

Monahan, J., & Powell, J. C. (2011). An embodied carbon and energy analysis of modern methods of construction in housing: A case study using a lifecycle assessment framework. *Energy and Buildings, 43*, 179–188.

Munns, A. K., & Al-Haimus, K. M. (2000). Estimating using cost significant global cost models. *Construction Management and Economics, 18*, 575–585.

Perera, S., & Victoria, M. (2017). The role of carbon in sustainable development. In P. Lombardi, G. Q. Shen, & P. S. Brandon (Eds.), *Future challenges for sustainable development within the built environment*. UK: Wiley's Publication.

Ramesh, T., Prakash, R., & Shukla, K. K. (2010). Life cycle energy analysis of buildings: An overview. *Energy and Buildings, 42*, 1592–1600.

RICS. (2014). *Methodology to calculate embodied carbon* (1st ed.). UK: RICS.

RICS. (2016). *Building cost information services (BCIS)*. UK: RICS.

Seeley, I. H. (1996). *Building economics: Appraisal and control of building design cost and efficiency*. Basingstoke: Macmillan.

Shafiq, N., Nurrudin, M. F., Gardezi, S. S. S., & Kamaruzzaman, A. B. (2015). Carbon footprint assessment of a typical low rise office building in Malaysia using building information modelling (BIM). *International Journal of Sustainable Building Technology and Urban Development, 6*(3), 1–16.

Tas, E., & Yaman, H. (2005). A building cost estimation model based on cost significant work packages. *Engineering, Construction and Architectural Management, 12*, 251–263.

Victoria, M., Perera, S., & Davies, A. (2015a). Developing an early design stage embodied carbon prediction model: A case study. In A. B. Raidén & E. Aboagye-Nimo (Eds.), *31st annual ARCOM conference, 7–9 September 2015*. Lincoln: Association of Researchers in Construction Management.

Victoria, M. F., Perera, S., Zhou, L., & Davies, A. (2015b). *Estimating embodied carbon: A dual currency approach. Sustainable buildings and structures*. UK: CRC Press.

Victoria, M., Perea, S., & Davies, A. (2016). *Design economics for dual currency management in construction projects*. RICS COBRA 2016. Retrieved from http://www.rics.org/lk/knowledge/research/conference-papers/design-economics-for-dual-currency-managementin-construction-projects/

Part III
Mitigation

Chapter 12
Applying Circular Economic Principles to Reduce Embodied Carbon

Danielle Densley Tingley, Jannik Giesekam, and Simone Cooper-Searle

Introduction

Through a series of case studies, this chapter explores the connections between the circular economy and the reduction of embodied carbon. Circular economic principles focus on maintaining the value of materials for as long as possible. A circular economy seeks to keep materials in circulation, removing the concept of waste from the system and the need for material extraction from primary sources. In a completely circular economy, all 'waste' outputs would equal system inputs; in Fig. 12.1 this means $M_{\mathrm{Eol}} + M_{\mathrm{w}} = M_{\mathrm{c}} + M_{\mathrm{m}}$. If the built environment is thought about in this way, as a system, then the inputs are construction materials; these materials accumulate in buildings, which can also be thought of as the stock. Demolition waste is the output flow of materials in this system.

The concept of a circular economy can also be extended to embodied carbon. Construction materials are input flows of embodied carbon. These emissions are new to the system. If demand for new buildings is reduced through retrofit and adaptation of buildings already in use, the input flow of embodied emissions would fall, and existing, already expended, embodied carbon would remain in stock. Another option to reduce flows of new embodied emissions is to channel demolition waste into material inputs. Steel sections, for example, can be extracted from a building at the end of its life and reused. This requires a shift in approach to building

D. Densley Tingley (✉)
Civil and Structural Engineering Department, University of Sheffield, Sheffield, UK
e-mail: d.densleytingley@sheffield.ac.uk

J. Giesekam
Sustainability Research Institute, School of Earth and Environment, University of Leeds, Leeds, UK

S. Cooper-Searle
Department of Engineering, University of Cambridge, Cambridge, UK

© Springer International Publishing AG 2018
F. Pomponi et al. (eds.), *Embodied Carbon in Buildings*,
https://doi.org/10.1007/978-3-319-72796-7_12

265

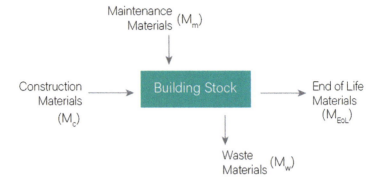

Fig. 12.1 The built environment as a system of stocks and flows

deconstruction and materials salvage. Some end-of-life buildings will be more suited to this approach than others, depending on the material composition and construction methods. Adapting the existing building stock and salvaging material at the end of a building's life will reduce, but not eliminate, the demand for new materials and the addition of embodied carbon to the building stock. Where new materials are required, strategies such as design for adaptability and deconstruction should facilitate longer building lifetimes and component and material reuse at end of life. This should enable a reduction in cradle-to-cradle embodied carbon. The following section outlines the state of the art in the area.

State of the Art

A review of the key literature on the circular economy in construction in general terms is included here, as well as an introduction and summary of some key literature for each of the four circular economic design strategies focused on in this chapter, namely, building reuse, material reuse, design for deconstruction and material reuse and design for adaptability.

The Circular Economy in Construction

Much of the literature to date that specifically focuses on the circular economy in construction or specifically buildings are industry-based reports rather than academic literature. This section provides a snapshot of some of the key documents in the area. The Ellen MacArthur Foundation has been key in increasing the profile of the circular economy, through their work to promote and progress the transition to a circular economy. Much of their work is on the benefits of the circular economy across a broad range of sectors. However, they have also compiled a report (CE100

2016), which documents a series of case studies from its CE100 network where the circular economy has been applied in the built environment; some of these case studies are explored in the case study analysis section of this chapter. Usefully this report also discusses the ReSOLVE framework in the context of the built environment. The framework sets out six strategies, which circular economic approaches can be gathered under; these are as follows:

1. *Regenerate*: which includes examples in the built environment of renewable energy, building on brownfield sites and resource recovery
2. *Share*: with examples of co-housing and office-sharing
3. *Optimise*: compact urban growth, energy and material efficiency
4. *Loop*: making use of end-of-life buildings/materials through repair, remanufacturing, upgrade and reuse
5. *Virtualise*: teleworking and smart appliances
6. *Exchange*: better performing materials and technologies (CE100 2016)

These six strategies won't necessarily reduce the embodied carbon of buildings in the design sense as some of them instead look to replace the need for additional buildings through more strategic use of existing space, e.g. teleworking. This approach would, however, reduce the embodied carbon of the broader built environment if less new buildings were required. The reduction of operational carbon is also targeted through some of the strategies, such as regenerate and virtualise.

A report on the potential of the circular economy in Dutch construction highlights four important conditions for a change to a more circular construction industry; these are as follows: minimising operational impacts, reuse of existing buildings/infrastructure, design of new buildings for circularity (including future deconstruction, adaptability and reuse) and selection of circular materials (durable, strong, light, non-toxic) (van Odijk and van Boven 2014). There are not only clear parallels with some of the strategies that will be explored in this chapter (points 2 and 3 particularly), but these ideas will also bring about reduced cradle-to-cradle embodied carbon through longer product lifetimes. The book by Cheshire (2016) explores some of these principles and the wider role of the circular economy in construction, weaving case studies throughout to demonstrate how principles can be integrated in practice. Cheshire also suggests alternative models which could be applied for additional value capture; this will likely be an important component if the circular economy is to be adopted at scale. Arup (2016) have also coalesced their ideas on the role of the circular economy in the built environment, discussing different strategies, which they categorise under the Ellen MacArthur Foundation's ReSOLVE framework. They also explore and discuss the opportunities, challenges and enablers across scales in the built environment from buildings to cities to the global scale. Another useful contribution of this report is the overlaying of the identified strategies with a layered building approach (following Brand 1994), meaning opportunities are identified across stuff, space, services, skin, structure, site and the wider system. Pomponi and Moncaster (2017a) consider the framing of the circular economy in the built environment, particularly exploring the literature that has helped to form this area. They propose a framework for circular economy research in the built environment, which considers six different dimensions: governmental, economic, environmental, behavioural, societal and technological. This brief review of literature on the circular economy in the built environment demonstrates that there are several strategies adhering to circular economic principles that would also bring about reductions in embodied carbon.

Building Reuse

Building reuse can be challenging to identify, in that buildings are frequently retrofitted to varying degrees, and there is potential for the terms building reuse and building retrofit to be used interchangeably. This chapter considers building reuse to be when a design team is presented with a site containing an existing building, and a decision is taken to reuse either all or part of the building, as opposed to demolishing the building and starting anew. Assefa and Ambler (2017) explore this, quantifying the potential life cycle impacts of repurposing and building system reuse compared to a demolition and rebuild scenario. They show an avoided impact of 33% global warming potential, purely in embodied carbon as the in-use impacts were not considered. However, in the broader context of building reuse, it is important that retrofit and improvement of the building fabric are integrated to ensure that operational carbon is also minimised. Andrews et al. (2017) debate the difficulties of energy efficient adaptive reuse of buildings, particularly considering the difficulties a change in use presents and the impacts this has on building service requirements. They suggest that regulation in the USA is a particular problem, and one potential solution would be an energy efficiency performance path rather than specific regulation in the case of building reuse. Laefer and Manke (2008) advocate building reuse, considering a number of benefits including cost savings, shorter construction times, waste management, minimised site disruptions and energy and material savings. The latter will likely result in embodied carbon savings. They propose an assessment method to apply to projects to consider the potential for reuse of an existing building's structure. In a similar vein, Langston et al. (2008) propose a framework in which to identify the adaptive reuse potential of different buildings; this considers the current age of the building as well as the predicted physical lifespan and then requires an assessment of a range of possible reasons for obsolescence, producing an index of reuse potential. The potential comparative nature of this approach would make it useful for owners with a significant property portfolio, or possibly local authorities, but it would likely be less useful for those involved in building design.

Material Reuse

Material reuse involves salvaging materials or components during retrofits or demolition. Architectural salvage occurs particularly for period pieces, although the potential for material and component reuse is much greater and could extend from floorboards and doors to structural elements such as steel or timber beams. Chau et al. (2015) have demonstrated the energy benefits of reuse at end of life, demonstrating in the modelling of a concrete frame, high-rise building that a maximum reuse scenario gives a 38.5% potential energy saving, which was higher than recycling. This, of course, translates to an embodied carbon saving as well. Pongiglione and Calderini (2014) specifically investigate the embodied carbon benefits of reusing structural steel, demonstrating a 138 tCO2e saving in the theoretical redevelopment of a train station in Italy, suggesting that 30% of the new steel could be replaced with reused steel sourced from a nearby building. This study again demonstrates the clear embodied carbon benefit of reusing materials. Inherently, if material reuse can displace the need for new material processing, then there will be an associated energy and greenhouse gas emission saving. It thus also follows that those materials with high-embodied carbon should be particularly targeted for reuse, as displacing the need for new materials of these types will yield a greater benefit. This is potentially best applied to durable materials such as steel and aluminium as they should have long enough lifespans to benefit from reuse, although they will also need to be utilised in forms where they can be salvaged with minimal damage for reuse.

There are, however, significant barriers to material reuse, particularly of structural materials; these have been explored in work by Addis and Schouten (2004), Densley Tingley and Davison (2011), Hosseini et al. (2015) and Densley Tingley et al. (2017). Commonly occurring and significant barriers include the potential cost of reuse, supply chain gaps or lack of integration (i.e. where do you source reused materials from), availability of reused materials, traceability (this is particularly an issue for structural materials where for future design, it is useful to know material properties such as steel grade) and a lack of client demand. The latter might be overcome if reduced embodied carbon becomes a client driver as reuse of materials would be an effective way to achieve this.

Design for Deconstruction and Material Reuse

Design for deconstruction and material reuse aims to facilitate future non-destructive dismantling of buildings to facilitate future reuse. Some of the barriers to this approach overlap with those outlined in the preceding section on material reuse, but another significant one is the uncertainty of the future – it is difficult to encourage clients and designers to incorporate a strategy which will see its main benefits in the future. However, work by Densley Tingley and Allwood

(2015) has shown that co-benefits such as faster construction times and greater flexibility in use can be better drivers for design for deconstruction than the environmental benefits of reuse and cradle-to-cradle embodied carbon savings. Design guides for approaches to design for deconstruction have been written by Addis & Schouten (2004) Morgan & Stevenson (2005), and Guy and Ciarimboli (n.d.), highlighting key areas such as reversible, accessible connections, non-composite design, deconstruction plans and material inventories.

Densley Tingley and Davison (2012) link design for deconstruction to embodied impacts, proposing a method to account for the embodied carbon benefits of design for deconstruction, whereby the impact is shared over the number of predicted lives. Module D of BS EN 15978:2011 Sustainability of Construction Works – Assessment of Environmental Performance of Buildings – Calculation Method would also allow for an estimate of a future benefit from reuse. Densley Tingley (2013) applies this method to series of case studies demonstrating embodied carbon savings, particularly for steel- and timber-framed structures. These construction types in particular lend themselves to reversible connections.

Delivering future reuse may require a fundamental change in perception from buildings as 'waste in waiting' (Giesekam et al. 2016) to valuable 'material banks'. This idea is being explored by the ongoing BAMB ('Buildings as Material Banks') project which is attempting to integrate 'material passports' with reversible building design to optimise circular industrial value chains. Luscuere (2017) explores how these material passports can retain critical context-specific information about material health, deinstallation, disassembly, location and reverse logistics to support greater reuse at end of life. Retaining such information could support stakeholders throughout the building value chain in preserving the economic value and embodied carbon of these stocked assets.

Design for Adaptability

Pinder et al. (2017) explore what practitioners in the construction industry perceive adaptability in buildings to mean, conducting a series of unstructured face-to-face interviews. They conclude that interpretation of adaptability and indeed the vocabulary associated with adaptability is varied and suggest that this lack of clear articulation could be a barrier to the development of adaptable buildings. However, the terms 'flexible' and 'resilient' are mentioned, and more generally the discussions allude to a building's ability to change over time. Manewa et al. (2013) discuss the role adaptable buildings play in creating a more sustainable future and explore design strategies, categorising these into themes: convertible (functions), scalable, moveable (location), adjustable (task), versatile (space) and refitable. This work demonstrates the broad nature of the term adaptable buildings.

For the purposes of this chapter, design for adaptability is taken to mean the design intention to increase the ability of buildings to change over time to suit different needs. This could be as simple as clear spans and increased floor to ceiling heights to accommodate internal reconfigurations and provision of additional services or could include sizing columns to allow for future vertical expansion of the building or increasing the design criteria for imposed loading to allow for different uses in the future. These options would likely increase the initial embodied carbon as hypothetically they will require additional material use; however, if they increase the lifespan of the building, the cradle-to-cradle embodied carbon could be reduced. There has however been little work to date which quantifies any increase in initial embodied carbon of buildings designed for adaptability and considers the potential lifetime extension that would be required to counterbalance this.

State-of-the-Art Conclusions

Of the four strategies being explored, only design for adaptability seems to have little work to date that connects this strategy with any consideration of embodied carbon, and this is an area for future work. The literature on building reuse, material reuse and design for deconstruction all demonstrate an embodied carbon saving on particular case studies. However, work on how this has been achieved in practice and project drivers has been limited. This additional knowledge would be useful in practice in order to increase uptake of circular economic strategies in the built environment.

Research Design

A series of case studies were selected to illustrate the respective strategies across a range of structure types. Each case study is used to provide practical insights on project processes, drivers and the perceived benefits and challenges of adopting circular economic approaches. These insights are drawn from semi-structured interviews with members of each design team, supplemented by supporting literature. Each case study briefly sets out the key characteristics of the project, the solutions delivered and the perceived impact on cradle-to-cradle embodied carbon. Although an attempt to precisely enumerate and compare these benefits across case studies would appear to be informative, in practice ensuring comparability of data and assumptions is challenging. As noted by De Wolf et al. (2017), Giesekam et al. (2016) and Pomponi and Moncaster (2017b), there are still numerous inconsistencies in the underlying product data and system boundaries adopted by different carbon assessors. Furthermore, although BS EN 15978:2011 Sustainability of Construction Works – Assessment of Environmental Performance of Building – Calculation Method defines a common set of life cycle stages and approaches, it does not prescribe many key parameters, such as expected lifespan, that are integral to ensuring comparability between assessments. Commonly used complementary guidance documents, such as RICS (2014), also offer minimal practical guidance on accounting for the benefits of adaptability and reuse in practice, meaning industry assessments often ignore these issues. Future guidance documents, such as the upcoming RICS professional statement 'Whole life carbon measurement: implementation in the built environment' will seek to standardise some aspects of these approaches. Without extensive retrospective gathering of comparable data, it is thus not possible to compare the strategies numerically. In the meantime, the authors have chosen to forgo quantitative evaluation of the embodied carbon benefits, in favour of qualitative statements about the life cycle stages affected.

Case Study Analysis

This section provides a series of 'good practice' case studies for each of the four strategies being discussed, exploring practical implementation of the strategy and discussing the potential scale of cradle-to-cradle embodied carbon saving. Where possible, project drivers are explored to understand if and why circular economy principles drove the approach or if other co-benefits were the primary driver, e.g. embodied carbon reduction or cost reduction. Any enabling conditions are also discussed in order to understand the key criteria for successful inclusion of these strategies in projects.

Building Reuse

A prominent example of building reuse is Quadrant 3, also known as Air W1, a mixed use complex which links Piccadilly Circus to Regent Street. Preservation of the architectural heritage was an important project driver with the building being located within a conservation area. However, with sustainability at the heart of this project for the Crown Estate, embodied carbon reduction was also targeted. Building reuse was part of the embodied carbon strategy, with over 1500 tonnes of existing steel retained in the structure. The original 1919 construction was steel frame, with the frame embedded within the facade. Three sections of the original neo-classical facades were retained and restored; the steelwork columns and beams that were attached to these facades were retained and new connections welded on so that the old steel could be connected to new steelwork. A portion of floor slab was also retained; again the steel was retained in its original form, but corrosion protection was added to certain required locations. The retention of the original facades was a key driver for the structural building reuse; certain areas in the building are Grade 2 listed; thus, an option to demolish the whole building and start again was rejected for this scheme which carefully contrasts the aesthetics of the neo-classical facades with contemporary facades. As the steelwork was embedded and supported these retained facades, there was a strong practical benefit in retaining it, beyond the environmental benefits. There were, however, also challenges when reusing the retained steelwork. In order to ensure the steelwork was of sufficient quality, thousands of 50 pence piece-sized samples were taken and tested to verify the material properties. Corrosion and lead-based paint coating also provided additional challenges. In the case of the former, certain extensively corroded elements had to be removed and replaced, and in the case of the latter, additional health and safety precautions had to be taken on-site when welding to the original steelwork. In addition, art deco interiors were restored for restaurant use, and granite kerbs and slates from the existing building were salvaged and donated to a community garden project. In total, over 1500 tonnes of embodied carbon were saved on the project, which also utilised cement replacements, designed outwaste and achieved a 20% recycled content in materials (by value) (Crown Estate 2017). As discussed in the state-of-the-art section, operational carbon is also an important consideration in building reuse projects, and this too was targeted in this project which achieved a BREEAM excellent rating (Crown Estate 2017).

There is an excellent example of the reuse of an entire structural system in a pedestrian bridge, which was on its fifth use at the time of interview with the contractor. The 30–40 tonne steel bridge was originally used by Colchester Garrison as a temporary structure to enable troop movement. By chance when it was being disassembled, Sir Robert McAlpines obtained it for reuse on the Olympic site. With original drawings, some adaptation and the addition of another tonne of steel, the bridge was redeployed in Stratford to enable pedestrians to cross a road during construction, saving time. It was later moved within the site for use during the Olympics to facilitate access to ceremonies, before then moving back to Colchester for a period, finally coming to rest in the North East as a pedestrian bridge over a dual carriage way. This is a great example of how the value of an asset was realised and maintained over a number of moves, also displacing the need for new materials to create the various temporary bridges that would have otherwise been required.

Material Reuse

It was earlier suggested that for maximum embodied carbon benefit, materials with a higher environmental impact should be targeted for reuse. Steel is a one such example, energy intensive in production and commonly used, but with a history of reuse. For this reason, steel reuse case studies are the focus of this section, although many of the benefits and barriers would also apply to other materials.

BedZED, Fig. 12.2, in the UK, is a good example of a project which prioritised reuse, with 98 tonnes of structural steel reused (Sergio and Gorgolewski n.d.). Assuming a carbon factor of 1.53 $kgCO_2e$ (Hammond and Jones 2011), this would displace 150 tonnes CO_2e that would be attributed if new steel (with an average 59% recycled content) was utilised. The majority of the steelwork was sourced locally from a redevelopment at Brighton train station, which had steel girders that were suitable for reuse. In this example, the salvage contractor stored

Fig. 12.2 BedZED (Photo Credit: Tom Chance – www.flickr.com/photos/tomchance/1008213420/)

the elements until they were required on-site. This overcomes a frequently discussed barrier to reuse, the storage of reused materials. Where and for how long materials must be stored is often perceived as a problem when reusing materials (Densley Tingley et al. 2017).

Identification of a reused material sourced during the detailed design stage can be helpful in order to ensure sufficient lead times and to take into account any potential alterations in size due to material availability. This could include reuse of elements salvaged from an existing building on-site, as in the case of the Ottawa Convention Centre, where 7, 4.5 m deep trusses were reused from the roof structure of the original Congress Centre (Canada Green Building Council 2010). The trusses weighed approximately 36 tonnes and were re-engineered to support the new roof design at a fabrication and storage facility. The enthusiasm of the building owner and engineer were both highlighted as key to achieving this component reuse.

The London 2012 main Olympic Stadium also featured steel reuse, although instead of being sourced from an existing construction site, the tubular steel was over-ordered, ex-pipeline tube. The steel had yet to be used but was considered a waste product disused in a storage yard. Five thousand tonnes of steel were reused in the roof structure. Sourcing disused pipeline steel resulted in shorter lead times than if the tube had been procured in a more traditional manner, which was perceived as a significant benefit of the approach. It is understood from discussions with those involved in the construction that testing and recertification of the tube was a relatively quick process. Coupon, sharpie hardness and chemical analysis tests were conducted on a sample from each piece of pipework. The retesting was necessary as a circumference strip of pipework would be tested to check resistance against pressure in its traditional use, whereas it needs to be tested along the length for structural use. The testing also demonstrated the quality of the steelwork, relieving fears from the client. Overall, the contractors, Sir Robert McAlpines, felt that the reuse was a positive experience and that others involved in the supply chain had similar feelings. From an environmental (displacing the need for new steel) and waste reduction perspective, it was seen as morally the right thing to do, not to mention the shorter lead times and cost savings to the project, which were significant benefits of the approach.

Design for Deconstruction and Material Reuse

Design for deconstruction and material reuse can be seen in a number of case studies around the globe, in both temporary event structures and permanent structures, including Brummen Town Hall and M&S's Cheshire Oaks store, which are explored within this chapter.

The Brummen Town Hall building in the Netherlands was designed to minimise total life costs via material and product reuse, rationalisation and renting. These actions can also deliver lower embodied carbon. BAM was the main contractor for the redevelopment of Brummen Town hall. The municipality was unsure if the building would be needed after 20 years and were looking to minimise costs and material impact. BAM, together with the architect, Thomas Rau, responded to this client brief by incorporating multiple material efficiency initiatives into the design of the building and the furnishings. As a consequence, the total costs, across the anticipated 20-year lifespan of the building were minimised, and almost all material inputs were reused and/or reusable in the future.

From the outset, the volume of construction material required was minimised by designing the new building around the existing structure. The roof was made almost entirely out of glass to minimise operational lighting costs and electricity use. This could also reduce operational carbon emissions if the electricity is generated from fossil fuels. Additional structural components were made of timber, which has a relatively lower embodied carbon than alternatives such as steel, and were designed for deconstruction. The building also included flexible walls and partitions to enable adaptability for future users. Symbolically, all materials used in the building were given their own 'passport' that details their material composition and plans for extraction at the end of the building's life (Fig. 12.3).

This commitment to reuse and cost minimisation for the building owner inspired new contracts with material suppliers. The ownership of key building components including timber, mechanical and electrical installations, lighting, tiles and flooring were retained by the manufacturer rather than the municipality. Instead, the municipality pays for the performance of these components, which goes beyond the traditional concept of product leasing. This novel approach necessitated a joint

Fig. 12.3 Brummen Town Hall during construction (Photo Credit: Royal BAM Group nv)

partnership between the designer and product manufacturers. BIM modelling was used to guarantee the performance of all components and to identify the material composition that minimised both capital and maintenance costs over the anticipated 20-year lifetime of the building since these were borne by the manufacturers rather than the building owner. An example of this type of service provision is the 'pay-per-lux' scheme from Philips (2017), where the client pays for lighting as a service and the manufacturer is responsible for maintaining the performance of the lighting – this potentially incentivises the manufacturer to ensure that the lighting operates at the lowest energy cost and thus reduces the operational carbon. Some evaluation of the capital cost of lighting replacement would be factored in here, but it is unclear if the trade-off and payback of embodied carbon would be factored into the decision (Fig. 12.4).

Since the building is still in use, it is unclear what will happen to the materials afterwards. However, the steps taken by BAM and Thomas Rau mean that there is the potential to reuse material and displace material from virgin sources. Lower demand for virgin materials can mean lower emissions generated during material production.

Marks and Spencer has introduced a number of initiatives with the aim of reducing the waste generated by product sales and construction projects. The latter aim led to a number of innovative approaches to build their 210,000 ft^2 development in Cheshire Oaks in 2012 (Datta 2012). One hundred percent of the construction waste was diverted from landfill (WRAP 2012). Many of the building materials have a high-recycled content. For example, the aluminium in the roof is 100%

Fig. 12.4 Brummen Town Hall (Photo Credit: Royal BAM Group nv)

recycled along with 60% of the high-grade aggregates. The roof is made out of glulam beams, held together with large bolts that allow for future deconstruction. Glulam has both a lower embodied energy and lower embodied carbon than steel and concrete and sequesters more carbon than is emitted in its manufacture (Hill and Dibdiakova 2016). Operational electricity use and emissions if electricity is from fossil fuels were reduced by maximising the amount of natural light in the stores. Additionally, an 80,000-l rainwater harvesting system meets around a third of the store's water requirements. The walls are cladded with a mixture of lime, water and hemp plant, known as Hemclad, which has both lower embodied carbon and delivers a better thermal performance. As a consequence of these and other initiatives, Cheshire Oaks store was estimated to be 42% more energy efficient and 40% more carbon efficient than a peer store (Faithful+Gould 2013). Marks and Spencer's commitment to 'zero waste' also extends to the end of the store's life. The building includes a disassembly and recycling guide, with details of the quantities of each "resource" element and instructions on how to reuse, resell or recycle each element. This provides an additional potential source of revenue for Marks and Spencer. The building was rated 'excellent' by BREEAM, scoring over 80% in the management, energy, water, pollution and materials and waste categories.

Lessons can also be learnt from temporary structures; two such examples were designed and constructed by ES Global for the London 2012 Olympics: the water polo arena and the shooting arena. These utilise a steel truss system, which forms part of ES Global's kit of parts for use in temporary events structures. These structures were leased for the games; in the case of the water polo arena, the lease cost amounted to approximately £1.2 million of a total £25 million project (Densley Tingley and Allwood 2015). This leasing scenario meant that the Olympic Committee had the required sports venues but removed the risk of empty, wasted stadia in the years after the games and meant that the components could then be reused again in future temporary events' structures by ES Global, lowering the wider embodied carbon arising from temporary events. With recurring sporting events being held frequently around the world, this type of approach, to design for deconstruction and reuse, is ideal to reduce the material demand and embodied carbon of the world's built environment.

Design for Adaptability

St John Bosco Arts College in Liverpool, UK used construction insights from outside the education sector to build an innovative flexible space at a lower cost than a conventional school. The new site for John Bosco Arts College in Liverpool opened in Sept. 2014. The architects, BDP, designed a three-storey, $11,100m^2$, column-free, single-span steel portal frame building. A large multifunctional assembly space is in the centre of the building, circled by two floors of classrooms with 2.7 m floor-to-ceiling heights. The interior structure of columns and floors are demountable and separate to the main frame, enabling removing and dismantling of semi-permanent features in the building as the school's needs evolve. There were few precedents of this building design for schools, so the architect's reference point came more from more open-plan workplaces and commercial shopping centres.

The main reported driver for this innovative design was cost. In 2010, John Bosco Arts College was selected to receive government funding to rebuild the school buildings as part of the 'Building Schools for the Future' (BSF) programme. BDP were selected as the architect; however, the government programme was cancelled before any money was allocated to the project. This had both short-term and long-term impacts on the design of the school. In the short term, BDP developed a cheaper building design for John Bosco Arts College. Costs were £11.91/m^2 compared with the £17.50/m^2 under the original BSF-funded design and have 15% more area than a traditional BB98 compliant school (Buxton 2015). Fortunately, Liverpool City Council was able to source the £18m needed to rebuild the school in the new design. The experience with the BSF funding also prompted considerations for the longer-term future needs of the school. Liverpool City Council wanted a highly flexible building with an adaptable-changeable configuration to provide the option of changing the building's use in the future for little cost. This reduces life cycle embodied carbon through two mechanisms. First, it helps to extend the building life by ensuring it meets the evolving needs of its users. Second, the inbuilt adaptability reduces the need for extra construction materials to modify the building over time.

In the end, the experience was reportedly very positive for the key stakeholders involved, and the school was awarded a RIBA Regional Award in 2015. The head teacher, Anne Pontifex, who was also awarded RIBA 'client of the year', noted in an interview with *Architects' Journal* (Pritchard 2015):

> Our partnership with the architects developed into an open, honest and creative relationship, which enabled ideas, theory and the concept of truly 'thinking outside the box' to take place. As a result, we have what I believe is a unique, exciting and stimulating school, which supports our already outstanding teaching and learning.

Adaptable Use

At the end of 2014, Marks and Spencer opened up their first community room to the public in their Wolstanton store (Horner 2014). This initiative allows local community groups to utilise the room when it is not being used as a meeting room by the staff, thereby increasing the productivity of their capital assets. A number of stores have since followed suit, providing a free space to local charities, youth and community groups, schools and small businesses.

This initiative is aligned with the company's Plan A commitment on being in touch with the local community. The decision to offer M&S facilities is driven by the store manager, but it requires initial coordination and planning from central office. For example, a multifunctional room requires a larger floor space, needs access to washrooms and water facilities and needs to be reachable via the front of the supermarket to manage security of the store. These considerations are more easily addressed when designing and building a new store but can also be delivered through retrofits. There may be minor additional capital outlays on store branding, furniture and television screens for the community users. Staff will also need to spend a couple of extra hours a week booking out and preparing the room. However, these additional expenditures are very small relative to the total operating costs of the store.

Even though the scheme is less than a year old, a number of benefits are already emerging. The use of store space allows Marks and Spencer to be more integrated in the community and alleviates pressure on local authorities to provide equivalent services to local residents. Employees have expressed a stronger connection with their customers, public relations have improved, and there are early indications of increased footfall in stores where this service is offered. By making some of their stores multi-use, Marks and Spencer are reducing demand for new construction projects to provide equivalent services to the community, reducing the need for new material inputs and embodied emissions.

Discussion

From the exploration of the literature and the case study investigation, an understanding of project drivers and incentives to incorporate the four discussed strategies can be derived. These are split initially into lessons for each of the four strategies, with overarching lessons learnt highlighted in the 'Common Lessons' section.

Building Reuse

It is clear that buildings of historical significance, whether listed or not, will likely be prime targets for building reuse on sites where revitalisation and redevelopment are strategic priorities. This can be seen not only in the case study example discussed, Quadrant 3, but also in projects where historic facades are retained and propped during redevelopment and the remainder of the building significantly redeveloped or in some cases partially demolished and rebuilt. Another key factor is the condition of the existing structure of a building and the inherent flexibility offered by the available space. For instance, warehouse conversions are common due to large clear spans and high floor to ceiling heights offering a wide range of options for adaptation.

In order to maximise building life, and reduce embodied carbon, two steps should be carried out on existing structures on a site where redevelopment is proposed. Firstly, an assessment of whether the existing structure is fit for purpose or what alternations/additions would be required to make it suitable. Secondly, consideration of the flexibility offered and whether it is suited to the potential new building use. Even where the existing building footprint is desired, an assessment could be made to explore if vertical expansion is possible under the current load capacity of the structure or what alterations would be required to achieve this. The environmental benefits across embodied carbon reduction and waste reduction for this approach to building reuse are clear, but a more significant project driver would be the potential programme savings, in turn achieving project cost savings and earlier functional use of the building for the client.

Material Reuse

If an assessment of the viability of building reuse is found to be unfavourable, then the next step in dealing with an existing building should be to assess the material salvage and reuse potential (in conjunction with the recycling and downcycling options), essentially carrying out a demolition audit. If there are materials available for reuse on-site, then the integration of these into the new proposed project should be explored, as demonstrated by the Ottawa Convention Centre.

Another option for new projects would be to identify particular elements which would see an embodied carbon benefit from reuse and to then explore if any local demolition sites have appropriate materials available. There are also a number of online reused material marketplaces opening up, for example, Planet Reuse (2017) in the USA and Enviromate (2017) in the UK, where materials wanted and for sale can be posted. These sites could be explored to source reused materials for projects.

The earlier in the project that the materials desired for reuse can be identified, then the earlier sourcing of appropriate materials can occur. Early consideration allows for easier integration of elements into an existing design and leaves open the possibility of design alterations. The London 2012 stadium demonstrated that in cases where specialised reused materials are available that this could result in shorter lead times than ordering of bespoke fabrication, which would be a significant project benefit. Aesthetics could also drive material reuse, as well as embodied carbon reduction, particularly for those clients that are now prioritising this.

Design for Deconstruction and Material Reuse

A key factor for this approach is the longer time periods over which the benefits must be measured; a cradle-to-cradle embodied carbon measurement should result in reductions as design for deconstruction facilitates future material reuse, thus lessening the demand for new materials in the future. A move to measuring whole life costs will also yield reductions, as the approach maintains the value of future material assets. Information transfer over time is critical. Whether it be disassembly guides or material passports, having this information will facilitate future deconstruction and material reuse. Ensuring this information is preserved, accessible and passed between building owners will be important to maximising the value from this future planning strategy.

Design for deconstruction and material reuse also presents opportunities for new business models such as leasing, as the asset can be held and sold on by the leaser. The examples show that this model can also have operational carbon benefits through the delivery of services in buildings as an alternative to buying individual products. Temporary structures, or projects which are predicted to have very short lifetimes, are also likely to see significant benefits from this leasing approach, as the return on investment should be quicker on assets leased for short time periods – for example, a year to 18 months in the case of rotating sporting events. It is unclear whether current legislation sufficiently incentivises or rewards construction companies to enable future reusability or sustainable collaboration/partnerships across the construction supply chain, and this is an area to explore going forward.

Design for Adaptability

There are parallels with design for deconstruction here, as the benefits of design for adaptability will also be seen over longer, whole life cycle, time periods. Inherently, the additional adaptability and flexibility in use provided by a building that is designed considering future uses and changing requirements will only be recognised during the life cycle of the building. There can be reduced life cycle costs as the building can be easily adapted during use. If the building does successfully adapt to changing needs, then it is likely that building's life will be extended, displacing the need for future materials and buildings, and thus embodied carbon. There is a potential tension between the probable increase in carbon emissions incurred in initially enabling adaptability and the urgent need for deep carbon reductions, consistent with pathways to keep global temperature rises to below 2 °C above pre-industrial levels. This timeline suggests that short-term strategies that yield an immediate embodied carbon reduction should be prioritised first. However, these longer-term and shorter-term strategies can be integrated and complementary, for example, incorporating material reuse into a project designed for adaptability. Deployment of both will be essential in meeting the long-term goal of achieving a balance of emissions sources and sinks.

Adaptable space that can be used by different stakeholders, enabling a more intensive use of space – as in the M&S community spaces – will also largely see benefits in use. Increasing the frequency with which other stakeholders access the space can produce co-benefits, such as an increased sense of community or increased footfall with potential sales implications. This approach again potentially displaces the need for additional buildings and thus materials. As the benefits are seen during the building's life, to date, the inclusion of this strategy has largely been client driven from a practicality of use perspective rather than by an embodied carbon reduction strategy, which could be considered more of a co-benefit.

Conclusion

The four design strategies that form the focus of this chapter all aim to maintain the value of material assets, either now or in the future, making them central to a circular economic approach to the built environment. The potential benefits from these strategies are wide ranging, with the main drivers being focused on cost reductions or flexibility of use and at end of life. The cost reductions are both capital, through shorter programme times or reduced material costs, and lifetime through maintained or recovered asset value. Embodied carbon reduction should be realised through the deployment of all of these strategies, when whole lifetimes are considered, but whilst the environmental benefit is recognised, it is perhaps secondary to these other project drivers. A move to whole life costing will help a greater number of projects to see the economic benefits of these strategies. The wider exploration of opportunities to add value to projects through these strategies, the exploitation of similar niche project circumstances to increase uptake and improved quantification of the co-benefits will ultimately drive the longer-term progression of these strategies into mainstream construction. The more these strategies are explored, the more the wider industry can respond in developing the required skill sets and underpinning supply chains to enable these practices to become business as usual in the sector. The wider potential economic benefits of these strategies are likely to drive their uptake, particularly in the current marketplace, when compared to embodied carbon reduction alone. Thus these two should be considered in tandem to help drive the sector to achieving a low embodied or indeed low whole life carbon built environment.

Acknowledgements The authors would like to sincerely thank all those who gave their time up for interviews. The contribution of the second author was supported by the Research Council UK Energy Programme through the Centre for Industrial Energy, Materials and Products (grant number EP/N022645/1). The third author was supported by an EPSRC PhD studentship, grant reference EP/L504920/1.

References

Addis, W., & Schouten, J. (2004). *Design for deconstruction. Principles of design to facilitate reuse and recycling.* London: CIRIA.

Andrews, C. J., Hattis, D., Listokin, D., Senick, J. A., Sherman, G. B., & Souder, J. (2017). Energy-efficient reuse of existing commercial buildings. *Journal of the American Planning Association, 82*(2), 113–133. https://doi.org/10.1080/01944363.2015.1134275.

Arup. (2016). *The circular economy in the built environment.* Available at: http://publications.arup.com/publications/c/circular_economy_in_the_built_environment. Accessed 27/04/17.

Assefa, G., & Ambler, C. (2017). To demolish or not to demolish: Life cycle consideration of repurposing buildings. *Sustainable Cities and Society, 28*, 146–153.

Brand, S. (1994). *How buildings learn: What happens after they are built.* New York: Viking Penguin.

Buxton, P. (2015). *St John Bosco Arts College, Liverpool – Article in RIBA journal.* Available at: http://www.steelconstruction.info/St_John_Bosco_Arts_College,_Liverpool. Accessed 02/05/17.

Canada Green Building Council. (2010). *Material reuse at the Ottawa Convention Centre. Green Space, Ottawa Region Chapter Newsletter, Materials Issue, 8, April 2010.* Available at: http://static1.1.sqspcdn.com/static/f/322919/7289577/1276203149047/CaGBC_ORC_Green_Space_Issue_8_short.pdf?token=QXNNcA5t2gT9HuhpYnr2uM4wOJ8%3D. Accessed 12/07/17.

CE100. (2016). *Circularity in the built environment: Case studies. A compilation of case studies from the CE100.* Ellen MacArthur Foundation. Available at: https://www.ellenmacarthurfoundation.org/assets/downloads/Built-Env-Co.Project.pdf. Accessed 27/04/17.

Chau, C. K., Xu, J. M., Leung, T. M., & Ng, W. Y. (2015). Evaluation of the impacts of end-of-life management strategies for deconstruction of a high-rise concrete framed office building. *Applied Energy, 185*, 1595–1603. https://doi.org/10.1016/j.apenergy.2016.01.019.

Cheshire, D. (2016). *Building revolutions: Applying the circular economy to the built environment.* London: RIBA Publishing. ISBN: 9781859466452.

Datta, M. (2012). *M&S Cheshire Oaks store.* Available at: http://corporate.marksandspencer.com/blog/stories/mands-cheshire-oaks-store. Accessed 02/05/17.

De Wolf, C., Pomponi, F., & Moncaster, A. (2017). Measuring embodied carbon dioxide equivalent of buildings: A review and critique of current industry practice. *Energy and Buildings, 140*, 68–80. Available at: http://linkinghub.elsevier.com/retrieve/pii/S0378778817302815.

Densley Tingley, D. (2013). *Design for deconstruction: An appraisal* (Doctoral dissertation). University of Sheffield.

Densley Tingley, D., & Allwood, J. M. (2015). The rise of design for deconstruction – A cradle to cradle approach for the built environment. In: *International Sustainable Development Research Society conference, the tipping point: Vulnerability and adaptive capacity,* 10th–12th July 2015, Geelong, Australia.

Densley Tingley, D., & Davison, B. (2011). Design for deconstruction and material reuse. *Proceedings of the ICE – Energy, 164*(4), 195–204. https://doi.org/10.1680/ener.2011.164.4.195.

Densley Tingley, D., & Davison, B. (2012). Developing an LCA methodology to account for the environmental benefits of design for de-construction. *Building and Environment, 57*, 387–395. https://doi.org/10.1016/j.buildenv.2012.06.005.

Densley Tingley, D., Cooper, S., & Cullen, J. M. (2017). Understanding and overcoming the barriers to structural steel reuse, a UK perspective. *Journal of Cleaner Production, 148*, 642–652.

Enviromate. (2017). Available at: https://www.enviromate.co.uk/. Accessed 14/07/17.

Faithful + Gould. (2013). *Marks & Spencer Cheshire Oaks building performance evaluation first year performance summary.* Available from: http://www.greencoreconstruction.co.uk/down loads/MS_Cheshire_Oaks_Building_Performance_Evaluation_-_Summary_-_Technical_a_. pdf

Giesekam, J., Barrett, J., & Taylor, P. (2016). Construction sector views on low carbon building materials. *Building Research & Information, 44*(4), 423–444. https://doi.org/10.1080/09613218.2016.1086872.

Guy, B., & Ciarimboli, N. (n.d.). *Seattle guide: Design for disassembly in the built environment.* Available at: http://your.kingcounty.gov/solidwaste/greenbuilding/documents/Design_for_Dis assembly-guide.pdf. Accessed 02/05/2017.

Hammond, G., & Jones, C. (2011). *Inventory of carbon and energy, Version 2.0.* Available at: http://www.circularecology.com/embodied-energy-and-carbon-footprint-database.html#. WWc7GojyuUk. Accessed 13/07/17.

Hill, C. A. S., & Dibdiakova, J. (2016). The environmental impact of wood compared to other building materials. *International Wood Products Journal.* https://doi.org/10.1080/20426445.2016.1190166.

Horner, R. (2014). *Our first community room – Marks and Spencers blog.* Available at: http://corporate.marksandspencer.com/blog/stories/our-first-community-room. Accessed 02/05/17.

Hosseini, M. R., Rameezdeen, R., Chileshe, N., & Lehmann, S. (2015). Reverse logistics in the construction industry. *Waste Management Research, 33*(6), 499–514.

Laefer, D. F., & Manke, J. P. (2008). Building Reuse Assessment for Sustainable Urban Recon-struction. *Journal of Construction Engineering and Management, 134*(3), 217–227. https://doi.org/10.1061/(ASCE)0733-9364(2008)134:3(217).

Langston, C., Wong, F. K. W., Hui, E. C. M., & Shen, L.-Y. (2008). Strategic assessment of building adaptive reuse opportunities in Hong Kong. *Building and Environment, 43*(10), 1709–1718. https://doi.org/10.1016/J.BUILDENV.2007.10.017.

Luscuere, L. M. (2017). Materials Passports: Optimising value recovery from materials. *Proceedings of the Institution of Civil Engineers: Waste Resource Management, 170*(1), 25–28. https://doi.org/10.1680/jwarm.16.00016.

Manewa, A., Pasquire, C., Gibb, A., Ross, A.,& Siriwardena. (2013). Adaptable buildings: Striving towards a sustainable future. *People and the planet 2013 conference: Transforming the future, RMIT University, Melbourne, Australia, 2–4 July.* Available at: http://global-cities.info/wp-content/uploads/2013/11/Adaptable-Buildings-Striving-Towards-a-Sustainable-Future1.pdf. Accessed 27/04/17.

Morgan, C., & Stevenson, F. (2005) *Design and detailing for deconstruction.* Available at: http://www.seda.uk.net/index.php?id=136. Accessed 02/05/17.

Philips. (2017). *Light as a service – UK student movement is a beacon of sustainability for wider society.* Available at: http://www.philips.com/a-w/about/sustainability/sustainable-planet/cir cular-economy/light-as-a-service.html. Accessed 13/07/17.

Pinder, J. A., Schmidt, R., Austin, S. A., Gibb, A., & Saker, J. (2017). What is meant by adaptability in buildings? *Facilities, 35*(1/2), 2–20.

Planet Reuse. (2017). Available at: https://planetreuse.com/. Accessed 14/07/17.

Pomponi, F., & Moncaster, A. (2017a). Circular economy for the built environment: A research framework. *Journal of Cleaner Production, 143*, 710–718. https://doi.org/10.1016/j.jclepro.2016.12.055.

Pomponi, F., & Moncaster, A. (2017b). Scrutinising embodied car-bon in buildings: The next performance gap made manifest. *Renewable and Sustainable Energy Reviews*. https://doi.org/10.1016/j.rser.2017.06.049.

Pongiglione, M., & Calderini, C. (2014). Material savings through structural steel reuse: A case study in Genoa. *Resources, Conservation and Recycling, 86,* 87–92.

Pritchard, O. (2015). *St John Bosco Arts College by BDP – Article in the architect's journal.* Available at: http://www.architectsjournal.co.uk/buildings/st-john-bosco-arts-college-by-bdp/8679827.article. Accessed 02/05/17.

RICS. (2014). *Methodology to calculate embodied carbon, RICS guidance note, Global 1st edition.* ISBN: 9781842197950 Available at: http://www.rics.org/Documents/Methodology_embodied_carbon_final.pdf.

Sergio, C., & Gorgolewski, (n.d.). *Reuse of structural steel at BedZED (Beddington Zero Energy Development).* Available at: http://www.reuse-steel.org/files/projects/bedzed/bedzed%20case%20study%205-5.pdf. Accessed 12/07/17.

The Crown Estate. (2017). *Quadrant 3, modern renaissance.* Available at: https://www.thecrownestate.co.uk/media/5249/urb-quadrant-3-brochure.pdf. Accessed 12/07/17.

Van Odijk, S., & van Bovene, F. (2014). *Circular construction: The foundation under a renewed sector.* ABN.AMRO. Available at: https://www.slimbreker.nl/downloads/Circle-Economy_Rapport_Circulair-Construction_05_2015.pdf. Accessed 27/04/17.

WRAP. (2012). *Marks and Spencers sustainable learning store.* Available at: http://www.wrap.org.uk/sites/files/wrap/Refurbishment%20Resource%20Efficiency%20Case%20Study_Retail_M&S.pdf. Accessed 02/05/17.

Chapter 13
Embodied Carbon of Sustainable Technologies

S. Finnegan

Introduction

It is recognised that buildings and their operation contribute to a large percentage of total energy end-use worldwide. They account for one-third of global greenhouse gases (of which CO_2 is a major contributor) with commercial and residential buildings alone accounting for 40% of the world's energy consumption (RICS 2015). Clearly there is a need to reduce this impact with the creation of more sustainable buildings that use significantly less or ideally zero energy. One method to achieve this goal is to consider the use of sustainable technology. In this chapter 'sustainable technology' refers to the use of equipment to power, ventilate, heat and/or cool a building that relies on resources that have no long-term adverse carbon dioxide equivalent (CO_2e) impact on the environment. One can calculate this environmental impact using a universally accepted technique called life cycle assessment (LCA). This is a methodology to calculate and evaluate the whole-life operational and embodied impacts of a product which can include, for example, energy, water, waste and emissions, during its creation, use and final disposal. In this chapter a review of the embodied carbon dioxide equivalent (CO_2e) emissions is considered.

There is a long list of new sustainable technologies on the market that are primarily used to reduce the energy use in buildings. They should only be considered for a building once the building fabric has been maximised. A 'fabric first' approach (Design Buildings Wiki 2016), following the Passivhaus (2016) design principles of creating well-insulated thermally efficient buildings, is always the first step in creating a sustainable building. Once the building has considered each fabric option, the next step is to consider the use of sustainable technology to heat, cool

S. Finnegan (✉)
Liverpool School of Architecture (LSA), University of Liverpool, Liverpool, UK
e-mail: S.Finnegan@liverpool.ac.uk

© Springer International Publishing AG 2018   287
F. Pomponi et al. (eds.), *Embodied Carbon in Buildings*,
https://doi.org/10.1007/978-3-319-72796-7_13

Table 13.1 Common sustainable technologies

Activities	Sustainable technologies
Heating	Condensing boiler
Ventilation air	Biomass boiler
Conditioning	Ground- and air-source heat pumps (GSHP/ASHPs)
	Heating, ventilation and air conditioning (HVAC)
	Biomass boiler
	Solar thermal
	Mechanical ventilation with heat recovery (MVHR)
Power	Solar photovoltaics (PV)
	Micro-combined heat and power (CHP) micro-wind generation
	Micro-hydro systems
	Voltage optimisers
Lighting	LED lighting
	Effective lighting controls

and/or power the building. Ideally, these technologies themselves should use less or zero energy in comparison with a conventional system. The aim of the fabric first approach and sustainable technology is to operate and maintain a level of thermal comfort (Thermal comfort in buildings n.d.) for the occupants, using the least amount of energy possible to do so. Table 13.1 shows the common sustainable technologies that are used for residential and commercial buildings in the European Union (EU).

A recent survey of over 200 UK residential housing associations by the NHBC Foundation (2015) concluded that *solar PV* is the most popular sustainable technology for housing associations. The survey also found that 82% of housing associations install solar PV in all new builds. In addition, 62% install *MVHR (mechanical ventilation with heat recovery)*, 61% install *solar thermal* and 56% install *air-source heat pumps (ASHPs)*.

There are a large number of examples of the use of sustainable technologies in commercial buildings, and Table 13.2 presents an overview of a sample of ten modern sustainable commercial buildings and the type of technology considered. Marked with a cross.

Based on the NHBC Foundation survey of residential buildings and the review of modern sustainable commercial buildings, this chapter is focused on the following selection of common sustainable technologies including:

– Solar PV
– Solar thermal
– Mechanical ventilation with heat recovery (MVHR)
– Air-source heat pumps (ASHPs)
– LED lighting

Typically, these technologies have a shorter lifespan than the building itself; however with advancements in, for example, building-integrated photovoltaics

Table 13.2 Common sustainable technologies for non-residential buildings

Building	Sustainable technology used			
	LED lighting	Micro-CHP	Solar PV	Solar thermal
Powerhouse Kjorbo (Norway) (Throndsen et al. 2015)			x	
The Edge (Netherlands) (The Edge 2017)	x		x	
DPR, San Francisco (USA) (DPR Construction Net-Zero Energy San Francisco Regional Office 2017)			x	x
IREA (UAE) (International Renewable Energy Agency (IREA) Headquarters 2017)			x	x
One Angel Square (Manchester)	x	x		
D&L Packard Foundation, (USA) (n.d.)			x	
GSW Headquarters (Berlin) (Oldfield et al. 2009)	x			
ARPT Headquarters (Algeria) (Autorité de Régulation de la Poste et des Télécommunications (ARPT 2017)			x	
Hive, Worcester (UK) (Simon 2013)		x		
Beddington Zero Emission Development (BedZED) (UK) (Chance 2009)	x	x	x	

(BIPV) (Peng et al. 2011), each technology is becoming more robust and connected to the building fabric. Once in place and in use on the building, these technologies indirectly emit CO_2e as each system requires power to operate. This power could come from the grid or from another sustainable technology, i.e. the electricity generated from solar PV system can be used to power an MVHR system. In each particular case, it is necessary to calculate the energy demands and engineer a system that works. This chapter is not investigating this operational CO_2e impact but the so-called 'embodied' impact, i.e. the energy required to create, manufacture, transport and dispose of the technology. In comparison with the operational impact and indeed the impact of conventional systems, the embodied CO_2e is a new area of research which may or may not be proven to have a significant impact.

As previously mentioned, the embodied impact of sustainable technologies in this chapter is limited to the CO_2e impact only. In order to fully understand this impact for each of the shortlisted technologies, we have investigated (1) the method upon which the environmental assessments have taken place, (2) the results of the investigation into the embodied impact, (3) the main findings and discussion points and (4) the conclusions that could be drawn.

The LCA Standard

There exists a common methodology to assess the environmental impact of any product or in this case sustainable technology, which includes an assessment of the embodied CO_2e impact. This method, as mentioned above, is life cycle assessment

(LCA) which is defined by the International Organisation for Standardisation (ISO) (2017) in accordance with two particular standards, ISO 14040 (2017) and ISO 14044 (2017). In the EU, the European Standards Committee published the TC350 standards (ECO Platform CEN/TC 350 standards 2017) to define the stages which should be included in a LCA (specifically for construction works, including products and buildings), further details of which are provided in De Wolf et al. (2017). In general, LCA consists of four stages: (1) goal and scope definition, (2) inventory analysis, (3) impact assessment and (4) interpretation as defined by ISO 14040 and 14044. For a thorough review of a sustainable technology, this is important as one would like to compare the various types on a like-for-like basis.

1. Goal and scope definition: ISO requires a clearly defined goal and scope which sets out the context and defines the scope. The goal of the assessment could be to calculate the amount of CO_2e emitted per energy unit used (i.e. kWh), per area covered (m^2), per (kg) or per unit. For sustainable technologies, it is common to consider the life cycle impact per unit of energy produced. This would typically take the form of $kgCO_2e/kWh$ or $kgCO_2e$ per unit. This measure is also known as the *functional unit*, i.e. a normalised indicator of environmental performance to compare different technologies. The scope also determines the physical system boundary, i.e. when to start and stop the assessment considering all of the stages involved.
2. Inventory analysis: On completion of the goal and scope, the ISO standards require a life cycle inventory (LCI) to be produced. This is an inventory of flows to and from the product. Inventory flows can include water, energy, raw materials and emissions to air, land and water. To develop the inventory, a flow model is constructed from which an impact assessment can be determined.
3. Impact assessment: This phase is aimed at evaluating the significance of environmental impacts based on the LCI flow results. ISO 14044 defines a series of compulsory stages with grouping and weighting optional. On completion, the user is then able to interpret the results and undertake comparative analysis.
4. Interpretation: This stage is crucial in assessing the true holistic impact of the product as the energy and carbon impact of each stage is interpreted. In order for the construction industry to be consistent in the approach to assessment, ISO 14040 states that the interpretation should include the identification of significant issues, consistency and sensitivity checks, conclusions, limitations and recommendations.

Across the EU, the Construction Products Regulation (CPR) (n.d.) provides further rules for the creation of construction products (including certain sustainable technologies). It requires that reliable information on the performance of products is available to the public and consumers. Where the declaration of the performance includes an assessment of the sustainable use of resources, a LCA (as described above) is required. A particular type of LCA is an Environmental Product Declaration (EPD). Their evolution has been explained by Ibáñez-Forés et al. (2016), and details of how they link to LCA are provided by Strazza et al. (2016). The guidelines for the creation of an EPD are outlined in EN 15804 (n.d.) and discussed in Achenbach et al. (2016). These declarations follow a set of Product Category Rules (PCR), which are explained in a paper by Gelowitz and McArthur (2016) who investigated the effect of them on a particular case study. Depending upon the country in question, EPDs are either mandatory or voluntary. For example, in France from the 1 July 2017, any technical equipment for use in buildings, i.e. electrical, electronic and heating, ventilation and air conditioning (HVAC), will require an EPD. A review of the current position in different European countries has been presented by Passer et al. (2015).

Results

Each of the sustainable technologies under consideration, solar PV, solar thermal, mechanical ventilation with heat recovery (MVHR), air-source heat pumps (ASHPs) and LED lighting, has been investigated to estimate the embodied CO_2e impact. The variance from a selection of UK and EU studies for each technology is presented in a common format, which is $kgCO_2e/kWh$, i.e. the amount of embodied CO_2e for every energy unit (kWh) generated. Such values have either been selected from an LCA study or derived through the following equation:

$$EC = M/E \qquad (13.1)$$

where EC is the amount of embodied CO_2e per unit of energy generated ($kgCO_2e/kWh$), M is mass of CO_2e (kg) and E is energy generated (kWh).

In the review of each sustainable technology below, if a value has been calculated, an explanation is provided.

Solar PV

A large amount of research and investigation into the embodied CO_2e impact of solar PV has been undertaken in different parts of the world by Nawaz and Tiwari (2006), Sherwania et al. (2010), Lamnatoua et al. (2014), Kristjansdottira et al. (2016), Jayathissaa et al. (2016), Yue et al. (2014), Fthenakis and Kim (2011) and Lin Lu and Yang (2013) with various and numerous conclusions drawn depending upon the types of system used. Solar PV systems can generally be categorised into three main types (IRENA and IEA ETSAP 2013):

- Silicon cells (first generation)
- Thin-film solar cells (second generation)
- Emerging and novel photovoltaic technologies (third generation)

The most important consideration is the market uptake and penetration of each type of system. In a 10-year review of PV usage worldwide from 1999 to 2009

(Redondo de la Mata 2014), approximately 85% of panels are known to be first-generation silicon cells, with the remainder a mixture of emerging second and third generation. Given this is the case, it is safe to assume that the majority of systems in use today and in the near future on residential and non-residential buildings in the EU are therefore monocrystalline or polycrystalline first-generation silicon cell.

A typical silicon cell PV will consist of silicon and other semiconductor materials such as germanium, gallium arsenide and silicon carbide (IRENA and IEA ETSAP 2013). The environmental impact of sourcing these materials, manufacturing the panels and disposing of the components at the end-of-life stage affects the environment in many different ways including water use, hazardous waste, land use and associated ecological impacts. As discussed, this chapter is focused solely on the total CO_2e impact, and numerous studies have estimated this impact for silicon solar PV panels. Most estimates from predominantly EU studies estimate the embodied carbon impact of silicon cell PV systems to be between 0.03 and 0.23 kg of CO_2e per kWh of energy created (Nawaz and Tiwari 2006; Alsema 2000; Pacca et al. 2007; Schaefer and Hagrdorn 1992; Niewlaar et al. 1996; Meier 2002; Battisti and Corrado 2005; Tripanagnostopoules et al. 2005; Hosenuzzaman et al. 2015). During operation, the panels themselves do not directly emit CO_2e and in the vast majority of cases are used to replace or supplement a grid-connected system. As a result, Hosenuzzaman et al. (2015) states that solar PV systems have a negligible effect on global CO_2 levels. They actually save 0.53 kg of CO_2e for every kWh of electricity produced in the EU (Cheng and Tu 2008) (Table 13.3).

According to the UK government (Department for Business 2016), using a kWh of 2016 grid electricity results in the output of 0.4 kg of CO_2e. Therefore, four UK solar PV panels (0.1 kg CO_2e per kWh) have the same impact.

Solar Thermal

Unlike solar PV systems, solar thermal systems capture heat from the sun and use it to heat water for use within the building. They are most commonly used for residential buildings and operate using a simple process. Further details and a breakdown of the component parts of a typical system can be seen in a study by Aini Masruroh et al. (2006). The variance from two studies can be seen in Table 13.4.

The study by Ardente et al. (2005) investigated the impact of an Italian solar thermal system, and in that study the EC can be estimated as 0.025 $kgCO_2e/kWh$; see Eq. 13.2.

$$0.025 \ kgCO_2e/kWh = 650 \ kgCO_2e/25{,}000 \ kWh \qquad (13.2)$$

Table 13.3 Solar PV studies

Location	Assumptions	kgCO$_2$e/kWh
Italy (Tripanagnostopoules et al. 2005)	A 3 kWp array in use for 30 years	0.227
UK (Hammond and Jones 2005)	A 2.1 kWp array in use for 25 years	0.100

Table 13.4 Solar thermal studies

Location	Assumptions	kgCO$_2$e/ kWh
Italy (Ardente et al. 2005)	One solar thermal collector used for 25 years	0.025
EU (Aini Masruroh et al. 2006)	One solar heating and storage unit used for 20 years	0.030

Over the life cycle of this solar thermal system, 1000 kWh of energy was generated per year for 25 years. The same study also reported an embodied CO$_2$e of 650 kg.

A further EU study by Aini Masruroh et al. (2006) found that for every kWh of energy created, the solar thermal system equivalent life cycle CO$_2$e was between the ranges of 0.023 and 0.036 kg, of which 99% is reported to be from the embodied CO$_2$e stages. A higher estimate of 0.03 kg CO$_2$e per kWh is used as an estimate.

Mechanical Ventilation with Heat Recovery (MVHR)

MVHR systems are becoming increasingly popular due to the fact that buildings are becoming better insulated or in many cases super insulated, and as a result they become airtight and lack the infiltration of fresh air and expulsion of stake air. In short an MVHR system can supply this fresh air and extract stale air with negligible heat loss in the process. Recent reviews of the embodied carbon impact of MVHR systems, following the ISO 14040 methodology, are sparse with Nyman and Simonson (2004), Goggins et al. (2016), ProAir 600LI Heat Recovery Ventilation Unit (2015), and Hernandez and Kenny (2011) providing the best detailed analysis. The variance from two of these studies can be seen in Table 13.5.

A ProAir MVHR system used in a semi-detached home in Ireland was studied using ISO 14040 in the study by Goggins et al. (2016); the results show that the total embodied kgCO$_2$e was 0.108; see Eq. 13.3.

$$0.108 \ \text{kgCO}_2\text{e/kWh} = 661 \ \text{kgCO}_2\text{e}/6148 \ \text{kWh} \qquad (13.3)$$

Over the lifetime use of this MVHR system, 6,157 kWh of energy was generated to meet the demands. This figure was highlighted in Goggins et al. (2016) and sourced from the ecoinvent database (Hischier et al. 2009). This value and the total embodied CO$_2$e was reported as 661 kg.

Table 13.5 MVHR studies

Location	Assumptions	kgCO$_2$e/ kWh
Ireland (Goggins et al. 2016)	An MVHR used in a semi-detached home used for an assumed 50 years	0.108
UK (Hernandez and Kenny 2011)	An MVHR used in a two-storey newly built detached house for an assumed 50 years	0.100

Further work by Hernandez and Kenny (2011) in the UK estimates the embodied kgCO$_2$e of MVHR systems to be 0.100 kgCO$_2$/kWh, based on the same energy generation figure as above and an estimated kgCO$_2$ of 600 kg; see Eq. 13.4.

$$0.100 \ kgCO_2e/kWh = 600 \ kgCO_2e/6148 \ kWh \quad (13.4)$$

Air-Source Heat Pump (ASHP)

An ASHP is commonly used to replace/supplement a conventional heating system. In short, they will absorb heat from the outside air to heat radiators, supplement underfloor heating systems and/or provide hot water. The ASHP can extract heat from the outside air even when temperatures are as low as −15°C. In a study by Johnson (2011), the total life cycle impact of a typical UK ASHP was assessed using the Publicly Available Standard (PAS) 2050 (BSI, Carbon Trust 2008). This is a specification for the assessment of individual goods and services based on ISO 14044. The results show that the embodied CO$_2$e impact of this specific type of ASHP used in a semi-detached home in the UK with cavity walls for 15 years is an estimated 0.028 kgCO$_2$e/kWh; see Eq. 13.5 and Table 13.6.

$$0.028 \ kgCO_2e/kWh = 1653 \ kgCO_2e/59,135 \, kWh \quad (13.5)$$

Over the life cycle use of this type of ASHP, 59,135 kWh of energy was generated, and the total embodied CO$_2$e was reported as 1,653 kg.

Further details in relation to the full life cycle impact including the contribution from the operational phase can be seen in Parliamentary Office of Science and Technology (2016) and Greening and Azapagic (2012). This later study stated that ASHPs release 0.276 kg of CO$_2$ per kWh of energy created, of which 95% is due to the use stage related to electricity generation and natural gas combustion. Therefore, 5% of this value would provide an estimate of the total embodied CO$_2$e. This figure as presented in Table 13.5 is 0.014 kg.

Table 13.6 ASHP studies

Location	Assumptions	kgCO$_2$e/ kWh
UK (Greening and Azapagic 2012)	Used for domestic space heating over 20 years	0.014
UK (Johnson 2011)	A 10 kW pump for space heating over 15 years	0.028

LED Lighting

It is universally accepted that light-emitting diode (LED) lighting uses less energy than conventional lighting and therefore would emit less CO_2e. Less is known of the embodied impact of this new type of lighting, and the potential life cycle impacts have been investigated in a number of studies in the EU and USA (Lim et al. 2011; US Department of Energy 2012; Principi and Fioretti 2014; Tähkämö et al. 2013). The findings from two of these studies for a US and EU LED are presented in Table 13.7.

A review of over 25 LED LCA studies was conducted by the US Department of Energy (2012). Using the ISO methodology for life cycle analysis, the US study considers the impact of an LED luminaire in use for 20 years using the SimaPro (Hischier et al. 2009) and ecoinvent (Hischier et al. 2009) databases. An LED luminaire used for 20 years would emit 0.32 kg CO_2 per lumen. In Tähkämö et al. (2013), it is estimated that a 2017 LED total lighting output (lumen) would be 824. Therefore, the total kg of CO_2e per light over 20 years is 264 kg, of which 98% is the operational CO_2e (259 kg) and 2% is the embodied CO_2e (5 kg). According to the US Department of Energy (2012), a typical 5 W LED light will consume approximately 9.13 kWh of energy per year. Over 20 years, this figure is therefore 182.6 kWh. The embodied CO_2e/kWh can be estimated as 0.027; see Eq. 13.6.

$$0.027 \ kgCO_2e/kWh = 5 \ kgCO_2e/182.6 \ kWh \qquad (13.6)$$

The embodied CO_2e figure presented in this example is based upon energy consumed rather than energy generated.

An additional investigation by Tähkämö et al. (2013) in the EU found that the whole-life CO_2e impact of a 20-year LED light was 167 kg. Assuming 2% is related to the embodied stage (3.34 kg) and using the same energy consumption figure as Eq. 13.6, the total embodied CO_2e can be estimated as 0.018; see Eq. 13.7.

$$0.018 \ kgCO_2e/kWh = 3.34 \ kgCO_2e/182.6 \ kWh \qquad (13.7)$$

Table 13.7 LED studies

Location	Assumptions	kgCO$_2$e/ kWh
USA (US Department of Energy 2012)	One LED luminaire used for 20 years	0.027
EU (Tähkämö et al. 2013)	One LED luminaire used for 20 years	0.018

Discussion

In this chapter, the embodied CO$_2$e of a selection of sustainable technologies used in residential and commercial buildings has been considered, based on readily available information from predominantly LCA studies. Although it is possible to make an educated estimate as to the impact of the average technology, it is by no means accurate. A full LCA using the ISO methodologies on each specific sustainable technology used in each building is necessary in order to provide an accurate and true measure. Detailed information is not readily available, and any studies that have been completed are specific to particular regions and/or focused on particular case studies as discussed above. This makes a general estimate difficult. However, based on the information available, it has been possible to compare the different impact of each technology. Table 13.8 presents the findings from a selection of the studies investigated. It is unwise to compare the technologies against each other as in many cases one cannot be used to replace another. However, they do show the relative impacts. It should also be noted that all kgCO$_2$/kWh values are based on the energy generated with the exception of LED lighting which is based upon energy consumed.

A comparison can be made to show the embodied kgCO$_2$e impact of these technologies, using a kWh of 2016 UK grid electricity for power and/or lighting results in the output of 0.4 kg of CO$_2$e (Department for Business 2016). A kWh of UK grid electricity is therefore equivalent to 4 UK solar PV or 13 solar thermal systems. According to a UK parliamentary publication (Parliamentary Office of Science and Technology 2016) on the carbon footprint of heat generation, a number of conventional and new technologies were compared. A new A or B-rated gas-fired boiler used to generate hot water for central heating has an equivalent CO$_2$e footprint of approximately 0.220 kgCO$_2$e/kWh. Older boilers have a higher figure of approximately 0.340 kgCO$_2$e /kWh. For every kWh of hot water replaced by a UK ASHP, the kgCO$_2$ is reduced by a factor of 8 for new boilers and 12 for older boilers. Moreover, new electric heaters are estimated to emit 0.370 kgCO$_2$/kWh. Therefore, for every kWh of solar PV used as a replacement, the kgCO$_2$ is reduced by a factor of approximately 4.

Table 13.8 Embodied carbon of sustainable technologies

Technology type	Study location	kgCO$_2$/kWh	Study	Year
Solar PV	UK	0.100	Hammond and Jones (2005)	2005
Solar thermal	EU	0.030	Aini Masruroh et al. (2006)	2006
MVHR	UK	0.100	Hernandez and Kenny (2011)	2011
ASHP	UK	0.028	Johnson (2011)	2011
LED lighting	EU	0.018	Tähkämö et al. (2013)	2013

Conclusion

The analysis has compared the full life cycle impact of a selection of sustainable technology systems commonly used in the EU. There remains a lack of accurate data and information on the whole-life impact of sustainable technologies. This is a significant gap as these technologies are becoming increasingly popular, and this research has shown that the impact of new sustainable technologies is significantly less than conventional systems, when compared on a whole-life basis. However, each development is unique with a combination of technologies used, and therefore generalisations should be avoided; for example, an airtight building would typically use solar PV and MVHR in combination. Clearly there are environmental benefits in the use of sustainable technologies. However, there remains a significant problem not considered thus far, that of finance and the extra capital expenditure required to install and maintain the technology which has been considered by Finnegan et al. (2015). It is recommended that it is made mandatory for all sustainable technology used on new or existing buildings to have an Environmental Product Declaration (EPD) to enable the decision maker to make the best-informed choice possible.

References

Achenbach, H., Diederichs, S., Wenker, J., & Rüter, S. (2016). Environmental product declarations in accordance with EN 15804 and EN 16485 – How to account for primary energy of secondary resources? *Environmental Impact Assessment Review, 60*, 134–138.

Aini Masruroh, N., Li, B., & Klemeš, J. (2006). Life cycle analysis of a solar thermal system with thermochemical storage process. *Renewable Energy, 31*(4), 537–548.

Alsema, E. (2000). Energy payback time and CO2 emissions of PV systems. *Progress in Photovoltaics Research and Application, 8*, 17–25.

Ardente, F., Beccali, G., Cellura, M., & Lo Brano, V. (2005). Life cycle assessment of a solar thermal collector. *Renewable Energy, 30*(7), 1031–1054.

Autorité de Régulation de la Poste et des Télécommunications (ARPT) Headquarters, Algeria. (n.d.). https://www.2degreesnetwork.com/groups/2degrees-community/resources/big-beautiful-and-sustainable-10-worlds-most-energy-efficient-offices/. Accessed 24 Jan 2017.

Battisti, R., & Corrado, R. (2005). Evaluation of technical improvements of photovoltaic systems through life cycle assessment methodology. *Energy, 30*, 952–967.

BSI, Carbon Trust. (2008). PAS 2050. Specification for the assessment of the life cycle greenhouse gas emissions of goods and services. Publicly available specification 2050.

Chance, T. (2009). Towards sustainable residential communities; the Beddington Zero Energy Development (BedZED) and beyond. *Environment and Urbanization, 21*, 527–544.

Cheng, J., & Tu, Y. (2008). Trust and knowledge sharing in green supply chains. *International Journal of Supply Chain Management, 13*, 283–295.

Construction Products Regulation (CPR). (n.d.). https://ec.europa.eu/growth/sectors/construction/product-regulation_en. Accessed 24 Jan 2017.

David & Lucile (D&L) Packard Foundation, California (USA). (n.d.). https://www.2degreesnetwork.com/groups/2degrees-community/resources/big-beautiful-and-sustainable-10-worlds-most-energy-efficient-offices/. Accessed 24 Jan 2017.

De Wolf, C., Pomponi, F., & Moncaster, A. (2017). Measuring embodied carbon dioxide equivalent of buildings: A review and critique of current industry practice. *Energy and Buildings, 140*, 68–80.

Department for Business, Energy and Industrial Strategy Greenhouse has reporting – Conversion factors 2016. (2016). https://www.gov.uk/government/publications/greenhouse-gas-reporting-conversion-factors-2016. Accessed 2 Feb 2017.

Design Buildings Wiki. (2016). Fabric first. https://www.designingbuildings.co.uk/wiki/Fabric_first. Accessed 24 Jan 2017.

DPR Construction Net-Zero Energy San Francisco Regional Office. (n.d.). http://www.dpr.com/projects/dpr-construction-san-francisco-regional-office. Accessed 24 Jan 2017.

ECO Platform CEN/TC 350 standards. (n.d.). http://www.eco-platform.org/cen-tc-350.html. Accessed 22 Feb 2017.

EN 15804. (n.d.). Sustainability of construction works. Environmental Product Declarations. Core rules for the product category of construction products https://www.en-standard.eu/csn-en-15804-a1-sustainability-of-construction-works-environmental-product-declarations-core-rules-for-the-product-category-of-construction-products/. Accessed 24 Jan 2017.

Finnegan, S., Dos-Santos, J. D., Chow, D. H. C., Yan, Q. O., & Moncaster, A. (2015). Financing energy efficiency measures in buildings – A new method of appraisal. *International Journal of Sustainable Building Technology and Urban Development, 6*(2), 62–70.

Fthenakis, V., & Kim, H. (2011). Photovoltaics: Life-cycle analyses. *Solar Energy, 85*, 1609–1628.

Gelowitz, M., & McArthur, J. (2016). Investigating the effect of environmental product declaration adoption in LEED® on the construction industry: A case study. *Procedia Engineering, 145*, 58–65.

Goggins, J., Moran, P., Armstrong, A., & Hajdukiewicz, M. (2016). Lifecycle environmental and economic performance of nearly zero energy buildings (NZEB) in Ireland. *Energy and Buildings, 116*, 622–637.

Greening, B., & Azapagic, A. (2012). Domestic heat pumps: Life cycle environmental impacts and potential implications for the UK. *Energy, 39*(1), 205–217.

Hammond, G., & Jones, C. (2005). *Inventory of carbon and energy*. Bath: University of Bath.

Hernandez, P., & Kenny, P. (2011). Development of a methodology for life cycle building energy ratings. *Energy Policy, 39*(6), 3779–3788.

Hischier, R., Weidema, B., Althaus, H., Bauer, C., Doka, G., Dones, R., Frischknecht, R., Hellweg, S., Humbert, S., Jungbluth, N., Köllner, T., Loerincik, Y., Margni, M., & Nemecek, T. (2009). *Implementation of life cycle impact assessment methods ecoinvent report No. 3, v2.1.* Dübendorf: Swiss Centre for Life Cycle Inventories.

Hosenuzzaman, M., Rahim, N., Selvaraj, J., Hasanuzzaman, M., Malek, A., & Nahar, A. (2015). Global prospects, progress, policies, and environmental impact of solar photovoltaic power generation. *Renewable and Sustainable Energy Reviews, 41*, 284–297.

Ibáñez-Forés, V., Pacheco-Blanco, B., Capuz-Rizob, S., & Bovea, M. (2016). Environmental product declarations: Exploring their evolution and the factors affecting their demand in Europe. *Journal of Cleaner Production, 116*, 157–169.

International Organization of Standardization. (n.d.). http://www.iso.org/iso/home.htm. Accessed 24 Jan 2017.
International Organization of Standardization (ISO 14040:2006). (n.d.). Environmental management. Life Cycle Assessment Principles and Framework http://www.iso.org/iso/catalogue_detail?csnumber=37456. Accessed 24 Jan 2017.
International Organization of Standardization (ISO 14044:2006). (n.d.). Environmental management. Life Cycle Assessment Requirements and guidelines http://www.iso.org/iso/home/store/catalogue_tc/catalogue_detail.htm?csnumber=38498. Accessed 24 Jan 2017.
International Renewable Energy Agency (IREA) Headquarters. (n.d.). https://www.2degreesnetwork.com/groups/2degrees-community/resources/big-beautiful-and-sustainable-10-worlds-most-energy-efficient-offices/. Accessed 24 Jan 2017.
IRENA & IEA ETSAP. (2013). Solar Photovoltaics. Technology Brief. Ed.11.
Jayathissaa, P., Jansen, M., Heeren, N., Nagy, Z., & Schluetera, A. (2016). Life cycle assessment of dynamic building integrated photovoltaics. *Solar Energy Materials and Solar Cells, 156*, 75–82.
Johnson, E. (2011). Air-source heat pump carbon footprints: HFC impacts and comparison to other heat sources. *Energy Policy, 39*(3), 1369–1381.
Kristjansdottira, T., Good, C., Inman, M., Schlanbusch, R., & Andresen, I. (2016). Embodied greenhouse gas emissions from PV systems in Norwegian residential Zero Emission Pilot Buildings. *Solar Energy, 133*, 155–171.
Lamnatoua, C., Notton, G., Chemisanaa, D., & Cristofari, C. (2014). Life cycle analysis of a building-integrated solar thermal collector, based on embodied energy and embodied carbon methodologies. *Energy and Buildings, 84*, 378–387.
Lim, S., Kang, D., Ogunseitan, O., & Schoenung, J. (2011). Potential environmental impacts of light-emitting diodes (LEDs): Metallic resources, toxicity and hazardous waste classification. *Environmental Science and Technology, 45*, 320–327.
Lin Lu, J., & Yang, H. (2013). Review on life cycle assessment of energy payback and greenhouse gas emission of solar photovoltaic systems. *Renewable and Sustainable Energy Reviews, 19*, 255–274.
Meier, P. J. (2002). *Life cycle assessment of electricity generation systems and applications for climate change policy analysis* (Doctoral dissertation). Fusion Technology Institute University of Wisconsin, Madison.
Nawaz, I., & Tiwari, G. (2006). Embodied energy analysis of photovoltaic (PV) system based on macro- and micro-level. *Energy Policy, 34*(17), 3144–3152.
NHBC Foundation. (2015). Sustainable Technologies – The experience of housing associations. Published by HIS BRE Press ISBN 978-84806-415-7.
Niewlaar, E., Alsema, E., & Van Engelenburg, B. (1996). Using life cycle assessments for the environmental evaluation of greenhouse gas mitigation options. *Energy Conversion and Management, 37*, 831–836.
Nyman, M., & Simonson, C. (2004). Life cycle assessment of residential ventilation units in cold climate. *Building and Environment, 40*, 15–27.
Oldfield, P., Trabucco, D., & Wood, A. (2009). Five energy generations of tall buildings: An historical analysis of energy consumption in high-rise buildings. *The Journal of Architecture, 14*, 591–613.
Pacca, S., Sivaraman, G., & Keoleian, G. (2007). Parameters affecting the life cycle performance of PV technologies and systems. *Energy Policy, 35*, 3316–3326.
Parliamentary Office of Science and Technology. (2016). Carbon footprint of heat generation. *POSTnote 523.* www.parliament.uk/post
Passer, A., Lasvaux, S., Allacker, K., De Lathauwer, D., Spirinck, C., Wittstock, B., Kellenberger, D., Gschosser, F., Wall, J., & Wallbaum, H. (2015). Environmental product declarations entering the building sector: Critical reflections based on 5 to 10 years' experience in different European countries. *The International Journal of Life Cycle Assessment, 20*(9), 1199–1212.
Passivhaus. (2016). http://www.passivhaus.org.uk/. Accessed 24 Jan 2017.
Peng, C., Huang, Y., & Wu, Z. (2011). Building-integrated photovoltaics (BIPV) in architectural design in China. *Energy and Buildings, 43*(12), 3592–3598.

Principi, P., & Fioretti, R. (2014). A comparative life cycle assessment of luminaires for general lighting for the office – Compact fluorescent lighting (CFL) vs light emitting diode (LED) – A case study. *Journal of Cleaner Production, 83*, 96–107.

ProAir 600LI Heat Recovery Ventilation Unit. (2015). http://www.proair.ie/products/proair-600/. Accessed 12 Feb 2017.

Redondo de la Mata, A. (2014). Life cycle and sustainable assessment of integrated photovoltaic roof systems. Universidad de Catabria Masters Thesis. https://repositorio.unican.es/xmlui/bitstream/handle/10902/5693/Redondo%20de%20la%20Mata%20Alfonso.pdf?sequence=1. Accessed 24 Jan 2017.

RICS. (2015). COP21: Built Environment crucial to attaining emissions targets. http://www.rics.org/uk/news/news-insight/press-releases/cop21-built-environment-crucial-to-attaining-emissions-targets/. Accessed 18 Jan 2017.

Schaefer, H., & Hagrdorn. (1992). Hidden energy and correlated environmental characteristics of P.V. power generation. *Renewable Energy, 2*(2), 159–166.

Sherwania, A., Usmani, J., & Varun. (2010). Life cycle assessment of solar PV based electricity generation systems: A review. *Renewable and Sustainable Energy Reviews, 14*(1), 540–544.

Simon, G. (2013). Sustainable buildings: BREEAM case studies. *Journal of Building Survey, Appraisal & Valuation, 2*, 7–15.

Strazza, C., Del Borghi, A., Magrassi, F., & Gallo, M. (2016). Using environmental product declaration as source of data for life cycle assessment: A case study. *Journal of Cleaner Production, 112*, 333–342.

Tähkämö, L., Bazzana, M., Ravel, P., Grannec, F., Martinsons, C., & Zissis, G. (2013). Life cycle assessment of light-emitting diode downlight luminaire – A case study. *The International Journal of Life Cycle Assessment, 18*(5), 1009–1018.

The Edge. (n.d.). Amsterdam, The Netherlands. http://www.plparchitecture.com/the-edge.html. Accessed 24 Jan 2017.

Thermal comfort in buildings. (n.d.). https://www.designingbuildings.co.uk/wiki/Thermal_comfort_in_buildings. Accessed 24 Jan 2017.

Throndsen, W., Berker, T., & Knoll, B.E. (2015). Powerhouse Kjorbo. Evaluation of construction process and early use phase. Zero Emission Buildings (ZEB) Project Report no 25. https://brage.bibsys.no/xmlui/handle/11250/2398994. Accessed 24 Jan 2016.

Tripanagnostopoules, Y., Souliotis, M., Battisti, R., & Corrado, A. (2005). Energy, cost and LCA results of PV and hybrid PV/T solar systems. *Progress in Photovoltaics: Research and Applications, 13*, 235–250.

US Department of Energy. (2012). Life cycle assessment of energy and environmental impacts of LED lighting products. Part 2: LED manufacturing and performance. Prepared for: Solid-State Lighting Program Building Technologies Program Office of Energy Efficiency and Renewable Energy U.S. Department of Energy Prepared by: Pacific Northwest National Laboratory N14 Energy Limited.

Yue, D., You, F., & Darling, S. (2014). Domestic and overseas manufacturing scenarios of silicon-based photovoltaics: Life cycle energy and environmental comparative analysis. *Solar Energy, 105*, 669–678.

Chapter 14
Accounting for Embodied Carbon Emissions in Planning and Optimisation of Transport Activities During Construction

Ahmed W. A. Hammad, Ali Akbarnezhad, and David Rey

Introduction

Tackling the issue of greenhouse gas (GHG) emissions in urban developments is of paramount importance nowadays, especially with governments and regulating authorities focusing more on the need to establish a sustainable environment within their cities. Urban regions are responsible for up to 80% of the total GHG emissions (United Nations Environment Programme 2010) with this figure expected to increase as the urbanisation rate continues to rise. The construction sector constitutes a large portion of the carbon emissions produced within urban environments. Current studies suggest that around 45–60% of total GHG emissions in urban regions are caused by construction activities (Orabi et al. 2012). Within construction industry, which encompasses areas such as road and infrastructure construction, the building sector is the largest contributor to GHG emissions (Intergovernmental Panel on Climate Change (IPCC) 2007). As such it seems reasonable to address the rising rates of emissions in urban environments through reduction of the embodied carbon of building construction activities.

In order to quantify environmental impacts of products or services, life cycle assessment (LCA) is deployed in the existing literature. In construction, it is commonly adopted for establishing the environmental impacts of buildings, with focus primarily on the maintenance and operational phase of buildings (Brocklesby and Davison 2000; Hong et al. 2012; Bilec et al. 2006; Park and Hong 2011). However, carbon emitted during the construction process has not received much attention, in comparison to the maintenance or operational phases (Monahan and Powell 2011; Pomponi and Moncaster 2017), due to its relatively smaller impact on overall carbon emission. Yet, as stated by IPCC, an encompassing strategy that

A. W. A. Hammad (✉) · A. Akbarnezhad · D. Rey
School of Civil and Environmental Engineering, UNSW, Sydney, NSW, Australia
e-mail: a.hammad@unsw.edu.au

© Springer International Publishing AG 2018
F. Pomponi et al. (eds.), *Embodied Carbon in Buildings*,
https://doi.org/10.1007/978-3-319-72796-7_14

aims to minimise embodied carbon in construction should also consider the GHG emissions from the construction phase (Metz et al. 2007). Additionally, when considering the short duration of the construction phase, relative to the operation and maintenance phases, the resulting carbon emitted, when conducting an analysis that cancels the effects of phase duration, reveals the critical nature of the construction stage on total carbon emissions (Hong et al. 2013). This has prompted many regulations to specify the need for extensive measures that address carbon emissions in buildings, which requires consideration of the construction phase (Core Writing Team et al. 2007; Greenhouse Gas Inventory and Research Center of Korea (GGIRCK) 2011).

One means of reducing GHG emissions in construction is through addressing the associated embodied carbon, defined as the total carbon emitted during the life cycle of the building process, excluding its operation phase. Mitigation measures recommended by the IPCC include the use of low-carbon buildings, which can be achieved through proper planning that is carried out at the onset of projects. Though methods have been extensively implemented for assessing and addressing the emissions at the operational phase of buildings, the same cannot be said for embodied carbon, as this is not commonly considered in the design and planning phases of buildings (Monahan and Powell 2011).

This chapter proposes to tackle the issue of embodied carbon emissions in building projects, through addressing the transportation cycle in building construction operations. The literature highlights that transportation activities account for almost 16.5% of the overall embodied emissions in building construction (Rodríguez Serrano and Porras Álvarez 2016). Other works have suggested that within the TC350 standards defining the cradle to grave impact of buildings, the transportation stage, which considers transport from factory gate to site, accounts for around 9% of the total embodied energy in residential dwellings (Moncaster and Symons 2013). This figure however does not account for transport that occurs during construction stage.

Two well-known problems with applications within construction are modelled and solved here to optimise their performance, hence leading to reductions in transport-related carbon emissions. The first problem is a site layout planning problem (SLPP) (Hammad et al. 2015, 2016; Zouein et al. 2002), where the location of tower cranes on construction sites is optimised, so that carbon emissions resulting from their operations are minimised. The selection of locations can considerably impact transport needs between facilities on construction sites and hence emissions. Factors including the transportation distance and mode of transportation, which are known to influence total embodied carbon (Akbarnezhad and Xiao 2017), are related to the location decisions taken within the presented models. The SLPP presented in this chapter is modelled combining principles from the facility location problem (FLP) and the facility layout problem (FLAP). In FLAP, the focus is on finding a layout for the facilities within a constrained area during the different phases of the project such that a certain objective function is minimised/ maximised, while in the FLP, the focus is placed on locating facilities in a relatively larger area. In particular, two relevant FLP formulations, adapted to the context of

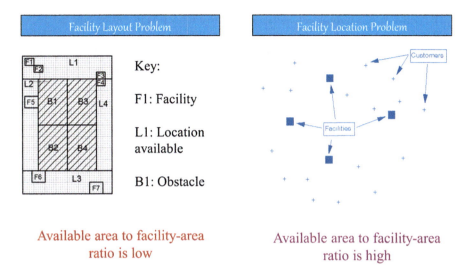

Fig. 14.1 Major difference between FLAP and FLP

the SLPP, are contrasted, namely, a set covering location model (Eiselt and Marianov 2009), where full coverage by the tower crane to all material demand points on a construction site is achieved, and the maximal coverage model (Karasakal and Karasakal 2004), where coverage to the demand points is maximised. In the set covering model, given a set of elements called U and a collection of finite sets labelled, S, the aim is to find the smallest subcollection of S whose union equals U. On the other hand, in the maximal coverage problem, it is not necessary to fully cover each element in U; the aim is to choose at most k sets to cover as many elements in U as possible. Figure 14.1 highlights the difference between FLAP and FLP.

The second problem type, the container loading problem (CLP), deals with off-site transportation activities. This is particularly relevant to the transportation activities involved with moving material from manufacturers to suppliers and also from suppliers to construction sites.

Although the problems presented in this chapter have been widely adopted in supply chain management, their use in construction for the purpose of minimising embodied carbon during construction activities has not been explored. Hence the optimisation framework proposed in this work is a novel contribution to the reduction of embodied carbon in construction through solving the tower crane location and the container loading problem.

The rest of the chapter is organised as follows. In the next section, a brief literature review of the problems from which SLPP is adopted, namely, FLAP and FLP, is given. In addition, a review on CLP is presented. Next, models are presented to locate the tower crane and associated material supply facilities in such a way that carbon emissions resulting from the crane's operation are minimised. Following that, a CLP model is described where the number of truck trips required to be performed is minimised, hence leading to a reduction in carbon emitted due to truck transport activities. Case examples are presented to highlight the applicability of the models. Lastly, concluding remarks are highlighted.

Literature Review

Facility Layout Problem

The FLAP is applicable where departments with known dimensions, examples of which are workshops and plants/machines, are to be located within a known area, such that the travel cost/material handling between the various departments is minimised (Heragu and Kusiak 1991; Kulturel-Konak and Konak 2013). Within FLAP, a facility is defined as an entity which facilitates the performance of a task (Drira et al. 2007). The solution to the FLAP is a block layout displaying the dimensions along with the relative positions of the departments within a given area (Chen et al. 2015). In engineering construction and management, an adaptation of the FLAP, called the SLPP, involves finding an appropriate physical arrangement for a given number of temporary facilities operating during the construction process, to the available planar region of the construction site (Andayesh and Sadeghpour 2013; Hammad et al. 2016; Khalafallah and El-Rayes 2011; Li and Love 2000; Zouein et al. 2002).

Facility Location Problem

The FLP, which involves locating single/multiple facilities in a wide planar region, forms a vital facet for many strategic planning applications (Drezner and Hamacher 2004). Applications of FLP in the literature include locating public facilities, such as fire stations, hospitals and emergency response units (Badri et al. 1998; Balcik and Beamon 2008; Batta et al. 2014; Hahn and Krarup 2001; Li et al. 2011), as well as private facilities, such as warehouses and manufacturing firms (Revelle et al. 1970; Klose and Drexl 2005; Thanh et al. 2008), to maximise delivery efficiency of services to customers. FLP can either be solved as an uncapacitated FLP, where each facility to be located is assumed to be able to supply all demand points (Ghosh 2003; Kratica et al. 2014; Verter 2011), or as a capacitated problem, where a limit is placed on the demand that can be supplied by each positioned facility (Boyaci et al. 2013; Fischetti et al. 2016).

Categorising the FLP is based on the type of model that is formulated to solve the problem. Two major problem types, adopted in the literature, include the set covering location problem and maximum covering location problem (Farahani et al. 2012). In the set covering problem (Eiselt and Marianov 2009; Gunawardane 1982; Rajagopalan et al. 2008; ReVelle et al. 1976), the aim is to satisfy all demand leading to a specified level of coverage, through minimising the average cost of locating the facilities. For maximal coverage problems, the aim is to maximise the demand coverage (Church and Velle 1974; Church et al. 1996; Hochbaum and Pathria 1998; Karasakal and Karasakal 2004; Murawski and Church 2009).

Container Loading Problem

Packing problems have been widely studied since the early 1960s. During this period, many variants have been developed, including knapsack loading (Gehring et al. 1990), bin-packing (Martello et al. 2000) and strip packing (Bischoff et al. 1995) problems. In this chapter, the focus is on the container loading problem, which involves a given set of items that need to be placed within a container of a specified size, so that the space of the container is fully utilised (Eley 2002). Two constraints need to be satisfied, namely, ensuring that the items are fully loaded within the container and that none of the items overlap with one another. The problem has numerous practical applications in the industry; this is due to its impact on transportation and shipping costs (Huang et al. 2016; Nowakowski 2016). In particular, an optimised packing pattern within the container will reduce the number of containers required to load the items and hence will also reduce the total number of trips that need to be conducted to transport the items to their destination. Given that transport activities contribute heavily to the embodied carbon in construction (British Standards Institution 2011; Moncaster and Symons 2013), addressing the CLP leads to an efficient utilisation of total number of trucks required to transport construction materials and hence the reduction of embodied carbon that is a result of such transport activities. Many of the inefficiencies in material transportation in construction are caused by incorrect loading operations of materials. This is specifically true whenever forklifts are deployed to load trucks from the top, as commonly occurs in construction (Nowakowski 2016). This in turn leads to the ill use of space in transport truck and trailers which entails extended transportation activities and hence an increase in the resulting carbon emissions released as a result.

Framework

In this chapter, an optimisation framework is presented to address the overall embodied carbon in construction transport activities. This is depicted in Fig. 14.2. As can be seen, two major optimisation problems are addressed, namely, a SLPP, which is solved to minimise the total carbon emissions resulting from the tower crane operations on site, and a CLP, which is solved such that the number of trips required to transport materials to the construction site is reduced, hence reducing carbon emissions due to material transportation to the construction site.

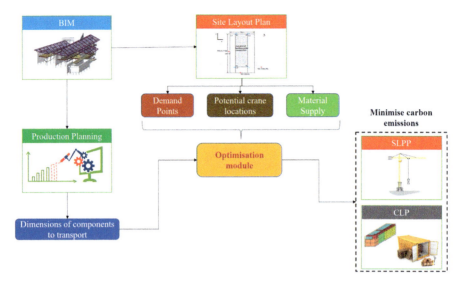

Fig. 14.2 Framework for reducing embodied carbon through optimising transport activities in construction

The framework is based on having a building information model (BIM) as the primary source of data regarding the building elements making up the construction project. From BIM, a preliminary site layout plan can be deduced, which is used to obtain information regarding demand point locations, possible supply points associated with the demand points and potential locations of cranes on site. All of this information is then directed to the optimisation module, where a site layout planning problem is solved. At the same time a production plan can be devised, which incorporates the various materials and element compositions that will be required to be delivered to the construction site during the various stages of construction. It is through the component production plan that data, regarding the dimensions of the material components to be delivered to the construction site, can be obtained. A CLP is then solved, which reduces the total number of trips required to be conducted between the material suppliers and construction site. Both optimisation problems are designed to minimise the total carbon emissions, aiming to provide significant reductions in embodied carbon of construction building materials.

Site Layout Planning Models

In the industry, the location of tower cranes is achieved through trial and error, with reference to loading charts. Practitioners rely mostly on experience to align the location of tower canes with the topology of the construction site (Zhang et al. 1999). Therefore, qualitative references are lacking for project engineers to rely on when making tower crane location decisions. In literature, the tower crane location problem is typically formulated as a SLPP, where studies focus on optimising the location of the tower cranes through minimising the crane operating cost (Zhang et al. 1996, 1999). Some studies have also focused on the assignment of a priori placed supply points to known demand regions as part of the tower crane location process (Huang et al. 2011; Lien and Cheng 2014; Nadoushani et al. 2016; Tam and Tong 2003). Yet to the authors' best knowledge, no attempt has been made to integrate modelling formulations from FLP to locate tower cranes on a construction site, for the sake of minimising embodied carbon emissions associated with crane operations.

In this section, two models are proposed to solve the SLPP, with a focus on minimising carbon emissions resulting from crane operation. The first model, labelled as the set covering tower crane location (SCTL) model, aims to minimise the total carbon emission associated with the tower crane operations while ensuring that all demand points on the construction site are serviced. The second model, labelled as the maximum covering tower crane location (MCTL) model, maximises the total number of demand points that are serviced given a restrained total number of tower cranes that can be installed on site. Both models address the issue of carbon emission reduction in a different way. Model SCTL tries to minimise emissions directly without sacrificing the servicing of all demand points. Model MCTL influences carbon emissions through limiting the number of tower cranes that can operate during the project while maximising the total demand points that are serviced by the located tower cranes. The optimisation models presented in this section are assumed to be applicable to medium/large construction sites, where significant savings in carbon can be realised through optimising the transport activities involved. Given that in large construction sites it is very common to see multiple tower cranes in operation, the proposed models are also applicable to such cases.

The representation of nodes making up the tower cranes, supply points and demand regions in both models is depicted in Fig. 14.3. Formulations for both models are presented next.

SCTL

Since carbon emissions associated with tower cranes on a construction site are impacted by both the number of operating tower cranes during the construction phase and the operating duration of the tower crane, an important parameter making up the objective function defined for Model SCTL is the crane's operating time parameter, T_{ij}^k. This is determined by the crane's tangential and radial trolley travel time, T_{ijk}^ω and T_{ijk}^r, respectively, and the horizontal and vertical hook travel time, T_{ijk}^h and T_{ijk}^v, respectively. The following equations, commonly adopted in the literature (Zhang et al. 1996), Eqs. 14.1, 14.2, 14.3, 14.4, 14.5, 14.6, 14.7, and 14.8, are used to derive the operation timing parameter of the tower crane:

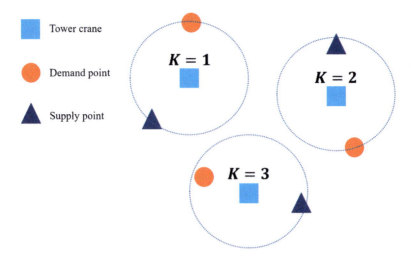

Fig. 14.3 Crane model representation, with the dashed circles highlighting the operating radius of the tower cranes, in accordance with their relevant loading charts

$$\rho_{ak} = \sqrt{(xc_a - xc_k)^2 + (yc_a - yc_k)^2} \quad \forall a \in I \cup J, \forall k \in K \qquad (14.1)$$

$$l_{ij} = \sqrt{(xc_i - xc_j)^2 + \left(yc_i - yc_j\right)^2} \quad \forall i \in I, \forall j \in J \qquad (14.2)$$

$$T^r_{ijk} = \frac{|\rho_{ik} - \rho_{jk}|}{V_r} \quad \forall i \in I, \forall j \in J, \forall k \in K \qquad (14.3)$$

$$\eta_{ijk} = \frac{l_{ij}^2 - \rho_{ik}^2 - \rho_{jk}^2}{2\rho_{ik}\rho_{jk}} \quad \forall i \in I, \forall j \in J, \forall k \in K \qquad (14.4)$$

$$T^\omega_{ijk} = \frac{1}{V_\omega} \cos^{-1}(\eta_{ijk}), \quad \begin{array}{l} \forall i \in I, \forall j \in J, \forall k \in K, \\ 0 \le \theta = \cos^{-1}(\eta_{ijk}) \le \pi \end{array} \qquad (14.5)$$

$$T^h_{ijk} = \max\left\{T^r_{ijk}, T^\omega_{ijk}\right\} + \alpha\min\left\{T^r_{ijk}, T^\omega_{ijk}\right\} \quad \forall i \in I, \forall j \in J, \forall k \in K \qquad (14.6)$$

$$T^v_{ij} = \frac{|zc_i - zc_j|}{V_v} \quad \forall i \in I, \forall j \in J \qquad (14.7)$$

$$T_{ijk} = \gamma_k\left(\max\left\{T^h_{ijk}, T^v_{ij}\right\} + \beta\min\left\{T^h_{ijk}, T^v_{ij}\right\}\right) \quad \forall i \in I, \forall j \in J, \forall k \in K \qquad (14.8)$$

Let I, J and K denote the sets of demand points, potential supply points' locations and potential tower cranes' locations, respectively. Notation for Eqs. 14.1, 14.2, 14.3, 14.4, 14.5, 14.6, 14.7, and 14.8 is highlighted in Fig. 14.3. Parameters α, β and γ_k define the degree of coordination of hook movement in tangential and radial direction, in vertical and horizontal planes, and the degree of difficulty in hook movement control for a crane located at k, respectively. For further explanation of Eqs. 14.1, 14.2, 14.3, 14.4, 14.5, 14.6, 14.7, and 14.8, the reader is referred to Zhang et al. (1996) (Fig. 14.4).

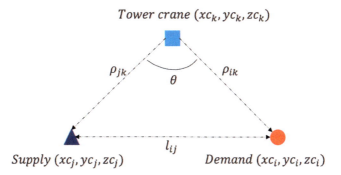

Fig. 14.4 Notation adopted for crane location models

The sets J_i and K_j are defined as: $J_i = \{j \in J : D_{ji} < \rho\}$ for $i \in I$, and $K_j = \{k \in K : D_{jk} < \rho\}$ for $j \in J$, respectively. The parameter ρ acts as an upper limit imposed on the distance D_{SW} between two nodes s *and* w in the network representation. This ensures that the supply points chosen, j, are positioned at a servicing distance away from the demand point, i, and crane point k. The carbon emissions per hour resulting from a single tower crane operation are denoted by E. An arbitrary large value is given as M. The binary variable, δ_k, is defined to equal 1 if a tower crane is positioned at location k, and zero otherwise, while the binary variable, x_{ij}, takes the value 1 if supply node j is assigned to a demand point i, and zero otherwise. An auxiliary binary variable, namely, ϕ_j, is adopted. For the SCTL the formulation is then given as:

$$\min \; E \sum_{i \in I} \sum_{j \in J} \sum_{k \in K} T_{ij}^k \delta_k x_{ij} \tag{14.9}$$

subject to

$$\sum_{j \in J_i} x_{ij} \geq 1 \quad \forall i \in I \tag{14.10}$$

$$\sum_{i \in I} x_{ij} \leq \phi_j M \quad \forall j \in J \tag{14.11}$$

$$\sum_{k \in K_j} \delta_k \geq \phi_j \quad \forall j \in J \tag{14.12}$$

$$x_{ij}, \phi_j, \delta_k \in \{0, 1\} \quad \forall i \in I, \forall j \in J, \forall k \in K \tag{14.13}$$

The objective function in SCTL, Eq. 14.9, is formulated as a quadratic function, which minimises the overall carbon emissions resulting from the operations of tower cranes on the construction site, which achieve full level of coverage to all demand points. Equation 14.10 requires that each demand point is serviced by at least one supply point that is close enough to permit the tower crane's jib to reach it. Equations 14.11 and 14.12 are defined to control the auxiliary variable ϕ_j, which linearises the condition that a supply point is assigned to one or more demand points and permits a link between the supply points assignment variables, and x_{ij} and the variable for locating a tower crane δ_k. This ensures that the supply point is positioned within operating distance of the tower crane. Equation 14.13 establishes the domain of the model variables.

MCTL

The second model, defined for solving the tower crane location problem, maximises the coverage of a maximum number of tower cranes that can be set up on the construction site. Through restraining the maximum number of tower cranes that can operate on the construction site, a limit is placed on the total carbon emissions that result from tower crane operations.

The **MCTL** model is presented as:

$$\max \sum_{i\in I} Q_i Z_i \tag{14.14}$$

subject to

$$\sum_{j\in J} \psi_j \le PS \tag{14.15}$$

$$\sum_{k\in K} \delta_k \le PT \tag{14.16}$$

$$\sum_{i\in I_j} x_{ij} \le \bar{M}\psi_j \quad \forall j\in J \tag{14.17}$$

$$\sum_{i\in I_j} x_{ij} \ge \psi_j \quad \forall j\in J \tag{14.18}$$

$$Z_i \le \sum_{k\in K} \delta_k \quad \forall i\in I \tag{14.19}$$

$$Z_i \le \sum_{j\in J_i} x_{ij} \quad \forall i\in I \tag{14.20}$$

$$\sum_{j\in J_i} x_{ij} \le 1 \quad \forall i\in I \tag{14.21}$$

$$x_{ij}, Z_i, \delta_k, \psi_j \in \{0,1\} \quad \forall i\in I, \forall j\in J, \forall k\in K \tag{14.22}$$

Where the set I_j is defined as $I_j = \{i \in I : D_{ij} < \rho\}$. Parameters PS and PT define the maximum number of supply points and tower cranes to be assigned, while parameter $\bar{M}$ denotes an arbitrary large value. The binary variable Ψj equals 1 if a supply point is activated at node j, while the binary variable Zi equals 1 if demand node is covered and 0 otherwise. The rest of the notation is described for Model SCTL.

The objective function, Eq. 14.14, maximises the number of demand points covered. Equations 14.15 and 14.16 state that the maximum number of supply points and tower cranes is to be less than the specified limits, PS and PT, respectively. Equations 14.17 and 14.18 link variables x_{ij} and Ψj. The coverage variable is activated only if a supply point is assigned to a demand point and a tower crane is placed in close proximity to the demand point; this is specified through Eqs. 14.19 and 14.20. Equation 14.21 requires that each demand point be associated with a maximum of 1 supply point, to avoid over allocation to the demand points. Finally, the domain of the variables is given by Eq. 14.22.

For the MCTL model, the following equation is used to assess the carbon emissions per unit material resulting from the site plan adopted, Eq. 14.23:

$$E \sum_{i \in I} \sum_{j \in J} \sum_{k \in K} T_{ij}^{k} {}_{i} \bar{\delta}_{k} \bar{x}_{ij} \tag{14.23}$$

Where $\bar{\delta}_k$ and $\bar{x}_{ij}$ are now both parameters whose values are obtained after solving the model. This then enables contrasting both SCTL and MCTL to find which results in the site layout with the least embodied carbon emissions.

Container Loading Problem Model

The second aspect contained within the highlighted framework of Fig. 14.2 addresses the embodied carbon resulting from transportation activities of the construction industry, through optimising the layout of materials within the shipping container used to transport materials to and from the construction site, in such a manner that container space utilisation is maximised. This ensures that the total trips to be performed is minimised. As a result, the model to be presented next solves a container loading problem, where materials are assumed to be represented as a convex polyhedron bounded by six quadrilateral faces, Fig. 14.5. All material types are assumed to be bounded by such polyhedron in the proposed model.

Let V_c denote the volume of item c, and let (x_n, y_n, z_n) be the bottom left corner coordinate of item c. Define l_{cs} and r_{cs} to equal 1 if item S is positioned to the left and right of item C, respectively. Also, define a_{cs} and bl_{cs} to equal 1 if item S is placed above and below item C, respectively. Finally, define f_{cs} and bh_{cs} to equal 1 if item S is placed in front of and behind item C, respectively. The notation for the CLP presented is given in Fig. 14.6.

To solve the CLP, the following model is presented, namely, **CL**:

$$\max \sum_{c \in C} w_c d_c h_c \zeta_z \tag{14.24}$$

Subject to

$$l_{cs} + r_{cs} + a_{cs} + bl_{cs} + f_{cs} + bh_{cs} = \zeta_s + \zeta_c - 1 \quad \forall s, c \in C : c < s \tag{14.25}$$
$$x_c - x_s \geq w_s l_{cs} - W(1 - l_{cs}) \quad \forall s, c \in C : c < s \tag{14.26}$$

Fig. 14.5 Steel reinforcement, enclosed within a cuboid for the CLP

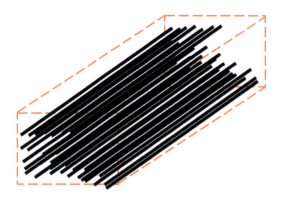

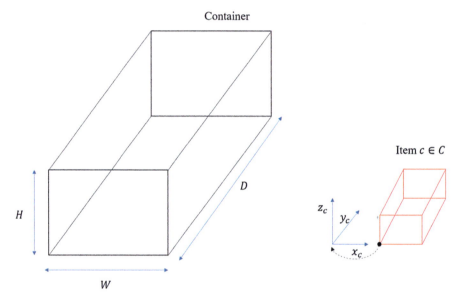

Container

Item $c \in C$

H

W

D

Fig. 14.6 Notation used in CLP model

$$x_s - x_c \geq w_c r_{cs} - W(1 - r_{cs}) \quad \forall s, c \in C : c < s \tag{14.27}$$
$$z_c - z_s \geq h_s \text{bl}_{cs} - H(1 - \text{bl}_{cs}) \quad \forall s, c \in C : c < s \tag{14.28}$$
$$z_s - z_c \geq h_c a_{cs} - H(1 - a_{cs}) \quad \forall s, c \in C : c < s \tag{14.29}$$
$$y_c - y_s \geq d_s f_{cs} - D(1 - f_{cs}) \quad \forall s, c \in C : c < s \tag{14.30}$$
$$y_s - y_c \geq d_c \text{bh}_{cs} - D(1 - \text{bh}_{cs}) \quad \forall s, c \in C : c < s \tag{14.31}$$
$$x_c, y_c, z_c \geq 0 \quad \forall c \in C \tag{14.32}$$
$$r_{cs}, l_{cs}, a_{cs}, \text{bl}_{cs}, f_{cs}, \text{bh}_{cs} \in \{0, 1\} \quad \forall s, c \in C : c < s \tag{14.33}$$

The objective function, Eq. 14.24, maximises the total volume utilised within the loaded container. Equation 14.25 requires one condition to hold, in order to ensure that the loaded items do not overlap within the container. These conditions are determined by Eqs. 14.26, 14.27, 14.28, 14.29, 14.30, and 14.31. In particular, Eqs. 14.26 and 14.27 state that item S is to the left or to the right of item C, respectively. Equations 14.28 and 14.29 state that item S is either below or above item C, respectively. Equations 14.30 and 14.31 state that item S is in front of or behind item C, respectively. Equations 14.32 and 14.33 define the domain of the variables within Model CL.

Once optimised solutions are obtained, to calculate the total embodied carbon, the following equation is used, as given in (Hong et al. 2013) Eq. 14.34:

$$\sum_{c \in C} \mathrm{TD} \cdot f \cdot Q_c \tag{14.34}$$

where TD is the total travel distance, f is the carbon emission factor of truck/trailer, and Q is the weight of material loaded into the truck/trailer, in tonnes.

Case Studies

In this section, the proposed models will be tested on two practical cases, obtained from a building contractor in Sydney, Australia. In particular, Case 1 involves a site layout planning problem for a medium-sized project, where tower cranes and material supply storage are desired to be located, while Case 2 represents a container loading problem for the same project that requires stocks to be delivered to the site to satisfy project requirements. The models proposed are implemented in AMPL (Fourer et al. 1993) and solved using CPLEX 12.6.0 (IBM Knowledge Center 2016). In all cases the units for embodied carbon is reported in kilogrammes carbon dioxide equivalent, $kgCO_2e$.

Case 1

In the first case study, the contractor needs to determine an appropriate site layout where supply points and tower cranes need to be located to service a set of predefined points. Both location models proposed above, SCTL and MCTL, are assessed for their relative impact on total carbon emissions resulting from the tower crane operations. The construction site has dimensions of 300 m by 200 m and is displayed in Fig. 14.7. A total of four tower crane locations are identified beforehand, along with four suitable supply point locations. The demand points correspond to areas within the constructed building where material is required. The site layout is analysed for a single phase of the construction process, namely, *Phase 2* of the project, involving the erection of the first and second floors of the steel building. The coordinates of predefined locations for the tower crane, supply points and existing demand regions are given in Table 14.1.

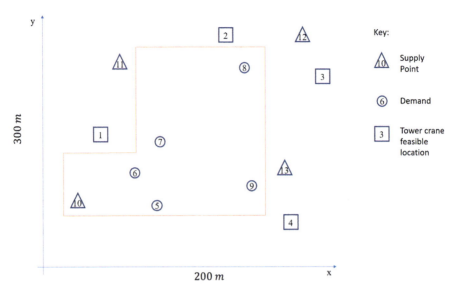

Fig. 14.7 Construction site layout for Case 1

Table 14.1 Coordinates for site layout of Case 1

	Nodes	Coordinates (xc_n, yc_n, zc_n)
Tower crane locations	1	(20,160,0)
	2	(110,290,0)
	3	(190,250,0)
	4	(160,100,0)
Demand points	5	(80,100,10)
	6	(50,150,6)
	7	(80,150,4)
	8	(100,260,7)
	9	(110,110,9)
Potential supply points	10	(10,100,0)
	11	(30,270,0)
	12	(160,290,0)
	13	(160,150,0)

The time needed for the tower crane operations per unit of material transfer in both models is calculated using Eqs. 14.1, 14.2, 14.3, 14.4, 14.5, 14.6, 14.7, and 14.8. The results obtained from both models are contrasted with the decision-making of a site planning engineer, whose solution is based solely on experience.

Table 14.2 displays the optimised results for Models SCTL and MCTL. A 32-ton electric tower crane is assumed for this example with an associated emission rate of 465 g/kWh (Hasan et al. 2013). Average tower crane load and total hours use are derived from Aurecon Australasia (2016). As can be noticed, the total embodied

Table 14.2 Optimised results for the site layout

	Models	
	SCLP	MCLP
Total carbon emissions $kgCO_2$ $KgCO_2 - e$	170	1240
Location of tower crane	@ node 1	@ node 1,2,3,4
Supply point allocation	Demand 5, 6, 7, 9 → supply node 10	Demand 5, 6, 9 → supply node 10
	Demand 8 → supply node 11	Demand 7, 8 → supply node 11

carbon of the site layout, measured as the carbon emissions released from tower crane operations per unit of material transported, produced by Model SCTL is 86 % less than that resulting from Model MCTL (2.83 kgCO$_2$e/unit material vs 21 kgCO$_2$e/unit material). This result is explained by the number of tower cranes deployed by both models, where for Model SCTL, only a single crane is utilised, whereas Model MCLP specifies the installation of four tower cranes. In both models, two supply points are placed, namely, at Nodes 10 and 11. The allocation of demand regions to these supply points is a bit different in both models, since in Model SCLP, demand regions 5, 6, 7 and 8 are assigned to supply point 10, while demand region 8 is assigned to supply point 11. The allocation is almost the same in Model MCLP, with demand region 8 this time assigned to supply region 11. Compared with the results of an experienced engineer, Models SCLP and MCLP produce a layout that is associated with 87% and 5% lower embodied carbon.

For the full duration of *Phase 2* of the project, a total of 734 units of material is required to be transported between the supply points and demand points. Considering the distribution of materials for this phase of the project, the bar chart of Fig. 14.8 is produced. Again, a significant saving in total carbon emissions is noticed in Model SCLT, in comparison to Model MCTL (2001 kgCO$_2$e vs 16,785 kgCO$_2$e). It is noted that the total carbon emissions associated with the project considered were measured at around 155,310 kgCO$_2$e.

Case 2

The problem in many construction transport activities is due to the inefficient use of available capacity for transporting materials. In this case study, a 15-tonne truck, with dimensions $7 \times 2.5 \times 2.5$ m and a total capacity of 43 m^3, is to be loaded with seven items whose dimensions are given in Table 14.3. The materials are to be transported to the construction site, located at a distance of 30 km away from the material manufacturer. An emission rate of 0.065 kgCO$_2$e/ton.km by the truck, as deduced from emission databases, like MOVES (US EPA 2016), is assumed.

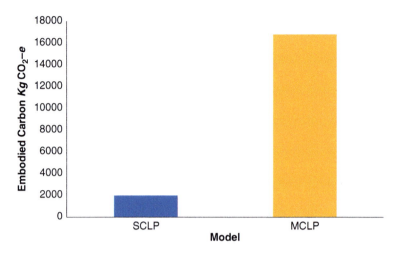

Fig. 14.8 Bar chart, showing total transport-related embodied carbon of site layout resulting due to tower crane operations, during *Phase 2* of the project

Table 14.3 Dimensions of items to be transported to construction site

Item	Width (m)	Depth (m)	Height (m)	Weight (tonnes)
1	2	1	2	2
2	2	4	2	1
3	2	2	1	1.5
4	2	2	1	2
5	2	2	2	1.5
6	2	1	2	2
7	2	1	2	2

The results of the model are then contrasted with the container layout decided on by the experienced engineer. The loading layout produced by the optimised model uses 65% of the truck, as opposed to only 40% when the expert opinion is sought. For the overall duration of the project, a total of 5530 kgCO$_2$e will be emitted from material transportation, if an expert opinion is sought, compared to 4095 kgCO$_2$e if the CL model is utilised. A 26% reduction in total embodied carbon is achieved in the material transportation mode if the proposed optimisation model is used, which equates to 1435 kgCO$_2$e in of carbon emissions. This number is around 9% of the total embodied carbon of the project considered. The use of the proposed CL model therefore achieves a significant reduction in embodied carbon associated with the building.

Conclusion

In this chapter, optimisation models were presented to address the embodied carbon resulting from transportation activities in the construction sector. In particular, the focus was on tower crane operations, through optimising the site layout plan and material transportation efficiency. The results indicated that considerable reduction in the transport-related embodied carbon of buildings can be achieved through optimising these operations.

For the site layout plans, two models were introduced, namely, SCLP, where a set covering problem is solved to minimise the total carbon emissions resulting from the movement of a tower crane between supply and demand points, and MCLP, where coverage of all demand points by a set of tower crane is maximised. When compared with a solution offered by an experienced engineer, both SCLP and MCLP produced a site layout with 46% and 21% less carbon emissions, respectively. In particular, comparisons between both of these models revealed that SCLP was able to save up to 86% of total carbon emissions from tower cranes.

In terms of addressing material transportation to construction sites, Model CL was introduced. The aim was to maximise the utilisation space of the truck used to transport materials to the construction site, hence minimising the total number of trips conducted and the total carbon emissions produced during the material hauling trip. Again, considerable improvements were noticed when utilising the model as opposed to solely relying on the expertise of an experienced practitioner, with a total carbon emission saving of 26%.

Model CL only accounts for space use within the container. The model can be expanded to encompass other factors such as the weight of materials and their impact on the resulting carbon emissions.

The models presented in this chapter can be used by decision-makers to reduce transport-related embodied carbon in construction. Future work will consider the overall transport activities that comprise a typical life cycle of a project, including various other equipment elements.

References

Akbarnezhad, A., & Xiao, J. (2017). Estimation and minimization of embodied carbon of buildings: A review. *Buildings, 7*, 5. https://doi.org/10.3390/buildings7010005.

Andayesh, M., & Sadeghpour, F. (2013). Dynamic site layout planning through minimization of total potential energy. *Automation in Construction, 31*, 92–102. https://doi.org/10.1016/j.autcon.2012.11.039.

Aurecon Australasia. (2016). *Appendix 1 greenhouse gas asssessment report*. New South Wales: Rail Operations Centre.

Badri, M. A., Mortagy, A. K., & Alsayed, C. A. (1998). EURO best applied paper CompetitionA multi-objective model for locating fire stations. *European Journal of Operational Research, 110*, 243–260. https://doi.org/10.1016/S0377-2217(97)00247-6.

Balcik, B., & Beamon, B. M. (2008). Facility location in humanitarian relief. *International Journal of Logistics Research and Applications, 11*, 101–121. https://doi.org/10.1080/13675560701561789.

Batta, R., Lejeune, M., & Prasad, S. (2014). Public facility location using dispersion, population, and equity criteria. *European Journal of Operational Research, 234*, 819–829. https://doi.org/10.1016/j.ejor.2013.10.032.

Bilec, M., Ries, R., Scott Matthews, H., & Sharrard, L. A. (2006). Example of a hybrid life-cycle assessment of construction processes. *Journal of Infrastructure Systems, 12*. https://doi.org/10.1061/(ASCE)1076-0342(2006)12:4(207).

Bischoff, E. E., Janetz, F., & Ratcliff, M. S. W. (1995). Loading pallets with non-identical items. *European Journal of Operational Research.*, Cutting and Packing, *84*, 681–692. https://doi.org/10.1016/0377-2217(95)00031-K.

Boyaci, B., Altinel, İ. K., & Aras, N. (2013). Approximate solution methods for the capacitated multi-facility Weber problem. *IIE Transactions, 45*, 97–120. https://doi.org/10.1080/0740817X.2012.695100.

British Standards Institution. (2011). *BS EN 15978, sustainability of construction works – Assessment of environmental performance of buildings – Calculation method.* London: British Standards Institution.

Brocklesby, M. W., & Davison, J. B. (2000). The environmental impacts of concrete design, procurement and on-site use in structures. *Construction and Building Materials, 14*, 179–188. https://doi.org/10.1016/S0950-0618(00)00010-6.

Chen, Y., Jiang, Y., Wahab, M. I. M., & Long, X. (2015). The facility layout problem in non-rectangular logistics parks with split lines. *Expert Systems with Applications, 42*, 7768–7780. https://doi.org/10.1016/j.eswa.2015.06.009.

Church, R., & Velle, C. R. (1974). The maximal covering location problem. *Papers in Regional Science, 32*, 101–118. https://doi.org/10.1111/j.1435-5597.1974.tb00902.x.

Church, R. L., Stoms, D. M., & Davis, F. W. (1996). Reserve selection as a maximal covering location problem. *Biological Conservation, 76*, 105–112. https://doi.org/10.1016/0006-3207(95)00102-6.

Core Writing Team, Pachauri, R. K., & Reisinger, A. (2007). *IPCC, climate change 2007: Synthesis report, Contribution of working groups I, II and III to the Fourth Assessment. Report of the Intergovernmental Panel on Climate Change.* Geneva: IPCC.

Drezner, Z., & Hamacher, H. W. (2004). *Facility location: Applications and theory.* Springer Science & Business Media.

Drira, A., Pierreval, H., & Hajri-Gabouj, S. (2007). Facility layout problems: A survey. *Annual Reviews in Control, 31*, 255–267. https://doi.org/10.1016/j.arcontrol.2007.04.001.

Eiselt, H. A., & Marianov, V. (2009). Gradual location set covering with service quality. *Socio-economic Planning Science.*, The contributions of Charles S. Revelle, *43*, 121–130. https://doi.org/10.1016/j.seps.2008.02.010.

Eley, M. (2002). Solving container loading problems by block arrangement. *European Journal of Operational Research, 141*, 393–409. https://doi.org/10.1016/S0377-2217(02)00133-9.

Farahani, R. Z., Asgari, N., Heidari, N., Hosseininia, M., & Goh, M. (2012). Covering problems in facility location: A review. *Computers and Industrial Engineering, 62*, 368–407. https://doi.org/10.1016/j.cie.2011.08.020.

Fischetti, M., Ljubić, I., & Sinnl, M. (2016). Benders decomposition without separability: A computational study for capacitated facility location problems. *European Journal of Operational Research, 253*, 557–569. https://doi.org/10.1016/j.ejor.2016.03.002.

Fourer, R., Gay, D., & Kernighan, B. (1993). Ampl. Boyd & Fraser.

Gehring, H., Menschner, K., & Meyer, M. (1990). A computer-based heuristic for packing pooled shipment containers. *European Journal of Operational Research.*, Cutting and Packing, *44*, 277–288. https://doi.org/10.1016/0377-2217(90)90363-G.

Ghosh, D. (2003). Neighborhood search heuristics for the uncapacitated facility location problem. *European Journal Operational Research, O.R Applied to Health Services, 150*, 150–162. https://doi.org/10.1016/S0377-2217(02)00504-0.

Greenhouse Gas Inventory & Research Center of Korea (GGIRCK). (2011). *Emissions prospect and three different reduction scenarios*. Korea: Greenhouse Gas Inventory & Research Center of Korea (GGIRCK).

Gunawardane, G. (1982). Dynamic versions of set covering type public facility location problems. *European Journal of Operational Research, 10*, 190–195. https://doi.org/10.1016/0377-2217 (82)90159-X.

Hahn, P. M., & Krarup, J. (2001). A hospital facility layout problem finally solved. *Journal of Intelligent Manufacturing, 12*, 487–496. https://doi.org/10.1023/A:1012252420779.

Hammad, A., Akbarnezhad, A., Rey, D., & Waller, S. (2015). A computational method for estimating travel frequencies in site layout planning. *Journal of Construction Engineering and Management, 4015102*. https://doi.org/10.1061/(ASCE)CO.1943-7862.0001086.

Hammad, A. W. A., Akbarnezhad, A., & Rey, D. (2016). A multi-objective mixed integer nonlinear programming model for construction site layout planning to minimise noise pollution and transport costs. *Automation in Construction, 61*, 73–85. https://doi.org/10.1016/j. autcon.2015.10.010.

Hasan, S., Bouferguene, A., Al-Hussein, M., Gillis, P., & Telyas, A. (2013). Productivity and CO2 emission analysis for tower crane utilization on high-rise building projects. *Automation in Construction, 31*, 255–264. https://doi.org/10.1016/j.autcon.2012.11.044.

Heragu, S. S., & Kusiak, A. (1991). Efficient models for the facility layout problem. *European Journal of Operational Research, 53*, 1–13. https://doi.org/10.1016/0377-2217(91)90088-D.

Hochbaum, D. S., & Pathria, A. (1998). Analysis of the greedy approach in problems of maximum k-coverage. *Naval Research Logistics NRL, 45*, 615–627. https://doi.org/10.1002/(SICI)1520-6750(199809)45:6<615::AID-NAV5>3.0.CO;2-5.

Hong, T., Ji, C., & Park, H. (2012). Integrated model for assessing the cost and CO2 emission (IMACC) for sustainable structural design in ready-mix concrete. *Journal of Environmental Management, 103*, 1–8. https://doi.org/10.1016/j.jenvman.2012.02.034.

Hong, T., Ji, C., Jang, M., & Park, H. (2013). Assessment model for energy consumption and greenhouse gas emissions during building construction. *Journal of Management in Engineering, 30*, 226–235.

Huang, C., Wong, C. K., & Tam, C. M. (2011). Optimization of tower crane and material supply locations in a high-rise building site by mixed-integer linear programming. *Automation in Construction, 20*, 571–580. https://doi.org/10.1016/j.autcon.2010.11.023.

Huang, Y.-H., Hwang, F. J., & Lu, H.-C. (2016). An effective placement method for the single container loading problem. *Computers and Industrial Engineering, 97*, 212–221. https://doi.org/10.1016/j.cie.2016.05.008.

IBM Knowledge Center. (2016). IBM ILOG CPLEX Optimization Studio V12.6.0 documentation [WWW Document]. URL http://www.ibm.com/support/knowledgecenter/SSSA5P_12.6.0/ilog.odms.studio.help/Optimization_Studio/topics/COS_home.html. Accessed 10.16.16.

Intergovernmental Panel on Climate Change. (2007). IPCC Fourth Assessment Report (AR4): Climate change 2007.

Karasakal, O., & Karasakal, E. K. (2004). A maximal covering location model in the presence of partial coverage. *Computers and Operations Research, 31*, 1515–1526. https://doi.org/10.1016/S0305-0548(03)00105-9.

Khalafallah, A., & El-Rayes, K. (2011). Automated multi-objective optimization system for airport site layouts. *Automation in Construction, 20*, 313–320. https://doi.org/10.1016/j. autcon.2010.11.001.

Klose, A., & Drexl, A. (2005). Facility location models for distribution system design. *European Journal Operational Research, Logistics: From Theory to Application, 162*, 4–29. https://doi.org/10.1016/j.ejor.2003.10.031.

Kratica, J., Dugošija, D., & Savić, A. (2014). A new mixed integer linear programming model for the multi level uncapacitated facility location problem. *Applied Mathematical Modelling, 38*, 2118–2129. https://doi.org/10.1016/j.apm.2013.10.012.

Kulturel-Konak, S., & Konak, A. (2013). Linear programming based genetic algorithm for the unequal area facility layout problem. *International Journal of Production Research, 51*, 4302–4324. https://doi.org/10.1080/00207543.2013.774481.

Li, H., & Love, P. E. (2000). Genetic search for solving construction site-level unequal-area facility layout problems. *Automation in Construction, 9*, 217–226. https://doi.org/10.1016/S0926-5805(99)00006-0.

Li, X., Zhao, Z., Zhu, X., & Wyatt, T. (2011). Covering models and optimization techniques for emergency response facility location and planning: A review. *Mathematical Methods of Operational Research, 74*, 281–310. https://doi.org/10.1007/s00186-011-0363-4.

Lien, L.-C., & Cheng, M.-Y. (2014). Particle bee algorithm for tower crane layout with material quantity supply and demand optimization. *Automation in Construction, 45*, 25–32. https://doi.org/10.1016/j.autcon.2014.05.002.

Martello, S., Pisinger, D., & Vigo, D. (2000). The three-dimensional bin packing problem. *Operations Research, 48*, 256–267.

Metz, B., Davidson, O. R., Bosch, P. R., & Dave, R. (2007). *IPCC, climate change 2007: Mitigation of climate change, Contribution of Working Group III to the 4th Assessment Report of the Intergovernmental Panel on Climate Change* (p. 13). Cambridge: Cambridge Int. J. Sustain. Eng.

Monahan, J., & Powell, J. C. (2011). An embodied carbon and energy analysis of modern methods of construction in housing: A case study using a lifecycle assessment framework. *Energy and Buildings, 43*, 179–188. https://doi.org/10.1016/j.enbuild.2010.09.005.

Moncaster, A. M., & Symons, K. E. (2013). A method and tool for "cradle to grave" embodied carbon and energy impacts of UK buildings in compliance with the new TC350 standards. *Energy and Buildings, 66*, 514–523. https://doi.org/10.1016/j.enbuild.2013.07.046.

Murawski, L., & Church, R. L. (2009). Improving accessibility to rural health services: The maximal covering network improvement problem. *Socio-Economic Planning Sciences.*, The contributions of Charles S. Revelle, *43*, 102–110. https://doi.org/10.1016/j.seps.2008.02.012.

Nadoushani, M., Sadat, Z., Hammad, A. W. A., & Akbarnezhad, A. (2016). Location optimization of tower crane and allocation of material supply points in a construction site considering operating and rental costs. *Journal of Construction Engineering and Management, 143*(1), 04016089.

Nowakowski, P. (2016). A proposal to improve e-waste collection efficiency in urban mining: Container loading and vehicle routing problems – A case study of Poland. *Waste Management.* https://doi.org/10.1016/j.wasman.2016.10.016.

Orabi, W., Zhu, Y., & Ozcan-Deniz, G. (2012). Minimizing greenhouse gas emissions from construction activities and processes. In: *Construction research congress 2012.* American Society of Civil Engineers, 1859–1868.

Park, J., & Hong, T. (2011). Maintenance management process for reducing CO_2 emission in shopping mall complexes. *Energy and Buildings, 43*, 894–904. https://doi.org/10.1016/j.enbuild.2010.12.010.

Pomponi, F., & Moncaster, A. (2017). Scrutinising embodied carbon in buildings: The next performance gap made manifest. *Renewable and Sustainable Energy Reviews*, https://doi.org/10.1016/j.rser.2017.06.049

Rajagopalan, H. K., Saydam, C., & Xiao, J. (2008). A multiperiod set covering location model for dynamic redeployment of ambulances. *Computer & Operations Research.*, Part Special Issue: New Trends in Locational Analysis, *35*, 814–826. https://doi.org/10.1016/j.cor.2006.04.003.

Revelle, C., Marks, D., & Liebman, J. C. (1970). An analysis of private and public sector location models. *Management Science, 16*, 692–707. https://doi.org/10.1287/mnsc.16.11.692.

ReVelle, C., Toregas, C., & Falkson, L. (1976). Applications of the location set-covering problem. *Geographical Analysis, 8*, 65–76. https://doi.org/10.1111/j.1538-4632.1976.tb00529.x.

Rodríguez Serrano, A. Á., & Porras Álvarez, S. (2016). Life cycle assessment in building: A case study on the energy and emissions impact related to the choice of housing typologies and construction process in Spain. *Sustainability, 8*, 287. https://doi.org/10.3390/su8030287.

Tam, C. M., & Tong, T. K. L. (2003). GA-ANN model for optimizing the locations of tower crane and supply points for high-rise public housing construction. *Construction Management and Economics, 21*, 257–266. https://doi.org/10.1080/0144619032000049665.

Thanh, P. N., Bostel, N., & Péton, O. (2008). A dynamic model for facility location in the design of complex supply chains. *International Journal of Production Economics.*, Special Section on Advanced Modeling and Innovative Design of Supply Chain, *113*, 678–693. https://doi.org/10.1016/j.ijpe.2007.10.017.

United Nations Environment Programme. (2010). *Global initiative for resource efficient cities.* Paris: United Nations (UN).

US EPA, O. (2016). MOVES and other mobile source emissions models [WWW document]. URL https://www.epa.gov/moves. Accessed 2.13.17.

Verter, V. (2011). Uncapacitated and capacitated facility location problems. In H. A. Eiselt & V. Marianov (Eds.), *Foundations of location analysis, International Series in Operations Research & Management Science* (pp. 25–37). New York: Springer US.

Zhang, P., Harris, F. C., & Olomolaiye, P. O. (1996). A computer-based model for optimizing the location of a single tower crane. *Building Research and Information, 24*, 113–123. https://doi.org/10.1080/09613219608727511.

Zhang, P., Harris, F. C., Olomolaiye, P. O., & Holt, G. D. (1999). Location optimization for a Group of Tower Cranes. *Journal of Construction Engineering and Management, 125*, 115–122. https://doi.org/10.1061/(ASCE)0733-9364(1999)125:2(115).

Zouein, P., Harmanani, H., & Hajar, A. (2002). Genetic algorithm for solving site layout problem with unequal-size and constrained facilities. *Journal of Computing in Civil Engineering, 16*, 143–151. https://doi.org/10.1061/(ASCE)0887-3801(2002)16:2(143).

Chapter 15
Design Strategies for Low Embodied Carbon in Building Materials

Marianne Kjendseth Wiik, Selamawit Mamo Fufa, and Inger Andresen

Introduction

The Norwegian zero emission building (ZEB) research centre has a series of concept and pilot buildings that investigate design strategies and material choices for low carbon footprints in buildings in order to achieve a net zero emission balance throughout the lifetime of a building. According to the centre's definition, a net ZEB can be achieved by offsetting greenhouse gas (GHG) emissions from the entire life cycle of the building through the generation of on-site renewable energy.

The ZEB research centre's definition is very ambitious; therefore, a stepwise approach has been developed, whereby the concept of a zero emission building is divided into a range of ZEB ambition levels. These ambition levels are depicted in Fig. 15.1. The lowest ZEB ambition level is ZEB-O÷EQ, whereby O stands for operational energy use and EQ stands for technical equipment. This ambition level means that all embodied carbon emissions from operational energy use, excluding technical equipment, will be compensated for through on-site, renewable energy generation. The highest ZEB ambition level is ZEB-COMPLETE. Here, C stands for the construction phase and includes the transport of building materials to the construction site and their installation into the building, and O stands for operational energy use. M stands for the production and replacement of building materials; PLET stands for use, maintenance and repair; and E stands for the end-of-life phase and includes the demolition, transport, waste processing and final disposal of the building. This ambition level means that all embodied carbon emissions from

M. K. Wiik (✉) · S. M. Fufa
SINTEF Building and Infrastructure, Oslo, Norway
e-mail: marianne.wiik@sintef.no

I. Andresen
Norwegian University for Science and Technology (NTNU), Department of Architecture, Trondheim, Norway

© Springer International Publishing AG 2018
F. Pomponi et al. (eds.), *Embodied Carbon in Buildings*,
https://doi.org/10.1007/978-3-319-72796-7_15

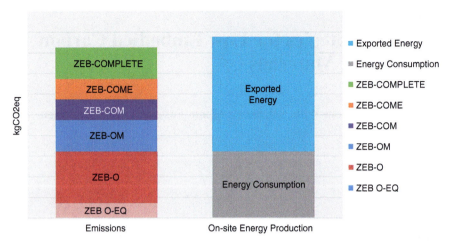

Fig. 15.1 Example of a ZEB balance (Wiik et al. 2017): on the left, the total CO_{2eq} emissions/m^2/year for each ZEB ambition level; on the right, the on-site energy generation from renewable energy sources

the entire life cycle of the building will be compensated for through on-site, renewable energy generation that is exported to the grid. To date, the most ambitious Norwegian pilot project built from the ZEB research centre has achieved ZEB-COM. A full overview of the ZEB definition and ambition levels can be found in Kristjansdottir et al. (2014) and Fufa et al. (2016). This balance is ambitious if embodied carbon from building materials is considered. Experiences collected from the research centre's pilot buildings demonstrate that a combination of carbon reduction design strategies and material choices are necessary to achieve a net ZEB.

Background

An analysis of some of the ZEB research centre's projects has been carried out to highlight design strategies for low embodied carbon in buildings. The projects include two conceptual studies or virtual building models (ZEB office building and ZEB single-family house) and six pilot buildings (Powerhouse Kjørbo, Campus Evenstad, Heimdal high school, Multikomfort house, Living Laboratory and Skarpnes). These pilot buildings are well documented and are described in more detail in the book published by the ZEB research centre (Hestnes and Eik-Nes 2017).

Figure 15.2 shows a series of images for the pilot projects described above. The images show that Norwegian ZEBs are not necessarily characterised by one architectonic profile but show that a range of design strategies can be implemented to achieve a net zero emission balance and low embodied carbon emissions. However, the range of pilot buildings do include some common traits. For example, the

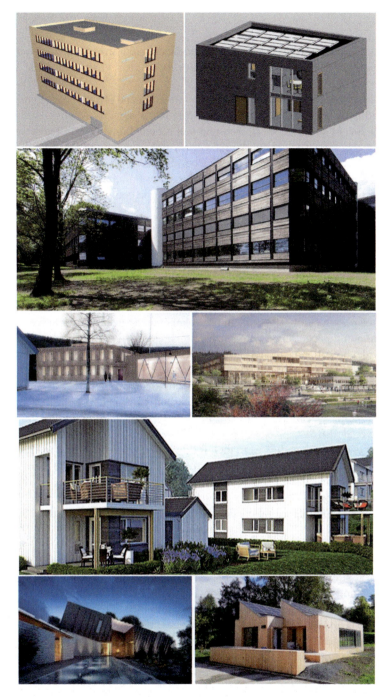

Fig. 15.2 Pilot projects from the Norwegian research centre for zero emission buildings (ZEB). Office concept study (Dokka et al. 2013a); single-family house concept study (Dokka et al. 2013b); Powerhouse Kjørbo (Powerhouse, Powerhouse Kjørbo – Trinn1 2017); Campus Evenstad administration and educational building, courtesy of Ola Roald Arkitekter AS, Heimdal high school courtesy of Skanska AS, Skarpnes residential building (Dokka and Rasmussen 2012); Multikomfort house courtesy of Snøhetta AS (Sørensen et al. 2017); and the Living Laboratory (Inman and Houlihan Wiberg 2015)

majority of pilot buildings are characterised by concrete strip foundations, a timber load-bearing structure with mineral wool insulation, LED lighting, hybrid ventilation, photovoltaic panels and heat pumps.

The first conceptual study was carried out in 2013 and is for an office building; it is theoretically located in Oslo and has a heated floor area of 1980 m^2. The building has a ZEB-OM ambition level, meaning that all embodied carbon emissions from operational energy use and building materials are compensated for through on-site, renewable energy production. The office building is of predominantly concrete and steel construction (Dokka et al. 2013a). The office building was later adapted to a timber frame structure for embodied carbon emission comparison (Hofmeister et al. 2015).

The second conceptual study was also carried out in 2013 and is for a single-family house, theoretically located in Oslo; it has a heated floor area of 160 m^2 and has a ZEB-OM ambition level. The concept building is characterised by a reinforced concrete slab foundation; a compact roof; timber-framed walls; a well-insulated building envelope, combined with solar façade-mounted thermal collectors; and an air-to-water heat pump (Dokka et al. 2013b). A roof-mounted and grid-connected photovoltaic (PV) system is implemented to generate enough electricity. The heat pump is combined with solar thermal collectors through a hot water tank to provide both domestic hot water (DHW) and space heating. Afterwards, a secondary study of the single-family house was carried out, whereby the original, generic embodied carbon emission factors from life cycle inventory databases were replaced with product-specific embodied carbon emission factors from Environmental Product Declarations (EPDs) for some of the main building components (Wiberg et al. 2015).

These conceptual studies were then followed by a series of pilot projects for an office building (Powerhouse Kjørbo), two school buildings (Campus Evenstad and Heimdal high) and three residential buildings (Multikomfort house, Living Laboratory and Skarpnes).

The office building is called Powerhouse Kjørbo and is located in Sandvika (Throndsen et al. 2015). The office building was originally built in the 1980s and was renovated to a ZEB-OM ambition level, which means that all embodied carbon emissions from operational energy use and building materials are compensated for through on-site, renewable energy production. The office building consists of two connected units, providing a heated floor space of 5180 m^2 (Fjeldheim et al. 2015). The foundations and load-bearing structure are reused directly in the refurbished building. The external walls were rebuilt with a timber frame construction, 300 mm of insulation and a charred wood cladding. The charred wood cladding is based on a Japanese technique called 'shou sugi ban' or 'yakisugi' and has good fire-, rot- and bug infestation-resistant qualities. The external walls have enlarged window openings to increase natural daylighting and external solar shading of dark grey textile screens. The roof has been upgraded to include 400 mm of rigid mineral wool insulation, and the basement exterior walls have been well insulated. Internally, 40% of the concrete ceilings are exposed as thermal mass to regulate internal temperature fluctuations, and the acoustics are regulated via sound-absorbing baffles. The energy system includes energy wells and 1560 m^2 of photovoltaic (PV) panels (Hestnes and Eik-Nes 2017).

The first school building is located at Campus Evenstad high school in Hedmark and is for an administration and educational building with 1141 m^2 of heated floor area (Selvig et al. 2017; Wiik et al. 2017). The building has a ZEB-COM ambition level, meaning that all embodied carbon emissions from operational energy use, building materials and the construction phase are compensated for through on-site, renewable energy production. This is the most ambitious of the ZEB pilot projects. The building is characterised by its solid wood construction, wood fibre insulation and untreated timber cladding. During the design phase, embodied carbon emissions were considered in conjunction with cost, building physics, acoustics and fire safety requirements in order to obtain an optimal and holistic design solution (Hestnes and Eik-Nes 2017). The energy system consists of a combined heat and power (CHP) unit that supplies both heat and electricity to the building and is powered by the gasification of wood chips (Selvig et al. 2017; Wiik et al. 2017).

In contrast, Heimdal high school represents a larger school development project and consists of a school building and sports hall, with approximately 26,356 m^2 of heated floor area. The building is located south of Trondheim and is currently under construction (Schlanbusch et al. 2017). A ZEB-O+20%M ambition level has provisionally been set, which means that all embodied carbon emissions from operational energy use, and 20% of embodied carbon emissions from building materials are compensated for through on-site, renewable energy production (Hestnes and Eik-Nes 2017). The school building will utilise ground source heat pumps, a biogas-based CHP unit and photovoltaic panels in its energy system. The building will also test electro-chromatic smart windows in its external facade (Schlanbusch et al. 2017).

The Multikomfort house is located in Larvik. This residential building is a demonstration and exhibition building and has a ZEB-OM ambition level. The house contains approximately 200 m^2 of heated floor area (Sørensen et al. 2017). The house is characterised by its sloping roof, glue-laminated timber structure, natural stone and timber external facades, brick internal walls and large glazing area. The ground floor is based on a timber and wood fibre plate construction resting on concrete strip foundations, and the crawl space underneath is lined with reflective foil to minimise heat loss (Hestnes and Eik-Nes 2017). The energy system includes an air-to-water and ground-to-water heat pump, grey water heat recovery, 16.8 m^2 of solar thermal collectors and 150 m^2 photovoltaic panels. Any excess electricity generated is stored in a battery bank and used to power the family's electric car.

The Living Laboratory is located at the Norwegian University of Science and Technology (NTNU) in Trondheim and is a small, demonstrational and experimental residential dwelling. The building is characterised by its timber-framed construction and has 102 m^2 of heated floor space and has a ZEB-OM ambition level. The Living Laboratory contains multiple technologies, such as ground source heat pumps, 4 m^2 solar thermal collectors, two south-facing arrays of photovoltaic panels at 79 m^2 and a 30° tilt, grey water heat recovery, phase change material and vacuum insulation panels in order to simulate a range of energy strategies (Inman and Houlihan Wiberg 2015; Wiik and Houlihan Wiberg 2016).

The residential development at Skarpnes, situated outside of Arendal, consists of 17 detached houses, 20 apartments and 3 terraced houses. Five of the buildings are designated as ZEBs and have a ZEB-O ambition level. These buildings have a heated floor area of approximately 154 m^2 each (Hestnes and Eik-Nes 2017). The development is characterised by predominantly timber-framed houses, providing a total heated floor area of between 1500 and 3000 m^2 (Dokka and Rasmussen 2012). The energy system comprises of ground source heat pump and four rows of 32 south-facing photovoltaic panels at 40 m^2 and a 32° tilt (Kristjansdottir et al. 2016).

The photovoltaic system at Skarpnes has been compared to the photovoltaic systems at the Multikomfort house and the Living Laboratory (Kristjansdottir et al. 2016), despite the Skarpnes development only having a ZEB-O ambition level. A ZEB-O ambition level means that only the carbon emissions from operational energy use are compensated for through renewable, local energy production. The photovoltaic systems for the three residential buildings have been compared in terms of their carbon emissions and payback times (Kristjansdottir et al. 2016).

From the range of pilot projects described above, five core design strategies to reduce embodied carbon impacts from building materials are identified. These include reducing area and material use, reusing and recycling materials, selecting low carbon building materials, sourcing local materials and adopting materials with high durability and a long service life.

LCA of a Zero-Energy Building

The Norwegian ZEB research centre has developed its own methodology for embodied carbon calculation and evaluation of ZEBs (Kristjansdottir et al. 2014; Fufa et al. 2016). In short, this complies with the international standard for life cycle assessment ISO 14040/44, the European standard for the assessment of the environmental performance of buildings EN 15978 and the European standard for Environmental Product Declarations (EPD) core rules for product category of construction products EN 15804 (ISO 14040/44 2006; EN 15978 2011; EN 15804: 2012 + A1 2013) and facilitates for the comparison of carbon emission results between buildings. The international standard for life cycle assessment ISO 14040/44 includes four main stages: definition of goal and scope, life cycle inventory, life cycle impact assessment and interpretation. This same methodology is used for each of the Norwegian pilot buildings, when assessing embodied carbon emissions from building materials.

The goal of a life cycle assessment (LCA) for a ZEB is to evaluate, quantify and provide an overview of the life cycle GHG emissions of the building based on the defined scope and intended use of the assessment. The scope of the ZEB ambition levels has been standardised in accordance with the life cycle modularity principle defined in EN 15978 (see Fig. 15.3). Essentially, operational energy use (O) corresponds to life cycle module B6 for operational energy use, and materials (M) correspond to life cycle modules A1–A3 for the production of building materials and life cycle module B4 for the replacement of said building materials. Construction (C) corresponds to life cycle modules A4 and A5, which represent transport from the factory to the construction site and the installation of building materials and other construction site activities. The end-of-life (E) phase corresponds to life cycle modules C1–C4 which include the demolition, transport, waste processing and final disposal of the building materials. The use (PLET) phase corresponds to the remaining life cycle modules, B1, B2, B3, B5 and B7, for use, maintenance, repair, refurbishment and operational water use. In all of the ZEB pilot buildings, a functional unit of '1 m^2 of heated floor area over a reference study period of 60 years' is used. This common functional unit allows for comparison between the various pilot projects. Alongside the functional unit, it is also required to state total embodied emissions (kgCO$_2$e) of the building and present embodied emission results according to building component and life cycle module.

The life cycle inventory data for material quantities is collected from architect's drawings, building information models (BIM) and information gathered from construction professionals involved in the projects. The building material inventory is structured according to the Norwegian Standard NS 3451: 2009 Table of Building Elements (NS 3451 2009). This standard provides a structured nomenclature checklist of building materials and components.

The replacement interval of building materials and components during the lifetime of the building is also considered. This is achieved by extracting reference service lifetimes from product literature, EPDs or SINTEF design guidelines for building component replacement intervals (BKS 700.320 2010). This information provides a basis for replacement scenario calculations.

Within the Norwegian ZEB research centre, global warming potential (GWP) is used as an environmental indicator. GWP is calculated in terms of carbon dioxide equivalents (CO$_{2eq}$). The GWP indicator converts the global warming potential of other greenhouse gases than CO$_2$ (such as methane) to its kg of CO$_2$ equivalence. This is achieved by using the Intergovernmental Panel on Climate Change (IPCC) GWP 100-year horizon method (IPCC 2014).

Embodied carbon emission factors are obtained from product-specific Environmental Product Declarations (EPDs), generic life cycle inventory databases such as Ecoinvent (2014; EPD-Norge 2017) and other LCA studies. These emission factors are multiplied with the corresponding material quantities to obtain the total embodied carbon emissions. Finally, the embodied carbon emission results are interpreted for the whole building in terms of a zero emission balance, as demonstrated in Fig. 15.1.

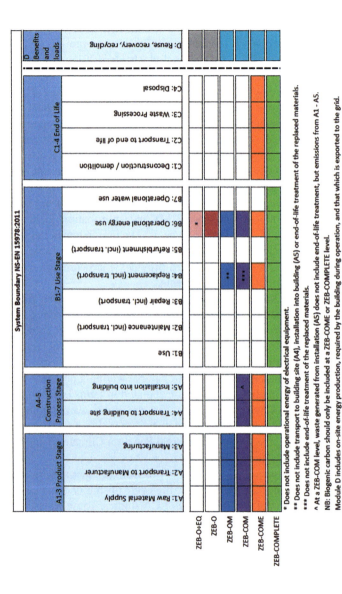

Fig. 15.3 Correlation between the ZEB ambition levels and the life cycle modules outlined in NS-EN15978: 2011. The life cycle stages (A1–A5, B1–B7 and C1–C4) mandatory for the different ZEB ambition levels (ZEB-O÷EQ, ZEB-O, ZEB-OM, ZEB-COM, ZEB-COME, ZEB-COMPLETE) are highlighted

Results

Reduction

One of the most effective mitigation strategies for low embodied carbon emissions may be demonstrated by the carbon emissions not released, in much the same way that low energy demand in energy efficient buildings is achieved primarily by reducing the energy demand of the building, with the adage that the most environmentally friendly kWh is the kWh not spent. This reasoning can be extrapolated to embodied carbon emissions and by association to building materials, whereby low embodied carbon emissions can be achieved through designing space-efficient, flexible, compact buildings through material reduction.

In the Living Laboratory, a sensitivity analysis of the functional unit was carried out using a range of definitions for area (Inman and Houlihan Wiberg 2015). This included heated floor area, gross floor area, net floor area and built-up area. Although the total life cycle phase emissions are proportional across the four definitions of area, there is almost a twofold variation in embodied carbon emissions (Inman and Houlihan Wiberg 2015). A similar trait was identified when the total embodied carbon emission results from the Living Laboratory were compared against the single-family house concept. Total embodied carbon emissions for the Living Laboratory are 23.5 $kgCO_2e/m^2$/year and 7.2 $kgCO_2e/m^2$/year for the single-family house concept. However, the Living Laboratory (102 m^2) is more compact than the single-family house concept (160 m^2) (Dokka et al. 2013b; Inman and Houlihan Wiberg 2015). This means that the Living Laboratory has a higher concentration of emissions per m^2 of heated floor area than the single-family house concept because of its more compact building form. It is thought that total embodied carbon emissions in buildings can be reduced by building compact building forms; however this will also result in a higher intensity of building materials and consequently embodied carbon per m^2 of heated floor area. Measures to counteract this may include designing slimmer wall constructions to maximise on heated floor space and designing slimmer floor constructions to maximise on the number of floors that can be implemented in high-rise buildings.

The Living Laboratory and Multikomfort use timber floors on top of concrete strip foundations, compared to the raft foundation design in the single-family house concept (Dokka et al. 2013b; Sørensen et al. 2017; Inman and Houlihan Wiberg 2015). This results in a 68–80% decrease in embodied carbon arising from reduced concrete use. The lower emissions from the Multikomfort house compared to Living Laboratory are due to the use of low carbon concrete. In Heimdal high school, the amount of concrete used has also been reduced by implementing thinner retaining walls in the sports hall (Schlanbusch et al. 2017). A lighter roof construction has also reduced the amount of structural steel required in the building (Schlanbusch et al. 2017).

Another method for reducing the amount of building materials used is centred on holistic, integrated design solutions, which avoid the double up of building material functions. Integrated design can be achieved by implementing a multidisciplinary design team. This strategy was used in the design of Powerhouse Kjørbo and has led to some interesting results. For example, the amount of ventilation channels and ducts installed in the building have been reduced by utilising the staircase as a ventilation shaft, allowing for the flow of both people and air (Hestnes and Eik-Nes 2017). Furthermore, the traditional suspended ceiling found in many office buildings has been replaced with 40% exposed concrete ceilings for thermal mass and acoustic baffles (Hestnes and Eik-Nes 2017). These integrated design measures not only save on building materials but also improve the indoor environment and reduce operational energy use. Finally, material reduction can also be achieved by reducing the amount of construction waste generated on site.

Reuse and Recycling

At least 50% of the Norwegian building stock in 2050 has already been built (Hestnes and Eik-Nes 2017). Therefore, it is of paramount importance to renovate buildings in a sensitive manner and reuse and recycle building materials wherever possible. Closing material loops by recycling materials minimise emissions by substituting environmentally intensive primary production with lower impact secondary production. This also helps to preserve non-renewable resources and minimise waste and the amount of landfill. In the design phase, it may be beneficial to consider the demountability of buildings, in terms of maintenance, repair, reuse and recycling of materials.

In terms of reuse and recycling, the Multikomfort house uses bricks reclaimed from a nearby derelict barn as the central massive wall of the house. This reuse strategy leads to a saving of 40% CO_{2eq} when compared to a new brick wall and 29% when compared to a concrete wall (Sørensen et al. 2017). In addition, the brick wall provides thermal mass for a good indoor climate. The reclaimed brick wall from the Multikomfort house is depicted in Fig. 15.4. Similarly, the renovated office building Powerhouse Kjørbo reuses the reinforced concrete foundations and load-bearing structure. In addition, the building reuses the green-tinted external laminated glass facade as internal glass partitions; this prolongs the service life of building materials and postpones emissions associated with end-of-life treatment, but most importantly it avoids the use of new materials for the partition walls (Throndsen et al. 2015). These appropriated internal glass partitions are depicted in Fig. 15.5. In addition, the sound-absorbing acoustic baffles are manufactured from recycled plastic bottles (Hestnes and Eik-Nes 2017).

In Campus Evenstad, the internal glass partition walls from Statsbygg's main office are reused as internal glass partition walls in the administration and educational building (Selvig et al. 2017; Wiik et al. 2017). The only embodied carbon emissions arising from these internal partitions arise from the transport and installation of the building elements.

Fig. 15.4 Photograph of the reclaimed brick wall in the Multikomfort house (Sørensen et al. 2017)

Fig. 15.5 Photograph of the glass internal partitions and staircase at Powerhouse Kjørbo (Powerhouse, Powerhouse Kjørbo – Trinn1 2017)

Low Carbon

In all of the pilot buildings, wood is considered as a key low carbon construction material and is used to replace materials with traditionally high-embodied emissions. In the office concept building (Dokka et al. 2013a), the change from the original concrete and steel structure to a timber structure of similar technical performance leads to a 30% reduction in weight and a 50% reduction in embodied carbon (Hofmeister et al. 2015). In Multikomfort and Powerhouse Kjørbo, glue-laminated beams and charred wood for the new facade cladding are used, to reduce the embodied emission from the buildings. The ZEB research centre has not considered the biogenic carbon uptake of wood in the ZEB-OM ambition level (e.g. Multikomfort house) or the ZEB-COM ambition level (e.g. Campus Evenstad high school). This is because the end-of-life stage is not taken into account (Fufa et al. 2016). It is assumed that the uptake and later release of biogenic carbon are zero over the lifetime of a building. Therefore, considering biogenic carbon at a ZEB-OM or ZEB-COM ambition level would only consider the uptake of biogenic carbon in the production phase and would present a negative carbon emission, providing an unfair advantage to wood products. However, biogenic carbon should be included in calculations at a ZEB-COME or ZEB-COMPLETE level, since the whole life cycle is considered and will show the temporary carbon storage of GHG emissions in wood products.

Other low embodied carbon building materials include low carbon concrete, as seen in the Living Laboratory and Multikomfort house. It should be noted that the ZEB research centre has not considered the carbonation of concrete in embodied carbon emission calculations and that the carbonation of concrete will undoubtedly lead to a reduction in embodied carbon emissions over time.

Research at the ZEB centre has identified that photovoltaic systems can contribute significantly to embodied carbon emissions (Kristjansdottir et al. 2016). A comparison of embodied carbon from the photovoltaic (PV) systems in Skarpnes, Multikomfort and Living Laboratory shows that emissions from PV modules and supporting systems vary considerably (Fig. 15.6) (Kristjansdottir et al. 2016). In the Multikomfort house, the PV modules included recycled components that lowered total embodied carbon considerably. On the other hand, the PV systems varied between being building integrated (Skarpnes), building adapted (Multikomfort) and in-roof (Living Laboratory) systems. The results in Fig. 15.6 show that the building adapted system has the lowest embodied carbon emissions, followed by the building integrated system and then the in-roof system. In terms of the roofing material being replaced by the different mounting systems, the biggest saving was made in Skarpnes, as the mounting system replaced concrete roofing tiles, followed by the Living Laboratory since roofing felt was replaced. No material savings were to be made from Multikomfort since the mounting system did not replace any roofing elements.

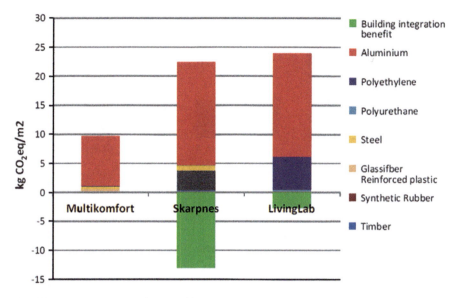

Fig. 15.6 Emission results for the different materials for the roof mounting structures and building integration benefits (Kristjansdottir et al. 2016)

Local

The implementation of locally sourced materials reduces the use of carbon-intensive materials and the embodied carbon from long-distance transportation. This also supports the development of local industries, which in turn provide local economic development. Evenstad high school considers embodied transport emissions and demonstrates the importance of selecting local materials. In Evenstad high school, excavated material is sourced from a local quarry, steel connections are formed by a local workshop, locally sourced timber and wood fibre insulation and other local manufacturers are selected to reduce embodied emission from the transportation of materials, construction machinery and workers (Selvig et al. 2017; Wiik et al. 2017). In addition, high-quality, locally produced wood chips have been sourced to fuel the renewable, combined heat and power (CHP) plant on campus. This measure supports and generates local energy production. Although symbolic, sculptures have also been made by students, from the construction waste generated on site.

A sensitivity analysis of different concrete hollow core slabs (HD1–HD4) and cross-laminated timber floors (CLT1–CLT3) was performed in Heimdal high school during the early design phase for material selection. Emission results for HD1 to HD4, and CLT1 to CLT3 correspond to EPD production and transport emission data from various manufacturers. Although concrete hollow core slabs typically have high-embodied carbon emissions, the cross-laminated timber floors require additional soundproofing and a thicker floor construction. When

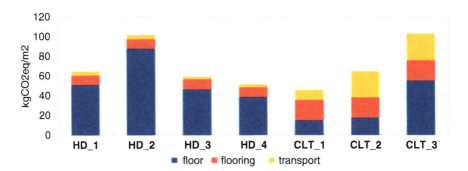

Fig. 15.7 Comparison of the embodied carbon emissions from various manufacturers for concrete hollow core slab (HD) and cross-laminated timber (CLT) floor constructions (Hestnes and Eik-Nes 2017)

considering embodied carbon emissions from transport, although the concrete hollow core slabs are heavier than the cross-laminated timber floors, it was not possible to find a local manufacturer for the cross-laminated timber floors, meaning that they had to be transported across a greater distance. The results from the production of materials (A1–A3) and transport to the construction site (A4) are given in Fig. 15.7 and show a high level of variation in emissions between manufacturers. This highlights the importance of a holistic evaluation when selecting low carbon building materials in order to avoid problem shifting (Schlanbusch et al. 2017).

Durability

Building components often require maintenance, repair or replacement during the service life of a building (Fufa et al. 2017). The total embodied carbon emission results from the product stage (A1–A3) and replacement phase (B4) of the ZEB single-family house; office building and Living laboratory show that the replacement of materials contributes to approximately 25%, 27% and 43% of total carbon emissions, respectively (Dokka et al. 2013a, b; Inman and Houlihan Wiberg 2015; Wiik and Houlihan Wiberg 2016). The Living Laboratory has high carbon emissions for the replacement phase because it is a temporary building and was not designed with durability in mind. It is thought that these replacement phase emissions could be reduced significantly if efforts were made to select building materials and components with a long reference service life and high durability (Fufa et al. 2017). For example, in Powerhouse Kjørbo, charred wood was used for the new facade cladding to increase the durability and service life of the wood and minimise the frequency of maintenance intervals.

Attributional LCA aims to find out the environmental properties of a product through its life cycle within a defined system boundary; therefore all carbon emissions are typically modelled as occurring at the start of the analysis period, and all calculations are carried out based on scenarios from today's best practice (Peuportier et al. 2011). However, embodied carbon emissions could be further reduced by considering future technology innovations and by integrating more dynamic scenarios into LCA calculations. At the ZEB research centre, the potential of future technological improvements of PV panels is considered. This includes an assumption of increased material efficiency, improved production processes and the transition to increased use of renewable energy in the future production process. Therefore, a 50% reduction in emissions for replaced PV panels, with an estimated service life of 30 years, has been considered in all pilot building emission calculation. This assumes that the PV system will be produced 50% better in 30 years' time, with half the amount of material emissions per m^2 (Kristjansdottir et al. 2016; Fufa et al. 2017). This assumption significantly affects the embodied material emission results, as PV modules are one of the main emission drivers in the Norwegian concept and pilot ZEBs.

During the design phase of Heimdal high school, the embodied emission from timber window frames with and without a protective aluminium cladding was evaluated for material selection. The aluminium cladding, despite its elevated embodied emissions, gives the frame a longer service life. This results in fewer replacements during the service life of the school. Over a 60-year calculation period, approximately 22 $kgCO_2e$/window is saved when the aluminium cladding is implemented (Schlanbusch et al. 2017; Fufa et al. 2017).

Conclusion

For successful embodied carbon mitigation, it is important to consider embodied carbon reduction strategies from an early design phase and throughout the entire design and building process (Pomponi and Moncaster 2016). The findings from the Norwegian ZEB research centre have also demonstrated the importance of assessing the whole life cycle of buildings to avoid shifting the embodied carbon emission burden to other parts of the building's life cycle. The results have also shown that by considering low embodied carbon building materials at an early design phase, considerable time, energy and cost savings can be made.

Some of the most efficient low embodied carbon design strategies, identified through the ZEB pilot projects, include reducing area and material use, applying reused and recycled materials, using materials with low embodied carbon, sourcing local materials and adopting materials with high durability and a long service life. Embodied carbon calculations from eight of the ZEB pilot projects provide an insight into the measured effect of low embodied carbon design strategies and material choices.

Acknowledgments The authors would like to acknowledge the Research Council of Norway and partners of the Norwegian research centre for zero emission buildings.

References

BKS 700.320. (2010). *Intervals for maintenance and replacement of building components (in Norwegian)*. Oslo: SINTEF Building and Infrastructure.

Dokka, T. M., Thyholt, H., & Rasmussen, R. (2012). *The Skarpnes residential development – A zero energy pilot project*. 5th Nordic Passivhus Conference (PHN 2012), Trondheim.

Dokka, T. H., Kristjansdottir, T., Time, B., Mellegård, S., Haase, M., & Tønnesen, J. (2013a). *A zero emission concept analysis of an office building, ZEB project report (8)*. Oslo: SINTEF Academic Press.

Dokka, T. H., Houlihan Wiberg, A., Georges, L., Mellegård, S., Time, B., Haase, M., Maltha, M., & Lien, A. G. (2013b). *A zero emission concept analysis of a single-family house, ZEB project report (9)*. Oslo: SINTEF Academic Press.

Ecoinvent. (2014). Ecoinvent database version 3.1. In E. Centre (Ed.), www.ecoinvent.org, Zurich, Switzerland.

EN 15804: 2012 + A1: 2013. (2013). *Sustainability of construction works – Environmental product declarations – Core rules for the product category of construction products*. Brussels: European Committee for Standardization.

EN 15978: 2011. (2011). *Sustainability of construction works – Assessment of environmental performance of buildings – Calculation method*. Brussels: European Committee for Standardization.

EPD-Norge. (2017). Environmental Product Declarations (EPDs) http://epd-norge.no/epder/. Accessed 20 Feb 2017.

Fjeldheim, H., Kristjansdottir, T., & Sørnes, K. (2015). *Establishing the life cycle primary energy balance for Powerhouse Kjørbo*. 7th Passivhus Norden Conference on Sustainable Buildings, Copenhagen.

Fufa, S. M., Schlanbusch, R. D., Sørnes, K., Inman, M., & Andresen, I. (2016). *A Norwegian ZEB definition guideline, ZEB project report (29)*. Oslo: SINTEF Academic Press.

Fufa, S. M., Wiik, M. K., Schlanbusch, R. D., & Andresen, I. (2017). *The influence of estimated service life on the embodied emission of ZEBs when choosing low-carbon building product*. In XIV International Conference on Durability of Building Materials and Components, 29–31 May 2017, Ghent.

Hestnes, A. G., & Eik-Nes, N. L. (2017). *Zero emission buildings*. Bergen: Fagbokforlaget.

Hofmeister, T. B., Kristjansdottir, T., Time, B., & Wiberg, A. H. (2015). *Life cycle GHG emissions from a wooden loadbearing alternative for a ZEB office concept, ZEB project report (20)*. Oslo: SINTEF Academic Press.

Inman, M. R., & Houlihan Wiberg, A. A.-M. (2015). *Life cycle GHG emissions of material use in the living laboratory ZEB project report (24)*. Oslo: SINTEF Academic Press.

IPCC. (2014). *Working group III, IPCC fifth assessment report: Mitigation of climate change*, Berlin.

ISO 14040/44: 2006. (2006). *Environmental management – Life cycle assessment – Principles and framework/requirements and guidelines*. Geneva: International Organization for Standardization.

Kristjansdottir, T., Fjeldheim, H., Selvig, E., Risholt, B., Time, B., Georges, L., Dokka, T. H., Bourrelle, J., Bohne, R., & Cervenka, Z. (2014). *A Norwegian ZEB definition: Embodied emissions, ZEB project report (17)*. Oslo: SINTEF Academic Press.

Kristjansdottir, T., Good, C. S., Schlanbusch, R. D., & Inman, M. R. (2016). A GHG emission balance for roof mounted PV systems installed in ZEB residential pilot buildings in Norway. *Solar Energy, 133*, 155–171.

NS 3451: 2009. (2009). *Table of building elements*. Oslo: Standard Norge.

Peuportier, B., et al. LoRe-LCA. (2011). Low resource consumption buildings and constructions by use of LCA in design and decision making. Deliverable D3.2. Guidelines for LCA, ARMINES, p. 54.

Pomponi, F. & Moncaster, A. (2016). Embodied carbon mitigation and reduction in the built environment: The evidence. *Journal of Environmental Management, 181,* 687–700.

Powerhouse, Powerhouse Kjørbo – Trinn1. (2017). http://www.powerhouse.no/prosjekter/kjorbo/. Accessed 20 Feb 2017.

Schlanbusch, R. D., Fufa, S. M., & Andresen, I. (2017). *ZEB pilot Heimdal high school and sports hall – Design phase report, ZEB project report (34).* Oslo: SINTEF Academic Press.

Selvig, E., Wiik, M. K., & Sørensen, Å. L. (2017). *Campus Evenstad. Jakten på nullutslippsbygget. ZEB-COM.* Oslo: Statsbygg.

Sørensen, Å. L., Andresen, I., Kristjansdottir, T., Amundsen, H., & Edwards, K. (2017). *ZEB pilot house Larvik – as built report, ZEB project report.* Oslo: SINTEF Academic Press.

Throndsen, W., Berker, T., Knoll, E. B., & Powerhouse Kjørbo. Evaluation of construction process and early use phase, ZEB project report (25) SINTEF Academic Press, Oslo, (2015).

Wiberg, A. H., Georges, L., Fufa, S. M., & Risholt, B. (2015). *A zero emission concept analysis of a single family house: Part 2 sensitivity analysis., ZEB project report (21).* Oslo: SINTEF Academic Press.

Wiik, M. K., & Houlihan Wiberg, A. (2016). *Life cycle GHG emissions of material use in the living laboratory.* Central Europe towards Sustainable Buildings Conference (CESB16), Praha.

Wiik, M. K., Sørensen, Å. L., Selvig, E., Cervenka, Z., Fufa, S. M., Andresen, I. (2017). *ZEB Pilot Campus Evenstad. Administration and educational building. As-built report. ZEB project report (36).* Oslo: SINTEF Academic Press.

Chapter 16
Embodied Carbon of Tall Buildings: Specific Challenges

Donald Davies and Dario Trabucco

Introduction

Tall buildings are becoming the predominant building typology in cities and megacities worldwide, with only a few regional exceptions. The number of 200+ meter buildings is increasing daily, and buildings between 100 and 200 m are now becoming a norm of urban construction, especially in dense Asia cities and US urban city cores. This chapter focuses on the unique issues related to the embodied carbon footprint of tall buildings, their place in our built environment, and the challenges and opportunities for how to build more sustainable tall buildings now and in the future. While a wide range of topics are presented and discussed, this is a limited overview of some of the more significant issues on this topic. The authors have hoped to provide insightful thoughts and challenges to the readers on where we can and should be focusing more of our efforts within the life cycle analysis of tall buildings, with a goal of reducing the carbon footprint of our built environment.

Tall buildings have unique construction characteristics, which impact their embodied carbon footprints over other building typologies. Like all buildings, their structural system is proportioned to account for multiple purposes, including horizontal forces (such as wind and seismic) and vertical loads such as gravity. These are typical of all buildings, but the load concentration in a tall building requires higher material concentrations in the lower floors and different system options than are sometimes used within low-rise construction (Khan 1969). In the more slender and tallest towers, there are also further material requirements to address human perception to motion criteria. MEP systems have to consider stack

D. Davies (✉)
Magnusson Klemencic Associates, Seattle, WA, USA
e-mail: ddavies@mka.com

D. Trabucco
Council on Tall Buildings and Urban Habitat/IUAV University of Venice, Venice, Italy

© Springer International Publishing AG 2018
F. Pomponi et al. (eds.), *Embodied Carbon in Buildings*,
https://doi.org/10.1007/978-3-319-72796-7_16

effects and typically require pumping water to greater heights. To maintain system efficiencies, these issues require a thoughtful management of the load paths and system layouts and often more intensive design efforts than might be acceptable for many low-rise building typologies.

Tall building fire and life safety systems also include more materially intensive and self-reliant strategies for occupant safety and evacuation, as well as the structural frames being protected for burnout of the building. Typically, fire ladder trucks can only service up to 26 m (85 ft) of building height. This limit is one of the major separators between low-rise and high-rise building code fire and life safety requirements. Facades are also distinctive in tall buildings, with large surface areas designed for higher wind pressures than might be required in shorter construction.

With these life safety and associated structural and MEP system impacts, when looked at in isolation, tall buildings will show a higher unit area of embodied carbon footprint to build than many low-rise buildings. However, if the boundaries conditions are expanded, including the longevity that also often goes with the more robust construction of tall buildings, as well as the overall site density of tall buildings built within an urban core, where residents can live close enough to walk to work and the population density allows for mass transit to be affordable and functional, the total carbon footprint of the population with time can tell a different conclusion.

This chapter does not attempt to answer the question of what is the lowest carbon footprint solution for the greatest percentage of the world's population or to defend or condemn tall buildings for their place in our population's efforts to address global warming. It is relatively easy to show the extremes of poor carbon footprint choices in both low-rise and high-rise construction. For both the suburban sprawl development, with its extensive roadways and consumptive land requirements, as well as some recent iconic "world's tallest" towers, neither are typically conceived with carbon footprint optimization in mind. For the greatest percentage of the world's population, optimal buildings for a lowest carbon footprint living are somewhere between these extremes. It is also a mistake to assume there is going to be a one-solution-fits-all answer in building typologies. Diversity and variety are important values to our society. This is reflected within our buildings and building choices as well.

Others have attempted to answer the question of the lowest carbon footprint living for our populations with enlightening but inconclusive findings, including a thoughtful paper by Smith Gill Architecture titled "The Environmental Impact of Tall vs. Small: A Comparative Study" (Drew et al. 2014). At this time, there are many places of missing or misleading data for defendable embodied carbon comparisons between whole-building systems and typologies, and much work is yet to happen in this space. That doesn't mean, though, that actionable date is not available within the LCA data we have to work with today or that quality studies are not happening. Much can be done to reduce the embodied carbon footprint of our built environment if the right questions are asked and optimization efforts focused on a lower carbon footprint are acted upon, based upon the findings that come from the answers to those questions.

Much of the focus of this chapter is on the "shell and core" of the tall building and those decisions the initial building design team will often face if they look to optimize the embodied carbon footprint of a project. However, as is demonstrated later in this chapter, the embodied carbon impacts of material finishes within the building and the frequency of their replacement over the life of a project can and often will become a dominant part of the whole-building cradle-to-grave embodied carbon footprint of what is built.

Commercial office buildings are often designed through a shell and core approach with the design team commissioned to design the structural system, primary MEP systems, and the exterior façade. Outside main lobby public spaces, interiors, lighting, etc. are often left to the tenant's designers and are also subject to frequent renovations and remodeling. Residential construction will often include a full fit out of the interior spaces, but due to changing tenant tastes, move-ins and move-outs, and the relatively low durability of many interior finishes and appliances, they are also subject to frequent renovations and remodeling. Since finishes are not unique to a tall building design and issues raised in this paper around the frequency of replacement may be applicable regardless of the height of a project, we have not focused on this topic as far as it could be, other than to highlight some of the issues whose influence should not be ignored.

With these qualifications noted, the authors believe the design team has great potential to influence the initial embodied carbon content of tall buildings when built, with a 400% spread (or more) being reported in potential embodied carbon footprints of the structural frames within tall buildings within the industry today (Cho et al. 2004). Focusing on where the embodied carbon is located within tall buildings, and how it can be reduced and optimized, is thus an important topic to consider when looking to reduce the embodied carbon footprint of our built environment.

Where Is the Carbon?

Tall buildings, as with most building types, have seen attempts to lower of their greenhouse gas emissions over the past 10–20 years, with awareness of the concentrated energy consumption within these buildings dating back to the late 1970s (Stein 1977). This pursuit to date has largely focused on operational improvements and has resulted in much more energy-efficient buildings (Oldfield et al. 2009). Less has been done from an embodied carbon perspective, though, which is a focus that is now receiving greater scrutiny.

The studies available in literature confirm a great variability in the embodied carbon and energy content of tall buildings (Oldfield 2012; Kofoworola and Gheewala 2009; Chang et al. 2012). These studies are rapidly increasing with the debate on this building type's sustainability now being transferred from "abstract" academic research to applied professional practice. Several studies are now available to seize the problem of embodied emissions and embodied energy content of tall buildings, though the results vary significantly from case to case. Buildings of various heights are compared against one another by Treloar (Treloar et al. 2001) with values of embodied energy progressively increasing with the building height. Similar results were obtained in a study (Foraboschi et al. 2014) based on a theoretical exercise that focused on building heights with different floor system solutions.

A life cycle assessment of tall building structural systems performed by the Council on Tall Buildings and Urban Habitat (CTBUH) (Trabucco et al. 2015), on which the authors of the present chapter worked, is, to date, the most comprehensive study on the embodied carbon and energy content that goes into the construction of tall buildings. Though this study focused on the structural components of the building frames, the results demonstrate several key findings.

For one, while embodied carbon footprints do increase with tower height, the greatest volume of consumptive materials, and therefore of the related embodied emissions due to material usage, are typically within the horizontal components of the structural frame (slabs and beams) (Kim et al. 2008). Only in the largest supertall buildings do the vertical components of the structural system (columns, shear walls, and braces) begin to be equally predominant in terms of materials usage. Assuming a tall building project starts with a rationally proportioned lateral system, the horizontal framing of a building, often regardless of building height, is where optimization efforts to reduce the materials, and correspondingly the embodied carbon footprint of a building, will often find the most potential.

Another notable finding of the CTBUH study is that the embodied carbon within a tower was less sensitive to the six different tower frames and their variations studied, than if the designers optimized their designs of those systems, or their assumptions of how and where materials were produced. This study considered steel, concrete, and hybrid floor and column framing options, as well as outriggered concrete cores, and steel diagrid lateral systems of variable steel strength (Trabucco et al. 2015).

This brings into focus design optimization and the design team's attention to detail. It also highlights the importance of embodied carbon and other life cycle data being considered at the time of material procurement for a project. Asking where and how materials are made and using that information to influence purchasing decisions can vary the embodied carbon footprint of a project significantly (Trabucco et al. 2015).

Finally, the literature shows that when considering embodied carbon impacts at a regional scale, perhaps one of the most fundamental conclusions is that proper land use planning and decision-making around where to thoughtfully place the building infrastructure makes an impact. The resiliency of that infrastructure to last over time, which is also tied to land use planning, is perhaps where the greatest impact to reducing the building carbon footprint of our built environment can occur (Du et al. 2016).

The Influence of Urban Density

While tall buildings represent a larger embodied carbon footprint at initial construction, several studies have tried to identify the benefits of the urban density made possible by tall buildings. The findings of these studies to date have been good at highlighting trends, but actionable data is still a work in progress.

Notably, a graph showing the decreased energy intensity of urban mobility as urban density grows, developed several decades ago (Newman and Kenworthy 1989), is still often used to demonstrate the benefits of high-density versus low-density urban settlements. The general assumption extrapolated from this study was, in reality, much narrower in scope. The authors' intention was to monitor transportation habits in some of the most populated urban environments which present a remarkable variability in terms of geographic location, climate conditions, economic status, and, not to be neglected, culture. The resulting graph clearly showed energy intensity per person trending downward with increased density, but the ideas derived from this have been given more significance than was initially intended. While this study and its graph are often used to "justify" the need for higher urban density and, in turn, more tall buildings, the variables considered from this study miss many important topics around the carbon footprint associated with urban living.

In more recent years, a study conducted (Du et al. 2016) has focused on the larger implications of urban versus suburban living, by comparing the real behavior/energy spending of a large group of families living in a low-rise suburban neighborhood and in two high-rise downtown condos, both in Chicago, Illinois.

This study more credibly confirms the reduced energy usage required of families and individuals living in urban areas, due to the reduced transport energy used to get to their desired locations. The study is also enlightening to the energy usage per capita of households for heating, appliances, air conditioning, etc. The study concludes that mobility and the energy expended around the personal choices of how we spend time and where we drive or fly to, more than anything else, control the carbon footprint consequences of modern living habits.

The evidence from the abovementioned study by Du et al. may be influenced (perhaps significantly) by the sample families chosen or by other local conditions, but they clearly put in perspective the influence of the building construction type on the embodied consumption/emissions of people.

A lower carbon footprint society needs to start with fewer cars on the roadways and people living close to the activities they choose to participate in the most, including their work and recreation. This may or may not be in a tall building, but urban development that includes enough density that transportation choices are other than being in a car for the greatest portion of the population is a key (Perkins et al. 2009). While not a one-size-fits-all answer, urban city cores with tall buildings become a fundamental part of creating the desirable locations for this density to exist (Norman et al. 2006).

Carbon Over Time

For most buildings today, tall or short, it isn't until 30–40 years past construction that the operational carbon footprint exceeds the CO_2 to build in the first place (Stein 2011). The Architecture 2030 Challenge for Products (2030 Inc. 2011) and the Carbon Leadership Forum's Time Value of Carbon studies (Strain 2017) try to address this topic and issues around the embodied and operational carbon impacts with time.

When tall buildings are considered compared to low-rise wood-framed residential buildings, the timing of when the operational carbon exceeds the embodied carbon to construct changes. A fundamental finding, though, is that the longer a building remains standing, the ratio of embodied to operational carbon will fall, reducing the overall influence of embodied carbon on the full carbon life cycle of the building. The sustainability focus on operational energy is thus only well spent if buildings are built to be around to the time when operational carbon use exceeds the embodied carbon to build. If designers are not thoughtful on where or how to build to begin with and if buildings are built with similar materials but discarded after shorter time frames such as 25 years, we are creating some of the most unsustainable infrastructure possible.

Tall buildings, because of the economic investments required to construct them initially and because of the infrastructure that usually goes around them, frequently become a long-term infrastructure investment. This is one of the benefits with urban cores and cities. The infrastructure and investments required to support these buildings necessitate long-term land use planning and coordination that typically is looking past a 25- or 50-year horizon, forcing considerations outside of shorter-term speculative developer-led economic decision-making windows.

Despite the number of tall buildings built since this typology was born almost 150 years ago, only a small number of them have been demolished (Trabucco 2016). While 50-year life spans are often used in design discussions, a 100-year life span for a tall building is not unusual and, if maintained, can often be much longer. The longevity of buildings is not unique to tall buildings only, but tall buildings certainly encourage a longer-lasting and more durable built environment infrastructure (Fig. 16.1).

As noted earlier, embodied carbon studies that look at just the up-front construction do not tell the full story of a tall building. To expand on one of these boundaries, the future renovation efforts within a tower's life span are considered below. Changes in the local market condition, the evolution of the technical equipment used by the tenants, or simply the need to "refresh" the building aesthetics to compete against a dynamic market will cause an ever-growing embodied carbon footprint of the building. Cradle-to-grave, instead of cradle-to-gate, discussions are necessary to capture these and similarly important changes with time.

To demonstrate this renovation impact and to highlight the importance of building longevity, this example is for a projected 100-year potential building life

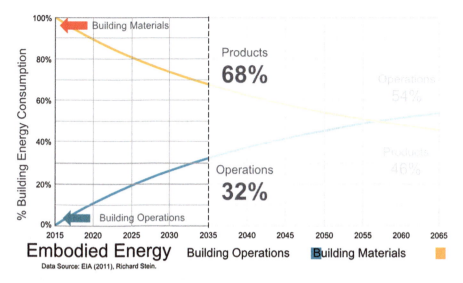

Fig. 16.1 Building energy split between operation and embodied after 20 years

span. It is based upon an unbuilt 25-story hotel (shown below), which consists of posttensioned concrete flat plate construction, concrete shear walls, glass, precast and aluminum cladding, and a centralized four-pipe fan coil cooling system (Fig. 16.2).

As with many LCAs today, full accounting of the embodied carbon percentages for this hotel is not known, and this study is not the conclusion of an exhaustive academic study. The ratios of the embodied carbon between each impact category were derived from reasonable material quantity estimates, after looking at the consumptive material volumes from several similar projects and the study by CTBUH (Trabucco et al. 2015) for the environmental carbon footprint of the materials going into a reference building. However, since it is presented as percentages of a total footprint only and it comes from reasonable material quantity estimates, for the purpose of highlighting a trend, this approximation is appropriate.

While the specific numbers of this chart can and should be debated further, and they will change with every project, this chart highlights where to spend time in a project's initial design to optimize the building's embodied carbon footprint at the time of construction.

While building system choices will vary these percentages, the three areas typically representing the largest initial construction embodied carbon footprint are, in order, the structure, envelope, and MEP systems (the building shell and core). Collectively these typically represent 60–70% of the total embodied carbon footprint to construct the project. A more detailed case study on the comparative percentage breakdown of embodied carbon within tall buildings can also be found in studies by Clark (2013).

348   D. Davies and D. Trabucco

Fig. 16.2 Example of reference building for longevity scenario

Fig. 16.3 Estimate of the embodied carbon within a 150 m tall hotel at initial construction

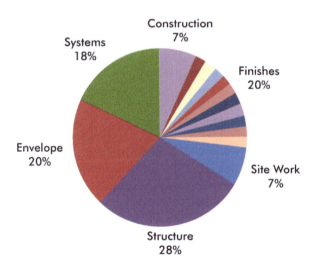

While the above percentages represent embodied carbon at the time of construction, considering the longevity assumptions for when systems will be replaced further changes these ratios in a notable way. For this 100-year life span example, we have further assumed that:

- Mechanical, electrical, and plumbing (MEP) systems will be updated about every 25 years.

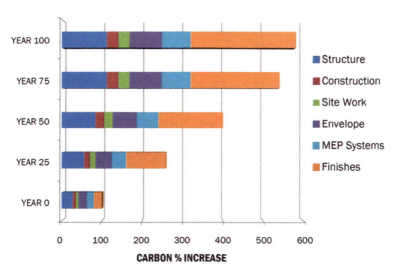

Fig. 16.4 Scenario #1: structure replacement every 25 years – carbon split over time

- Facades typically last between 20 and 70 years. For this example, we have assumed cladding systems will be updated every 50 years.
- Interiors are often rejuvenated more frequently for a hotel, often around every 5–15 years. For this example, we have assumed a 10-year average for the changing of secondary finishes within the building.

Given the above assumptions, when considering up to a 100-year time span and if the building has to be fully rebuilt multiple times over that 100 years, or if the primary structural frame stays and only the secondary systems are upgraded, the embodied carbon footprint of the site changes significantly. This provides an enlightening view of the cumulative embodied carbon impacts of a building with time (Figs. 16.4, 16.5, and 16.6).

If looking to reduce embodied carbon footprint impacts over time, the longevity and durability of the structural frame are important (Pomponi and Moncaster 2016). Minimizing the use of and replacement frequency of secondary finishes is also a vital aspect of reducing a building's overall embodied carbon footprint, but much of that may be irrelevant if the structure is scrapped and rebuilt due to functional or life safety obsolescence over a shorter building life. Also to this point, architectural quality which gives the potential for buildings to achieve landmark status over time, where restoration and preservation over demolition and replacement are desirable, matters.

As this example shows, improving a building's durability, and the related topic of resiliency, is a beneficial strategy in both embodied carbon and economic terms, to the long-term needs of the built environment. Similar realizations have led to

SCENARIO #2 – Structure Replacement in 50 Years

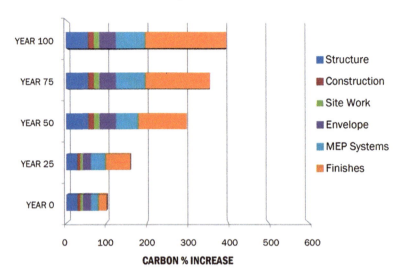

Fig. 16.5 Scenario #2: structure replacement every 50 years – carbon split over time

SCENARIO #3 – Structure Replacement in 100 Years

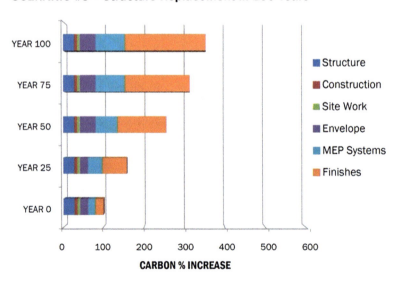

Fig. 16.6 – Scenario #2: structure replacement every 100 years – carbon split over time

thoughtful efforts by others to quantify resiliency benefits beyond a replacement cycle date as used in the example above (US Resiliency Council 2015; Almufti and Willford 2013).

Building Performance Objectives

Tall building longevity qualifications for embodied carbon reduction discussions, and how they can be influenced, need to also include an understanding of how we design for and predict a building's structural life safety performance with time. Current US and multiple international building codes follow earthquake and wind criteria specified in ASCE 7–10 (ASCE 2013 examples are shown for reference). Note that ASCE7–16 (ASCE 2017) tables, the newest version of this code, will soon be available for use, but were not available at the timing of this publication. The discussion on this topic, though, is still relevant for either model code.

For ordinary buildings, the underpinnings of today's building code are focused on risk-targeted maximum considered earthquake (MCE_R) demands of a 1% in 50-year collapse risk. For the majority of the USA, this is approximately a once in every 2000 years event. A "design earthquake" is then defined as 2/3 of these MCE_R demands and, on average, has a 10% probability of exceedance in 50 years (475-year mean recurrence interval). The data scatter for this design earthquake varies from 200 to 800 years (Figs. 16.7 and 16.8).

For wind, the underpinnings of the ASCE building code are a 700-year return period based on a single element in the structural system reaching its limit state, which is roughly targeted at a 10% probability of exceedance in 50 years.

In the above references, for both wind and seismic, there is a 50-year criterion from which risk probabilities are established, based upon event recurrence intervals that are hundreds of years apart. But as noted earlier, most tall buildings are likely to be in existence for 100 years or more.

This 50-year criterion, though, establishes a frame of reference for many that we are only designing our building inventories for 50-year durability, after which it is likely to be replaced. If properly maintained, our tall building structural frames can and should last much longer than 50 years. Tall buildings and their longevities could be improved if we extended this mind-set, but what would that actually mean to what is built today?

For earthquakes, buildings built to ASCE 7–10 (or similar) criteria have generally performed as expected, with repairable damage in many cases if it was a code-level or lesser event. Some of the biggest earthquake vulnerabilities today, when considering a code-level "design-basis" event, typically involve older buildings or buildings that do not meet current ASCE 7–10 or equivalent standards (Harris 2015). For say 100-year longevity, given the uncertainties around earthquake frequencies, increasing the current design-basis earthquake ground shaking levels for most locations may not be warranted (unless there is a goal to move to a higher standard of building performance). This is a worthy resiliency debate, but assuming that is not the objective for a moment, simply changing the terminologies to keep the design-basis earthquake event the same but to start thinking in a 100-year mind-set for the rest of the building could be significant for the discussions around the quality and longevity of cladding and other secondary systems. Note that a 2%

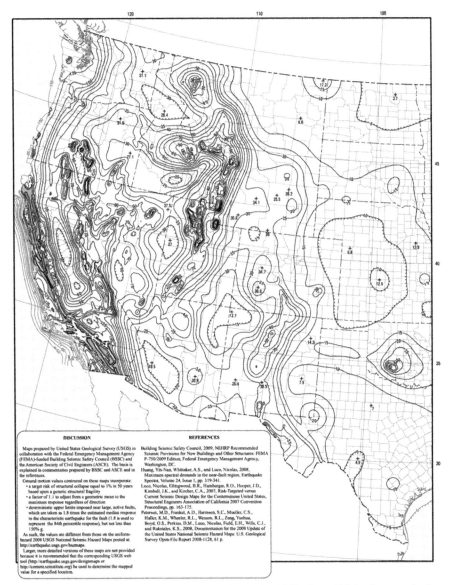

FIGURE 22-1 S_S **Risk-Adjusted Maximum Considered Earthquake (MCE$_R$) Ground Motion Parameter for the Conterminous United States for 0.2 s Spectral Response Acceleration (5 % of Critical Damping), Site Class B.**

Fig. 16.7 ASCE seismic design criteria – Western US American Society of Civil Engineers 7–10 (Reprinted with permission from ASCE)

probability in 100-year earthquake is approximately the same 2000-year mean return period earthquake as a 1% probability in 50-year event. Changing the mind-set to be thinking about 100-year exterior enclosure systems, or an

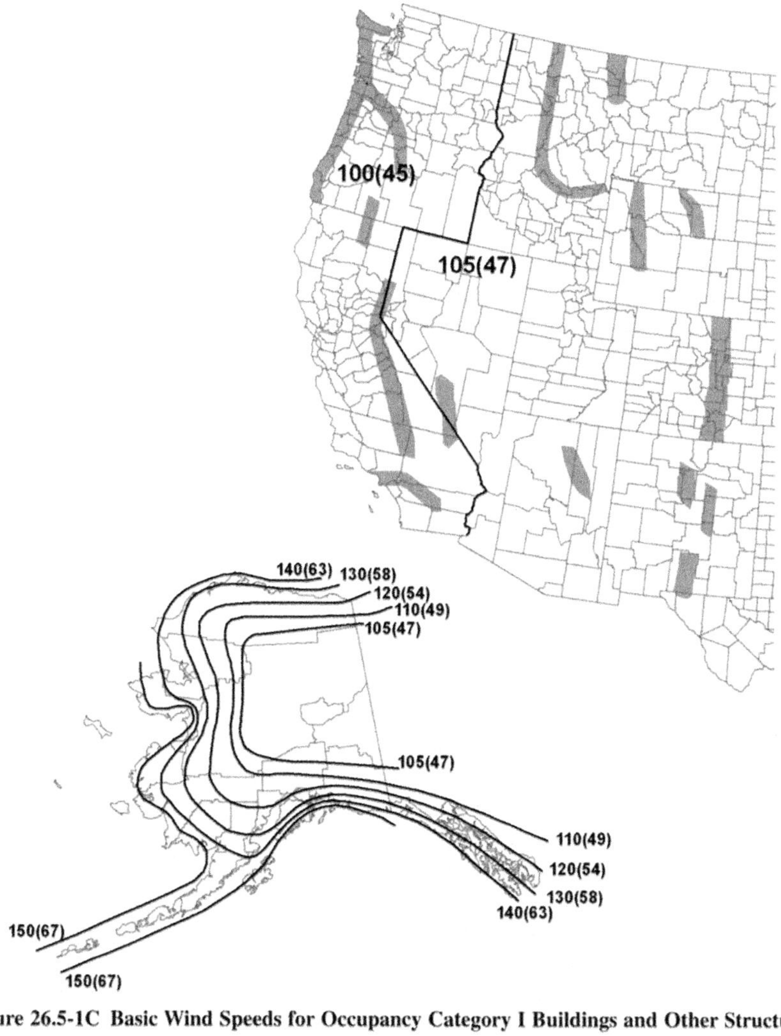
100(45)
105(47)
140(63) 130(58)
120(54)
110(49)
105(47)
105(47)
110(49)
120(54)
130(58)
140(63)
150(67)
150(67)
Figure 26.5-1C Basic Wind Speeds for Occupancy Category I Buildings and Other Structures.

Fig. 16.8 ASCE wind design criteria – Western US American Society of Civil Engineers 7–10 (Reprinted with permission from ASCE)

economical way to repair or remove and replace a building's exterior enclosure over a build's life, is fundamental to getting to longer-lasting buildings.

Similar considerations apply to wind design, but the implications may be more profound when moving to say a 100-year criterion. In many coastal locations of the world (locations like Florida being an exception), the influence of hurricanes and typhoons is only nominally considered as the events have historically had a probability of less than once every 50 years. When the design criterion is extended to 100 years in many coastal regions, though, wind design impacts may grow significantly.

For tall buildings, wind tunnel specialists are frequently engaged during a building's design, so site- and region-specific wind data from best available science can be uniquely brought to each design. But today, if the building code does not require consideration of 100-year events, especially for cladding system design, it is only the more resiliently minded owners that will, by choice, upgrade their buildings for the true impacts from a hurricane or typhoon. Hurricane and typhoon probabilities, their increasing frequency, and how building codes react to these events are a currently evolving topic.

In response to the above, performance-based wind design is a growing topic of attention (Griffis et al. 2012), as is the use of damping systems to reduce energy input into a building from a windstorm. Damping both improves occupant comfort and can potentially reduce damage impacts from within. These systems and their advancements are not a detailed consideration of this chapter, but more information can be found from resources such as the CTBUH. Damping systems, though, in most cases can help improve building performance and thus have the potential to reduce long-term embodied carbon footprint requirements of a building.

Embodied Carbon Reductions Through Longevity

Earthquakes

Statistically, approximately 70% of earthquake losses over time are due not to collapse but damage to nonstructural cladding and building systems and contents (Arroyo et al. 2015). The resulting embodied carbon impacts are significant as nonstructural system replacements of cladding and MEP systems, as well as finishes, as shown in Fig. 16.3 above, represent approximately 60% of the embodied carbon of a building at the time of construction. Possibly as or even more significant, if the economic damage to the building exceeds replacement costs, or if the building receives a reputation as being a bad building to be in for an earthquake, the full building and its structure are likely to be scrapped and replaced instead of renovated and repaired. Reducing damage from an earthquake is thus significant to long-term embodied carbon reductions within a building. Where do these nonstructural elements pose the most problems? Any heavy building components or water tanks that can topple over are a good starting point.

Most significantly for tall buildings, though, the proper jointing of cladding systems for movement, especially at corners, requires special attention for improved survivability and to reduce the embodied carbon impacts associated with repairs. These systems need to be explicitly designed, detailed, and jointed for the true movements that the building will experience, without damage during service-level earthquakes (Court et al. 2012; FEMA 2006). While this is a topic for both curtain walls and infill window wall construction, most new tall buildings within first world cities will attempt to appropriately account for these movements, although multiple examples exist showing there is still work to do (Court et al. 2012).

Unfortunately though, for a wide collection of new tall buildings still being constructed today, especially residential towers in emerging markets, there is inconsistency in what happens for the building enclosures, with both life safety and embodied carbon impacts. Some of these towers still rely on infill systems with minimum engineering and historical trade practices such as infill brick and hollow clay tile enclosures with a plaster exterior that are not appropriate for the building movements that will likely occur within the building's life. There can be a lack of understanding of the anticipated building movements and major wind or seismic events to often bring significant damage to these buildings. The resulting repairs may exceed the value of the building when they occur, and the buildings degrade rapidly, with a service life that is significantly reduced (FEMA 2012). This leads to earlier building demolition and replacement that might otherwise not be necessary.

Water infiltration at cracks in these systems affects long-term durability, and when proper embedded or backup steel support is not provided, earthquakes bring a high probability of bricks "raining" onto the street. As shown below, the "primary" structural frames of these buildings may be fairly well designed. Significant damage will occur, though, in even a moderate earthquake due to infill brick falling off the building exterior. This exterior infill system is now banned in new construction for US high seismic regions and many other countries for obvious life safety reasons (Figs. 16.9 and 16.10).

Well-engineered curtain wall and window wall systems today can generally achieve the necessary building movements. Even brick, when jointed appropriately and used in a rain screen application, designed as reinforced brick panels, or when appropriately anchored to a backup system that gives it ductility, can perform well. Infill exterior brick systems such as those shown above, though, represent a significant long-term liability in any location subject to earthquakes. The resulting economic and embodied carbon costs to repair and replace the building exteriors for these shortcomings, or at worst to replace the buildings in total due to the extent of damage incurred, are significant. This may seem obvious, but it is still one of the most common mistakes in the design of residential buildings in international emerging markets, which leads to significant unnecessary economic and embodied carbon expenditures to remedy.

Wind

Wind design is another area where the choices made around building systems have a significant embodied carbon impact for the infrastructure repairs and replacements that have to happen following major storms. This topic is getting increased attention, especially given the increasing frequency of hurricanes, typhoons, and tornados due to global warming.

Fig. 16.9 Brick infill
cladding with improper
movement joints (Images
courtesy of Maffei
Engineering)

Fig. 16.10 Brick infill
cladding with improper
movement joints (Images
courtesy of Maffei
Engineering)

As an example, low-rise communities in the US Midwest, with their propensity for lightweight construction, are very vulnerable to tornados. Because of their historical infrequency, most building codes currently do not address designing for tornados, so the losses seen in areas where they unfortunately do occur within communities are not unexpected (Zhou et al. 2002). There are, however, inherent winners and losers in a major windstorm. Below is Joplin, Missouri, following a tornado in May 2011. The image is very telling regarding the survivability of lightweight construction compared to heavier-weight buildings that are more similar to many tall building construction techniques, even though neither was specifically designed to resist a tornado. The life safety differences between building options are notable, but strictly from an embodied carbon perspective which is the focus of this chapter, there is much to be said for the building that survives with nonstructural repairs being required and the building that has to be fully scrapped and replaced (Fig. 16.11).

This does not mean tall buildings don't have problems in extreme wind events, especially since the winds often intensify with height. Proper design for both building movements and the impacts associated with high winds will often control tall building structural designs, as well as the glass and façade design. Below is an example of a failed exterior façade of a tall building after a hurricane (Fig. 16.12).

Even in areas where structures and tall buildings are designed to resist the impact of wind-borne debris such as in Florida (International Code Council 2014), the general understanding is that code-complying facades must provide sufficient lifesaving measures of occupants finding shelter within the building (Judah and Cousins 2015).

Although this is certainly the most important objective, from an embodied carbon perspective, as noted when earthquake serviceability objectives are considered, carbon footprints may be reduced even further and building longevity greatly improved, if minimizing damage is an objective beyond life safety goals (Patterson et al. 2014). It requires an initial higher investment in embodied carbon and materials, but facades that can also maintain water-tight capacities at moderate storms and that can be easily and efficiently evaluated and repaired are a long-term project value. If a window is fully broken and missing, water penetration during high winds can spray water within a building's interiors resulting in mold growth and other types of water damage. In hurricane-impacted buildings, even when cladding damage is limited, the remediation for mold is often the most expensive repair, and it can result in the need to replace a majority of the interior finishes at a significant economic and embodied carbon cost (Trabucco et al. 2017).

The same problem – thought usually at a smaller scale – also happens during lower-intensity wind events that do not cause damage to facades through wind-borne debris or excessive building movements. In these circumstances, if the cladding system had too small of joints to begin with and/or maintenance is overly deferred, water can still find its way through decayed silicon joints, producing water patches and mold concerns in the building interiors.

Through better design, possible water infiltration warning instrumentation, and/or continuous maintenance of façade elements and façade joints, water

Fig. 16.11 Joplin, Missouri – May 2011 – Charlie Riedel, Associate Press

Fig. 16.12 Failed façade after a hurricane (Source: United Public Adjusters & Appraisers, Inc. (2016) http://www. unitedpublicadjustersny. com/commercial-hurricane-damage/)

penetration during windstorms can be minimized (Patterson et al. 2012). If water damage does occur, an immediate response of the building maintenance crew is fundamental to prevent mold growth.

Avoiding Obsolescence

Another, not to be trivialized, part of longer-lasting buildings is a project's adaptability over time. Although the resiliency of a tall building has generally referred to the structure's performance during an extreme natural event such as an earthquake or hurricane, the ability to change a building's program is also important. Buildings have to face the continuous evolution of the market they exist in. They need to be able to adapt to the evolution of the market requirements, or they will be demolished and replaced sooner than necessary, increasing our embodied carbon expenditures for a built environment. Office towers have successfully been converted into residential buildings, residential buildings into hotels, or vice versa. Several notable changes of function and upgrades have occurred in recent years for iconic older tall towers, such as the Empire State Building in New York (Rode and Lin 2010), Taipei 101 (Tsai and Chou 2016), or the Willis Tower of Chicago. But tall buildings from all periods have been modified to house new functions with the vast number of recent examples being office towers being converted in hotels or residential condos (Trabucco and Fava 2013). Hotel and residential functions typically require shorter structural spans and less MEP system depths than offices, allowing for an easier restoration process of old office buildings, where the typical core to window depth does not exceed 9 m and the interstory height is between 4.3 m and 4.6 m (Space Management Group 2006).

A change in a building's function, while preserving the structure and the façade system, can significantly extend a building's useful life while reducing the embodied carbon equivalence of a replacement building. Several boutique hotel brands, such as Kimpton and Joie de Vivre, have made the renovation and repurposing of buildings a core part of their brand. Examples would include Kimpton's repositioning of numerous office projects into hotels or even a prison into the Liberty Hotel in Boston.

Other historic towers have kept their original function while still undergoing significant renovation processes to meet present-day energy efficiency standards and other requirements for building systems, comfort, and sustainability. The Empire State Building is an example of an older tower whose iconic value became the main driver of its preservation and eventual renovation. The half-billion-dollar revamp of the tower started in 2010 and included a great number of modifications to improve the energy efficiency of the building, but it has remained primarily as an office tower, its original intended function (Malkin 2015).

A few tall building examples embrace the use of flexible and adaptable construction systems that, through the extensive use of accessible and pre-envisioned mechanical connections, achieves unique pre-assembly advantages during the construction process and faster construction speed. Such systems when designed appropriately can be inherently easier to modify and also offer unique advantages for maintenance, which can have future economic and embodied carbon benefits.

The Lloyd's building in London, completed in 1978, was conceived as an "assembly-kit" of prefabricated components. The idea started from the necessity for a fast construction process. At the same time, by the need to possibly enlarge the building's program with time, the vast interior atrium of the building includes the ability to be partially filled in with more floor plates as a future project phase. The design of the building has a very clear subdivision of service and serviced areas, with all "back of house" functions (toilets, stairs, elevators, etc.) built with modular construction.

The Blue Cross Blue Shield building in Chicago provided future adaptability in a different way, having been designed with the capacity for a future vertical expansion that would double the original project's size. A central atrium was provided within the original tower that envisioned future elevator banks. A network of transfer trusses at the phase 1 roof also both allowed for the future phase 2 construction above and for the MEP systems to service the future much larger building. While the original building was built in 1997, the vertical expansion occurred in 2010, taking advantage of this pre-envisioned adaptability capacity (Fig. 16.13).

In both building cases above, functional obsolescence was reduced within the original project designs due to a thoughtful consideration of the future and how the building could be adapted with time.

Other taller buildings often benefit from the use of modular construction and prefabrication, such as the Hong Kong and Shanghai Bank Corporation Headquarters in Honk Kong or the Torre Mapfre tower in Barcelona. These projects all offer, if required, an advantage in terms of future adaptability. These examples can be also seen as "designed for expansion and deconstruction," implying that the solutions that have been adopted are highly reversible and can allow, if needed, an easier dismantling process and an improved recyclability potential for the building elements.

From an embodied carbon perspective, though, there is a trade-off that can come with such pre-envisioned adaptability provisions. It often means greater economic and embodied carbon investments up front in order to build in the flexibility for a future, which may or may not happen. An up-front expansion investment is only realized if that future adaptability is taken advantage of. Because of this, the best examples for future adaptability are actually those that offer pre-envisioned strategies for how to expand in the future, with minimal additional infrastructure or additional material to achieve that expansion.

Fig. 16.13 Blue Cross Blue
Shield building vertical
expansion

Conclusion

Tall buildings represent an important typology for the future of urban areas. They
have often been considered a carbon- and energy-intensive typology due to the
materials that are needed for their construction. But for much of this building type
(excluding the extreme examples at both ends of the short and tall built environment
spectrum), the height of the building is not the most significant variable in the
embodied carbon of the built environment. The embodied carbon of initial construction certainly needs to be considered, but infrastructure that offers lower
carbon personal living and mobility choices, longevity, and optimized designs
can create lower carbon footprint impacts more than height considerations alone.

Embodied carbon reduction also requires greater resiliency and longevity of the built environment, perhaps one of the most important realizations.

Commercial properties are often designed by following a highly speculative approach, with the developer only focusing on 10–20-year investment periods. This leads to a tendency for many buildings to have a relatively low durability and design life. Tall building design, though, due to its intensified use of resources and the infrastructure requirements around these projects, often leads to more thoughtful planning and longer-lasting buildings.

Longevity, both from a material and a functional point of view, is fundamental to reducing the embodied content of any building. Thoughtful design of the building components, and in particular of the structural system and façade, can successfully reduce damage in areas where natural hazards such as hurricanes and earthquakes frequently occur. In "calm zones" where natural hazards are not a significant threat, functional obsolescence and accelerated aesthetic aging are the main drivers for building obsolescence or more frequent invasive renovation efforts. The design team should constantly pursue measures to prolong the expected life of a building and of its internal finishes. The design team should also attempt to incorporate system flexibilities that envision the inevitable future modifications that will be required to "refresh" the building or to change its function. Ultimately, construction technologies that consider "designing for disassembly" principles can facilitate the final dismantling process at the end of a building life and allow an easier recyclability of its materials.

Tall building designs represent an important element of modern cities. Where they are appropriate, they can represent an important part of a lower overall carbon lifestyle for a significant portion of the urban population.

References

2030 Inc. (2011). Architecture 1. Products challenge. Retrieved from http://www.architecture2030.org/files/2030products_cp.pdf

Almufti, I., & Willford, M. (2013). REDiTM rating system. Resilience-based earthquake design initiative for the next generation of buildings, Verison 1.0. Arup. https://www.arup.com/publications/research/section/redi-rating-system?query=redi

Arroyo, D., Ordaz, M., & Teran-Gilmore, A. (2015). Seismic loss estimation and environmental issues. Earthquake Spectra, 31(3), 1285–1308. https://doi.org/10.1193/020713EQS023M.

ASCE American Society of Civil Engineers. (2013). 7–10 minimum design loads for buildings and other structures. Reston: American Society of Civil Engineers.

ASCE American Society of Civil Engineers. (2017). 7–16 minimum design loads and associated criteria for buildings and other structures. Reston: American Society of Civil Engineers.

Chang, Y., Ries, R. J., & Lei, S. (2012). The embodied energy and emissions of a high-rise education building: A quantification using process-based hybrid life cycle inventory model. Energy and Buildings, 55, 790–798.

Chen, T. Y., Burnett, J., & Chau, C. K. (2012). Analysis of embodied energy use in the residential building of Hong Kong. Energy and Buildings, 55, 790–798.

Cho, H. W., Roh, S. G., Byun, Y. M., & Yom, K. S. (2004). Structural quantity analysis of tall buildings. Proceedings of the CTBUH 2004 Seoul Conference.

Clark, D. (2013). *What colour is your building?: Measuring and reducing the energy and carbon footprint of buildings*. London: Royal Institute of British Architects.

Court, A., Simonen, K., Webster, M., Trusty, W., & Morris, P. (2012). *Linking next-generation performance-based seismic design criteria to environmental performance*. 2012 Structures Congress, March 29–31, Chicago.

Drew, C., Ova, K. F., & Fanning, K. (2014). *The environmental impacts of tall vs. small: A comparative study*. Proceedings of: CTBUH 2014 Shanghai Conference.

Du, P., Wood, A., & Stephens, B. (2016). Empirical operational energy analysis of downtown high-rise vs. suburban low-rise lifestyles: A Chicago case study. *Energies, 9*, 445.

FEMA. (2006). *Designing for earthquakes: A manual for architects, providing protection to people and buildings*. Retrieved: https://www.fema.gov/media-library/assets/documents/8669

FEMA. (2012). *Seismic performance assessment of buildings, Vol. 1 – Methodology*, Redwood. Retrieved: https://www.fema.gov/media-library-data/1396495019848-0c9252aac91dd1854 dc378feb9e69216/FEMAP-58_Volume1_508.pdf

Foraboschi, P., Mercanzin, M., & Trabucco, D. (2014). Sustainable structural design of tall buildings based on embodied energy. *Energy and Buildings, 68*, 254–269.

Griffis, L., Patel, V., Muthukumar, S., & Baldava, S. (2012). A framework for performance based wind engineering. ATC & SEI Conference on Advances in Hurricane Engineering, 2012, Oct. 24.

Harris, J. (2015). Building seismic safety council (BSSC) colloquium, 2015, Feb. 11, Seismic performance objectives.

International Code Council. (2014). *Florida building code – Buildings*, 5th Edition. https://codes. iccsafe.org/public/collections/FL

Judah, I., & Cousins, F. (2015). *The resilient urban skyscraper as refuge*. In: CTBUH Conference Proceeding of CTBUH 2015 New York Conference, pp. 230–237

Khan, F. (1969). *Recent structural systems in steel for high-rise buildings*. Proceedings of: The British Constructional Steelwork Association Conference on Steel in Architecture London S. W.1, 24th to 26th November, 1969.

Kim, S. B., Lee, Y. H., & Scanlon, A. (2008). Comparative study of structural material quantities of high-rise residential buildings. *Structural Design of Tall and Special Buildings, 17*, 217–229.

Kofoworola, O. F., & Gheewala, S. H. (2009). Life cycle energy assessment of a typical office building in Thailand. *Energy and Buildings, 41*, 1076–1083.

Malkin, A. (2015). A landmark sustainability program for the empire state building. Interchanges: Resurgence of the Skyscraper City Conference, pp. 364, 369.

Newman, P. W. G., & Kenworthy, J. R. (1989). Gasoline consumption and cities. *Journal of the American Planning Association, 55*(1), 24–37.

Norman, J., MacLean, H. L., & Kennedy, C. A. (2006). Comparing high and low residential density: Life-cycle analysis of energy use and greenhouse gas emissions. *Journal of Urban Planning and Development, 132*, 10–21.

Oldfield, P. (2012). Embodied carbon and high-rise. In Asia Ascending: Age of the Sustainable Skyscraper City – A collection of state-of-the-art, multi-disciplinary papers on tall buildings and sustainable cities, Proc. of the CTBUH 9th World Congress, pp. 614–622

Oldfield, P., Trabucco, D., & Wood, A. (2009). Five energy generations of tall buildings: An historical analysis of energy consumption in high-rise buildings. *The Journal of Architecture, 14*(5), 591.

Patterson, M., Martinez, A., Vaglio, J., & Noble, D. (2012). New skins for skyscrapers: anticipating Façade Retrofit Proceedings of: CTBUH 2012 9th World Congress, Shanghai.

Patterson, M., Silverman, B., & Casper, J. (2014). Curtainwall lifecycles: Evaluating durability and embodied energy Proceedings of CTBUH 2014 Shanghai Conference Proceedings; Mic Patterson, Ben Silverman & James Casper.

Perkins, A., Hamnett, S., Pullen, S., Zito, R., & Trebilcock, D. (2009). Transport, housing and urban form: The life cycle energy consumption and emissions of city centre apartments compared with suburban dwellings. *Urban Policy and Research, 27*, 377–396.

Pomponi, F., & Moncaster, A. M. (2016). Embodied carbon mitigation and reduction in the built environment: The evidence. *Journal of Environmental Management, 181*, 687–700.

Rode, P., & Lin, H. (2010). Talking tall: Greening supertalls: The green retrofit programs of the empire state building and Taipei 101. *CTBUH Journal, 2010*(3), 46–50.

Space Management Group. (2006). *Promoting space efficiency in building design*. Retrieved: http://www.smg.ac.uk/reports.html

Stein, R. G. (1977, 2011). Energy cost of building construction. *Energy and Buildings*. Richard G. Stein and Associates, *1*(1), 27–29.

Strain, F. A. I. A. (2017, May). Time value of carbon. *Carbon Leadership Forum*. http://www.siegelstrain.com/wp-content/uploads/2017/09/Time-Value-of-Carbon-170530.pdf

Trabucco, D. (2016). End of life of a tall building. In N. Clark & B. Price (Eds.), *Tall buildings a strategic design guide* (pp. 123–129). Newcastle upon Tyne: Riba Publishing.

Trabucco, D., & Fava, P. (2013). Confronting the question of demolition or renovation. *CTBUH Journal, 2013*, 38–43, ISSN: 1946–1186.

Trabucco, D., Wood, A., Vassart, O., Popa, N., & Davies, D. (2015). *Life cycle assessment of tall building structural systems*. Chicago: CTBUH.

Trabucco, D., Mejorin, A., Miranda, W., Nakada, R., Troska, C., & Stelzer, I. (2017). *Cyclone resistant glazing solutions in the Asia-Pacific region: A growing market to meet present and future challenges*. Proceedings of GPD Glass Performance Days 2017, Tampere.

Treloar, G. J., Fay, R., Ilozor, B., & Love, P. E. D. (2001). An analysis of the embodied energy of office buildings by height. *Facilities, 19*(5/6), 204–214.

Tsai, F., & Chou, J. (2016). *A perspective on Taipei' 101's decision to upgrade recertification to LEED O+M v4*. Proceedings of the 2016 CTBUH Shenzhen – Guangzhou – Hong Kong Conference.

U.S. Resiliency Council. (2015). Implementation Manual 2015. https://codes.iccsafe.org/public/collections/FL

Zhou, Y., Kijewski, T., & Kareem, A. (2002). Along-wind load effects on tall buildings: Comparative study of major international codes and standards. *Journal of Structural Engineering, 128*, 788–796.

Part IV
Approaches Across Global Regions

Chapter 17
Managing Embodied Carbon in Africa Through a Carbon Trading Scheme

N. Kibwami and A. Tutesigensi

Introduction

Need for Embodied Carbon Management in Africa

Africa is in its earliest stages of economic development, a situation characterized by industrialisation, mechanisation of agriculture, rapid urbanisation and high rates of construction activities. Unfortunately, Africa's development is to come at a cost. Hypotheses, such as the environmental Kuznets curve, suggest that at such early stages of economic growth, environmental degradation and pollution increase at an increasing rate because of increased energy demand and concomitant carbon emissions (Osabuohien et al. 2014; Esso and Keho 2016; Asane-Otoo 2015). Moreover, unlike the developed world where most buildings that will be operating in decades to come are already built, in the developing world, such buildings are either being built or yet to be built. In order not to repeat the same mistakes made in the developed world, Africa should tread a low-carbon path to development. This calls for, among other things, integrating carbon management strategies into policies of economic development, so as to promote sustainable development. This call is supported by recent research (e.g. Chen et al. 2015; Lam et al. 2015) which suggest that carbon trading approaches could be a viable strategy to promoting sustainable buildings in developing countries.

Although consideration of embodied carbon continues to attract attention, especially in the developed world, little is known about the same in Africa. This

N. Kibwami (✉)
School of the Built Environment, Department of Construction Economics and Management, Makerere University, Kampala, Uganda
e-mail: knathan@cedat.mak.ac.ug

A. Tutesigensi
School of Civil Engineering, University of Leeds, Leeds, UK

observation is supported by studies suggesting that research related to this topic with regard to Africa is scanty (Cabeza et al. 2014; Kibwami and Tutesigensi 2016), yet embodied energy of buildings in developing countries can be large, compared to operating energy, as the latter is quite low (Levine et al. 2007). For instance, a study found the average embodied energy consumed in brick production in Uganda to be over five times more than that in developed countries (Hashemi et al. 2015). This is partly because the technologies used in the Ugandan manufacturing industry are highly energy intensive, inefficient and associated with high levels of pollution (Okello et al. 2013). Although some initiatives for curbing emissions have been realized in a number of sectors in Africa, such as forestry, waste management and energy industries (Buchholz et al. 2012; Zanchi et al. 2013; Henry et al. 2011; Kees and Feldmann 2011), little has been realized in the building sector. In particular, there is little or no information about the applicability of existing carbon management policies, such as carbon trading schemes, with regard to managing embodied carbon emissions in Africa.

Carbon Trading Schemes: The CDM

The Kyoto Protocol, a leading framework in managing global carbon emissions, provides for three carbon management mechanisms – emissions trading, joint implementation and the Clean Development Mechanism (CDM) (UNFCCC 2013). Of these, the CDM is the only mechanism implemented between the developed (Annex 1 countries) and developing countries (non-Annex 1 countries). In CDM initiatives, the Annex 1 countries can purchase certified emissions reduction (CER) credits, each equivalent to one ton of emissions avoided. The purchased CER credits can then be used to offset emission reduction targets in the Annex 1 countries. The CER credits must have been generated from emission reduction activities (e.g. planting of trees, renewable energy projects, energy efficiency measures etc.) undertaken in the participating developing countries. While the CDM has not been without its challenges and controversies (Gillenwater and Seres 2011; Hinostroza et al. 2007; Winkelman and Moore 2011), it has arguably remained the best available concerted global effort of tackling climate change by establishing a market for carbon emission reductions. Since most of the developing countries are located in Africa (UNCTAD 2011), the CDM could equally present the best opportunity, as a carbon trading scheme, in managing embodied carbon emissions in Africa.

Unfortunately, although energy-related CDMs constitute the biggest proportion of the hitherto registered CDMs (UNFCCC 2016), attempts to address buildings' emissions through CDMs are still meagre. By February 2006, a year after establishing the CDM, less than 5% of the total registered CDMs were related to buildings, with none in pipeline for registration (Novikova et al. 2006). By May 2008, of the 3000 CDMs in pipeline then, only six were related to buildings (Cheng et al. 2008). Moreover, the few registered CDMs related to buildings are focused on

Table 17.1 Some registered CDMs in Uganda and extent of emission reduction

No.	Project title and registration date	Total reductions (tCO$_2$eq)	Annual reductions (tCO$_2$eq)	Operation period (years)	Sector[a]
1	West Nile Electrification Project (WNEP); 10th February 2007	760,417	36,210	21	E/R
2	Uganda Nile Basin Reforestation Project No.3; 21st August 2009	111,798	5564	20	A/R
3	Bugoye 13.0 MW Run-of-River Hydropower Project; 1st January 2011	510,740	51,074	10	E/R
4	Kachung Forest Project: Afforestation on Degraded Lands; 4th April 2011	547,373	24,702	20	A/R
5	Uganda Nile Basin Reforestation Project No. 4; 29th August 2011	79,395	3969	20	A/R
6	Bujagali Hydropower Project; 7th October 2011	6,007,211	858,173	7	E/R
7	Mpererwe Landfill Gas Project; 20th January 2012	182,612	18,261	10	W/D
8	Buseruka Mini Hydro Power Plant; 21st May 2012	314,679	31,468	10	E/R
9	Namwasa Central Forest Reserve Reforestation Initiative; 31st January 2013	226,564	11,328	20	A/R

Source: UNFCCC (2016)
[a]*E/R* Energy industries renewable/nonrenewable, *A/R* Afforestation and reforestation, *W/D* Waste handling and disposal

operational energy, that is, energy demand in the operation phase of buildings – none addresses embodied energy. Although some studies, such as Mok et al. (2014), suggest that implementing CDM projects during the construction phase of buildings is feasible, the practicality of such suggestions is yet to be tested. Even in Uganda, a country that has hosted a variety of CDM projects (See Table 17.1), including the first forest-related CDM in Africa (World Bank 2009), building-related CDMs are unheard of. As such, little is known about how the CDM concept could be a possible strategy in managing embodied carbon of buildings in Africa. This chapter presents a lesson to demonstrate how, in the context of Uganda, the CDM can be extended to buildings, based on integration of accounting for embodied carbon in the building development process.

Methods

Integrating EC in the Development Approval Process

Budding practices suggest that embodied carbon accounting should be integrated in building projects, although the scope often used limits broader integration. These practices, whether voluntary (see Franklin and Andrews 2013; RICS 2012) or mandatory (see Brighton and Hove 2013) put emphasis on the cradle-to-gate boundary. While this boundary arguably presents the least complications in accounting for embodied carbon (EC), it does not give a complete picture of a building project since activities like construction are excluded. Consideration of extended boundaries such as cradle-to-grave requires making difficult assumptions (energy-use behaviour, number of renovations, demolition activities, etc.) about the operation phase of a building (Hong et al. 2015) and thus uncertainties in EC can manifest. Integrating EC accounting in building projects in a manner that does not underrepresent activities and yet minimises uncertainties requires consideration of the cradle-to-construction completion boundary. The appropriate boundary to use in creating a CDM should be one where it is possible to optimise the greatest comprehensiveness with the greatest certainty. The cradle-to-construction completion boundary, the authors argue, would be the most appropriate for that purpose. However, this boundary necessitates consideration of the whole buildings' development approval process, that is, from planning permission to commissioning. Since development approval regimes vary by country, development of country-based EC accounting procedures is necessary.

With a case of Uganda, the present authors considered an EC accounting procedure previously proposed (Fig. 17.1). The procedure introduces EC accounting into various processes associated with the development approval process of building projects. As part of environmental impact assessment (EIA), EC accounting is presented as a requirement for environmental approvals. Similarly, as part of the development permit application (DP) process, EC accounting of prospective projects is included as a requirement for issuing building and occupation permits (see 'Account for carbon 1' and 'Account for carbon 2' linkages in Fig. 17.1). With regard to building projects (BP), preliminary EC estimates are made during preliminary design stage, detailed EC estimates during detailed design stage and interim EC estimates during the construction stage.

Evaluation of the Procedure

The proposed procedure (Fig. 17.1) was evaluated using structured interviews with a sample of 120 construction professionals registered and working in Uganda in 2014. The professionals consisted of 30 architects, 30 engineers, 30 quantity surveyors and 30 environmental impact assessors. Kibwami and Tutesigensi

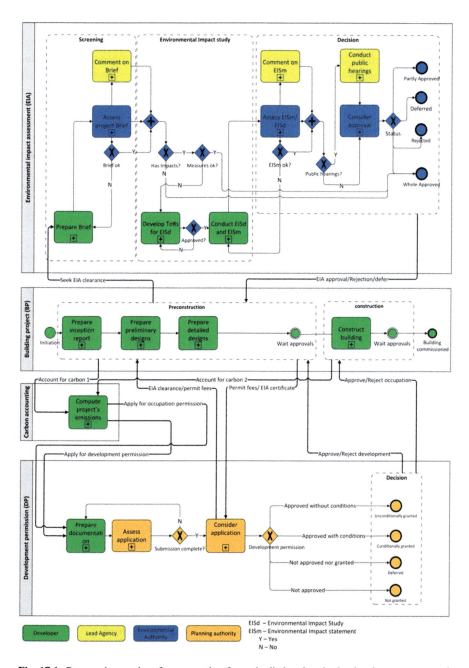

Fig. 17.1 Proposed procedure for accounting for embodied carbon in the development approval process of building projects in Uganda (Source: Kibwami and Tutesigensi 2015)

(2016) demonstrate that these professionals' level of experience and understanding of sustainability aspects in relation to the environment, and consequently EC, were suitable enough to generate meaningful opinions about the procedure. The evaluation was carried out to assess how the procedure was perceived with regard to its feasibility based on the principles of implementing environmental policy. The principle criteria for evaluating environmental policies, as specified in the IPCC fourth assessment report, were considered. These included cost implications, the extent to which the procedure can achieve its goals at minimum cost; distributional considerations, which largely relate to how the consequences of a procedure are distributed; and institutional feasibility, the extent to which the procedure is likely to be embraced given existing frameworks (Gupta et al. 2007). It was therefore hypothesised that a policy on introducing an environmentally related procedure of accounting for EC in Uganda should address such criteria. The following hypotheses were broadly considered:

- *Hypothesis H1*: null hypothesis $H1_0$, the procedure has cost implications, and alternative hypothesis $H1_1$, the procedure has no cost implications.
- *Hypothesis H2*: null hypothesis $H2_0$, the procedure does not address distributional considerations, and alternative hypothesis $H2_1$, the procedure addresses distributional considerations.
- *Hypothesis H3*: null hypothesis $H3_0$, the procedure is not institutionally feasible; and alternative hypothesis $H3_1$, the procedure is institutionally feasible.

The details on how data for testing these hypotheses were collected are discussed hereunder.

Perceptions of Cost Implications of the Procedure

Literature suggests that policies that (1) require new institutions in order to implement them, (2) are difficult to understand and (3) contribute to little or no benefits usually have high cost implications (Boardman et al. 2013; Mickwitz 2003; Gupta et al. 2007). Therefore, evaluation of cost implications (hypothesis H1) involved checking whether new institutions are required to implement the procedure, whether the procedure is easy to understand and whether it can contribute to other benefits, such as carbon trading (see Table 17.2).

Perceptions of Distributional Considerations of the Procedure

Table 17.3 summarises how perceptions of distributional considerations were evaluated so as to test hypothesis H2. This included assessing the procedure's legitimacy, fairness and transparency (Gupta et al. 2007; Mickwitz 2003; Huitema et al. 2011). Legitimacy was interpreted in terms of willingness to use the procedure, whereas fairness was directly interpreted as the extent to which the procedure

Table 17.2 Assessing perceptions on cost implications

Item	Descriptors and coding references				
The procedure contributes to other benefits[a]	1	2	3	4	5
	Strongly disagree	Disagree	Undecided	Agree	Strongly agree
The procedure is easy to understand[b]	1	2	3	4	5
	Not easy at all	Not easy	Undecided	Somewhat easy	Very easy
Institutional changes required to implement the procedure[bc]	1	2	3	4	0
	New institutions	Significant modification	Minor modification	Existing suffice	Don't know

[a]1 and 2 = does not have benefits (Failure), 4 and 5 = has benefits (Success)
[b]1 and 2 = requires new institutions (Failure), 3 and 4 = does not require new institutions (Success)
[c]1 and 2 = Difficult (Failure), 4 and 5 = easy (Success)

Table 17.3 Assessing perceptions on distribution considerations

	Descriptors and coding references				
Item	1	2	3	4	5
Willingness to use procedure (Legitimacy)[a]	Extremely unlikely	Unlikely	Neither likely nor unlikely	Likely	Extremely likely
Fairness of the procedure (Fairness)[b]	Very unfair	Unfair	Neither fair nor unfair	Fair	Very fair
Clarity of intentions (Transparency)[c]	Very unclear	Unclear	Neither clear nor unclear	Clear	Very clear

[a]1 and 2 = not legitimate (Failure), 4 and 5 = Legitimate (Success)
[b]1 and 2 = not fair (Failure), 4 and 5 = fair (Success)
[c]1 and 2 = not transparent (Failure), 4 and 5 = transparent (Success)

was perceived to be fair to the various stakeholders. Lastly, transparency was interpreted in terms of clarity of the procedure's intentions.

Perceptions of Institutional Feasibility

In assessing institutional feasibility (hypothesis H3), the procedure was evaluated for its legal acceptance, compatibility, persistence and predictability (Huitema et al. 2011; Mickwitz 2003) (see Table 17.4).

Analysis of the Perceptions

A chi-square test of independence was used to assess the extent of variation of responses across the four categories of professionals (i.e. architects, engineers, quantity surveyors and environmentalists) (Bryman and Cramer 2011). This was

Table 17.4 Assessing perceptions on institutional feasibility

Criterion	Question
Legal acceptance	Does the procedure fit into existing regulatory framework?
Compatibility	Is the procedure compatible with national priorities?
Persistence	Would you consider the procedure a persistent solution to minimising emissions?
Predictability	Are the impacts resulting from implementing the procedure foreseeable?

done in order to establish whether analysis could proceed based on a combined sample of the four categories or on a category by category basis. Where significant variations in the responses were found, hypothesis tests were carried out per category. The three hypotheses mentioned above were tested by comparing the observed proportion of cases appearing in the specified two response options (e.g. no (failure) and yes (success)) with threshold proportions selected as 75% and 25% to correspond to the null and alternative hypotheses, respectively (Green and Salkind 2005). Responses of 'undecided' and 'don't know' were taken to be non-substantive and excluded to improve validity of responses (Foddy 1993). The null hypotheses ($H1_0$, $H2_0$, and $H3_0$) were to be rejected only if the observed proportion was significantly different ($p < 0.05$) from the respective threshold proportion. The hypotheses were tested using the chi-square test (Bryman and Cramer 2011; Green and Salkind 2005), based on the threshold proportion and the resulting effect size. The effect size d was calculated using the Eq. 17.1 below (Green and Salkind 2005, p. 359):

$$d = \frac{x^2}{(n)(r - 1)} \tag{17.1}$$

where x^2 is the chi-square value, n is the sample size and r is the number of response options. The result was classified according to Cohen's criteria (Cohen 1988).

Managing EC Using CDM

Measurement of Embodied Carbon

From the old management adage, what cannot be measured cannot be managed; management of EC requires measurement. However, most mathematical models for measuring EC can be described as aggregated models as they depend on universal parameters such as generic carbon emission factors (Kibwami and Tutesigensi 2014). The use of generic parameters does not facilitate consideration of EC at the micro level such as at project level. Yet, CDMs are project-based programs by design. This has led to the need for disaggregated models which are refined mathematical models that seek to employ project-level parameters in the quantification of EC so as to facilitate CDMs. Disaggregated mathematical models can enable refined analysis and exploration of 'what-if' scenarios in specific project contexts, thereby enabling consideration of a broad range of carbon reduction opportunities. Moreover, disaggregation enables the specific sources of energy to bear on the quantification of EC in a manner that enables distinguishing the sources of the resulting EC. Thus the computation of EC in this work was based on a mathematical model, previously proposed by the authors (see Kibwami and Tutesigensi 2014), as it accommodates disaggregation of EC. The equations relevant for this work were extracted from the model as follows.

Manufacture and Transportation of Materials

Emissions from manufacturing and transporting n construction materials, using e different sources of energy, are given by Eqs. 17.2 and 17.3, respectively:

$$\text{EC}_{m1} = \sum_i^n \rho_i \left(\sum_j^e V_{ij} C_j^a \theta_j^a + S_i \right) \tag{17.2}$$

$$\text{EC}_{m2} = \sum_i^n X_j^a \left(\sum_j^e C_j^b \alpha_j^a \right) \tag{17.3}$$

where EC_{m1} is the total emissions from manufacturing materials (in $kgCO_2$); ρ_i is the quantity of material type i (in kg); V_{ij} is the quantity of energy j to manufacture a unit of material i (in kWh/kg); C_j^a is the carbon emission factor (in $kgCO_2$/kWh) per unit energy j used; θ_j^a is a disaggregation factor in manufacturing material(i.e. the proportion of energy derived from energy source j) i; S_i is a constant for process emissions (i.e. emissions from chemical reactions that lead to the material) per unit of material i (in $kgCO_2$/kg); EC_{m2} is the total emissions from transporting materials (in $kgCO_2$/kg); W_{ij} is the quantity of energy j to transport a unit of material i per unit distance (in kWh/kgkm); X_j^a is the transport distance for material i (in km); α_j^a is a disaggregation factor in transporting materials; C_j^b is the carbon emission factor per unit distance (in $kgCO_2$/km) with respect to the corresponding transportation energy j.

Transportation of Workforce

Emissions from transporting workforce for duration r, using e different sources of energy were given by Eq. 17.4.

$$EC_l = \sum_f^r \beta_f L_f X_j^c \left(\sum_j^e C_j^d \alpha_j^d \right) \tag{17.4}$$

where EC_l is the total emissions from transporting workforce (in $kgCO_2$); β_f is the duration of f workforce used (in days); L_f is the number of people in the workforce required; X_j^c is the distance travelled by a person per duration (in km/day); C_j^d is the carbon emission factor per person per unit distance depending on the mode (e.g. bus, train, cycle) of transport used (in $kgCO_2$/personkm); and α_j^d is a disaggregation factor for the mode used in transportation.

The total EC can be computed using Eq. 17.5 below:

$$EC_T = EC_{m1} + EC_{m2} + EC_l \tag{17.5}$$

where EC_T is the total emissions (in $kgCO_2$) from cradle to practical completion of the building.

Although emissions from using equipment ought to be included (Cole 1998; Kibwami and Tutesigensi 2014), the activity of constructing the walls was assumed to be entirely carried out by human workforce without need for powered tools/ equipment.

Demonstration of the CDM

Basic Assumptions

For purposes of demonstrating how CDM can be used to manage EC, a typical dwelling unit (see Table 17.5) which was to be constructed was considered. Two options of constructing the dwelling's walls were assessed using the disaggregated mathematical model, based on the evaluated procedure shown in Fig. 17.1: (1) a baseline constructed using typical materials, and workforce, and (2) a 'green' alternative constructed using provisions to reduce EC. Thus for the entire dwelling unit, potential EC reductions were associated with construction of its walls only, similar to CDM-related proposals by UNFCCC (2013b). Given that consideration of a single project as a CDM would be unrealistic because of the financial implications, among other things, the results were projected to a wider scale using a factor of 28,000 so as to generate substantial emission reductions. This was based on a United Nations report which highlights that addressing housing shortage in the capital city of Uganda requires constructing over 28,000 houses annually for at least 10 years (UN-HABITAT 2010, p. 37). Therefore, the assumption for the proposed CDM was that it involves mitigating EC arising from constructing 28,000 similar dwelling units over a period of 10 years.

For a typical dwelling unit, energy sources were diesel, biomass, heavy fuel oil, biodiesel and grid electricity, since these are either predominantly used or have a great potential (UBOS 2013). The emission factors (see Table 17.6) were taken

Table 17.5 Information about the house

Parameter	Description
Building type	Typical two-bedroom residential house
Construction type	Traditional: masonry burnt mud bricks
Floor to wall plate height	3 m
Number of bedrooms	2 no.
Number of floors	1 no.
Internal floor area	103 m^2
Total wall area	223 m^2
Wall width (unplastered)	0.107 m (based on 228 × 107 × 69 mm bricks)[a]
Openings areas: doors	21 m^2
Windows	24 m^2
Roof type and structure	Corrugated iron sheets on timber roof truss structure
Total cement required (walling only)	2.23 tons (assuming 0.01 tons per m^2, stretcher bond)[a]
Total bricks required (walling only)	11,147 bricks (50 bricks per m^2)
Total sand required (walling only)	1 trip, 6 ton truck

[a]Source: UNFCCC (2013b)

Table 17.6 Emission factors for common energy sources in Uganda

Fuel/energy source	Emissions factor[a]	Conversion to MJ 1kWh = 3.6 MJ
Diesel (100% mineral diesel) for vehicles	0.545 kgCO$_2$/km	N/A
Diesel (electricity)	0.68 KgCO$_2$/kWh	0.189 KgCO$_2$/MJ
Heavy fuel oil (HFO) for electricity	0.71 KgCO$_2$/kWh	0.197 KgCO$_2$/MJ
Biomass	0 kgCO$_2$	N/A
Grid electricity (diesel, HFO and hydroelectricity mix)	0.14 kgCO$_2$/kWh	0.039 KgCO$_2$/MJ

[a]Source: UNFCCC (2010)

from UNFCCC (2010) which is a country-related source and thus considered to be representative of the context. These factors, such as that of grid electricity, often change year on year depending on the availability of hydroelectricity, which is greatly affected by prolonged dry seasons. Various proportions of energy required for the baseline and alternative options were considered so as to facilitate the disaggregation concept (see Table 17.7). The proportions for the baseline option were based on typical energy use in Uganda. For instance, energy used in the cement industry comes from diesel, biomass, heavy fuel oil and grid electricity; in some factories, biomass accounts for 30% of the total energy used (Lafarge 2012). The alternative option was based on the goal of Uganda's renewable energy policy: dependence on 61% renewable energy, with biofuel blends of up to 20% in the transport sector (The Republic of Uganda 2007). Therefore, for manufacture of materials in the alternative option, the manufacturers involved in the CDM

Table 17.7 Proportion of energy used

Energy sources	Material (cement) manufacture		Transportation (material or workforce)	
	Base line	Alternative	Base line	Alternative
Diesel	0.35	0.10	1.00	0.80
Non-fossil	0.30	0.60	0.00	0.20
Heavy fuel oil	0.05	0.00	N/A	N/A
Electricity	0.30	0.30	N/A	N/A
Total	1.00	1.00	1.00	1.00

initiative were assumed to source 60% of the energy from non-fossil renewable energy, whereas 20% biofuel blend was assumed in all transportation activities. The overall EC computed arose from manufacture and transportation of materials and transportation of workforce.

Assumptions regarding computation of EC from manufacture (and transportation) of materials and transportation of workforce were posed. For cement manufacture, which causes both energy (46%) and process-related (54%) emissions, the energy requirement was taken as 4.9 MJ/kg (Worrell et al. 2001, p. 321). The country's two largest cement producers 'Hima' (in the West) and 'Tororo' (in the East) are located approximately 350 km and 209 km, respectively, from the capital city (based on Google Maps); a 560 km average roundtrip was used, based on a 6-ton diesel truck (UNFCCC 2010). According to typical brick manufacturing practices in Uganda (i.e. wood-fired kilns), the associated emissions were cautiously taken as zero, similar to Pooliyadda and Dias (2005), since wood fuels are generally assumed to be green (i.e. with zero emissions). Also, no production emissions were considered for sand, as it is a naturally occurring material that is usually unprocessed, though requires transportation. Bricks and sand are usually sourced not very far from construction sites; a 50 km roundtrip was used in each case, based on a 6-ton diesel truck. For emissions from transportation of workforce, in order to come up with emissions per person, a typical 14-passenger public transportation vehicle was considered. Similar to Cole (1998), no vehicle-sharing was assumed and thus each person travelled individually to and from the site using public transport. Emissions per person per unit distance were obtained as: $0.545kgCO_2/km \div 14 = 0.0390kgCO_2$. Each person was assumed to travel a 20 km round trip per day and thus emissions per person per day were $0.039 \times 20 = 0.780 \ kgCO_2$. A total workforce of four people was presumed: two masons, each with an assistant. Since a mason can construct 3.17 m^2/day (Nalumansi and Mwesigye 2011), yet 223 m^2 of walls were to be constructed, the total construction duration was obtained as: 223 $m^2 \div 3.17 \ m^2$/day $\div 2 = 35$ days.

Computation of EC

Based on Eq. 17.2, EC from manufacture of materials was computed by multiplying the total energy required to manufacture a unit of material, with the proportion of energy source used, with the emission factor of that energy source and with the total quantity of material required. For instance, considering diesel emissions in manufacturing cement, the baseline and alternative options were computed as 4.9 MJ/Kg $\times$ 0.35 $\times$ 0.189 gCO$_2$/MJ $\times$ 2230 Kg $=$ 722 kgCO$_2$ and 4.9 MJ/Kg $\times$ 0.10 $\times$ 0.189 KgCO$_2$/MJ $\times$ 2230 Kg $=$ 207 kgCO$_2$, respectively. This calculation process was repeated for other energy sources but with varying proportions (as per Table 17.7) of energy sources used.

Based on Eq. 17.3, EC from transporting materials were computed by multiplying the distance of transporting materials, with the proportion of energy source used, with the emissions emitted per unit distance for that energy source. Taking an example of transporting cement, the baseline and alternative options were computed as: 560 km $\times$ 1.00 $\times$ 0.545 kgCO$_2$/km $=$ 305 kgCO$_2$ and 560 km $\times$ 0.80 $\times$ 0.545 kgCO$_2$/km $=$ 244 kgCO$_2$, respectively. A similar calculation was applied for bricks and sand.

As per Eq. 17.4, EC from transporting workforce were computed by multiplying the emissions per person per day, with the proportion of energy source used, with the total workforce required for the activity and with the total duration of the activity. Thus, the baseline and alternative options were computed as: 0.780 kgCO$_2$/person/day $\times$ 1.00 $\times$ 4 people $\times$ 35 days $=$ 110 kgCO$_2$ and 0.780 kgCO$_2$/person/day $\times$ 0.80 $\times$ 4 people $\times$ 35 days $=$ 88 kgCO$_2$, respectively.

Based on Eq. 17.5, the total emissions were obtained by summing up results from Eqs. 17.2, 17.3, and 17.4.

Results and Discussion

Professionals' Perceptions on the Proposed Procedure

Overview of Responses

By the end of the data collection process, which lasted for 9 weeks, all the potential respondents in the initial sample of 120 (16% of the research population) individuals had been contacted, indicating a contact rate of 100%. However, of the 120 potential respondents, data were successfully collected from 85 of them, indicating a response rate of 71% (see Table 17.8), which was found to be acceptable, according to Kervin (1992, p. 422). This achieved sample, having been generated from a randomized stratified population, was large enough to generate meaningful statistical analyses.

Summarised results pertaining to how the procedure was perceived with regard to the various criteria are presented in Table 17.9. As can be seen, the procedure was largely perceived to be feasible with exception that most respondents thought significant modification of existing institutions or new ones will be required to implement it. In that case, the corresponding null hypothesis regarding to cost implication of the procedure was accepted and equally, the effect size was too

Table 17.8 Response rates

Professionals	Population size (No.)	Initial sample size (No.)	Response Achieved sample (No.)	%	Non response Not achieved (No.)	%
Architects	163	30	20	67	10	33
Engineers	405	30	21	70	9	30
Quantity surveyors	42	30	21	70	9	30
Environmentalists	144	30	23	77	7	23
Overall	754	120	85	71	35	29

Table 17.9 Summary of the results

Criteria	Success/failure	P-value	Effect size	Null hypothesis
Cost implications of the procedure				
Need for institutions	Failure	0.61	0.003	Accept
Simplicity of the procedure	Success	0.0005	1.85	Reject
Contribution to other benefits	Success	0.0005	2.39	Reject
Distributional considerations of the procedure				
Willingness to use the procedure	Success	0.0005	2.72	Reject
Fairness of the procedure	Success	0.0005	2.06	Reject
Transparency of the procedure	Success	0.0005	2.62	Reject
Institutional feasibility of the procedure				
Legal acceptance	Success	0.0415	3.15	Reject
Compatibility	Success	0.0005	7.48	Reject
Persistence	Success	0.0005	1.19	Reject
Predictability	Success	0.0005	2.36	Reject

small to reject the null hypothesis. A detailed explanation of the statistical implications and how the various conclusions were arrived at is presented in the following sections.

Cost Implications of the Procedure

Need for Institutions

It is stated in IPCC (2007) that cost implications broadly manifest as direct and indirect costs. Whereas direct costs are associated with implementation, indirect costs result from intended or unintended effects of implementation. A chi-square test for independence indicated no significant differences in the responses among the professionals, x^2 (3, $n = 84$) = 6.23, $p = 0.10$. The dichotomised results from descriptive statistics show that 27% of the responses thought new institutions are not required, whereas 72% indicated new institutions are required. This implies that majority of the respondents considered the procedure to have high cost implications with regard to implementation, since significant modification or need for creating new institutions was envisaged. A one-sample chi-square test confirmed that the difference between the observed 72% and the threshold of 75% was not significant, x^2 (1, $n = 84$) $= 0.25$, $p = 0.61$. The effect size of 0.003 obtained was classified as small. As such, the null hypothesis (H1$_0$) was accepted, suggesting that the procedure could have high cost implications with regard to the need for institutions. Therefore, EC management was envisaged to have high direct costs.

Simplicity of the Procedure

One of the ways in which cost implications of a policy can be kept low is by ensuring that implementation procedures are simple (Gupta et al. 2007). A chi-square test for independence indicated no significant differences in the responses, x^2 (3, $n = 81$) $= 1.17$, $p = 0.76$. The dichotomised results showed that 15% found the procedure difficult, whereas 81% considered it to be easy. A one-sample chi-square test confirmed that the difference between the observed 15% and the threshold of 75% was significant, x^2 (1, $n = 81$) $= 150.13$, $p < 0.0005$. The effect size of 1.85 obtained was classified as large. As such, the null hypothesis (H1$_0$) was rejected in favour for the alternative hypothesis (H1$_1$) which suggested that the cost implications with regard to ease of understanding are low. Therefore, it will be necessary to strike a balance between the high cost implications associated with the need for new institutions and the low cost implications associated with the ease of understanding, such that the overall cost implications can be offset.

Contribution to Other Benefits

Upon dichotomising the responses, results showed that 8% disagreed, whereas 81% agreed that the procedure can contribute to other benefits. A chi-square test for independence indicated no significant differences in the responses among the professionals, x^2 (3, $n = 75$) $= 4.13$, $p = 0.25$. A one-sample chi-square test confirmed that the difference between the observed 8% and the threshold of 75% was significant, x^2 (1, $n = 75$) $= 179.56$, $p < 0.0005$. The effect size of 2.39 obtained was classified as large. As such, the null hypothesis (H1$_0$) was rejected in favour for the alternative hypothesis (H1$_1$), inferring that the procedure can contribute to benefits such as carbon trading. This finding corroborates claims such as those made by UK's Embodied Carbon Industry Task Force (2014) that EC accounting can contribute to several benefits. Therefore, cost implications of the procedure can be low if its benefits are exploited in order to offset its cost implications. One of the ways to achieve this is by augmenting the procedure with a carbon trading scheme, as argued in this work.

Distributional Considerations of the Procedure

A chi-square test for independence indicated no significant differences in responses among the professionals with regard to distributional considerations: willingness to use the procedure, x^2 (3, $n = 83$) $= 1.69$, $p = 0.64$; fairness of the procedure, x^2 (3, $n = 62$) $= 1.01$, $p = 0.80$; and transparency of the procedure, x^2 (3, $n = 81$) $= 2.67$, $p = 0.45$. Further analyses are detailed below.

Willingness to Use the Procedure

Upon dichotomising the responses about willingness to use the procedure, 95% of the respondents were willing to accept it in comparison with 4% who were not. A one-sample chi-square test confirmed that the difference between the observed 4% and the threshold of 75% was indeed significant, x^2 (1, $n = 83$) $= 225.58$, $p < 0.0005$. The effect size of 2.72 obtained was classified as large. The null hypothesis (H2$_0$) was rejected in favour for the alternative hypothesis (H2$_1$). Therefore, the findings suggested that the procedure can address distributional considerations related to legitimacy.

Fairness of the Procedure

Since fairness concerns how the outcomes of a measure are distributed (Mickwitz 2003; Huitema et al. 2011), the procedure's outcomes were envisaged to be fairly distributed in Uganda. This was so because the dichotomised responses of 'generally fair' and 'generally unfair', showed that 64% considered the procedure to be generally fair, whereas 10% thought it was generally unfair. A one-sample chi-square test confirmed that the difference between the observed 10% and the threshold of 75% was significant, x^2 (1, $n = 62$) $= 127.51$, $p < 0.0005$. The obtained effect size of 2.06 was classified as large. In that regard, the null hypothesis (H2$_0$) was rejected in favour of the alternative hypothesis (H2$_1$).

Transparency of the Procedure

Results suggested that the procedure was perceived to address distributional considerations in regard to transparency. On the whole, the dichotomised responses revealed that 91% considered procedure's intentions to be clear, compared with 5% who indicated that its intentions were not clear. A one-sample chi-square test confirmed that the difference between the observed 5% and the threshold of 75% was significant, x^2 (1, $n = 81$) $= 212.05$, $p < 0.0005$. The effect size of 2.62 obtained was classified as large. As such, the null hypothesis (H2$_0$) was rejected in favour for the alternative hypothesis (H2$_1$). Since transparency is one of the criteria for democratic accountability (Huitema et al. 2011), this finding suggests that the procedure upholds principles of democratic accountability.

Institutional Feasibility of the Procedure

Legal Acceptance

In assessing legal acceptance, the extent to which the procedure was compliant to the laws and regulations of Uganda was probed. A chi-square test for independence indicated significant variations in the responses across the four professions, x^2 $(3, n = 78) = 9.18, p = 0.03$. A one-sample chi-square test per category returned the following results:

- Architects: Observed $= 53\%, x^2 (1, n = 19) = 5.07, p = 0.024$, effect size $= 0.27$
- Engineers: Observed $= 44\%, x^2 (1, n = 18) = 8.96, p = 0.003$, effect size $= 0.50$
- Quantity Surveyors: Observed $= 50\%, x^2 (1, n = 18) = 6.00, p = 0.014$, effect size $= 0.33$
- Environmentalists: Observed $= 13\%, x^2 (1, n = 23) = 8.96, p = 0.0005$, effect size $= 2.05$

These results show that for each profession category, the differences between the observed percentages and the threshold of 75% were significant ($p < 0.05$). The corresponding effect sizes were also large and as such, the null hypothesis (H3$_0$) was rejected in favour for the alternative hypothesis (H3$_1$). It turned out that the significant variation in profession categories has nothing to do with the perceptions as to whether the procedure is legally acceptable or not; all shared similar perceptions that the procedure is legally acceptable, albeit with significantly varying levels of consensus.

Compatibility with National Priorities

Similar to legal acceptance, responses regarding the compatibility of the procedure with national priorities significantly varied across the professions, $x^2 (3, n = 79) = 9.55, p = 0.02$. A one-sample chi-square test conducted per profession category revealed the following:

- Architects: Observed $= 35\%, x^2 (1, n = 17) = 14.29, p = 0.0005$, effect size $= 0.84$
- Engineers: Observed $= 5\%, x^2 (1, n = 19) = 49.28, p = 0.0005$, effect size $= 2.59$
- Quantity Surveyors: Observed $= 25\%, x^2 (1, n = 21) = 29.35, p = 0.0005$, effect size $= 1.4$
- Environmentalists: Observed $= 4\%, x^2 (1, n = 22) = 58.24, p = 0.0005$, effect size $= 2.65$

These results show that for each professional category, the differences between the observed percentages and the threshold of 75% were significant ($p < 0.05$). The corresponding effect sizes were also large and as such, the null hypothesis ($H3_0$) was rejected in favour for the alternative hypothesis ($H3_1$). This implied that the procedure is compatible with Uganda's priorities. Thus, the significant variation in profession categories noted above has nothing to do with the perceptions as to whether the procedure is compatible with national priorities or not; all professionals agreed that it is compatible albeit with significantly varying levels of consensus. Since it is crucial in Uganda that any proposed measure, especially in the context of addressing climate change, is compatible with Uganda's priorities (Olsen 2006; Bwango et al. 2000), the procedure demonstrates how national priorities can be reconciled with the mitigation of climate change.

Persistence

Mickwitz suggests that for a policy to be persistent, its impacts (intended or unintended) should be long lasting (Mickwitz 2003). With regard to whether the procedure is persistent, a one-sample chi-square test confirmed that the difference between the observed 24% and the threshold of 75% was significant, x^2 (1, $n = 72$) $= 85.63$, $p < 0.0005$. The effect size of 1.19 obtained was classified as large. As such, the null hypothesis ($H3_0$) was rejected in favour for the alternative hypothesis ($H3_1$). Therefore, the procedure was perceived to be institutionally feasible in regard to persistence.

Predictability

For a policy whose effects are predictable, its outcomes can be foreseen and thus it is possible for those affected to prepare in advance and take into account the implications (Mickwitz 2003). Regarding predictability of the impacts resulting from implementation of the procedure, a one-sample chi-square test confirmed that the difference between the observed 8% and the threshold of 75% was significant, x^2 (1, $n = 82$) $= 193.19$, $p < 0.0005$. The effect size of 2.36 obtained was classified as large. As such, the null hypothesis ($H3_0$) was rejected in favour for the alternative hypothesis ($H3_1$). It was therefore confirmed that the procedure is institutionally feasible in regard to predictability. Therefore, for the procedure, policy makers will have foresight ahead of its implementation because its impacts are predictable.

Overall Perception of the Procedure

Integrating the accounting for EC in the development approval process of building projects is feasible, as it largely addresses the criteria for implementing environmental policy, namely, cost implications, institutional feasibility and distributional considerations. Significant levels of consensus among the professionals were registered regarding the procedure's legal acceptability and compatibility with national priorities although this did not affect the overall in perception of the procedure in those aspects. Potentially, the high cost implications with regard to need for new institutions required to implement the procedure could, among other things, be an issue that can be addressed if the procedure is augmented by a CDM initiative as discussed in this work. On the whole, the results from the evaluation exercise suggested that the procedure can be used as a basis for demonstrating how EC can be managed using CDM.

EC of a Residential House

Results pertaining to application of the procedure on a typical residential house are as follows.

Baseline Option

The total emissions for the baseline option were 2550 $kgCO_2$, representing 11 $kgCO_2/m^2$ of wall (see Table 17.8). With respect to manufacture, diesel contributed the most (75%) energy-related emissions. The amount of emissions was highly sensitive to heavy fuel oil, as it had the largest emission factor (0.71 $kgCO_2/kWh$) amongst the fuels considered. Transportation emissions (including materials and workforce) were 18% of the total emissions, implying that at 82%, the manufacture of materials contributed the most emissions. This was not surprising since materials are known to constitute the biggest proportion of buildings' EC (Chang et al. 2012, p. 794; Nässén et al. 2007, p. 1599; Scheuer et al. 2003, p. 1057).

Alternative Option

For the alternative option, the total emissions were 1834 $kgCO_2$, which translated into 8 $kgCO_2/m^2$ of wall (see Table 17.10). This represented a reduction of 27% from the baseline option. The total energy-related emissions for manufacturing materials reduced from 957 to 334 $kgCO_2$, representing a reduction of 65%. Workforce and material transportation emissions reduced by 20%. The alternative option therefore demonstrates how a certain construction practice can deviate from the baseline practices and consider alternative greener options (e.g. by sourcing materials from manufacturers who use renewable energy, using biofuels in transporting materials and/or workforce, etc.) in order to reduce EC.

Table 17.10 Emissions from baseline and alternative options

		Baseline (kgCO$_2$)	Alternative (kgCO$_2$)
Manufacture of materials	Diesel	722	207
	Non-fossil	0	0
	Heavy fuel oil	108	0
	Electricity	127	127
	Nonfuel-related emissions (54%)	1124	1124
	Subtotal	2081	1458
Transportation of materials	Cement	305	244
	Bricks	27	22
	Sand	27	22
	Subtotal	359	288
Transportation of workforce	Diesel-vehicle	110	88
Grand total		2550	1834

Managing EC Using CDM

As per the earlier assumption to upscale the results so as to generate a single CDM composed of constructing 28,000 houses, for 2550 kgCO$_2$ per house, constructing (walls of) 28,000 houses would result into baseline EC of 71 ktCO$_2$ (i.e. 2550 × 28,000) annually. However, for the alternative 'greener' scenario, the annual EC would be 51 ktCO$_2$ (i.e. 1834 × 28,000), resulting in reductions of 20 ktCO$_2$ annually. If a duration of 10 years is considered, a total of 200 ktCO$_2$ would be avoided. These figures are comparable to those of other CDMs in Uganda that are not related to the building sector (see Table 17.1). Therefore, creating a CDM related to managing EC (EC-CDM) can generate emission reductions of an appropriate order of magnitude, and considering the prevailing CDM modalities, it would be classified under small-scale CDM types which have emission reductions of up to 60 kt per year (UNFCCC 2014, p. 40). However, as demonstrated in the bottom-up projection, in order to achieve substantial emission reductions, the initiative would require covering a substantial geographical area providing opportunity for substantial numbers of dwellings. In this chapter, we have demonstrated that a capital/large city would be sufficient for consideration as a single CDM project.

Operation of the EC-CDM

The structure of the CDM that is based on EC is presented in Fig. 17.2. Since building projects are usually geographically spread, yet individually lead to small

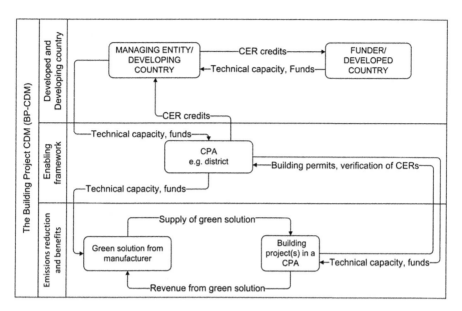

Fig. 17.2 Suggested structure of the CDM related to embodied carbon of buildings

emission reductions, a programme of activities (PoA) type of CDM would be appropriate, contrast to a typical mono-project CDM whose set-up costs would be rather prohibitive. In PoA CDMs, several projects sharing similar goals can be registered as a single CDM (UNFCCC 2014). Since the project sites in a PoA can be located in various parts of a country (Fenhann and Hinostroza 2011), this can similarly relate to building projects which are usually scattered in a given geographical area. To manage the geographical spread of building projects, existing local government administrative authorities such as districts, can be used. Each local authority (e.g. Kampala Capital City Authority) would be taken as a component project activity (CPA) of the PoA. A CPA is technically defined as 'a single measure, or a set of interrelated measures under a PoA, to reduce emissions or result in net removals, applied within a designated area" (UNFCCC 2014, p. 22). In operationalising the EC-CDM, the CPAs would keep up-to-date official records (e.g. of EC factors) specific to the geographical region concerned. Upon building permit applications, as specified in the procedure shown in Fig. 17.1, baseline emissions would be assessed as per the baseline option demonstrated in this work. Alternative options such as one indicated in this work can then be considered. The investors (e.g. clients, contractors) who opt in for alternative options can then be advised of 'greener' choices such as which manufacturers to buy materials from. On completing construction, before issuing occupation permit, a reassessment could be done, and the extent of deviations from the baseline revealed. If positive (i.e. EC reduced), a verification check can be carried out to assess where the EC reductions were achieved (e.g. whether manufacturer, contractor, client or workforce) in order to apportion incentives appropriately. These ideas suggest that market-based mechanisms like CDM can be used to manage EC in buildings at project level.

The operation of the EC-CDM can be structured into three levels (developed and developing country, enabling framework and emissions reduction and benefits) as illustrated in Fig. 17.2, each with various actors and responsibilities. In the top level, the developed country offers technical capacity and funds to implement a 'green' solution and in return, receives CERs from the developing country. Technical capacity and funds are extended to the CPAs (see middle level of the diagram) which also extend the same to the implementers of the green solution, who might be manufacturers or building projects. When manufacturers supply 'green' materials to the building project, they receive revenue. If manufacturers have obtained funds from the CPAs in order to manufacture 'green' materials, they can be tasked to offer the materials at lower competitive prices. But, if manufacturers do not claim funds from CPAs, and therefore sell materials at premium prices, the building projects could then redeem the premium from the CPAs. With such incentives, manufacturers can be tasked to be more innovative in search for low-EC solutions since the demand will be available. These incentives could help to mitigate the high cost implications with regard to implementing EC accounting management in the development approval process. For building projects, this could prompt stakeholders to adopt practices that are less EC intensive. In so doing, this could translate into a market-based mechanism of managing EC.

Conclusions

Africa is in its early stages of economic development, a situation characterised by activities that lead to carbon emissions. Addressing this issue, as argued in this work, requires integrating carbon management strategies in development activities. This chapter has demonstrated how a carbon management strategy based on CDM, an emission trading scheme, can be used to manage embodied carbon (EC) at building project level. A procedure previously proposed by the authors which integrates EC accounting in the prevailing development approval process of building projects in Uganda was considered. The feasibility of implementing the procedure, based on the criteria of considering environmental policy, was evaluated. Results suggested that construction professionals perceived the procedure to (1) have high cost implications with regard to implementation, (2) address distributional considerations and (3) be institutionally feasible. The procedure was then used as a basis for demonstrating how EC can be managed using CDM. The demonstration, which involved assessing EC resulting from constructing a typical dwelling unit, showed that through a bottom-up analysis, 20 $ktCO_2$ of emissions could be avoided annually within Kampala, Uganda's capital city, if an EC-based CDM was to be set up.

A working structure of the CDM was presented and discussed. Since promotion of sustainable development is one of CDMs' objectives, if suggestions in this chapter are adopted, construction processes in Uganda and developing countries alike can follow a low-carbon path and thus contribute to sustainable development. In such an African country like Uganda, where resources are chronically insufficient, the benefits associated with the suggested CDM could help to mitigate on the high cost implications related to implementing EC management. In addition, the perspective to incentivising lower EC, as presented in this work, shows that EC reductions can be obtained by varying fuel parameters without recourse to alternative materials that may be associated with higher social, technical and economic implications. This perspective, however, extends the responsibility for reducing EC from the construction industry to the manufacturing sector, an aspect that further corroborates the argument that management of EC should not be an isolated but, rather, a collaborative effort. That said, more studies that consider more aspects of the building fabric, apart from walling, are necessary to corroborate the findings presented in this chapter.

In furthering the contribution made by this work, there is a need to engage various CDM and built environment stakeholders in Uganda such as funders (e.g. World Bank), managing entities (e.g. ministries), local authorities, manufacturers and built environment professionals in order to assess the practicality of using CDM to manage EC. With the recognised need for reducing carbon emissions associated with buildings, Africa nevertheless still lags behind; using CDM to manage EC could be a plausible option for sustainable management of construction activities in developing countries, especially in Africa.

References

Asane-Otoo, E. (2015). Carbon footprint and emission determinants in Africa. *Energy, 82*, 426–435.

Boardman, A., Greenberg, D., Vining, A. R., & Weimer, D. (2013). *Cost-benefit analysis: Pearson New International Edition*. Harlow: Pearson Education.

Brighton and Hove. (2013). Brighton and Hove City council's local development framework [Online]. Available: http://goo.gl/KI888c. Accessed 28 Feb 2015.

Bryman, A., & Cramer, D. (2011). *Quantitative data analysis with IBM SPSS 17, 18 and 19: A guide for social scientists*. London: Routledge.

Buchholz, T., Da Silva, I., & Furtado, J. (2012). Power from wood gasifiers in Uganda: A 250 kW and 10 kW case study. *Proceedings of the Institution of Civil Engineers: Energy, 165*, 181–196.

Bwango, A., Wright, J., Elias, C., & Burton, I. (2000). Reconciling national and global priorities in adaptation to climate change: With an illustration from Uganda. *Environmental Monitoring and Assessment, 61*, 145–159.

Cabeza, L. F., Rincón, L., Vilariño, V., Pérez, G., & Castell, A. (2014). Life cycle assessment (LCA) and life cycle energy analysis (LCEA) of buildings and the building sector: A review. *Renewable and Sustainable Energy Reviews, 29*, 394–416.

Chang, Y., Ries, R. J., & Lei, S. (2012). The embodied energy and emissions of a high-rise education building: A quantification using process-based hybrid life cycle inventory model. *Energy and Buildings, 55*, 790–798.

Chen, Y., Jiang, P., Dong, W., & Huang, B. (2015). Analysis on the carbon trading approach in promoting sustainable buildings in China. *Renewable Energy, 84*, 130–137.

Cheng, C., Pouffary, S., Svenningsen, N., & Callaway, M. (2008). *The Kyoto Protocol, the clean development mechanism and the building and construction sector – A report for the UNEP Sustainable Buildings and Construction Initiative*. Paris: United Nations Environment Programme.

Cohen, J. (1988). *Statistical power analysis for the behavioral sciences*. Hillsdale: L. Erlbaum Associates.

Cole, R. J. (1998). Energy and greenhouse gas emissions associated with the construction of alternative structural systems. *Building and Environment, 34*, 335–348.

Embodied Carbon Industry Task Force. (2014). Proposals for standardised measurement method and recommendations for zero carbon building regulations and allowable solutions [Online]. Available: http://goo.gl/OLv6gt. Accessed 27 Mar 2015.

Esso, L. J., & Keho, Y. (2016). Energy consumption, economic growth and carbon emissions: Cointegration and causality evidence from selected African countries. *Energy, 114*, 492–497.

Fenhann, J., & Hinostroza, M. (2011). *CDM information and guidebook*. Roskilde: UNEP Risoe Centre.

Foddy, W. (1993). *Constructing questions for interviews and questionnaires: Theory and practice in social research*. Cambridge, UK: Cambridge University Press.

Franklin and Andrews. (2013). CESMM4 Carbon & Price Book 2013 [Online]. Available: http://goo.gl/LCH280. Accessed 23 Mar 2015.

Gillenwater, M., & Seres, S. (2011). *The clean development mechanism. A review of the fist international offset program*. Virginia: Pew Center on Global Climate Change.

Green, S. B., & Salkind, N. J. (2005). *Using SPSS for windows and Macintosh: Analyzing and understanding data*. Upper Saddle River: Pearson Prentice Hall.

Gupta, S., Tirpak, D. A., Burger, N., Gupta, J., Höhne, N., Boncheva, A. I., Kanoan, G. M., Kolstad, C., Kruger, J. A., Michaelowa, A., Murase, S., Pershing, J., Saijo, T., & Sari, A. (2007). Climate change 2007: Mitigation. In B. Metz, O. R. Davidson, P. R. Bosch, R. Dave, & L. A. Meyer (Eds.), *Policies, instruments and co-operative arrangements. Contribution of working group III to the fourth assessment report*. Cambridge, UK: Intergovernmental Panel on Climate Change.

Hashemi, A., Cruickshank, H., & Cheshmehzangi, A. (2015). Environmental impacts and embodied energy of construction methods and materials in low-income tropical housing. *Sustainability, 7*, 7866.

Henry, M., Maniatis, D., Gitz, V., Huberman, D., & Valentini, R. (2011). Implementation of REDD plus in sub-Saharan Africa: State of knowledge, challenges and opportunities. *Environment and Development Economics, 16*, 381–404.

Hinostroza, M., Cheng, C.-C., Zhu, X., Fenhann, J., Figueres, C., & Avendaño, F. (2007). *Potentials and barriers for end-use energy efficiency under programmatic CDM*. Roskilde: UNEP Risø Centre.

Huitema, D., Jordan, A., Massey, E., Rayner, T., Asselt, H., HAUG, C., Hildingsson, R., Monni, S., & Stripple, J. (2011). The evaluation of climate policy: Theory and emerging practice in Europe. *Policy Sciences, 44*, 179–198.

IPCC. (2007). *Climate change 2007: Working group III: Mitigation of climate change [Online]*. Available: http://www.ipcc.ch/publications_and_data/ar4/wg3/en/ch1s1-3.html. Accessed 23 Nov 2013.

Hong, J., Shen, G. Q., Feng, Y., Lau, W. S.-t., & Mao, C. (2015). Greenhouse gas emissions during the construction phase of a building: a case study in China. *Journal of Cleaner Production, 103*, 249–259.

Kees, M., & Feldmann, L. (2011). The role of donor organisations in promoting energy efficient cook stoves. *Energy Policy, 39*, 7595–7599.

Kervin, J. B. (1992). *Methods for business research*. New York: HarperCollins publishers.

Kibwami, N., & Tutesigensi, A. (2014). Mathematical modelling of embodied carbon emissions of building projects. In A. B. Raiden & E. Aboagye-Nimo (Eds.), *Procs 30th Annual ARCOM*

Conference, 1–3 September 2014. Portsmouth: Association of Researchers in Construction Management.

Kibwami, N., & Tutesigensi, A. (2015). Exploring the potential of accounting for embodied carbon emissions in building projects in Uganda. In A. B. Raiden & E. Aboagye-Nimo (Eds.), *Procs 31st Annual ARCOM Conference, 7–9 September 2015*. Lincoln: Association of Researchers in Construction Management.

Kibwami, N., & Tutesigensi, A. (2016). Enhancing sustainable construction in the building sector in Uganda. *Habitat International, 57*, 64–73.

Lafarge. (2012). *Energy consumption and resource management [Online]*. Available: http://goo.gl/JlfvCE. Accessed 10 Dec 2013.

Lam, P. T. I., Chan, E. H. W., Yu, A. T. W., Cam, W. C. N., & Yu, J. S. (2015). Applicability of clean development mechanism to the Hong Kong building sector. *Journal of Cleaner Production, 109*, 271–283.

Levine, M., Ürge-Vorsatz, D., Blok, K., Geng, L., Harvey, D., Lang, S., Levermore, G., Mongameli Mehlwana, A., Mirasgedis, S., Novikova, A., Rilling, J., & Yoshino, H. (2007). Climate change 2007: Mitigation. In B. Metz, O. R. Davidson, P. R. Bosch, R. Dave, & L. A. Meyer (Eds.), *Residential and commercial buildings. Contribution of working group III to the fourth assessment report*. Cambridge, UK: Intergovernmental Panel on Climate Change.

Mickwitz, P. (2003). A framework for evaluating environmental policy instruments: Context and key concepts. *Evaluation, 9*, 415–436.

Mok, K. L., Han, S. H., & Choi, S. (2014). The implementation of clean development mechanism (CDM) in the construction and built environment industry. *Energy Policy, 65*, 512–523.

Nalumansi, J., & Mwesigye, G. (2011). Determining productivity of masons for both stretcher and header bonding on building sites. In J. A. Mwakali & H. M. Alinaitwe (Eds.), *Second international conference on advances in engineering and technology: Contribution of scientific research in development*. Entebbe: Macmillan Uganda Ltd.

Nässén, J., Holmberg, J., Wadeskog, A., & Nyman, M. (2007). Direct and indirect energy use and carbon emissions in the production phase of buildings: An input–output analysis. *Energy, 32*, 1593–1602.

Novikova, A., Ürge-Vorsatz, D., & Liang, C. (2006). *The "Magic" of the Kyoto mechanisms: Will it work for buildings?* California: American Council for an Energy Efficient Economy Summer Study 2006; Central European University.

Okello, C., Pindozzi, S., Faugno, S., & Boccia, L. (2013). Development of bioenergy technologies in Uganda: A review of progress. *Renewable and Sustainable Energy Reviews, 18*, 55–63.

Olsen, K. H. (2006). National ownership in the implementation of global climate policy in Uganda. *Climate Policy, 5*(6), 599–612.

Osabuohien, E. S., Efobi, U. R., & Gitau, C. M. W. (2014). Beyond the environmental Kuznets curve in Africa: Evidence from panel cointegration. *Journal of Environmental Policy & Planning, 16*, 517–538.

Pooliyadda, S. P., & Dias, W. P. S. (2005). The significance of embedded energy for buildings in a tropical country. *The Structural Engineer, 83*, 34–36.

RICS. (2012). Methodology to calculate embodied carbon of materials [Online]. RICS. Available: http://goo.gl/4YZwvt. Accessed 19 Oct 2012.

Scheuer, C., Keoleian, G. A., & Reppe, P. (2003). Life cycle energy and environmental performance of a new university building: Modeling challenges and design implications. *Energy and Buildings, 35*, 1049–1064.

The Republic of Uganda. (2007). *The renewable energy policy for Uganda*. Kampala: Ministry of Energy and Mineral Development.

UBOS. (2013). *Statistical abstract*. Kampala: Uganda Bureau Of Statistics.

UNCTAD. (2011). *The least developed countries report 2011*. Geneva: United Nations Conference on Trade and Development (UNCTAD).

UNFCCC. (2010). *PoA 2956: Uganda municipal waste compost programme – POA design document [Online]*. Available: https://goo.gl/MVASox. Accessed 26 Nov 2013.

UNFCCC. (2013). *Kyoto protocol [Online]*. Available: http://unfccc.int/kyoto_protocol/items/
2830.php. Accessed 29 Nov 2013.

UNFCCC. (2013b). *AMS-III.BH: Small-scale methodology, Version 01.0 [Online]*. Available:
http://goo.gl/KPUoL1. Accessed 18 Nov 2013.

UNFCCC. (2014). *CDM methodology booklet* (5th ed.). Bonn: United Nations Framework Convention on Climate Change.

UNFCCC. (2016). *Project cycle search [Online]*. Available: https://cdm.unfccc.int/Projects/
projsearch.html. Accessed 4 Dec 2016.

UN-HABITAT. (2010). *Uganda urban housing sector profile*. Nairobi: United Nations Human
Settlements Program.

Winkelman, A. G., & Moore, M. R. (2011). Explaining the differential distribution of clean
development mechanism projects across host countries. *Energy Policy, 39*, 1132–1143.

World Bank. (2009). *Uganda registers first forestry project in Africa to reduce global warming
emissions [Online]*. Kampala/Bangkok: World Bank; Press Release No: 2010/093/AFR.
Available: http://goo.gl/s3WtXb. Accessed 22 Mar 2013.

Worrell, E., Pric, L., Martin, N., Hendriks, C., & Meida, L. O. (2001). Carbon dioxide emissions
from the global cement industry. *Annual Review of Energy and the Environment, 26*, 303–329.

Zanchi, G., Frieden, D., Pucker, J., Bird, D. N., Buchholz, T., & Windhorst, K. (2013). Climate
benefits from alternative energy uses of biomass plantations in Uganda. *Biomass & Bioenergy,
59*, 128–136.

Chapter 18
Embodied Carbon in Buildings: An Australian Perspective

Robert H. Crawford, André Stephan, and Monique Schmidt

Introduction

This chapter explores the current status of the methodological developments, policy and industry implementation associated with embodied carbon in Australia. It provides insight into the contribution that Australia has made to the development of embodied carbon data and assessment methods. It also draws upon existing studies and building projects to highlight the extent to which embodied carbon is being addressed in Australia, as well as where Australia sits in relation to the management and mitigation of embodied carbon in the global context. The degree to which policy has been driving the consideration of embodied carbon within the Australian building sector is also discussed.

Our understanding of embodied carbon in buildings, including its magnitude, key contributing factors, ideal approaches for measuring it and the most effective measures for its mitigation, is still very much in a state of development. Many of the aspects covered in this chapter are ongoing work and progressing at quite a rapid pace. They are primarily focused on advancing this understanding, improving the way in which we quantify embodied carbon and supporting decision-making around how best to go about mitigating embodied carbon within our buildings. As we delve further into our understanding of the ways in which buildings contribute to our broader embodied carbon footprint, priorities and areas for further work will emerge. The chapter concludes by highlighting some of the areas that will need to be explored in our strive towards 'net-positive life cycle carbon' buildings.

R. H. Crawford (✉) · A. Stephan · M. Schmidt
Faculty of Architecture, Building and Planning, The University of Melbourne,
Parkville, VIC, Australia
e-mail: rhcr@unimelb.edu.au

© Springer International Publishing AG 2018
F. Pomponi et al. (eds.), *Embodied Carbon in Buildings*,
https://doi.org/10.1007/978-3-319-72796-7_18

Data and Methods for Embodied Carbon Assessment in Australia

This section discusses the role that Australia has played in the development of embodied carbon data and assessment methods. The past three decades have seen considerable advancements in the way embodied carbon is measured. Australia has played a significant role in many of these developments, including the creation of multi-region input-output (MRIO) models and the hybridisation of life cycle inventory data. These recent methodological advancements in embodied carbon assessment in Australia and their drivers are explained below. This includes an overview of AusLCI and BP LCI process databases, Eora and IELab input-output databases, structural path analysis (SPA) and path exchange (PXC) hybrid analysis.

The Development of Embodied Carbon Data in Australia

Process Data

As is the case within many other regions of the world, Australian researchers and practitioners rely on a range of physical and financial data when quantifying the embodied carbon of buildings. With the advent of life cycle assessment in the 1990s, a concerted effort was made by the Australian life cycle assessment (LCA) community to collect physical data on the environmental flows associated with the production of a range of Australian products and processes, including building and packaging materials, energy and transport. This included life cycle inventory (LCI) data for timber, concrete, steel, PVC, aluminium and glass which were compiled into the national life cycle inventory (LCI) database.

While initially funded by the federal government and state EPAs, ongoing data collection efforts were quite limited due to a lack of sufficient resources. However, at the time, this data was considered to form the most comprehensive database of product environmental flows for Australia. It began to be used widely within the LCA community for quantifying embodied environmental flows associated with buildings, materials and other products. The database continued to expand to include data on agriculture, fuels, food, raw materials and waste management. A renewed effort to update the national LCI database began with the launch of AusLCI in 2006. This Australian Life Cycle Inventory Database (www.auslci.com.au) was a new initiative of the Australian Life Cycle Assessment Society (ALCAS) and the Commonwealth Scientific and Industrial Research Organisation (CSIRO). The aim of AusLCI was to further expand the range of products and processes contained within the original database and establish more robust data collection protocols. The database now includes data on a range of products across various sectors, collected using a consistent framework and covering a broad range of resource inputs and outputs, including emissions of carbon dioxide. While quite limited in the coverage of construction materials, efforts to expand the database to cover a broader range of products are ongoing.

In parallel, the Building Products Innovation Council (BPIC) established the Building Products Life Cycle Inventory (BP LCI) (www.bpic.asn.au), a database of physical process data for over 100 different building materials and products, including concrete, concrete blocks, concrete and terracotta roof tiles, bricks, gypsum board, steel, timber and timber products, windows, glass and insulation materials. For each material, data on inputs such as fuels, raw materials, water as well as emissions of waste and pollutants are provided. Unlike AusLCI, which also includes products other than building materials and is open access, BPIC covers only building materials but also restricts access to registered users.

Environmentally Extended Input-Output (EEIO) Data

The scope limitations typically inherent in physical life cycle inventories have led to a significant body of work on the establishment of environmental inventories based on financial data. In this approach, input-output (IO) tables, containing information on monetary transactions between sectors of an economy, are combined with national environmental accounts (e.g. energy, greenhouse gas emissions (GHGE) and water). The resulting environmentally extended IO (EEIO) data provides information on the embodied environmental flows per monetary value of output from a particular sector (e.g. tonnes of GHGE per dollar value of construction). As this data is based on an economy-wide system boundary, it is considered to be systemically complete. However, the applicability of the use of this national average data in accounting for the environmental flows associated with a particular good or service is one of its major drawbacks, amongst others (Lenzen 2000).

Input-output tables are produced in over 100 countries but vary considerably in terms of their level of sectoral and regional disaggregation and the frequency by which they are compiled. While the use of IO data for environmental assessment goes back to at least the 1960s (Isard et al. 1968), most applications of IO data for this purpose have occurred since the mid-1990s (Hoekstra 2010).

The Australian Bureau of Statistics (www.abs.gov.au) has produced IO tables for Australia going back to 1958. These are usually produced on an annual basis with the latest tables covering 114 industry groups (ABS 2016b). The earliest evidence of the use of these tables in accounting for environmental flows is by Karunaratne (1981) who used them to estimate the primary energy demand associated with fossil fuels. EEIO data for Australia has been made available in a number of databases, including Eora and IELab, as discussed below.

Hybrid Data

In an effort to combat data gaps in AusLCI and BP LCI and deal with the non-specificity of EEIO data, researchers have also combined both sources of data to produce what is known as hybrid data. This data aims to maintain the reliability and comprehensiveness of the respective original data sources and is most typically provided in the form of resource or emission coefficients for different materials and products. Hybrid data for Australian construction materials is provided by Crawford and Treloar (2010).

Methods for quantifying embodied carbon use embodied carbon data in a variety of ways. Both physical process data and EEIO data can be used independently, or they can be combined as part of a hybrid analysis. A hybrid analysis can take numerous forms depending on how the data is combined. In essence, its main goal is to avoid the limitations inherent in process and IO analysis (Lenzen 2000) by capturing the strengths of each individual approach. Australia has played a key role in the development of hybrid analysis methods, as outlined in the following section.

Embodied Carbon Assessment Methods: The Australian Contribution

Two of the most significant contributions made to the field of embodied carbon modelling by Australian researchers are in the fields of multi-region input-output (MRIO) analysis and hybrid analysis.

Multi-region Input-Output Analysis

Historically, the use of IO analysis to quantify carbon, as well as other environmental flows, in the form of an EEIO analysis, has occurred at a single-region level (Tukker and Dietzenbacher 2013). However, environmental flows are very rarely confined within the boundaries of a single region, especially with the ever-increasing trade of goods and services between countries. In recognition of this, multi-region input-output (MRIO) analysis can be used to trace the interregional flows of goods and services in order to attribute the environmental effects of the entire supply chain to final demand, in a consumption-based accounting approach. This environmentally extended MRIO analysis can be performed at both a subnational and global (country) scale by linking IO data for multiple regions.

MRIO analysis has been used since the 1950s, initially for economic accounting (Isard 1951). However, the most rapid advancements have occurred within the last decade (Wiedmann 2017). One of the major reasons for its slow uptake was the time and complexity involved in bringing together often disparate, inconsistent data sources. In recent times there have been a number of initiatives aimed at compiling large-scale global MRIO databases, including IDE-JETRO, EXIOBASE, GLIO, GTAP, OECD, WIOD and Eora. These are described in more detail by Murray and Lenzen (2013).

Researchers at the University of Sydney were responsible for conceiving and developing the Eora MRIO database (http://www.worldmrio.com). This came about by the lack of geographical and sectoral detail, continuous time series and information on reliability and uncertainty provided by existing MRIO databases. Eora was designed to provide a disaggregation of IO data into countries and sectors at the highest possible level of detail, improving the accuracy of environmental life cycle and footprint-type assessments. Eora provides data for 187 countries across 15,909 sectors, more than any other MRIO database. It also allows for the creation of a historical time series back to 1990, greater flexibility, increased transparency, reliability and uncertainty analysis and regular data updates (Lenzen et al. 2013).

During Eora's development, the research team identified the need for a collaborative effort in the creation of a global MRIO database, through the establishment of an international collaborative research platform. Data could then be pooled and shared and MRIO tables released in a regular and timely manner (Lenzen et al. 2013). The Australian Industrial Ecology Virtual Laboratory (IELab) is a first attempt to establish and test this collaborative approach to MRIO compilation, initially developed as an Australian subnational MRIO database and analytical toolbox (https://ielab-aus.info). IELab is a collaborative cloud-based platform for compiling large-scale, high-resolution, economic, social and environmental accounts based on MRIO tables. A user-friendly GUI aims to improve access to MRIO analysis. The IELab uses a spatial classification based on the Australian Statistical Geography Standard (ASGS) which includes a Statistical Area Level 2 (SA2) subdivision of Australia into 2196 geographical entities (ABS 2010), each containing an average population of 10,000 persons. The Input-Output Product Categories (IOPC) sectoral classification is used, which distinguishes 1284 product groups (ABS 2012). Theoretically, it could be used to model embodied carbon for any product group or subnational region, including temporal changes based on time series data.

Wiedmann et al. (2013) provide a summary of the range of studies in which the IELab has been used, which includes assessing the production of biofuel and analysing the carbon footprint of cities and electricity supply. The IELab has also been used to assess the embodied carbon of construction materials (Teh et al. 2017) and can act as a useful tool for providing an initial estimate of the embodied carbon of the construction sector or of sectors supplying goods or services to the construction sector. Most of these studies would not have been possible without the functionality of the IELab, strengthened by the cloud-based format that enables easy updating, as well as improved accessibility (Wiedmann et al. 2013).

Path Exchange Hybrid Analysis (PXC)

Until the 1970s, methods for embodied carbon accounting tended to rely on the use of physical process data (in the form of a process analysis) or, in much rarer cases, EEIO data (in the form of an EEIO analysis). In 1978, Bullard et al. published a handbook for combining process and input-output analysis. This hybrid approach to accounting for environmental flows was developed in order to address some of the limitations inherent in the two separate methods. Process analysis suffers from truncation issues due to time and cost constraints in data collection, while input-output analysis suffers from a lack of specificity in regard to individual products (Lenzen 2000).

In the mid-1990s, Associate Professor Graham Treloar, then a PhD researcher at Deakin University in Geelong, Australia, identified a need for disaggregating input-output data into discrete paths or nodes[1] (Treloar 1997). Treloar observed that changing the transaction coefficient for a particular node in an input-output matrix would affect all supply chain paths that contain that node, even if the changed coefficients applied only to a particular path. Treloar developed a technique for extracting individual nodes from the IO data in order to avoid such undesired 'global' effects inherent to previous hybrid methods. For example, the direct purchase of cement for concrete by the residential building sector can be identified, as well as cement for the indirect purchase of concrete through the concrete product sector. Treloar termed this the 'path extraction technique' and demonstrated how it could be applied to the analysis of embodied energy (Treloar 1997). This approach is now commonly referred to as 'structural path analysis' (SPA) (Lenzen 2007).

Treloar demonstrated how a hybrid approach to accounting for environmental flows, such as carbon, could be achieved using SPA. The first step is to mathematically disaggregate the IO matrix into a series of mutually exclusive nodes. Specific nodes are then modified using process data that corresponds to the particular transaction. The modifications can affect either the value of the transaction, if identified as different for the particular good or service under study, or the environmental flow associated with the transaction, if specific process data is available. Using this approach, process data can be integrated for individual nodes rather than by replacing a direct coefficient in the IO matrix, which would otherwise flow through to all instances of transactions between two sectors at every tier of the supply chain. For example, if process data for the greenhouse gas emissions associated with the production of cement purchased by the residential building sector are available, this particular transaction or node in the SPA can be replaced with this data. Nodes relating to the production of cement elsewhere in the supply chain can then remain unchanged.

The approach of conducting a hybrid analysis using a SPA as the basis for the integration of process data was originally termed an 'input-output-based hybrid analysis', by Treloar (1997). It has more recently been referred to as the 'path exchange method' (PXC) (Lenzen and Crawford 2009). Treloar and fellow researchers have applied the PXC method within a range of studies, mostly in relation to embodied energy, carbon and water in the construction sector (inter alia

---

[1]A node represents a good or service provided by a particular sector within an input-output matrix. Node-to-node connections represent a transaction between input-output sectors, i.e. the purchase of a good or service from one sector by another. A series of nodes, corresponding to a chain of transactions leading to the sector being assessed, is referred to as a path or pathway.

Treloar et al. 2002; McCormack et al. 2007; Baboulet and Lenzen 2010; Crawford 2011b; Crawford and Pullen 2011). However, Treloar unfortunately did not have the opportunity to develop a general methodology, to solve a number of problems and to undertake further research in relation to the PXC method as he was struck down by a terminal illness in his prime. Lenzen and Crawford (2009) continued this research, presenting a general methodology for the PXC method and illustrating the relationship between the PXC method and other methods. While the PXC method is regularly referred to in publications using hybrid methods, its application is rare and often limited to those researchers involved in its development. The complexity of this method and the amount of data to be manipulated is one reason explaining why its use has not yet become common practice.

As with most other methods for quantifying embodied carbon, using the PXC method is most commonly applied utilising material-based carbon coefficients. In this case, coefficients for building materials are compiled using the PXC method, using available process data for material production, and data from a SPA of the relevant IO sector producing the material. These coefficients cover the complete system boundary for the individual material and are multiplied by physical material quantities to determine their total embodied carbon. However, further work is needed to determine the direct and indirect embodied carbon associated with the building construction process, involving the integration of a range of materials. In this process, a SPA of the construction sector is used to identify and subtract the pathways representing the materials for which embodied carbon has already been quantified, from the total carbon intensity of the construction sector. What remains is the total embodied carbon associated with the construction process as well as any other minor materials not previously covered (Crawford 2011a).

Studies using this method have demonstrated the degree to which alternative methods that rely on a more limited system boundary may underestimate environmental flows. For example, an embodied energy analysis using the PXC method of a range of different building types showed that on average 64% of energy inputs would be excluded with the use of a process analysis (Crawford 2008).

There is a growing awareness of the benefits that a hybrid approach to accounting for embodied carbon (as well as other environmental flows) brings and hybrid analyses are becoming more common. However, there is still much work to do in its development and broad accessibility. There is a need to address the remaining limitations of hybrid analysis, many of which are inherited from process and IO approaches. There is also a range of conflicting interpretations of the hybrid analysis definition. This does nothing but create further confusion about how a hybrid analysis can be used to address the limitations of traditional embodied carbon accounting methods. A consistent, widely recognised and used definition and framework for the different hybrid methods are therefore urgently needed.

Future Direction

Australian researchers continue to be at the forefront of the development of MRIO analysis and hybrid analysis. A large group of researchers from a range of Australian research institutions continue to develop and strengthen the capabilities of the Australian IELab. This work has already led to the establishment of work on an IELab for China and Indonesia (Faturay et al. 2017). This is part of a larger collaborative effort to establish a global IELab covering the entire world. This global IELab will integrate data from existing MRIO databases, such as WIOD, EXIOBASE and Eora, providing high-resolution, time series, automated updating, hybridisation and analytical tools. The functional capabilities will be expanded as well as the number of countries covered, and product- and process-level resolution will be improved. Irrespective of this, process data is still considered to be more reliable, and AusLCI and BP LCI databases will need to continue to evolve with new materials and processes being added as data becomes available.

A number of projects are underway that will help with improving access to, and the application of, hybrid embodied carbon analysis, particularly using the PXC method (Crawford et al. 2017b). A key component of this work is an attempt to automate the path exchange process to help with both the speed and usability of the PXC method (Crawford et al. 2017a). By linking this to the IELab, it will also enable an analysis of embodied carbon at a much higher sectoral resolution than is currently possible. Automatic updating of process data used within the model will ensure the latest data is being used. Work is also being done to compile a detailed list of hybrid embodied carbon coefficients for Australian construction materials based on the latest available process and IO data that is also able to be easily updated in a semi-automated fashion. Given the global trade of materials, links to global databases, such as ecoinvent, are also integral to this model.

Researchers are also in the process of defining a set of consistent terminologies and mathematical notations for the different approaches to hybrid analysis. The universal acceptance and use of these definitions may help with easing some of the confusion around existing hybrid methods and support their broader use.

The Role of Policy and Voluntary Certifications in Building Embodied Carbon Mitigation in Australia

As the operation of buildings represents over one third of final energy use, globally (IEA 2015), and 27% of total energy use in Australia (Brandon and Lombardi 2011), building regulations and certifications have been predominately focused on reducing operational carbon, particularly in relation to energy demand for heating and cooling. Energy efficiency regulations generally date back to the aftermath of the 1973 oil crisis (Nässén and Holmberg 2005) and have since evolved dramatically, both in requirements (e.g. *PassivHaus* Contributors of passipedia.passiv.de

2013) and in geographic coverage (OECD/IEA 2015). However, despite continuous improvement in these regulations and certifications, and their contribution to reducing operational energy and associated greenhouse gas emissions, they still broadly fail to consider embodied carbon (García-Casals 2006; Szalay 2007; Blengini and Di Carlo 2010; Stephan and Crawford 2014, 2016).

In this section, current building energy efficiency regulations and certifications in Australia are reviewed with a focus on those attempting to capture embodied requirements. Guidelines for possible future life cycle energy and carbon regulations are then discussed.

Current Status

Australia's current building energy efficiency regulations are part of the Building Code of Australia (BCA), which enforces maximum heating and cooling energy demands (per square metre of useful area) for new buildings and significant modifications to existing buildings (ABCB 2016). For residential buildings, compliance with the regulatory requirements is most commonly demonstrated through the Nationwide House Energy Rating Scheme (NatHERS), which provides a 'star rating' for a building, ranging from zero stars (very poor) to ten stars (zero thermal operational energy). The current minimum performance level that must be achieved is six stars, which allows a maximum of 114 MJ/(m$^2 \cdot$ a) for heating and cooling in Melbourne's temperate climate (NatHERS 2003). Apart from its mandatory nature, the strengths of this scheme include a very high climatic resolution and the fact that it corrects for house size.

The heating and cooling demands for a building depend largely on the climate zone in which it is located. Unlike other energy efficiency schemes that typically rely on a limited number of climate zones to define maximum energy use thresholds, e.g. Germany, a single climate zone; France, two climate zones; and Spain, five climate zones (Rodríguez-Soria et al. 2014), Australia relies on 69 different climate zones that provide a very high climatic resolution. This is one of the key strengths of the current Australian operational energy efficiency regulations.

Secondly, the star rating scheme for residential buildings takes into account building size when determining heating and cooling requirements (Delsante 2005). This is critical in order to avoid the typical effect of smaller buildings being penalised, as larger buildings will often result in lower energy use per square metre. If anything, smaller buildings should be encouraged as they tend to result in much lower overall energy use and embodied energy and carbon per capita, as demonstrated by Stephan and Crawford (2016). The majority of operational energy efficiency regulations do not correct for building size.

However, despite these two strong attributes, the star rating scheme is far from being able to effectively reduce the life cycle energy demand and carbon of buildings. Crawford et al. (2016) quantified the life cycle energy benefits resulting from improving the star rating of typical houses in the Australian cities of

Melbourne and Brisbane, either through improved thermal performance of the building envelope or through improved design. They found that simply increasing the thermal performance of the envelope can result in increased life cycle energy demand due to the embodied energy in insulation and high-performance glazing. This supports the findings of Stephan et al. (2013a) related to a *PassivHaus* in Belgium and makes a case for including embodied energy and carbon in building energy efficiency regulations and certifications. Recent developments in the Australian building industry tend to point towards this same conclusion, notably the Materials Life Cycle Impacts credit in the Green Star certification and the planned National Carbon Offset Standard for Buildings.

Green Star is a voluntary green building certification scheme (see GBCA 2015a for more information), developed and managed by the Green Building Council of Australia (GBCA). Like its more famous US (Leadership in Energy and Environmental Design – LEED) and UK (Building Research Establishment Environmental Assessment Method – BREEAM) counterparts, it uses a score-based system to rank the environmental performance of construction projects across a range of categories. Green Star ranks a building across nine categories: management, indoor environmental quality, energy, transport, water, materials, land use, ecology and emissions. One of the most recent features of Green Star is its incorporation of a life cycle assessment (LCA) credit and its use of Environmental Product Declarations (EPD) as a tool to prove the environmental credentials of materials.

Since 2015, Green Star has included a Material Life Cycle Impacts credit in its scoring system, providing up to 7–8 points of a possible 100. It includes six points if a comparative LCA of the building is conducted, comparing it to a standard building of the same typology. This LCA must (1) include six impact categories (including global warming potential), (2) be conducted by a professional LCA practitioner, (3) be peer-reviewed and (4) be performed according to the International Standard 14040 (2006). Including five additional environmental impact categories can earn the applicant an additional point on top of the six base points for the comparative LCA. Alternatively to the comparative LCA, three points can be awarded for a reduced use of concrete and Portland cement, one point for reducing steel use and four other points for reusing existing materials on site where possible (i.e. retaining a façade or structure). This can bring the total points of the Materials Life Cycle Impacts credit to 8. In addition to this credit, up to three points can be earned for using responsible building materials. This credit focuses mostly on the use of recycled materials (such as steel), of certified and sustainably sourced timber and of permanent formwork that is free of polyvinyl chloride (PVC).

The total possible points awarded for LCA-related attributes in a Green Star building project therefore totals 10–11% which is significant and underlines the importance of LCA in the eyes of the GBCA. This is further supported by their statement:

> The LCA and EPD initiatives in Green Star will be a catalyst for greater LCA use and the generation of life cycle data, both of which will improve the sustainability outcomes in the built environment. (GBCA 2015b)

However, while the inclusion of LCA-based credits in a green building certifi-cation may help support the uptake of the consideration of embodied carbon, it is not enough to address the issue. Certified green buildings represent a negligible fraction of total construction activity, in Australia and elsewhere. For instance, the reported 21 million m^2 of floor area of certified Green Star buildings since 2004 (GBCA 2017) represent less than 7% of the built floor area of new houses built in Australia since 2004 (excluding apartments and non-residential buildings) (ABS 2016a). This figure is much lower when all construction activity is considered. Furthermore, the voluntary nature of green building certifications does not help support a broad market penetration. In addition, the methods prescribed by the GBCA typically rely on process analysis which systematically underestimates embodied carbon (as highlighted in the previous section). A more comprehensive approach to the consideration of embodied carbon within the construction industry is needed.

Beyond the star rating scheme and the GBCA's LCA credits which are devel-oped solely for buildings, other certifications can also play a role in reducing carbon in the Australian construction industry. This is the case for voluntary 'carbon-neutral' certifications that were originally developed for consumer products and are being adapted to buildings. Such certifications typically include embodied carbon although the proposed Australian National Carbon Offset Standard for Buildings does not at this stage.

The Australian National Carbon Offset Standard overrides its predecessor, the Greenhouse Friendly scheme (DEE 2017), which promoted low-carbon products in Australia from 2001 to 2010. The draft standard for buildings, which is currently in its final draft phase (DEE/NCOS 2016), states that Scope 1–3 greenhouse gas emissions (WRI/WBCSD 2011) should be considered. These are Scope 1, direct emissions within the building's boundary; Scope 2, emissions associated with energy production outside the boundary but linked to activities in the building (typically electricity demand and heating); and Scope 3, all other indirect emis-sions. For instance, the standard mentions that emissions from transportation should be considered but that this is currently not possible due to a lack of data. Including user transport-related emissions is praiseworthy as these typically represent one third of the life cycle emissions of a residential building (Stephan and Crawford 2014) or precinct (Stephan et al. 2013b). However, while Scope 3 emissions would consider the embodied carbon in building materials, the draft standard explicitly states that:

> A building's embodied emissions (including energy associated with materials introduced through renovation, fit out or upgrade) are not considered part of a building's operational carbon account and is not covered by the Standard. Embodied energy may be considered for future versions of the standard that apply to building construction. (DEE/NCOS 2016)

In other words, embodied carbon in building materials is not considered by a voluntary standard that aims to lead to 'carbon-neutral' buildings. This could be addressed by simply including embodied carbon in the standard and considering the entire life cycle at once, instead of separating the construction stage from the operational stage.

Towards Embodied Carbon Regulations for Australia

Developing a building regulation that includes embodied carbon is a challenging endeavour, highlighted by a lack of existing schemes, globally. One of the rare (if not the only) regulations targeting embodied environmental requirements in buildings can be found in the Netherlands, as discussed by de Klijn-Chevalerias and Javed (2017). This pilot regulation converts embodied environmental requirements into a so-called shadow cost expressed in euros/m^2 of floor area. A maximum value for the shadow cost of new buildings (similar to the maximum energy use per square metre set by some operational energy efficiency standards) is specified. While the calculation of the shadow cost relies on weighting different life cycle impact categories and on a database of underestimated environmental impacts of materials that is derived using process analysis, this pilot scheme sets a precedent in terms of implementing regulations that target embodied requirements. However, the lack of a standard life cycle inventory technique and system boundaries (Dixit et al. 2013) further complicates the widespread adoption of regulations addressing embodied carbon. Despite these limitations, multiple existing studies could be used to inform the development of a pilot life cycle energy efficiency and carbon regulation for buildings. This regulation would need to be developed in consultation with relevant industry stakeholders and further improved based on continuous feedback, as has been the case with operational energy efficiency regulations.

The work of the CEN/TC 350 and the multiple associated standards (notably European Standard 15643-2 2011; European Standard 15978 2011) should be capitalised upon in any new regulation. However, these standards do not mandate the use of a particular life cycle inventory technique. The use of a PXC hybrid analysis would provide the most comprehensive system boundaries while using the most reliable data where possible (Treloar 1997; Suh et al. 2004; Crawford 2008; Majeau-Bettez et al. 2011; Dixit et al. 2013). In addition, multiple functional units should be used to provide a variety of perspectives, including $kgCO_2$-e/m^2 for efficiency, $kgCO_2$-e for total global warming potential and $kgCO_2$-e per occupant/user to capture the lifestyle/behaviour of occupants/users, based on recommendations from Calwell (2010) and Stephan et al. (2012). Furthermore, Stephan and Crawford (2016) have revealed the significance of considering house size in regulations. Notably, vertical and horizontal construction assemblies contribute differently to the embodied carbon of a building, depending on its size, and should therefore be targeted accordingly within regulations addressing embodied carbon. Regulations should also go beyond material choices to support design decisions that improve a building's life cycle carbon performance, as demonstrated by Crawford et al. (2016). The National Carbon Offset Standard for Buildings proposal by the

Australian government could easily be adapted to include these aspects and factor in consideration of embodied carbon.

Creating a market value for a reduction in embodied carbon could also provide an incentive for reducing building embodied carbon (Langston and Langston 2008; Ariyarante and Moncaster 2014; Wu et al. 2016). A carbon tax or trading system could be possible mechanisms for achieving this.

Regulations that focus on the life cycle environmental performance of buildings, including embodied carbon, are still in their infancy. More research is needed to better understand the relationship between design, building geometry, material choices and embodied carbon. More importantly, a robust and comprehensive database of embodied carbon coefficients for construction materials, based on a consistent, systemically complete assessment framework, is needed. Tools for systematically, robustly and easily quantifying the life cycle carbon of buildings are also needed to help support industry in their carbon reduction efforts. The next section describes how the Australian construction industry is currently approaching the implementation of embodied carbon reduction.

What Is the Australian Construction Industry Doing to Reduce Embodied Carbon?

With buildings accounting for almost one quarter of Australia's emissions, they represent one of the largest and most attractive opportunities to reduce emissions (Climate Works 2010). So what is Australia, one of the largest emitters of greenhouse gases per capita in the world (Climate Council 2015), currently doing to address these emissions? The previous section highlighted the fact that current Australian policies and energy efficiency regulations deal almost exclusively with operational carbon, leaving embodied carbon largely ignored. Embodied carbon has been demonstrated to represent between 10% (Ibn-Mohammed et al. 2013) and 70% (ASBP 2014) of a building's total life cycle carbon and approximately 20% of all carbon emissions in Australia (Schinabeck et al. 2016). Thus, it is critical that we consider both embodied and operational carbon in our efforts to improve the environmental performance of buildings. Several data sources and assessment methods exist to help support these efforts (as discussed earlier in this chapter) and are constantly evolving to further improve their reliability, usefulness and accessibility. This section will provide more detail about what Australia's construction industry is currently doing to reduce embodied carbon and the barriers hindering its more widespread consideration. Four case study buildings that have used a variety of strategies for reducing embodied carbon are briefly discussed.

The State of Embodied Carbon Assessment in Australia's Construction Industry

When it comes to reducing greenhouse gas emissions, the Australian construction industry is predominantly focused on the operational carbon of buildings. This is highlighted in a recent survey, completed by Fouché and Crawford (2015) for the CRC for Low Carbon Living (www.lowcarbonlivingcrc.com.au), which found that over 85% of construction industry consultants focus on providing operational energy/carbon assessment services. In total, 60% of the survey respondents, consisting predominately of LCA practitioners and sustainability consultants, provided a form of embodied carbon assessment. For the organisations that did not provide this service, almost 70% said that they would consider providing embodied carbon assessment as part of their services in the future. This demonstrates that even though the existing building stock is far from achieving carbon neutrality (Schinabeck et al. 2016), there is an increasing awareness of the need to address more than just operational carbon mitigation.

Compared to operational carbon assessment, 30% more of the respondents outsourced their embodied carbon assessment. This was found to be due to a lack of tools specific to the Australian context, lack of in-house expertise and concerns about reliability of data. When asked what software tools get used for embodied carbon assessment, eToolLCD (etoolglobal.com), a building-specific LCA tool developed in Australia, was only 4% less commonly used than SimaPro (simapro. com), the most popular LCA tool. There seemed to be an interest for locally developed tools and the need to address the weaknesses in existing data and tools available for embodied carbon assessment. These weaknesses include a lack of Australian-specific data, inconsistent methodologies, time-intensive assessments, a need for expert knowledge, a lack of benchmarks and compatibility with building information modelling (BIM). This emphasises the importance of the work discussed earlier in developing more detailed MRIO databases and hybrid embodied carbon coefficients for Australia. When asked to indicate the top features desired in new or improved embodied carbon assessment tools, over 80% selected 'material cost' as the most beneficial feature followed by data on recycled materials (62%) and the source of materials (57%). Other recommendations included adherence to Australian regulations and standards, options for quick analysis, integration with existing tools, ease of updating and more transparency and consistency. This survey highlighted that the Australian construction industry seems to be interested in reducing the embodied carbon of buildings. The next four exemplar case studies provide practical examples of buildings where embodied carbon has been a key consideration.

5 × 4 Hayes Lane Project, Melbourne

The 5 × 4 Hayes Lane Project is an inner city dwelling, completed in June 2015, set on a small footprint of 5 × 4 m in East Melbourne (Fig. 18.1). One of the key design strategies was to minimise the building's life cycle carbon, with the help of both passive and active measures. Driven by the One Planet Living principles (Bioregional 2015), which encouraged, amongst other things, the use of low embodied energy materials that are locally sourced and made from renewable or waste sources, an environmentally exemplar dwelling was created. The embodied carbon of 11 different floor assemblies and 52 different wall assemblies was analysed over a 100-year period with the use of the PXC hybrid method. The results were used to inform the selection of these building elements and identify strategies for further embodied carbon reductions. The project made use of phase-change materials, timber from sustainably harvested forests, and focused on sourcing local materials where possible. Most of the superstructure was prefabricated off-site, which can help reduce the time, waste and cost of a project (Moncaster and Song 2012). This, in turn, can lead to embodied carbon reductions due to greater production and material efficiencies (Sturgis and Roberts 2010). This building acts as an educational tool with information publically available (the occupant's operational carbon emissions are monitored and shared online – www.5x4.com.au) and can be used as a vehicle to showcase, demonstrate and inform the practical incarnation of a carbon-conscious building (Johnson 2014).

Fig. 18.1   5 × 4 Hayes Lane (Source: Ralph Alphonso.)

Forte, Melbourne

Forte is a 10-storey apartment building, containing 23 apartments located in Melbourne's Docklands precinct (Fig. 18.2). At 32.2 m tall, and constructed of cross-laminated timber (CLT), it was until 2017 the world's tallest modern timber apartment building (Wood Solutions 2013). It was also the first CLT building in Australia. While local production of CLT did not exist at the time and the CLT panels had to be imported from Austria, which increased the transport-related embodied carbon, timber has a lower embodied carbon content than other materials such as concrete and steel, which would typically be used for a building of this type (Milne and Reardon 2013). The lighter weight of the structure also reduced the size of the footing system, leading to further embodied carbon reductions for the building.

Melbourne School of Design, Melbourne

The Melbourne School of Design building is an education building completed in 2014 at the centre of the University of Melbourne's Parkville campus (Fig. 18.3). This six-star Green Star-rated building's entire roof is constructed from laminated veneer lumber (LVL) with a long span of 22 m across the atrium. The use of LVL, instead of typical roofing materials such as steel, delivers significant environmental benefits, reducing the embodied carbon of the building. Timber has been used extensively elsewhere in the building also, for wall linings, staircases and floor finishes. Reducing embodied carbon was also a consideration in the selection of the external shading system. Perforated zinc was chosen due to its lower embodied carbon compared to the typical alternative of aluminium (Irwinconsult 2013). One of the keys to the building achieving the highest Green Star rating for an

Fig. 18.2 Forte (Source: Lend Lease 2012)

Fig. 18.3 Melbourne School of Design (Source: John Gollings/Peter Bennetts)

educational building was it being awarded the maximum number of points available for the Materials Life Cycle Impacts credit.

Barangaroo, Sydney

Barangaroo is a 22-ha precinct development on a former container terminal on the western edge of Sydney Harbour in which embodied carbon mitigation has been a major focus (Fig. 18.4). The project aims to reduce embodied carbon across the entire site by 20% compared to standard construction (Boral 2014). The aim is for the precinct to be climate positive with the stated ambition of being 'Australia's first carbon-neutral community'. As concrete represents over a quarter of the embodied carbon of the project and the cement binder component of this concrete is one of the key contributors (Lend Lease 2014), reduction in the concrete-related embodied carbon was achieved through the use of supplementary cementitious materials (SCMs), such as fly ash and ground-granulated blast-furnace slag. The concrete manufacturer also built a tailored onsite batching plant to reduce the transport-related carbon emissions (Lend Lease 2014). The embodied carbon of steel was also reduced with a 20% reduction target in carbon intensity of the reinforcing steel. Another interesting aspect to note was that the tender process also included an embodied carbon awareness element. The embodied carbon performance of differ-ent concrete mixes and supply options was assessed, and the information was used to inform the selection process. This can be an innovative way of selecting the most appropriate suppliers. In addition, the project has a commitment of carrying out a

Fig. 18.4 Barangaroo (Source: Title Magazine 2015)

life cycle assessment on the top 20 materials, in terms of volume, used on site. The project has some ambitious goals as highlighted by the developer:

> This project is to act as a catalyst for change in the wider industry to help incentivise suppliers to examine the life cycle impacts of their products, encourage the publication of EPD and to support greater collaboration between builders and product suppliers. (Lend Lease 2014)

These case studies demonstrate that from a small residential scale (5×4) to a larger precinct scale (Barangaroo), the Australian construction industry is starting to incorporate strategies for improving the embodied carbon performance of buildings. The next section describes some of the constraints preventing the widespread implementation of more projects of this type.

Current Barriers and Future Direction

Even though Australia has been at the forefront of developments in the field of embodied carbon assessment tools, methodologies and research, the uptake of embodied carbon considerations in practice has been slow. Some of the key barriers affecting this uptake within the construction industry were highlighted in a report published by ASBP (2014) that placed consistency of method at the top of the list followed by availability of comparable data and mandatory legislation. The inconsistency and lack of availability of comprehensive embodied carbon data are often quoted as key barriers affecting both embodied carbon and life cycle assessment (Ariyarante and Moncaster 2014; Dixit et al. 2015; Schinabeck et al. 2016) and were discussed in detail in the first section of this chapter. The lack of mandatory legislation was described in the previous section. The Australian CRC for Low Carbon Living survey (Fouché and Crawford 2015) identified several other critical barriers such as a lack of project budget (the most prevalent barrier with 60% of the respondent votes), client disinterest and no clear profit incentive. The effect of budgetary constraints on the uptake of embodied carbon considerations was also identified by Ariyarante and Moncaster (2014), Langston and Langston (2008) and Wu et al. (2016). These studies emphasise that the cost of embodied carbon reduction is not well understood and more research is required to gain further understanding as to the role of financial cost in embodied and life cycle carbon reduction.

Conclusions

This chapter has provided an overview of Australia's contribution to addressing embodied carbon, from a methodological, policy and implementation perspective. Despite the relative insignificance of building embodied carbon in Australia, in a global context, the country has played a pivotal role in the methodological developments underpinning some of the more sophisticated approaches for quantifying embodied carbon and our growing understanding of its significance. An increase in the uptake of more comprehensive embodied carbon assessment methods, such as MRIO analysis and the PXC hybrid method, is needed and will help to ensure that our carbon mitigation efforts are appropriately targeted. These developments also highlight the need for further work such as improving the quality and completeness of data and improving access to information and tools that support embodied carbon mitigation.

While, historically, policy- and regulatory-based approaches to mitigating carbon in Australia have focused on building operation, recent developments in green building certifications and policy-based discussions have started to consider embodied carbon. The National Carbon Offset Standard for Buildings that is currently being developed is a much needed framework for carbon accounting for buildings. Its contribution could be significantly enhanced with the inclusion of embodied carbon, something that it does not yet address. Whether it is this standard or something else, there is a pressing need for a mandatory scheme that enforces robust designs that aim to reduce the life cycle carbon of buildings. With no regulatory drivers, embodied carbon considerations are currently entirely voluntary within the Australian construction industry. The lack of awareness of embodied carbon amongst construction clients and construction industry professionals does nothing to encourage more than what is currently a very rare and piecemeal approach to mitigating embodied carbon.

Exemplar buildings are starting to emerge, much the way they did when addressing building operational carbon became a priority. Regulations, education, financial incentives and improved industry practices have made the consideration of operational carbon an integral part of the building design process, in most cases. These same strategies are likely to be needed to ensure embodied carbon mitigation is seen as an equally important part of the building design process.

While Australian researchers are contributing to world-leading methodological advancements in embodied carbon assessment, its application and the implementation of initiatives targeting embodied carbon reduction generally lag behind many other regions of the world. In light of this, a case exists for greater sharing and collaboration across global boundaries to accelerate the management and mitigation of embodied carbon within Australia's buildings. International collaborations are also critical to ensure consistency in data collection and embodied carbon assessment but also to help facilitate the integration of data, enabling modelling of global supply chains.

Even though there is still much to be learnt about embodied carbon in buildings, we are no longer able to ignore it. There is enough evidence to show that limiting our efforts to operational carbon will not lead to the considerable carbon savings that are needed to address the pressing environmental challenges of our day. A more holistic approach is needed, one that has the ultimate aim of creating buildings that are 'net-positive life cycle carbon'.

References

ABCB. (2016). *Building code of Australia (BCA) 2016*. Canberra: Australian Building Codes Board.
ABS. (2010). *Australian statistical geography standard: Design of the statistical areas level 2.*, , ABS Cat. No. 8731.0. Canberra: Australian Bureau of Statistics.
ABS. (2012). *Australian national accounts, input-output tables (product details), 2008–09.*, , ABS Cat. No. 5209.0. Canberra: Australian Bureau of Statistics.
ABS. (2016a, Jun). 8752.0 – Building activity, Australia. Australian Bureau of Statistics. http://www.abs.gov.au/AUSSTATS/abs@.nsf/DetailsPage/8752.0Jun%202016?OpenDocument#Time
ABS. (2016b). *Australian national accounts: Input-output tables – 2013–2014.*, , ABS Cat. No. 5209.0. Canberra: Australian Bureau of Statistics.
Ariyarante, C. I., & Moncaster, A. (2014). Stand alone calculation tools are not the answer to embodied carbon assessment. *Energy Procedia, 62*, 150–159.
ASBP. (2014). *Embodied carbon industry task force proposals*. London: The Alliance for Sustainable Building Products. http://www.asbp.org.uk/news/detail/?nId=87.
Baboulet, O., & Lenzen, M. (2010). Evaluating the environmental performance of a university. *Journal of Cleaner Production, 18*(12), 1134–1141.
Bioregional. (2015). The 5x4 project bioregional development group. http://bioregional.com.au/the-5x4-project/
Blengini, G. A., & Di Carlo, T. (2010). The changing role of life cycle phases, subsystems and materials in the LCA of low energy buildings. *Energy and Buildings, 42*(6), 869–880.

Boral. (2014). *Barangaroo South: Large scale concrete, demanding pump environment, sustainability focused mixes*. Australia: Boral. http://www.boral.com.au/major-projects/barangaroo/Boral-Construction-Materials_Major-Project-Case-Study_Barangaroo-South.pdf.

Brandon, P. S., & Lombardi, P. L. (2011). *Evaluating sustainable development in the built environment*. Chichester: Wiley-Blackwell.

Bullard, C. W., Penner, P. S., & Pilati, D. A. (1978). Net energy analysis: Handbook for combining process and input-output analysis. *Resources and Energy, 1*, 267–313.

Calwell, C. (2010). *Is efficient sufficient? The case for shifting our emphasis in energy specifications to progressive efficiency and sufficiency*. Stockholm: European Council for an Energy Efficient Economy.

Climate Council. (2015). New report reveals that Australia is among the worst emitters in the world. Climate Council. https://www.climatecouncil.org.au/new-report-reveals-that-australia-is-among-the-worst-emitters-in-the-world

ClimateWorks Australia (2010) Australian Carbon Trust Report: Commercial buildings emissions reduction oppurtunities, Available at: https://www.climateworksaustralia.org/sites/default/files/documents/publications/climateworks_commercial_buildings_emission_reduction_opportunities_dec2010.pdf

Contributors of passipedia.passiv.de. (2013). What is a passive house?. Passipedia.org. http://passipedia.passiv.de/passipedia_en/doku.php?id=basics:what_is_a_passive_house&rev=1383724060

Crawford, R. H. (2008). Validation of a hybrid life cycle inventory analysis method. *Journal of Environmental Management, 88*(3), 496–506.

Crawford, R. H. (2011a). *Life cycle assessment in the built environment*. London: Taylor and Francis.

Crawford, R. H. (2011b). Towards a comprehensive approach to zero-emissions housing. *Architectural Science Review, 54*(4), 277–284.

Crawford, R. H., & Pullen, S. (2011). Life cycle water analysis of a residential building and its occupants. *Building Research and Information, 39*(6), 589–602.

Crawford, R. H., & Treloar, G. J. (2010). *Database of embodied energy and water values for materials*. Melbourne: The University of Melbourne.

Crawford, R. H., Bartak, E. L., Stephan, A., & Jensen, C. A. (2016). Evaluating the life cycle energy benefits of energy efficiency regulations for buildings. *Renewable and Sustainable Energy Reviews, 63*, 435–451.

Crawford, R. H., Bontinck, P.-A., Stephan, A., & Wiedmann, T. (2017a). Towards an automated approach for compiling hybrid life cycle inventories. *Procedia Engineering, 180*, 157–166. https://doi.org/10.1016/j.proeng.2017.04.175.

Crawford, R. H., Wiedmann, T., & Stephan, A. (2017b). *Improving the environmental performance of Australian construction projects*. Melbourne School of Design. https://msd.unimelb.edu.au/improving-environmental-performance-construction

de Klijn-Chevalerias, M., & Javed, S. (2017). The Dutch approach for assessing and reducing environmental impacts of building materials. *Building and Environment, 111*, 147–159.

DEE. (2017). *Greenhouse friendly*. Department of the Environment and Energy. http://www.environment.gov.au/climate-change/carbon-neutral/greenhouse-friendly#Related_information-0

DEE/NCOS. (2016). *Draft – National carbon offset standard for buildings*. Canberra: Department of Environment and Energy and National Carbon Offset Standard. http://www.environment.gov.au/climate-change/publications/draft-buildings-standard.

Delsante, A. E. (2005). Is the new generation of building energy rating software up to the task? - A review of AccuRate. In: *CSIRO, manufacturing and infrastructure technology, highett, editor/s*. ABCB Conference 'Building Australia's Future 2005'; 11–15 Sept 11-15, 2005; Gold Coast, Qld. 2005.

Dixit, M. K., Culp, C. H., & Fernández-Solís, J. L. (2013). System boundary for embodied energy in buildings: A conceptual model for definition. *Renewable and Sustainable Energy Reviews, 21*, 153–164.

Dixit, M. K., Culp, C. H., & Fernandez-Solis, J. L. (2015). Embodied energy of construction materials: Integrating human and capital energy into an IO-based hybrid model. *Environmental Science & Technology, 49*(3), 1936–1945.

European Standard 15643-2. (2011). *Sustainability of construction works – Assessment of buildings – Part 2: Framework for the assessment of environmental performance.* Brussels: European Committee for Standardization (CEN), March 31st.

European Standard 15978. (2011). *Sustainability of construction works – Assessment of environmental performance of buildings – Calculation method.* Brussels: European Committee for Standardization (CEN), November 17th.

Faturay, F., Lenzen, M., & Nugraha, K. (2017). A new sub-national multi-region input–output database for Indonesia. *Economic Systems Research, 29*(2), 234–251. https://doi.org/10.1080/09535314.2017.1304361.

Fouché, M., & Crawford, R. H. (2015). In R. H. Crawford & A. Stephan (Eds.), *The Australian construction industry's approach to embodied carbon assessment: A scoping study.* Melbourne: The Architectural Science Association and The University of Melbourne.

García-Casals, X. (2006). Analysis of building energy regulation and certification in Europe: Their role, limitations and differences. *Energy and Buildings, 38*(5), 381–392.

GBCA. (2015a). Green star. Green Building Council of Australia. http://new.gbca.org.au/green-star/

GBCA. (2015b). Life cycle assessment in green star. Green Building Council of Australia. https://www.gbca.org.au/green-star/materials-category/life-cycle-assessment-in-green-star/

GBCA. (2017). Green star project directory. Green Building Council of Australia. http://www.gbca.org.au/project-directory.asp

Hoekstra, R. (2010). *(Towards) A complete database of peer-reviewed articles on environmentally extended input-output analysis.* Sydney, Australia: 18th International Input-Output Conference of the International Input-Output Association (IIOA), 20–25 June 2010.

Ibn-Mohammed, T., Greenough, R., Taylor, S., Ozawa-Meida, L., & Acquaye, A. (2013). Operational vs. embodied emissions in buildings—A review of current trends. *Energy and Buildings, 66*, 232–245.

IEA. (2015). Sustainable buildings. http://www.iea.org/topics/energyefficiency/subtopics/sustainablebuildings/

International Standard 14040. (2006). *Environmental management – Life cycle assessment – Principles and framework.* Geneva: International Organization for Standardization (ISO), July 1st.

Irwinconsult. (2013). Melbourne School of Design: A world class building for educating the building professionals of the future. Irwinconsult Pty Ltd. http://www.irwinconsult.com.au/experiences/msd-mufabp/

Isard, W. (1951). Interregional and regional input-output analysis: A model of a space-economy. *The Review of Economics and Statistics, 33*(4), 318–328.

Isard, W., Bassett, K., Choguill, C., Furtado, J., Izumita, R., Kissin, J., Romanoff, E., Seyfarth, R., & Tatlock, R. (1968). On the linkage of socio-economic and ecologic systems. *Papers in Regional Science, 21*(1), 79–99.

Johnson, N. (2014). Tiny project, enormous potential: 5x4m project sets the precedent for sustainable apartment design. *Architecture & Design.* http://www.architectureanddesign.com.au/news/tiny-project-enormous-potential-5x4-project-sets-t

Karunaratne, N. D. (1981). An input-output analysis of Australian energy planning issues. *Energy Economics, 3*(3), 159–168.

Langston, Y. L., & Langston, C. A. (2008). Reliability of building embodied energy modelling: An analysis of 30 Melbourne case studies. *Construction Management and Economics, 26*(2), 147–160.

Lend Lease. (2012). Forte. http://www.victoriaharbour.com.au/live-here/forte-living

Lend Lease. (2014). Barangaroo South: Sustainability report 2014. Lend Lease.

Lenzen, M. (2000). Errors in conventional and input-output-based life cycle inventories. *Journal of Industrial Ecology, 4*(4), 127–148.

Lenzen, M. (2007). Structural path analysis of ecosystem networks. *Ecological Modelling, 200* (3–4), 334–342.

Lenzen, M., & Crawford, R. H. (2009). The path exchange method for hybrid LCA. *Environmental Science & Technology, 43*(21), 8251–8256.

Lenzen, M., Moran, D., Kanemoto, K., & Geschke, A. (2013). Building Eora: A global multi-region input-output database at high country and sector resolution. *Economic Systems Research, 25*(1), 20–49.

Majeau-Bettez, G., Strømman, A. H., & Hertwich, E. G. (2011). Evaluation of process- and input-output-based life cycle inventory data with regard to truncation and aggregation issues. *Environmental Science & Technology, 45*(23), 10170–10177.

McCormack, M., Treloar, G. J., Palmowski, L., & Crawford, R. (2007). Modelling direct and indirect water requirements of construction. *Building Research and Information, 35*(2), 156–162.

Milne, G., & Reardon, C. (2013). Embodied energy, your home. http://www.yourhome.gov.au/materials/embodied-energy

Moncaster, A., & Song, J.-Y. (2012). A comparative review of exisitng data and methodologies for calculating embodied energy and carbon of buildings. *International Journal of Sustainable Building Technology and Urban Development, 3*(1), 26–36.

Murray, J., & Lenzen, M. (2013). *The sustainability practitioner's guide to multi-regional input-output analysis*. Champaign: Common Ground Publishing.

Nässén, J., & Holmberg, J. (2005). Energy efficiency—A forgotten goal in the Swedish building sector? *Energy Policy, 33*(8), 1037–1051.

NatHERS. (2003). *Star band criteria*. Canberra: Nationwide House Energy Rating Scheme.

OECD/IEA. (2015). *Building energy performance metrics – Supporting energy efficiency progress in major economies*. Paris: International Energy Agency. http://www.iea.org/publications/freepublications/publication/building-energy-performance-metrics.html.

Rodríguez-Soria, B., Domínguez-Hernández, J., Pérez-Bella, J. M., & del Coz-Díaz, J. J. (2014). Review of international regulations governing the thermal insulation requirements of residential buildings and the harmonization of envelope energy loss. *Renewable and Sustainable Energy Reviews, 34*, 78–90.

Schinabeck, J., Wiedmann, T., & Lundie, S. (2016). Assessing embodied carbon in the Australian built environment. Fifth Estate. http://www.thefifthestate.com.au/columns/spinifex/assessing-embodied-carbon-in-the-australian-built-environment/81887

Stephan, A., & Crawford, R. H. (2014). A multi-scale life-cycle energy and greenhouse-gas emissions analysis model for residential buildings. *Architectural Science Review, 57*(1), 39–48.

Stephan, A., & Crawford, R. H. (2016). The relationship between house size and life cycle energy demand: Implications for energy efficiency regulations for buildings. *Energy, 116*(Part 1), 1158–1171.

Stephan, A., Crawford, R. H., & de Myttenaere, K. (2012). Towards a comprehensive life cycle energy analysis framework for residential buildings. *Energy and Buildings, 55*, 592–600.

Stephan, A., Crawford, R. H., & de Myttenaere, K. (2013a). A comprehensive assessment of the life cycle energy demand of passive houses. *Applied Energy, 112*, 23–34.

Stephan, A., Crawford, R. H., & de Myttenaere, K. (2013b). Multi-scale life cycle energy analysis of a low-density suburban neighbourhood in Melbourne. *Australia, Building and Environment, 68*, 35–49.

Sturgis, S., Roberts, G. (2010). Redefining zero: Carbon profiling as a solution to whole life carbon emission, RCIS Research Report, Available at: http://sturgiscarbonprofiling.com/wp-content/uploads/2010/05/RICS.RedefiningZero.pdf

Suh, S., Lenzen, M., Treloar, G. J., Hondo, H., Horvath, A., Huppes, G., Jolliet, O., Klann, U., Krewitt, W., Moriguchi, Y., Munksgaard, J., & Norris, G. (2004). System boundary selection in life-cycle inventories using hybrid approaches. *Environmental Science & Technology, 38*(3), 657–664.

Szalay, A. Z.-Z. (2007). What is missing from the concept of the new European building directive? *Building and Environment, 42*(4), 1761–1769.

Teh, S. H., Wiedmann, T., Castel, A., & de Burgh, J. (2017). Hybrid life cycle assessment of greenhouse gas emissions from cement, concrete and geopolymer concrete in Australia. *Journal of Cleaner Production, 152*, 312–320.

Title Magazine. (2015). Weekends out: Barangaroo. *Title Magazine.* http://www.titlemagazine. com.au/weekends-out-barangaroo/

Treloar, G. J. (1997). Extracting embodied energy paths from input-output tables: Towards an input-output-based hybrid energy analysis method. *Economic Systems Research, 9*(4), 375–391.

Treloar, G. J., Ilozor, B. D., & Crawford, R. H. (2002). *Modelling greenhouse gas emissions associated with commercial building construction* (pp. 533–540). Geelong: Australia and New Zealand Architectural Science Association Conference.

Tukker, A., & Dietzenbacher, E. (2013). Global multiregional input–output frameworks: An introduction and outlook. *Economic Systems Research, 25*(1), 1–19.

Wiedmann, T. (2017). An input–output virtual laboratory in practice – survey of uptake, usage and applications of the first operational IELab. *Economic Systems Research, 29*(2), 296–312. https://doi.org/10.1080/09535314.2017.1283295.

Wiedmann, T., Crawford, R. H., Seo, S., & Giesekam, J. (2013). *The Industrial Ecology Virtual Laboratory and its application to sustainability and environmental engineering – The case of low carbon living.* Barton: Engineers Australia, 18–20 September 2013.

Wood Solutions. (2013). Forte living. https://www.woodsolutions.com.au/Inspiration-Case-Study/forte-living

WRI/WBCSD. (2011). *Product life cycle accounting and reporting standard.* Geneva: World Resources Institute and World Business Council for Sustainable Development.

Wu, H., Crawford, R. H., Warren-Myers, G., Dave, M., & Noguchi, M. (2016). The economic value of low-energy housing. *Pacific Rim Property Research Journal, 22*(1), 45–58. https://doi. org/10.1080/14445921.2016.1161869.

Chapter 19
Current Approaches for Embodied Carbon Assessment of Buildings in China: An Overview

Jingke Hong, Geoffrey Qiping Shen, and Miaohan Tang

Introduction

As the world's largest carbon contributor, China was responsible for 29% of global carbon emissions in 2015 (Le Quéré et al. 2016). To mitigate this situation, the central government of China signed an agreement in the China-US Joint Presidential Statement on Climate Change, promising to reduce CO_2 emissions per unit of GDP by 60–65% in 2030 in comparison to the level of 2005. From a global viewpoint, CO_2 emissions generated from buildings increased at an average of 2.7% per year from 1999 to 2004 (Metz et al. 2007). In China, the building sector, as a pillar industry of China's modern economy, plays a vital role in generating carbon emissions. According to Kuo et al. (2016), it was responsible for 40% of the national total carbon emissions.

In general, embodied carbon emissions include initial embodied carbon that occurs during material production and construction processes, recurrent embodied carbon that refers to the emissions from replacement materials for maintenance and refurbishment purposes, and demolition carbon emissions that are generated from building's deconstruction process. However, the major attention in China has been paid on the initial embodied carbon of buildings, which is defined as the sum of all carbon emissions that occur along the processes associated with the production of a building, from the mining and processing of natural resources to material

J. Hong (✉) · M. Tang
School of Construction Management and Real Estate, Chongqing University, Chongqing, China
e-mail: hongjingke@cqu.edu.cn; teie@foxmail.com

G. Q. Shen
Department of Building and Real Estate, The Hong Kong Polytechnic University, Hung Hom, Hong Kong
e-mail: geoffrey.shen@polyu.edu.hk

© Springer International Publishing AG 2018   417
F. Pomponi et al. (eds.), *Embodied Carbon in Buildings*,
https://doi.org/10.1007/978-3-319-72796-7_19

manufacturing, transport, administration, and building delivery. The simplest expression of a carbon account is the activity data (AD) during building construction and emission factor (EF), shown as Eq. (19.1) below.

$$Account = AD * EF \qquad (19.1)$$

There are numerous other types of emissions that have the potential to cause global warming. As the impacts of different types of carbon emissions may vary, they are accounted for by a group of conversion coefficients to establish a bridge between the different gases. The global warming potential (GWP), which translates the emission of a specific gas to carbon dioxide equivalent (CO_2e), was used for embodied carbon assessment. The Intergovernmental Panel on Climate Change (IPCC) has developed three sets of GWP to account for the impact of a particular gas with the same amount of CO_2 according to a set time horizon (TH). In this context, GWP is the integral of the global warming effect of gases compared with that of CO_2 in the same time interval. Three THs are commonly calculated, namely, 20 years, 100 years, and 500 years. IPCC's First Assessment Report (1990) quoted an atmospheric life span of CO_2 to be between 50 and 200 years. Therefore, it is common to use the IPCC 100 TH GWP. Regarding the other gases with global warming potential, it is the standard practice to convert them to an amount of CO2 which would cause an equivalent impact. For methane, the conversion coefficient is 25 and for nitrous oxide is 298. See Eq. (19.2).

$$CO_2e = AD * EF * GWP \qquad (19.2)$$

Literature Review

A Review of Carbon Assessment of Buildings from a Global Perspective

Based on different research objectives and research scopes, two approaches have generally been used for embodied carbon assessment of buildings: (1) macro-level based on the whole construction industrial chain and (2) micro-level based on a specific building project. At the macro level, previous studies mainly adopted an input-output (I-O) life cycle assessment (LCA) method to quantify carbon emissions in the whole building sector within the system boundary of the whole economy (Acquaye and Duffy 2010; Chen et al. 2011; Nässén et al. 2007). At the micro level, hybrid and process-based LCA methods were frequently used in the evaluation of carbon emissions from a certain building or construction project (Hong et al. 2015; Kim et al. 2011, 2012a, b; Salazar and Meil 2009; Yan et al. 2010). More importantly, recent research suggests that an increasing array of technologies is being applied to assess carbon emissions in the building sector. Melanta et al. (2013) used the carbon footprint estimation tool (CFET) to evaluate a transportation construction project. Barandica et al. (2013) developed a management information system to in-depth analyze the GHG emissions from road projects in Spain. Wong et al. (2013) applied virtual prototyping technology to predict onsite carbon emissions for a construction project. Tang et al. (2013) used interactive simulation-based method to select appropriate construction management strategies in controlling carbon emissions from unexpected disruptive events. However, in spite of these theoretical developments, high-efficient technologies, and a variety of environmentally friendly policies applied in the building sector, reducing embodied carbon emissions is proving hard to achieve.

A Review of Carbon Assessment of Buildings in China

Embodied carbon emissions generated from building construction have been also extensively studied in China. At the national level, the carbon emissions from the building sector have been quantified using a series of macro-level analysis techniques, such as input-output (I-O) analysis and structural path analysis (Chang et al. 2010, 2014; Chen and Zhang 2010; Liu et al. 2012). At the project level, numerous studies have investigated the carbon emissions from different types of buildings. As the major building type, residential buildings play a significant role in case studies for carbon emission assessment. Gao (2012) measured the embodied carbon footprint of residential buildings by conducting an empirical study of 17 buildings in Jiangsu province. Liu et al. (2009) quantified the life cycle CO_2 emissions of residential communities in China. Regarding office buildings, Xing et al. (2008) made a comparative analysis between buildings with steel and concrete structure in terms of life cycle energy consumption and carbon emissions. Their results showed that compared to concrete-framed building, steel-framed building only generated approximately half of CO_2 emissions per square meter. Wu et al. (2012) conducted a process-based LCA model to quantify the energy use and CO_2 emissions of an office building and emphasized the importance of environmental impacts from building operational phase. Zhang et al. (2006) evaluated the GWP of an office building by establishing a building environmental performance analysis system (BEPAS). Wang et al. (2016) employed two case studies to illustrate the current GHG emission reduction performance of Chinese green buildings. Yao (2013) developed a benchmark for the carbon emissions from office buildings based on life cycle assessment theory. Table 19.1 summarizes a number of building cases for initial embodied carbon assessment in the context of China. The quantitative results show that variation occurred in the reported values of carbon intensity owing to the inconsistent building types and structures, incomplete system boundary, and inaccurate inventory data.

Table 19.1 Initial embodied carbon emissions of different types of buildings in China

Reference	Building type	Structure	Floor area (m^2)	Method	kgCO$_2$e/ m^2
Hong et al. (2015)	Residential building	Concrete	11,508	Process	756
Dong and Ng (2015)	Residential building	Concrete	96,800	Process	637
Li et al. (2016)	Residential building	Masonry concrete	1838	Process	337
Wu et al. (2017)	Residential building	Frame	23,129	Process	400–450
		Frame-shear	13,896		920–950
		Shear wall	7928		940–970
		Frame	2831		470–490
Zhang and Wang (2017)	Residential building	Shear wall	17,558	Process	570
Wang et al. (2016)	Office building	Reinforced concrete frame	25,023	Process	1131
Xing et al. (2008)	Office building	Steel	46,240	Process	315
		Concrete	34,620		606
Wu et al. (2012)	Office building	Reinforced concrete	36,500	Process	803
Yan et al. (2010)	Commercial building	Reinforced concrete	43,210	Process	525

Current Policies and Industry Initiatives for Embodied Carbon Reduction

To relate to the current policy practice in China, a set of important national energy conservation and emission reduction policies promulgated after 2000 were reviewed (see Table 19.2). The central government has enacted a wide range of policies to mitigate carbon emissions of buildings in China.

Apart from the top-down carbon reduction policies, the central and local government of China has also launched a number of pilot initiatives and projects for demonstration purpose. Some typical initiatives with significant economic and environmental influences are summarized in Table 19.3.

Apparently, the major focus of the present policies and pilot projects for nationwide carbon reduction is still on the operational phase rather than the embodied phase of buildings in China. Although it is true that operational carbon emissions are more dominant from a life cycle perspective, the attention on embodied energy use should keep increasing both at the national and industrial levels due to its significant role in the creation of sustainability in the building sector, particularly with the increasing emergence of more energy-efficient buildings in China.

Table 19.2 National policies for energy conservation and emission reduction in China

Year	Name	Authority	Related content
2004	Special planning for medium and long-term energy conservation	National Development and Reform Commission	Residential and public buildings should reduce 50% of total energy consumption during 11th Five-Year Plan
			The local government should conduct green retrofit[a] in existing residential and public buildings with a proportion of total floor area at 25% in mega cities, 15% in medium cities, and 10% in small cities
2011	The working program for energy conservation and emission reduction during the 12th Five-Year Plan period	The State Council of the People's Republic of China	Develop low-carbon building materials, construction technologies, and energy-efficient buildings
2013	The 12th Five-Year Plan for the development of green building and ecological urban district	The State Council of the People's Republic of China	Build 100 pilot ecological urban district
			The completed floor area under green retrofit[a] for existing residential and public heating system should reach more than 400 million square meters and 120 million square meters
2013	Action plan for green buildings (2013)	The State Council of the People's Republic of China	Build 1 billion square meters of green buildings during 12th Five-Year Plan
			By the end of 2015, the share of green buildings should reach 20% in newly built buildings
			The completed floor area under green retrofit[a] for existing residential and public heating should reach more than 400 million square meters and 120 million square meters
			By the end of 2020, the green retrofit[a] for existing residential heating system in north China should be completed
2014	National plan for addressing climate change (2014–2020)	National Development and Reform Commission	Develop key technologies for energy conservation and carbon reduction in green buildings
			Promote light wood framework structure
			Improve quality of buildings
			Extend service life of buildings

(continued)

Table 19.2 (continued)

Year	Name	Authority	Related content
2015	China-US Joint Presidential Statement on Climate Change	China and the US government	Promote low-carbon buildings, with the share of green buildings reaching 50% in newly built buildings in cities and towns by 2020
2016	The working plan for controlling greenhouse gas emissions during the 13th five-year paln	The State Council of the People's Republic of China	Promote green construction and building industrialization
			Build ecological cities
			Develop pilot demonstrations of zero carbon buildings

[a]"Green retrofit" can be defined as the incremental improvement of the fabric and systems of a building with the primary intention of improving energy efficiency and reducing carbon emissions. It can also refer to other terms, such as refurbishment, rehabilitation, modernization, renovation, improvements, adaptation, additions, repairs, and renewal on existing buildings (Ali and Rahmat 2009; Liang et al. 2015, 2016)

Current Methods Used in Embodied Carbon Assessment of Buildings

With the booming of carbon assessment studies in China, there are multiple methods in common usage. This section will make an explicit introduction of these carbon assessment methods.

Embodied Carbon Assessment of Buildings from a Macro Perspective

National Level

The single-region input-output (SRIO) analysis was introduced by Leontief and completed in 1970. It has served as a validation method to analyze "externalities" of products or services by quantifying the interindustrial interdependence relationship in the entire economic system using publicly available data (Leontief 1970). Such analytical tools have been an efficient tool and technique to measure environmental impacts from a top-down perspective for many years (Chang et al. 2011; Joshi 1999; Wiedmann 2009; Wiedmann et al. 2007). A number of scholars have conducted national-level investigation for embodied carbon emissions of buildings in China (Chang et al. 2010, 2011, 2014, 2015; Hong et al. 2017a). In general, the input-output analysis can be expressed as:

Table 19.3 Typical pilot initiatives and projects launched in China

Projects	Location	Scale	Indicators	
Shenzhen International Low Carbon City	Shenzhen	1.8×10^6 m^2	Achievements	
				The demonstration project is awarded three-star green building[a] with a reduction of CO_2 by 10^6 kg per year
			Goals	
				Energy consumption per 10^4 GDP should be reduced 30% in comparison to the current level of Shenzhen, which should be less than 0.29 tce in 2020
				The total carbon emissions in 2020 should be reduced by 30% in comparison to the level of 2010
Sino-Singapore Tianjin Eco-City	Tianjin	1.3×10^5 m^2	Achievements	
				The share of green buildings has reached 100% in all newly built buildings
				The floor area of three-star green buildings[a] has reached 2.538 million square meters
				The demonstration project saved 30% of operation energy in comparison to similar buildings in Tianjin with a reduction of 4.27×10^5 kg CO_2 per year
				The share of renewable energy consumption has reached more than 20% in all energy sources
Qingdao Sino-German Ecopark	Qingdao	7.6×10^5 km^2	Achievements	
				Reduce CO_2 emissions by 64.6% in comparison to similar buildings
				The share of renewable energy has reached 84.6% in all energy sources
				The floor area of two-star green buildings[a] has reached 1.02 million square meters
			Goals	
				Energy consumption per 10^4 GDP should be less than 0.23 tce in 2018
Yujiapu Financial District	Tianjin	9.5×10^6 km^2	Achievements	
				The demonstration project has reduced 6×10^5 kg CO_2 per year with an average energy saving rate at 23.4%
				Nine buildings were awarded the two-star green building[a] label, covering more than 70% of the total floor area in the pilot district
				The share of green buildings has reached 100% in newly built buildings
			Goals	
				The total carbon emissions from buildings in 2020 will be reduced by 45% in comparison to the level of 2005

[a]Evaluation Standard for Green Building (GB/T 50378-2014, the second version) is an official and authoritative rating system for green buildings in China. A comprehensive weighted rating system is elaborated to label buildings as one-star, two-star, and three-star with the total score higher than 50, 60, and 80 in the aspects of site, energy, water, material, indoor environment, operation, and construction. Moreover, China has become the second largest market for LEED applications outside of the United States with more than 1000 LEED-certified projects until 2017

$$e_i x_i = \sum_{j=1}^{n} e_j a_{ji} x_i + c_i x_i \qquad (19.3)$$

where x_i represents the monetary value of the total output of sector i, a_{ji} represents the direct monetary coefficient from sector j to sector i, e is the embodied carbon emission intensity of products from a certain sector, and c_i is the direct carbon emission intensity of sector i, which is calculated based on Eq. (19.2) by using the direct energy consumption data reported from statistics.

Note that n equations are established under the whole economy, vectors and matrices that can therefore be introduced to simplify the mathematical expression.

$$E^T = \begin{bmatrix} e_1 \\ \cdots \\ e_n \end{bmatrix} C^T = \begin{bmatrix} c_1 \\ \cdots \\ c_n \end{bmatrix} X = \begin{bmatrix} x_1 & \cdots & 0 \\ \cdots & \cdots & \cdots \\ 0 & \cdots & x_n \end{bmatrix} A = \begin{bmatrix} a_{11} & \cdots & a_{1n} \\ \cdots & \cdots & \cdots \\ a_{n1} & \cdots & a_{nn} \end{bmatrix}$$

where E^T and C^T are the embodied carbon emission intensity vector and direct carbon emission intensity vector with n dimension, X is the diagonal matrix with n entries, and the coefficients in matrix X are equal to the total economic output. A is the intermediate input matrix in the input-output table. For the whole economic system, the above group of equations can be expressed in the form of a matrix.

$$XE^T = XAE^T + XC^T \qquad (19.4)$$

This equation can be further transformed as:

$$E = C(I - A^T)^{-1} \qquad (19.5)$$

However, SRIO was conducted based on the assumption that the manufacturing technology in the domestic production process is the same as the technology used in foreign regions, which failed to reflect specific regional characteristics such as the variations in the climate, geographical location, natural resources, and underlying economic activities that directly determine the cross-regional environmental shifting. This oversight presents the challenge of investigating strategies from a regional and industrial sector perspective.

Regional Level

To address the problems arisen from the SRIO for embodied carbon assessment of the building sector, the multiregional input-output (MRIO) model has been regarded as a systematic method to measure environmental impacts by considering regional disparities and technological differences (Chen and Chen 2013; Lenzen et al. 2004; Peters et al. 2004). The complexity and diversity in China's regional economy build an environment for development and implementation of this method. This is mainly because regional variations such as productivity and energy efficiency may have an indirect but significant effect on the embodied carbon assessment of buildings. Especially in China, regions may differ in terms of building materials, construction processes, and modes of transportation. Consequently, the regional productivity of a specific building being studied may be very different from the national average level. Such difference may further execrate the errors between the simulated value and the actual emissions. Therefore, measuring the differences in the production technology is crucial in obtaining robust estimations. To address this issue, Hong et al. (2016a, b, d, 2017b) conducted a series of studies focusing on the regional-level investigation for embodied energy assessment of buildings in China, which indicated that a large gap existed in terms of energy intensity and total energy consumption among different regions. By using this method, they also designed the corresponding strategies to achieve a more equitable energy reduction policy in China's building sector.

Similar with the SRIO, the carbon balance of sector i in region r in the MRIO table can be expressed as:

$$e_i^r x_i^r = \sum_{k=1}^{m} \sum_{j=1}^{n} e_j^k a_{ji}^{kr} x_i^r + c_i^r x_i^r \tag{19.6}$$

where x_i^r represents the monetary value of the total output of sector i in region r, and it is assumed that there are m regions and each region has n sectors; a_{ji}^{kr} represents the monetary input from sector j in region k as intermediate use for per unit produced from sector i in region r, e_i^r is the embodied carbon emission intensity of products from sector i in region r, and e_j^k is the embodied carbon emission intensity of products from sector j in region k. c_i^r is the direct carbon emissions intensity of sector i in region r, which is calculated based on Eq. (19.2) by using the direct energy consumption data reported from regional statistics.

Note that $m \times n$ equations are established under the whole economy, vectors and matrices that can therefore be introduced to simplify the mathematical expression.

$$\text{Nominate} \quad E^T = \begin{bmatrix} \begin{pmatrix} e_1^1 \\ \cdots \\ e_n^1 \end{pmatrix} \\ \begin{pmatrix} e_1^m \\ \cdots \\ e_n^m \end{pmatrix} \end{bmatrix} \quad C^T = \begin{bmatrix} \begin{pmatrix} c_1^1 \\ \cdots \\ c_n^1 \end{pmatrix} \\ \begin{pmatrix} c_1^m \\ \cdots \\ c_n^m \end{pmatrix} \end{bmatrix} \quad X = \begin{bmatrix} x_1^1 & 0 & \cdots & 0 \\ 0 & x_2^1 & \cdots & 0 \\ \cdots & \cdots & \cdots & \cdots \\ 0 & 0 & \cdots & x_n^m \end{bmatrix}$$

$$A = \begin{bmatrix} \begin{pmatrix} a_{11}^{11} & \cdots & a_{1n}^{11} \\ \cdots & \cdots & \cdots \\ a_{n1}^{11} & \cdots & a_{nn}^{11} \end{pmatrix} & \cdots & \begin{pmatrix} a_{11}^{1m} & \cdots & a_{1n}^{1m} \\ \cdots & \cdots & \cdots \\ a_{n1}^{1m} & \cdots & a_{nn}^{1m} \end{pmatrix} \\ \begin{pmatrix} a_{11}^{m1} & \cdots & a_{1n}^{m1} \\ \cdots & \cdots & \cdots \\ a_{n1}^{m1} & \cdots & a_{nn}^{m1} \end{pmatrix} & \cdots & \begin{pmatrix} a_{11}^{mm} & \cdots & a_{1n}^{mm} \\ \cdots & \cdots & \cdots \\ a_{n1}^{mm} & \cdots & a_{nn}^{mm} \end{pmatrix} \end{bmatrix}$$

where E^T and C^T are the embodied carbon emission intensity vector and direct carbon emission intensity vector with $m \times n$ dimension, X is the diagonal matrix with $m \times n$ entries, and the coefficients in matrix X are equal to the total economic output. A is the intermediate input matrix in the input-output table with $m \times n$ entries. For the whole economic system, the above group of equations can be expressed in the form of a matrix.

$$XE^T = XAE^T + XC^T \tag{19.7}$$

Similar with the SRIO, this equation can be further transformed as:

$$E = C(I - A^T)^{-1} \tag{19.8}$$

Based on the embodied carbon emission intensity calculated in Eq. (19.8), the carbon emission embodied in the final demand can be simply deduced for a general situation. However, as only the building sector is the focus of this study, further clarification is needed to help understand the embodied carbon emission intensity and interregional carbon emissions transfer of the regional building sector. Embodied carbon emission intensity can measure the direct and indirect carbon emissions within the entire supply chain of the building sector, which represents the total carbon emissions per unit monetary value. The carbon emission intensity vector for the regional building sector $E_b = \left[e_b^1, e_b^2, \ldots, e_b^m\right]$ can be extracted from vector E.

In this section, a supply chain-based method for structural analysis of embodied carbon emissions in the building sector will be also introduced. Structure path analysis (SPA) explores environmental transmissions within the entire economic system by decomposing the direct and indirect effects from interconnections in the upstream supply chain. This method helps to identify the key paths and sectors in the production chain where the economic interactions with other sectors lead to significant influences on the final output (Acquaye et al. 2011; Defourny and Thorbecke 1984; Roberts 2005). It reveals linkages and indirect transactions between the exogenous final demand and total output by tracing the transmissions among different upstream processes.

The total embodied carbon emissions F can be expressed as:

$$F = EV = C(I - A)^{-1}V \tag{19.9}$$

which can be further expanded based on a power series approximation theory as:

$$F = C(I - A)^{-1}V = CIV + CAV + CA^2V + CA^3V + CA^4V + \ldots \tag{19.10}$$

V is the set of commodities in the building sector. CA^tV represents the carbon emissions generated from the production process in the tth stage. Given this equation, this study establishes an infinite tree that is based on the computational algorithm implied in the SPA. The tree paths explore the inter-sectoral and interregional connections of different tiers in the upstream production process. Each node in a connected graph is a certain sector in a specific region within the economic system. It represents individual carbon emissions induced from the corresponding final demand. According to Eq. (19.10), it is understandable that the number of nodes is exponential via the growth of tiers.

The paths tree is based on the entire supply chain and can be further inspected from the horizontal and vertical perspective. Horizontally, the paths diagramed between two stages in the supply chain represent the direct carbon emissions of a certain tier. Consequently, the sum of the total environmental impact of the building sector in all tiers can be expressed as:

$$t = 0 : c_b \sum_{k=1}^{m} v_b^k$$
$$t = 1 : \sum_{j=1}^{m \times n} c_j a_{jb} \sum_{k=1}^{m} v_b^k$$
$$t = 2 : \sum_{l=1}^{m \times n} \sum_{j=1}^{m \times n} c_l a_{lj} a_{jb} \sum_{k=1}^{m} v_b^k$$
$$\ldots \ldots$$

(19.11)

where v_b^k represents the set of commodities in the building sector from region k. From the vertical perspective, paths connecting different stages represent the carbon linkage between the producers in the higher-order stages of the upstream process to the final consumer. The carbon emissions embodied in a certain path of the building sector in region k can be described as:

$$\overbrace{c_b v_b^k}^{Stage0} + \overbrace{c_j a_{jb} v_b^k}^{Stage1} + \overbrace{c_k a_{kj} a_{jb} v_b^k}^{Stage2} + \ldots$$

(19.12)

The adoption of Eqs. (19.11) and (19.12) enables us to trace the supply chain intuitively. This allows investigation of the carbon emissions from the production view. For each energy path that is extracted in this study, the starting point is the final demand v_b^k (consumption) of the building in region k. The end point represents carbon emissions generated from a given production. This carbon emission transfer process established a linkage between the final demand purchased and its corresponding production (Peters and Hertwich 2006). Therefore, examining the total carbon emissions from both a consumption and production perspective can effectively establish a holistic map of carbon emission interactions in the building sector. This may help policy makers to achieve a fair and equitable carbon reduction policy.

It is effective to focus on a specific degree of order within a sound number of paths because the number of nodes increases exponentially whereas the value of the node decreases sharply with the growth of the path length. Therefore, an optimized algorithm is commonly used in previous research to cut the redundant paths with negligible value. This optimization not only makes the computational process time-efficient but also helps decision makers extract and identify important paths with large environmental improvements. In general, the iterative recalculation of the (m*n)*(m*n) matrices results in an exponential increase of nodes. It is therefore

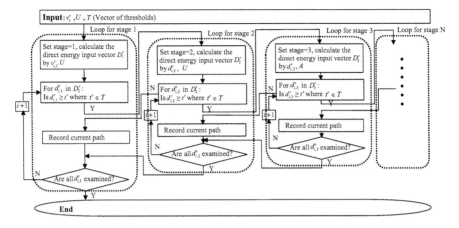

Fig. 19.1 Iterative computational processes for path extracting (Source: Hong et al. 2016c)

impossible to manually manage the energy path tree when higher-order upstream processes are considered (e.g., maximum stage = 10). In addition, the exploration of the carbon paths from such computational processes also makes the mathematical operation more challenging. Consequently, a backtracking algorithm based on a depth-first searching strategy can be implemented to carry out tests and extract the carbon paths of this research. The algorithm enumerates all the possible paths by setting sectors in paths in a trial-and-error way for certain regional construction industries. During enumeration, a vector of thresholds can be set to prune the branches with negligible embodied carbon emissions. Figure 19.1 shows the basic iterative computational processes for the building sector in a certain region r.

Embodied Carbon Assessment of Buildings from a Micro Perspective

The quantification models for embodied carbon assessment at the building level mainly include process-based model, input-output (I-O) model, and hybrid model. As aforementioned in the section of literature review, the use of process-based method for embodied carbon assessment of buildings is dominant in China. However, there is also a trend that relevant studies gradually shift their focus from process-based individual cases to a more hybrid and macro sense. This is mainly because there is an urgent need for China's government to ensure the emission reduction targets both at the national and provincial levels. On the other hand, the built environment of China is comparatively huge where an imbalance of economy exists between the eastern coast and the western interior. An insufficient number of case studies are unable to provide a holistic picture for the current distribution of buildings' carbon emissions in China.

Process-Based LCA

Process-based analysis quantifies the detailed resource and energy consumption from direct input of the manufacturing process to the indirect input with significant environmental contributions in the upstream and downstream process of the supply chain. Although the case-specific process data to some extent improve the accuracy of the calculation result, this model is time and cost intensive. In addition, the intuitive determination of system boundary is subject to truncation errors and thereby results in variations (Rowley et al. 2009).

The process-based LCA for building embodied carbon assessment can be expressed as:

$$\text{Account} = \sum_{i=1}^{n} (CO_2e)_i = \sum_{i=1}^{n} (AD_i * EF_i * GWP_i) \tag{19.13}$$

where AD represents a specific elementary flow data during building construction phase. In the traditional construction process, the carbon sources involved in the initial embodied carbon mainly include the following categories:

- Fuels used by construction equipment (direct emissions)
- Electricity consumption (direct and indirect emissions)
- Onsite assembly and miscellaneous works (direct emissions)
- Building material production (indirect emissions)
- Transportation (direct and indirect emissions)
- Construction-related human activities (direct and indirect emissions)

The basic procedures regarding the embodied carbon assessment of buildings are described in Fig. 19.2.

Hybrid LCA

To eliminate the truncation errors and guarantee the specificity in the environmental assessment process, hybrid analysis has been developed to provide more accurate assessment of environmental loadings. In general, three models have been commonly used in previous literature: tiered hybrid, input-output (I-O)-based hybrid, and integrated hybrid model. The first two models are commonly used in China for embodied energy and carbon assessment of buildings (Chang et al. 2012, 2013; Hong et al. 2016a, 2017b), while the integrated hybrid model is rare in China due to its high dependency for process-based specific data.

The tiered hybrid model was firstly proposed by Bullard et al. (1978). The scientific basis of this model is to employ process-based data at important lower-order upstream processes, usage phase, and downstream processes while supplementing I-O data for indirect impacts with negligible contributions from higher-order upstream processes. Such manipulation to large extent maximizes the

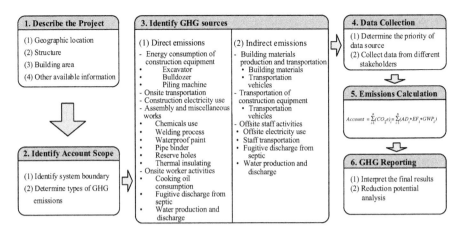

Fig. 19.2 Framework of embodied carbon assessment (Source: Hong et al. 2015)

accuracy and reliability of calculated results. However, the direct integration of process and I-O model may probably result in double counting. It is therefore important to subtract the process-based flows from the I-O model to represent only the cutoff inventory. Although the application of I-O-derived data improves the completeness of the system boundary in the upstream process, truncation errors may still arise in use and downstream phases due to the limitations in data availability for details processes. More importantly, the interface of system boundaries between process-based and I-O-based model is flexible, which depends on the research purpose, accuracy requirement, and time restriction.

I-O-based hybrid model is a top-down method which aims to further modify or disaggregate the direct supply chain of the sector in the I-O table that the product being investigated belongs. It allows the incorporation of the case-specific process data into I-O direct coefficient matrix, which provides the analyst with access to detailed process information within complete system boundary. However, according to Joshi (1999) and Suh et al. (2004), as the basic procedure for I-O-based hybrid analysis, the disaggregation is restrained due to the overdependence of the detailed data of input and sale information for the new hypothetical sector. Treloar (1997) proposed I-O-based hybrid approach in a different way by substituting the most energy-intensive paths with the process-based inventory data. A number of studies have been conducted under this hybrid framework (Crawford 2008; Crawford and Pullen 2011; Lenzen and Treloar 2002; Treloar et al. 2000, 2004). According to Treloar (1997) and Crawford (2008), I-O-based hybrid model was designed to substitute the process-based data for the most energy-intensive paths that were extracted from the I-O analysis. Treloar et al. (2001) summarized the basic procedures for this hybrid model, including:

1. Calculation of the initial total environmental burden of the product being studied by using I-O analysis

2. Disaggregation of the complex upstream process based on I-O analysis and determination of the key paths with significant environmental impact
3. Modification of key paths with both delivered quantity and energy intensity data derived from process-based inventory
4. Subtraction of the corresponding I-O value of the key paths represented in the process inventory from the initial total environmental impact calculated by I-O model
5. Integration of modified energy paths derived from process-based analysis into remaining unmodified I-O framework

The integrated hybrid model integrates the I-O model with a matrix representation of the physical process flows of a particular product, which makes the computational framework consistent (Bilec 2007). It incorporates the physical quantities of process-based data into the I-O model directly. However, because of its higher requirement in detailed data, it is time and cost intensive and more complicated to practical application.

In summary, tiered and I-O-based hybrid model is more dependent on budget information because only monetary value can be modeled in the I-O analysis for further environmental effect assessment. The results obtained from these two approaches are in the higher level of aggregation because the computational process is based on the sectoral framework derived from I-O model. In contrast, the integrated model is prioritized to incorporate physical unit and monetary transactions. According to Suh et al. (2004), it is difficult to determine the most suitable hybrid model intuitively in a certain application, which is to a large extent based on the actual data availability and accuracy requirement.

Uncertainty Analysis for Embodied Carbon Assessment of Buildings

Given the current methods employed for embodied carbon assessment of buildings in China, the source of uncertainties mainly comes from two aspects: the sector aggregation in I-O analysis and the data quality in the process-based LCA model.

Uncertainty in the I-O Analysis

The uncertainties in the input-output analysis come from two major sources. One is the theoretical assumptions before analysis including the assumption of proportionality, homogeneity, and identity of production technology. The other is due to transformation and reconstruction in the compilation of input-output tables. In general, the methodological uncertainties are mostly unavoidable and hard to estimate in the computational process. In contrast, uncertainties from subjective compilation can be quantified and improved based on uncertainty analysis. Moreover, Weber (2009) also emphasized that the level of sector aggregation has a direct impact on results accuracy—this is the most critical factor influencing the uncertainty in structural decomposition. Unfortunately, very few studies have discussed this problem systematically due to a lack of public data.

Broadly speaking, sector aggregation is the result of the trade-off between the level of detail in analysis and the availability of environmental data from the statistical yearbook. Generally, the I-O table compiled by the National Bureau of Statistics is more specific on the detailed monetary flow data. However, direct energy input data are recorded at a more aggregate level, where the sector classification standard is not consistent with the I-O table. Moreover, due to the improvement of sector classification standards in the past two decades, the compilation of input-output table has been changed across different years. Sector aggregation has to be performed to keep the consistency in table format across the time series.

The level of sector aggregation thus directly affects the final results of the I-O analysis because the number of sectors has been predetermined in structural decomposition analysis. As a result, the sector aggregation strategy needs to be implemented to match these two systems (Su et al. 2010). One strategy is to aggregate the I-O sectors to match the energy sector data. This strategy not only guarantees the accuracy of sector aggregation but also avoids extra assumptions. The other approach disaggregates energy consumption data to match the I-O table that retains all economic information while bearing the drawbacks on the subjective estimation of energy use among the subsectors. In the construction practice of China, a common method to examine the effect from the sector aggregation is to conduct a scenario analysis by employing multi-scale I-O tables (Hong et al. 2017a).

Uncertainty in the Process-Based LCA Model

Assessment tools vary in terms of uncertainty types, and they can generally be divided into qualitative and quantitative approaches. The data quality index (DQI) is the most commonly used qualitative assessment method because of its high applicability and feasibility. The data quality evaluation matrix (Weidema 1998) and the transformation matrix (Kennedy et al. 1996) are the two most efficient tools used in DQI assessment. However, DQI remains limited in terms of its assessment accuracy due to the subjective determination of data quality. Although the quantitative analysis techniques are complementary to the current qualitative uncertainty assessment methods to minimize variations, the results still tend to be underestimated as specified by Coulon et al. (1997). Considering the aforementioned limitations in the application of qualitative approaches in uncertainty analysis, quantitative approaches have been introduced based on data availability. In general, fuzzy logic analysis, interval theory, and possibility uncertainty analysis are commonly used for LCA uncertainty evaluation given that the data are insufficient for further processing (Chevalier and Le Téno 1996; Tan et al. 2002, 2007).

By contrast, stochastic methods such as MCS can be adopted if a large amount of actual data can be observed and collected (Canter et al. 2002; Geisler et al. 2004; Lo et al. 2005; Venkatesh et al. 2010). In particular, stochastic methods are superior to other methods due to their inherent advantages in capturing the variability and uncertainty in LCA. Tan et al. (2002) applied possibility theory to assess the uncertainties in LCIA and clarified that possibilistic methods have advantages in computational efficiency when compared with probabilistic approaches. André and Lopes (2012) held the same view by comparing possibilistic and probabilistic approaches and identified that the latter is limited by a relatively slow computational process but can provide sufficient uncertainty information. Lloyd and Ries (2007) asserted that stochastic methods were mostly used (67%) in previous research, followed by scenario analysis (29%) and fuzzy data sets (17%).

The estimation of uncertainty in LCA analysis is not only associated with uncertainty sources and assessment methods but also with the research scope and objectives. Given that this report focuses on the embodied carbon assessment of buildings, we should consider some basic characteristics relevant to the building sector, especially for building construction process. First, a specific building is distinctive due to its characterized building profile such as design parameters and construction structure. Therefore, elementary flows may vary among different buildings. This specificity determines that all input parameters involved in a certain building are unique. In this case, sufficient data can hardly be collected to describe the probability distribution of the elementary flows of buildings. However, such procedure can be partially mitigated by improving the data collection quality and by enhancing the measurement method during the building construction phase. Second, unlike the manufacturing industry that is characterized by reproducibility and mass production, the inventory data are rarely reported from private and public sources. The probable reason behind this condition is the confidentiality requirement between the client and the contractor, which results in an insufficient application of uncertainty analysis in real building cases in the construction industry.

In China, several uncertainty studies have been conducted for process-based embodied carbon assessment of buildings (Hong et al. 2016b; Wang and Shen 2013; Zhang and Wang 2017). Hong et al. (2016b) developed an uncertainty analysis framework by combining both DQI and probabilistic method to address the lack of actual data in the building sector. Based on this framework, Zhang and Wang (2017) conducted a comparative analysis between the deterministic and stochastic emissions from building construction. The present studies in China analyzed the uncertainty mainly from the aspects of measurement method, data source, geographic representativeness, technical representativeness, and temporal representativeness.

Parameter uncertainties stem from the lack of knowledge about the true value of a parameter (Huijbregts et al. 2003). A probabilistic approach can be used to reflect the most probable value of the objective. Traditionally, the goodness of fit between data samples and probability distributions can be examined to identify the most appropriate probability distribution for parameters based on a certain number of observations. However, the basic features of building products result in a limited

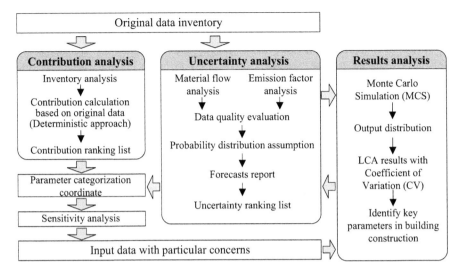

Fig. 19.3 A multi-method-based uncertainty analysis framework (Source: Hong et al. 2016b)

amount of actual data. In this case, expert judgment can be used as a substitute for the lack of knowledge to estimate uncertainty ranges. Simultaneously, DQI should be used as the supplement for uncertainty evaluation. However, the pure DQI method has its weakness of subjective evaluation; the latest improvements in uncertainty analysis can facilitate the comprehensive assessment of the parameter uncertainty. Therefore, combining DQI and probabilistic method can address the lack of actual data in the building sector.

Figure 19.3 shows the fundamental procedure of the uncertainty analysis. Contribution analysis and DQI assessment can be conducted simultaneously to determine the importance of input data based on the compilation of the original inventory. Sensitivity analysis further verifies the critical input parameters with particular concerns. Finally, according to the assumed probability distribution for each construction activity, the overall coefficient of variation and the final cumulative results are calculated with Monte Carlo simulation. The following sections will make a detailed introduction of each part in this uncertainty analysis framework.

DQI Assessment Method

A priority system should be established to reflect the reliability by ranking different data sources. Comprehensively considering the conclusions of previous research (Wang and Shen 2013; Weidema and Wesnæs 1996), we identify five types of data quality indicators, namely, data measurement method, source of data, geographic representativeness, technical representativeness, and temporal representativeness. The score for each indicator ranges from one to five. The value represents the

Table 19.4 Data quality index (DQI) evaluation system

Quality score	Data quality indicators				
	Measurement method	Data source	Geographic factor	Technical factor	Temporal factor
5	Consistently measured data	Verified data from independent source	Field survey/ measure data	Data from process studied of the enterprise with same technology	Less than 3 years
4	Regularly measured data	Verified data from interested party	Data from an area with similar production condition	Data from process studied of the enterprise with similar technology	Less than 6 years
3	Data estimated based on measurements	Unverified data from independent source	Regional data	Data from enterprises with different technology	Less than 10 years
2	Data estimated partly based on assumptions	Unverified data from irrelevant enterprise	National data	Data from process-related enterprises with similar technology	Less than 15 years
1	Subjective estimated data	Unverified data from interested party	International data	Data from process-related enterprises with different technology	More than 15 years or unknown

Source: Hong et al. 2016b

Table 19.5 Different data source in the building sector

Quality score	General data source	Specific in building sector
5	Verified data from independent source	Accounting receipt
4	Verified data from interested party	Stakeholder's report
3	Unverified data from independent source	Bill of quantity
2	Unverified data from irrelevant enterprise	Material use application record
1	Unverified data from interested party	Secondary data from the procurement agency

Source: Hong et al. 2016b

uncertainty of each category from low to high. In consideration of the basic characteristics of the building sector, a DQI matrix was established (see Table 19.4) by Weidema and Wesnæs (1996) and Wang and Shen (2013). Table 19.5 shows the difference of data sources from the general situation and building construction.

Contribution Analysis

The contribution analysis determines the importance of each construction activity to the final cumulative results. The use of original process-based inventory data enables the calculation of the deterministic contributions for each construction activity. However, the result of the deterministic analysis is relatively imprecise because the uncertainty of all coefficients contributes to the overall uncertainty value. Such deterministic contribution can be regarded as the basic reference and cornerstone for identifying the important parameters, which substantially contribute to the final output. Maurice et al. (2000) emphasized that the percentage of data contributions may vary from one model to another. Therefore, the scenario analysis is conducted to verify and validate the reliability of the selected key parameters.

Monte Carlo Simulation (MCS)

MCS is a numerical method used to sample a probability distribution for the concerned factors to produce thousands of possible outcomes. The results of MCS are further analyzed to obtain the probabilities of different occurring outcomes (Shen et al. 2011; Xu 1985). MCS is a useful tool for measuring the total uncertainty aggregated by various uncertainty factors with nonlinear relationship. In this chapter, the total embodied carbon emission of buildings is the aggregation of carbon emissions in various activities in the construction phase. If the carbon emissions generated in each activity is taken as a stochastic factor considering the uncertainty of parameters, then the total carbon emissions should also be a stochastic factor. The statistical features of the total carbon emissions can be determined through its probability distribution obtained by MCS. MCS first determines the corresponding probability distribution for each input factor. The probability distribution for the concerned factor can be obtained either by fitting a large number of existing data or by experts' view. Based on the established relationship between the observed and input factors, MCS can be run with Crystal Ball software to obtain the probability distribution of the observed factor. To identify the extent in which uncertainty can be considered, the coefficient of variation (CV) is used to describe the degree of uncertainty based on the MCS results. Considering the constraints and deficiencies of data availability in the building sector, β distribution is assumed to describe the possible value for each construction activity. Canter et al. (2002) and Wang and Shen (2013) thoroughly explored the reasons and rationale for applying beta function in uncertainty analysis. Kennedy et al. (1996) established a transformation matrix to define four unknown parameters involved in β distribution, namely, α, β, lower point, and upper point. MCS is conducted in Crystal Ball software with 10,000 iterations. The number of iterations directly determines the stability of the final output. The result will be convergent with the increasing numbers of simulations, and the excessive iterations are regarded as a less efficient means to run MCS. Therefore, the examination of convergence is a valid method to confirm the appropriate number of iterations (Cowles and Carlin 1996; Shen et al. 2011). Numerous mean values and standard deviations derived from simulations with different iterations are then generated to examine the trend of convergence.

Conclusions

To mitigate carbon emissions of buildings in China, a set of national emission reduction policies along with pilot low-carbon initiatives were launched by China's government. Regardless of the evident progress that has been achieved through these therapies, the major focus of concern in nationwide carbon reduction is still on the operational emission reduction rather than that in the embodied phase of buildings. Such bias mismatches the increasing role of embodied carbon reduction in the creation of sustainability in the building sector. In China, a vast body of work is conducted for embodied carbon assessment at different levels. The SRIO analysis, MRIO analysis, and SPA are commonly applied for embodied carbon assessment at the macro level with the focus on the building sector, while the process-based and hybrid LCA models are dominant in micro-level analysis with their emphases on a specific project. A comprehensive review indicates that the process-based method is most frequently used for embodied carbon assessment of buildings in China. A clear trend can be observed that the relevant studies gradually shift their focus from process-based individual cases to a more hybrid and macro sense in China. Moreover, a number of studies concentrating on the uncertainty analysis at both macro and micro levels allowed for a more accurate assessment of embodied carbon in China.

Acknowledgments The authors wish to express their sincere gratitude to the Research Grants Council of Hong Kong (Project No.15276916), Chongqing Social Science Planning Project (No. 2017BS30), and the Fundamental Research Funds for the Central Universities (Project No. 0204005201043) for funding this research. The whole chapter is organized referring to the work done by Hong et al. (2015, 2016a, b, c, d, 2017a, b).

References

Acquaye, A. A., & Duffy, A. P. (2010). Input–output analysis of Irish construction sector greenhouse gas emissions. *Building and Environment, 45*, 784–791.
Acquaye, A. A., Wiedmann, T., Feng, K., Crawford, R. H., Barrett, J., Kuylenstierna, J., Duffy, A. P., Koh, S. L., & McQueen-Mason, S. (2011). Identification of 'carbon hot-spots' and quantification of GHG intensities in the biodiesel supply chain using hybrid LCA and structural path analysis. *Environmental Science & Technology, 45*, 2471–2478.
Ali, A. S., & Rahmat, I. (2009). Methods of coordination in managing the design process of refurbishment projects. *Journal of Building Appraisal, 5*, 87–98.
André, J. C., & Lopes, D. R. (2012). On the use of possibility theory in uncertainty analysis of life cycle inventory. *The International Journal of Life Cycle Assessment, 17*, 350–361.

Barandica, J. M., Fernández-Sánchez, G., Berzosa, Á., Delgado, J. A., & Acosta, F. J. (2013). Applying life cycle thinking to reduce greenhouse gas emissions from road projects. *Journal of Cleaner Production, 57,* 79–91.

Bilec, M. M. (2007). *A hybrid life cycle assessment model for construction processes.* University of Pittsburgh: Pennsylvania, United States.

Bullard, C. W., Penner, P. S., & Pilati, D. A. (1978). Net energy analysis: Handbook for combining process and input-output analysis. *Resources and Energy, 1,* 267–313.

Canter, K. G., Kennedy, D. J., Montgomery, D. C., Keats, J. B., & Carlyle, W. M. (2002). Screening stochastic life cycle assessment inventory models. *The International Journal of Life Cycle Assessment, 7,* 18–26.

Chang, Y., Ries, R. J., & Wang, Y. (2010). The embodied energy and environmental emissions of construction projects in China: An economic input–output LCA model. *Energy Policy, 38,* 6597–6603.

Chang, Y., Ries, R. J., & Wang, Y. (2011). The quantification of the embodied impacts of construction projects on energy, environment, and society based on I–O LCA. *Energy Policy, 39,* 6321–6330.

Chang, Y., Ries, R. J., & Lei, S. (2012). The embodied energy and emissions of a high-rise education building: A quantification using process-based hybrid life cycle inventory model. *Energy and Buildings, 55,* 790–798.

Chang, Y., Ries, R. J., & Wang, Y. W. (2013). Life-cycle energy of residential buildings in China. *Energy Policy, 62,* 656–664.

Chang, Y., Ries, R. J., Man, Q., & Wang, Y. (2014). Disaggregated IO LCA model for building product chain energy quantification: A case from China. *Energy and Buildings, 72,* 212–221.

Chang, Y., Huang, Z., Ries, R. J., & Masanet, E. (2015). The embodied air pollutant emissions and water footprints of buildings in China: A quantification using disaggregated input–output life cycle inventory model. *Journal of Cleaner Production, 113,* 274–284.

Chen, Z.-M., & Chen, G. Q. (2013). Virtual water accounting for the globalized world economy: National water footprint and international virtual water trade. *Ecological Indicators, 28,* 142–149.

Chen, G. Q., & Zhang, B. (2010). Greenhouse gas emissions in China 2007: Inventory and input–output analysis. *Energy Policy, 38,* 6180–6193.

Chen, G. Q., Chen, H., Chen, Z. M., Zhang, B., Shao, L., Guo, S., Zhou, S. Y., & Jiang, M. M. (2011). Low-carbon building assessment and multi-scale input–output analysis. *Communications in Nonlinear Science and Numerical Simulation, 16,* 583–595.

Chevalier, J.-L., & Le Téno, J.-F. (1996). Life cycle analysis with ill-defined data and its application to building products. *The International Journal of Life Cycle Assessment, 1,* 90–96.

Coulon, R., Camobreco, V., Teulon, H., & Besnainou, J. (1997). Data quality and uncertainty in LCI. *The International Journal of Life Cycle Assessment, 2,* 178–182.

Cowles, M. K., & Carlin, B. P. (1996). Markov chain Monte Carlo convergence diagnostics: A comparative review. *Journal of the American Statistical Association, 91,* 883–904.

Crawford, R. H. (2008). Validation of a hybrid life-cycle inventory analysis method. *Journal of Environmental Management, 88,* 496–506.

Crawford, R. H., & Pullen, S. (2011). Life cycle water analysis of a residential building and its occupants. *Building Research and Information, 39,* 589–602.

Defourny, J., & Thorbecke, E. (1984). Structural path analysis and multiplier decomposition within a social accounting matrix framework. *The Economic Journal, 94,* 111–136.

Dong, Y. H., & Ng, S. T. (2015). A life cycle assessment model for evaluating the environmental impacts of building construction in Hong Kong. *Building and Environment, 89,* 183–191.

Gao, X. (2012). *Assessment methodology and empirical analysis of embodied carbon footprint of building construction, management science and engineering.* Beijing: Tsinghua University.

Geisler, G., Hellweg, S., & Hungerbühler, K. (2004). *Uncertainties in LCA of plant-growth regulators and implications on decision-making, complexity and integrated resources management*. Proceedings of the 2nd Biennial Meeting of the International Environmental Modelling and Software Society.

Hong, J., Shen, G. Q., Feng, Y., Lau, W. S.-T., & Mao, C. (2015). Greenhouse gas emissions during the construction phase of a building: A case study in China. *Journal of Cleaner Production, 103*, 249–259.

Hong, J., Shen, G. Q., Mao, C., Li, Z., & Li, K. (2016a). Life-cycle energy analysis of prefabricated building components: An input–output-based hybrid model. *Journal of Cleaner Production, 112*, 2198–2207.

Hong, J., Shen, G. Q., Peng, Y., Feng, Y., & Mao, C. (2016b). Uncertainty analysis for measuring greenhouse gas emissions in the building construction phase: A case study in China. *Journal of Cleaner Production, 129*, 183–195.

Hong, J., Shen, Q., & Xue, F. (2016c). A multi-regional structural path analysis of the energy supply chain in China's construction industry. *Energy Policy, 92*, 56–68.

Hong, J., Shen, G. Q., Guo, S., Xue, F., & Zheng, W. (2016d). Energy use embodied in China 's construction industry: A multi-regional input–output analysis. *Renewable and Sustainable Energy Reviews, 53*, 1303–1312.

Hong, J., Li, C. Z., Shen, Q., Xue, F., Sun, B., & Zheng, W. (2017a). An overview of the driving forces behind energy demand in China's construction industry: Evidence from 1990 to 2012. *Renewable & Sustainable Energy Reviews, 73*, 85–94.

Hong, J., Zhang, X., Shen, Q., Zhang, W., & Feng, Y. (2017b). A multi-regional based hybrid method for assessing life cycle energy use of buildings: A case study. *Journal of Cleaner Production, 148*, 760–772.

Huijbregts, M. A., Gilijamse, W., Ragas, A. M., & Reijnders, L. (2003). Evaluating uncertainty in environmental life-cycle assessment. A case study comparing two insulation options for a Dutch one-family dwelling. *Environmental Science and Technology, 37*, 2600–2608.

Joshi, S. (1999). Product environmental life-cycle assessment using input-output techniques. *Journal of Industrial Ecology, 3*, 95–120.

Kennedy, D. J., Montgomery, D. C., & Quay, B. H. (1996). Stochastic environmental life cycle assessment modeling: A probabilistic approach to incorporating variable input data quality. *The International Journal of Life Cycle Assessment, 1*, 199–207.

Kim, B., Lee, H., Park, H., & Kim, H. (2011). Greenhouse gas emissions from onsite equipment usage in road construction. *Journal of Construction Engineering and Management, 138*, 982–990.

Kim, B., Lee, H., Park, H., & Kim, H. (2012a). Estimation of greenhouse gas emissions from land-use changes due to road construction in the Republic of Korea. *Journal of Construction Engineering and Management, 139*, 339–346.

Kim, B., Lee, H., Park, H., & Kim, H. (2012b). Framework for estimating greenhouse gas emissions due to asphalt pavement construction. *Journal of Construction Engineering and Management, 138*, 1312–1321.

Kuo, C. F. J., Lin, C. H., & Hsu, M. W. (2016). Analysis of intelligent green building policy and developing status in Taiwan. *Energy Policy, 95*, 291–303.

Le Quéré, C., Andrew, R. M., Canadell, J. G., Sitch, S., Korsbakken, J. I., Peters, G. P., Manning, A. C., Boden, T. A., Tans, P. P., Houghton, R. A., Keeling, R. F., Alin, S., Andrews, O. D., Anthoni, P., Barbero, L., Bopp, L., Chevallier, F., Chini, L. P., Ciais, P., Currie, K., Delire, C., Doney, S. C., Friedlingstein, P., Gkritzalis, T., Harris, I., Hauck, J., Haverd, V., Hoppema, M., Klein Goldewijk, K., Jain, A. K., Kato, E., Körtzinger, A., Landschützer, P., Lefèvre, N., Lenton, A., Lienert, S., Lombardozzi, D., Melton, J. R., Metzl, N., Millero, F., Monteiro, P. M. S., Munro, D. R., Nabel, J. E. M. S., Nakaoka, S. I., O'Brien, K., Olsen, A., Omar, A. M., Ono, T., Pierrot, D., Poulter, B., Rödenbeck, C., Salisbury, J., Schuster, U., Schwinger, J., Séférian, R., Skjelvan, I., Stocker, B. D., Sutton, A. J., Takahashi, T., Tian, H., Tilbrook, B., van der Laan-Luijkx, I. T., van der Werf, G. R., Viovy, N., Walker, A. P., Wiltshire, A. J., & Zaehle, S. (2016). Global carbon budget 2016. *Earth System Science Data, 8*, 605–649.

Lenzen, M., & Treloar, G. (2002). Embodied energy in buildings: Wood versus concrete—Reply to Börjesson and Gustavsson. *Energy Policy, 30*, 249–255.

Lenzen, M., Pade, L.-L., & Munksgaard, J. (2004). CO2 multipliers in multi-region input-output models. *Economic Systems Research, 16*, 391–412.

Leontief, W. (1970). Environmental repercussions and the economic structure: An input-output approach. *The Review of Economics and Statistics, 52*, 262–271.

Li, D., Cui, P., & Lu, Y. (2016). Development of an automated estimator of life-cycle carbon emissions for residential buildings: A case study in Nanjing, China. *Habitat International, 57*, 154–163.

Liang, X., Geoffrey Qiping, S., & Li, G. (2015). Improving management of green retrofits from a stakeholder perspective: A case study in China. *International Journal of Environmental Research & Public Health, 12*, 13823–13842.

Liang, X., Peng, Y., & Shen, G. Q. (2016). A game theory based analysis of decision making for green retrofit under different occupancy types. *Journal of Cleaner Production, 137*, 1300–1312.

Liu, N., Wang, J., & Li, R. (2009). Computational method of CO_2 emissions in Chinese urban residential communities. *Journal of Tsinghua University (Science and Technology), 9*, 000.

Liu, Z., Geng, Y., Lindner, S., & Guan, D. (2012). Uncovering China's greenhouse gas emission from regional and sectoral perspectives. *Energy, 45*, 1059–1068.

Lloyd, S. M., & Ries, R. (2007). Characterizing, propagating, and analyzing uncertainty in life-cycle assessment: A survey of quantitative approaches. *Journal of Industrial Ecology, 11*, 161–179.

Lo, S.-C., Ma, H.-W., & Lo, S.-L. (2005). Quantifying and reducing uncertainty in life cycle assessment using the Bayesian Monte Carlo method. *Science of the Total Environment, 340*, 23–33.

Maurice, B., Frischknecht, R., Coelho-Schwirtz, V., & Hungerbühler, K. (2000). Uncertainty analysis in life cycle inventory. Application to the production of electricity with French coal power plants. *Journal of Cleaner Production, 8*, 95–108.

Melanta, S., Miller-Hooks, E., & Avetisyan, H. G. (2013). Carbon footprint estimation tool for transportation construction projects. *Journal of Construction Engineering and Management, 139*, 547–555.

Metz, B., Davidson, O. R., Bosch, P. R., Dave, R., & Meyer, L. A. (2007). *Contribution of Working Group III to the fourth assessment report of the Intergovernmental Panel on Climate Change.* Cambridge, UK: Cambridge University Press.

Nässén, J., Holmberg, J., Wadeskog, A., & Nyman, M. (2007). Direct and indirect energy use and carbon emissions in the production phase of buildings: An input–output analysis. *Energy, 32*, 1593–1602.

Peters, G. P., & Hertwich, E. G. (2006). The importance of imports for household environmental impacts. *Journal of Industrial Ecology, 10*, 89–109.

Peters, G. P., Briceno, T., & Hertwich, E. G. (2004). *Pollution embodied in Norwegian consumption.* Trondheim: Norwegian University of Scienece and Technology.

Roberts, D. (2005). The role of households in sustaining rural economies: A structural path analysis. *European Review of Agricultural Economics, 32*, 393–420.

Rowley, H. V., Lundie, S., & Peters, G. M. (2009). A hybrid life cycle assessment model for comparison with conventional methodologies in Australia. *The International Journal of Life Cycle Assessment, 14*, 508–516.

Salazar, J., & Meil, J. (2009). Prospects for carbon-neutral housing: The influence of greater wood use on the carbon footprint of a single-family residence. *Journal of Cleaner Production, 17*, 1563–1571.

Shen, L., Lu, W., Peng, Y., & Jiang, S. (2011). Critical assessment indicators for measuring benefits of rural infrastructure investment in China. *Journal of Infrastructure Systems, 17*, 176–183.

Su, B., Huang, H. C., Ang, B. W., & Zhou, P. (2010). Input–output analysis of CO2 emissions embodied in trade: The effects of sector aggregation. *Energy Economics, 32*, 166–175.

Suh, S., Lenzen, M., Treloar, G. J., Hondo, H., Horvath, A., Huppes, G., Jolliet, O., Klann, U., Krewitt, W., & Moriguchi, Y. (2004). System boundary selection in life-cycle inventories using hybrid approaches. *Environmental Science & Technology, 38*, 657–664.

Tan, R. R., Culaba, A. B., & Purvis, M. R. (2002). Application of possibility theory in the life-cycle inventory assessment of biofuels. *International Journal of Energy Research, 26*, 737–745.

Tan, R. R., Briones, L. M. A., & Culaba, A. B. (2007). Fuzzy data reconciliation in reacting and non-reacting process data for life cycle inventory analysis. *Journal of Cleaner Production, 15*, 944–949.

Tang, P., Cass, D., & Mukherjee, A. (2013). Investigating the effect of construction management strategies on project greenhouse gas emissions using interactive simulation. *Journal of Cleaner Production, 54*, 78–88.

Tegart, W. J. M., Sheldon, G. W., & Griffiths, D. C. (1990). *Report prepared for intergovernmental panel on climate change by working group II*. Camberra: Australian Government Publishing Service.

Treloar, G. J. (1997). Extracting embodied energy paths from input–output tables: Towards an input–output-based hybrid energy analysis method. *Economic Systems Research, 9*, 375–391.

Treloar, G., Love, P., Faniran, O., & Iyer-Raniga, U. (2000). A hybrid life cycle assessment method for construction. *Construction Management and Economics, 18*, 5–9.

Treloar, G. J., Love, P. E., & Holt, G. D. (2001). Using national input/output data for embodied energy analysis of individual residential buildings. *Construction Management and Economics, 19*, 49–61.

Treloar, G. J., Love, P. E., & Crawford, R. H. (2004). Hybrid life-cycle inventory for road construction and use. *Journal of Construction Engineering and Management, 130*, 43–49.

Venkatesh, A., Jaramillo, P., Griffin, W. M., & Matthews, H. S. (2010). Uncertainty analysis of life cycle greenhouse gas emissions from petroleum-based fuels and impacts on low carbon fuel policies. *Environmental Science & Technology, 45*, 125–131.

Wang, E., & Shen, Z. (2013). A hybrid data quality indicator and statistical method for improving uncertainty analysis in LCA of complex system: Application to the whole-building embodied energy analysis. *Journal of Cleaner Production, 43*, 166–173.

Wang, T., Seo, S., Liao, P.-C., & Fang, D. (2016). GHG emission reduction performance of state-of-the-art green buildings: Review of two case studies. *Renewable and Sustainable Energy Reviews, 56*, 484–493.

Weber, C. L. (2009). Measuring structural change and energy use: Decomposition of the US economy from 1997 to 2002. *Energy Policy, 37*, 1561–1570.

Weidema, B. P. (1998). Multi-user test of the data quality matrix for product life cycle inventory data. *The International Journal of Life Cycle Assessment, 3*, 259–265.

Weidema, B. P., & Wesnæs, M. S. (1996). Data quality management for life cycle inventories—An example of using data quality indicators. *Journal of Cleaner Production, 4*, 167–174.

Wiedmann, T. (2009). A review of recent multi-region input–output models used for consumption-based emission and resource accounting. *Ecological Economics, 69*, 211–222.

Wiedmann, T., Lenzen, M., Turner, K., & Barrett, J. (2007). Examining the global environmental impact of regional consumption activities—Part 2: Review of input–output models for the assessment of environmental impacts embodied in trade. *Ecological Economics, 61*, 15–26.

Wong, J. K. W., Li, H., Wang, H., Huang, T., Luo, E., & Li, V. (2013). Toward low-carbon construction processes: The visualisation of predicted emission via virtual prototyping technology. *Automation in Construction, 33*, 72–78.

Wu, H. J., Yuan, Z. W., Zhang, L., & Bi, J. (2012). Life cycle energy consumption and CO2 emission of an office building in China. *The international journal of life cycle assessment, 17*, 105–118.

Wu, X., Peng, B., & Lin, B. (2017). A dynamic life cycle carbon emission assessment on green and non-green buildings in China. *Energy and Buildings, 149*, 272–281.

Xing, S., Xu, Z., & Jun, G. (2008). Inventory analysis of LCA on steel-and concrete-construction office buildings. *Energy and Buildings, 40*, 1188–1193.

Xu, Z. J. (1985). *The method of Monte Carlo*. Shanghai: Science and Technology.

Yan, H., Shen, Q., Fan, L. C., Wang, Y., & Zhang, L. (2010). Greenhouse gas emissions in building construction: A case study of One Peking in Hong Kong. *Building and Environment, 45*, 949–955.

Yao, X. (2013). *Study on calculation of carbon emissions baseline of public building based on LCA, Environmental Engineering*. Wuhan: Huazhong University of Science & Technology.

Zhang, X., & Wang, F. (2017). Stochastic analysis of embodied emissions of building construction: A comparative case study in China. *Energy and Buildings, 151*, 574–584.

Zhang, Z., Wu, X., Yang, X., & Zhu, Y. (2006). BEPAS—A life cycle building environmental performance assessment model. *Building and Environment, 41*, 669–675.

Chapter 20
Embodied Carbon Measurement, Mitigation and Management Within Europe, Drawing on a Cross-Case Analysis of 60 Building Case Studies

A. M. Moncaster, H. Birgisdottir, T. Malmqvist, F. Nygaard Rasmussen, A. Houlihan Wiberg, and E. Soulti

Introduction

This chapter discusses the current state of the art of both the knowledge of and actions towards measuring and reducing embodied carbon within Europe. It draws on several sources of data, principal among which is a cross-case analysis of 60 case studies within nine European countries. The case studies and their original analysis were developed as part of an international project, carried out under an implementing agreement with the International Energy Agency Energy in Buildings and Communities Programme (IEA EBC). The Annex 57 project (Seo et al. 2016) ran from 2011 to 2016, and the authors of this chapter formed subtask 4 (ST4). The full reports from the project are available online (International Energy Agency 2016), and a number of journal articles have recently been published recently which are referenced within the chapter text. Birgisdottir et al. (2017) provides an overview of the full project. The analysis of these case studies is compared with a systematic review of over 100 published international papers conducted by Pomponi and Moncaster (2016).

A. M. Moncaster (✉)
School of Engineering and Innovation, Open University, Milton Keynes, UK
e-mail: alice.moncaster@open.ac.uk

H. Birgisdottir · F. Nygaard Rasmussen
Danish Building Research Institute, Aalborg University, Copenhagen, Denmark

T. Malmqvist
KTH Royal Institute of Technology, Stockholm, Sweden

A. Houlihan Wiberg
National Technical University of Norway (NTNU), Trondheim, Norway

E. Soulti
Building Research Establishment, Watford, UK

© Springer International Publishing AG 2018   443
F. Pomponi et al. (eds.), *Embodied Carbon in Buildings*,
https://doi.org/10.1007/978-3-319-72796-7_20

To set the context for the chapter, it is important to explain the characteristics of the European context, compared with other regions of the world. One distinguishing feature is that Europe includes a far higher percentage of old (>50 years) and very old (>100 years) buildings than more recently developed and developing regions. Due to the maturity of existing development, the European construction sector has low rates of new construction and is therefore responsible for less energy and lower greenhouse gas emissions than the equivalent construction sectors in rapidly developing nations (Yokoo et al. 2015). However, the other effect of the mature building stock is that a considerable proportion of the emissions from the European construction sector are due to refurbishment activities.

While there is little published work which considers the age profile of the whole building stock of Europe, a broad overview of the residential building stock in the member states of the EU can be found in Uihlein and Eder (2010), while Mata et al. (2014) have characterised residential and non-residential buildings in France, Germany, Spain and the UK, and Ballarini et al. (2014) have created a set of model reference buildings for each of 13 European countries. The age of buildings in China is discussed by Yand and Kohler (2008), while for North America, see Aktas and Bilec (2012), and for UK buildings, see DCLG (2017) and Summerfield et al. (2015).

The relatively recent work on characterising the building stock in Europe stems from the focus on reducing operational energy use from heating and lighting, and increasingly from cooling in Southern Europe in particular. The focus on reducing operational energy use in buildings is the outcome of the European Energy Performance of Buildings Directive (EPBD) of 2002 and subsequent recasts (European Commission 2010), which in turn has defined European member state building regulations. However, the EPBD continues to omit any consideration of embodied energy (Zöld-Zs and Szalay 2005).

While there is a certain rationale for keeping calculations of operational and embodied impacts of buildings separate, within industry practice one unintended consequence has been to provide an argument for developers to increase demolition of existing buildings and then replace them with new ones (Baker et al. 2017). This argument, based on reducing operational carbon and energy costs as required by national building codes, while ignoring the embodied carbon and energy costs of both the demolition of the existing buildings and the construction of the new buildings, has the potential result of increasing the whole-life energy and carbon (Andresen 2017). This is compounded by the fact that the actual reductions of operational energy have been repeatedly shown to be lower than the modelled reductions (Bordass et al. 2001; Lowe 2007; Sunikka-Blank and Galvin 2012), as well as that the measures taken to decrease operational energy usually require more materials consequently leading to an increase in embodied energy.

There are therefore two issues to be considered within the European context. First, as in other areas of the world, is the importance of minimising the whole-life energy and carbon cost of new build. However, of equal or even greater importance is the necessity to understand the embodied impacts of the multiple developments taking place on brownfield (previously developed) sites, through assessment of the whole-life impact of the two alternative solutions of demolish and build new buildings or retain and refurbish existing buildings.

The following section discusses the Annex 57 collection of case studies and the methodological approach. The subsequent three sections then focus on measurement, mitigation and management (in that order). The measurement section provides a general quantification of embodied carbon and energy in European buildings, for different life cycle stages and building components. The mitigation section identifies and discusses a number of approaches to reducing embodied impacts of buildings which are particularly suitable in the European context. The section on management then discusses some specific current policy and industry initiatives. These include a brief overview of the European standards for the life cycle analysis of buildings, which were published in 2011 and 2012 and form a useful framework for defining life cycle stages of buildings, and an overview of the approach to embodied carbon within BREEAM, an environmental assessment tool from the UK Building Research Establishment in widespread use. This chapter ends with a summary and concluding section, which identifies the particular need for further research on the whole-life environmental impacts of demolition versus refurbishment.

Developing Conclusions from a Comparative Analysis of Multiple Case Studies

The majority of the data used in the next two sections of this chapter comes from the Annex 57 project described in the introduction. The project team included participants from 19 countries, of which 6 were from outside Europe (Australia, Brazil (as an observer), China, Japan, Korea and the USA) and 13 were from within Europe (Austria, Czech Republic, Denmark, Finland, Germany, Italy, Sweden, the Netherlands, Norway, Portugal, Spain, Switzerland and the UK). Subtask 4 (ST4) of the Annex led the work on identification of design and construction strategies for the reduction of embodied carbon from buildings. The group was headed by the authors of this chapter, who work in and come from several European countries. The chosen method of ST4 was the collection and cross-case analysis of a large number of single-building case studies.

The use of such case studies to analyse the lifetime environmental impacts of a new building is common when developing policy advice and in informing industry initiatives. However, case studies are rarely definitive; that is to say, very few will follow the recommendations for full analysis as set out in the European TC350 standards (European Committee for Standardization 2011, 2012) and discussed in more detail later in the chapter, and even fewer are likely to be based on full and accurate input data for materials and components. There are in addition an almost infinite number of variations in construction methods, materials and building types and designs, with these variations accentuated across different geographical regions and cultures. This means that a single case study has little to offer of relevance for any building other than the one it is focused on. Any policy or industry initiatives which are based on outputs of single case studies are therefore unlikely to be valid for embodied carbon of buildings in general.

There is nevertheless considerable value to be found in a comparative analysis of a large collected set of case studies, such as that developed by the Annex 57 project. One important question is how large the set needs to be for adequate generalisations to be drawn. A big enough set will be one in which the most common variations in methodology can be identified and their impact on the results assessed. Once these impacts have been accounted for, the set of studies then needs to provide sufficient data for a number of valid and meaningful quantitative conclusions to be drawn. This data needs to provide enough replication of results for validity, and ideally to also cover variations of building types and designs for applicability to a range of projects.

The purpose of subtask 4 of the IEA Annex 57 was not therefore to calculate embodied carbon and energy of individual buildings to a high degree of accuracy, but to collect enough of such calculations to be able to draw useful conclusions about the relative embodied impacts of different design and construction choices, and thereby determine appropriate measures for reduction. The focus was therefore on 'the issues faced in practice by designers and policy makers' (Birgisdottir et al. 2016b) rather than on adding to the increasing body of case study information.

Annex 57 ST4 initially identified several key areas of interest and variation they wished to consider within the case studies. These included: the objective of the case study; the potential stakeholders who were likely to be interested in this type of case study; and whether the case study illustrated mitigation strategies, focused on significance of different life cycle stages, compared the impacts of methodology, compared the impacts of different components of materials, provided a national-level picture, or illustrated decision-making processes which led to reduction of embodied impacts. ST4 sent three repeated 'calls for case studies' (2013, 2014 and 2015) to all members of the Annex, asking for cases which identified or illustrated the specific defined aspects. All studies provided were based on published peer-reviewed articles or on postgraduate dissertations, and the details of these were included for each case study to provide the maximum transparency and data access. More detailed information on the case study collection and analysis method is given in Malmqvist et al. (2014) and Birgisdottir et al. (2016b).

The European case studies collected for this purpose are shown in Table 20.1; complete details of the case studies and the underpinning published research for each are given in Birgisdottir et al. (2016a). There were a total of 66 of these individual quantitative studies from Europe, including modelled variations of the same building. There were also 6 qualitative case studies from Europe, and an additional 12 quantitative case studies collected from Japan and Korea, which are not included in the analysis presented in this chapter. The European case studies collected provide a fair example of the range and quality of analyses being undertaken within different national contexts, at a building level, and within the academic context. These were then cross-analysed in order to identify the most

Table 20.1 Annex 57 European quantitative case studies

Case study	Database	RSP	Product stage			Construction process		Use stage					End of life				Next product	Main concept	Type
			A1 Raw material	A2 Transport	A3 Manufacturing	A4 Transport to building site	A5 Installation into building	B1 Use	B2 Maintenance	B3 Repair	B4 Replacement	B5 Refurbishment	C1 Deconstruction	C2 Transport to EoL	C3 Waste processing	C4 Disposal	D Reuse, recovery or recycling potential	Main concept	Type
AT1	Baubook eco2soft	100	X	X	X						X							New	Office
AT2	Baubook eco2soft	100	X	X	X						X							New	Residential
AT3	Baubook eco2soft	100	X	X	X						X							New	Office
AT4	EcoBat	60	X	X	X						X		X		X	X		Refurbishment	Residential
AT5	Baubook eco2soft	100	X	X	X						X							New	Residential
AT6	Ökobau 2009	50	X	X	X						X							New	Office
AT7	Ecoinvent 2.2	100	X	X	X						X				X	X		New	Residential
CH1	Ecoinvent 2.2	60	X	X	X						X				X	X		Refurbishment	School
CH2	Ecoinvent 2.2	60	X	X	X						X				X	X		Refurbishment	School
CH3	Ecoinvent 2.2	60	X	X	X						X				X	X		Refurbishment	School
CH4	Ecoinvent 2.2	60	X	X	X						X				X	X		Refurbishment	School
CH5	Ecoinvent 2.2	60	X	X	X						X				X	X		Refurbishment	School
CH6	Ecoinvent 2.2	60	X	X	X						X				X	X		New	School
CH7	Ecoinvent 2.2	60	X	X	X						X				X	X		New	School
CH8	Ecoinvent 2.2	60	X	X	X						X				X	X		Refurbishment	Residential
CH9	Ecoinvent 2.2	60	X	X	X						X				X	X		Refurbishment	Residential
CH10	Ecoinvent 2.2	60	X	X	X						X				X	X		New	Residential
CH11	Ecoinvent 2.2	60	X	X	X						X				X	X		Refurbishment	Residential
CH12	Ecoinvent 2.2	60	X	X	X						X				X	X		Refurbishment	Residential
CH13	Ecoinvent 2.2	60	X	X	X						X				X			Refurbishment	Residential
CH14	Ecoinvent 2.2	60	X	X	X						X			X		X		New	Residential
CH15	Ecoinvent 2.2	60	X	X	X						X			X		X		New	Residential
CZ1	Envimat	60	X	X	X													New	Residential
CZ2	Ecoinvent 2.2	100	X	X	X	X	X						X	X	X	X	X	–	Material
DE1	Ökobau 2011	50	X	X	X						X			X	X	X		New	School

(continued)

Table 20.1 (continued)

Case study	Database	RSP	Product stage			Construction process		Use stage					End of life				Next product	Main concept	Type
			A1 Raw material	A2 Transport	A3 Manufacturing	A4 Transport to building site	A5 Installation into building	B1 Use	B2 Maintenance	B3 Repair	B4 Replacement	B5 Refurbishment	C1 Deconstruction	C2 Transport to EoL	C3 Waste processing	C4 Disposal	D Reuse, recovery or recycling potential		
DE2	Ökobau 2009	50	X	X	X						X				X	X	X	New	School
DE3	Ökobau 2009	50	X	X	X						X				X	X	X	New	Residential
DE4	Ökobau 2009	50	X	X	X						X				X	X	X	New	Office
DK1	PE int	50	X	X	X						X				X	X	X	New	Office
DK2	PE int	50	X	X	X													New	Residential
DK3a	ESUOCO/Ökobau	150	X	X	X						X				X	X	X	New	Residential
DK3b	ESUOCO/Ökobau	150	X	X	X						X				X	X	X	New	Residential
DK3c	ESUOCO/Ökobau	50	X	X	X						X	X			X	X	X	New	Residential
DK3d	ESUOCO/Ökobau	50	X	X	X						X				X	X	X	New	Residential
DK3e	ESUOCO/Ökobau	50	X	X	X						X	X			X	X	X	New	Residential
DK4a	ESUOCO/Ökobau	50	X	X	X						X				X	X	X	New	Office
DK4b	ESUOCO/Ökobau	50	X	X	X						X				X	X	X	New	Office
DKc	ESUOCO/Ökobau	50	X	X	X						X				X	X	X	New	Office
DK4d	ESUOCO/Ökobau	50	X	X	X						X				X	X	X	New	Office
DK4e	ESUOCO/Ökobau	50	X	X	X						X				X	X	X	New	Office
DK4f	ESUOCO/Ökobau	50	X	X	X						X				X	X	X	New	Office
DK4g	ESUOCO/Ökobau	50	X	X	X						X			X	X	X	X	New	Office
IT1	Various	–	X	X	X	X	X		X				X	X	X	X	X	–	Material
IT2	Ecoinvent	50	X	X	X				X		X		X	X	X	X	X	New	Residential
IT2a	Ecoinvent	50	X	X	X				X		X		X	X	X	X	X	Refurbishment	Residential
IT3	Ecoinvent	70	X	X	X	X	X		X			X		X			X	New	Residential
IT4	(Not specified)	–	X	X	X	X												–	Material
N01	Ecoinvent	60	X	X	X						X							New	Residential
N02	Ecoinvent	60	X	X	X						X							New	Office
N04	EPD	60	X	X	X	X												New	Residential

ID	Data source	Number																Phase	Sector
NO8	Ecoinvent	60	X	X	X						X							Refurbishment	Office
NO9	Ecoinvent	60	X	X	X						X							New	Residential
SE1	Ecoinvent	1	X	X	X				X	X	X	X						–	Sector
SE2	Swedish IO data	50	X	X	X	X	X		X									New	Residential
SE2b	Ecoinvent, BECE	50	X	X	X													New	Residential
SE3	Ecoinvent, BECE	50	X	X	X													New	Residential
SE4	Ecoinvent	50	X	X	X													New	Residential
SE5	Ecoinvent	50	X	X	X													New	Office
SE6	EPD, Ökobau 2013, Ecoinvent, KBOB	1									X							Refurbishment	Office
SE7	Ecoinvent, KBOB,	50	X	X	X	X	X			X		X	X	X	X	X		New	Residential
UK2	Bath ice, ECEB	N/A	X	X	X	X	X						X	X				Refurbishment	Residential
UK4	Bath ice, ECEB	68	X	X	X	X	X			X	X	X	X	X	X	X		New	School
UK5	USLCI	20	X	X	X	X	X											New	Residential
UK7	Bath ice	60	X	X	X	X	X		X				X	X	X	X	X	New	Sports hall
UK9	EPD, ELCD, industry	–		X	X	X	X						X	X	X	X	X	New	Residential
UK12	Guide to	60	X	X	X	X	X		X	X	X	X	X	X	X	X	X	Refurbishment	Residential
	Total number		65	65	65	13	11	0	7	5	49	7	11	13	41	44	22		

Adapted from Birgisdottir et al. (2016a)

common variations in methodological approaches, and then the energy and green-house gas impacts of buildings of different typologies and materials, for individual life cycle stages and for the whole life and for different building components.

Measurement of Embodied Carbon and Energy in the Annex 57 European Case Studies

Embodied carbon and energy have been calculated for several decades and across all continents (Treloar 1998; Thormark 2002; Hammond and Jones 2008; Dixit et al. 2010; Chang et al. 2012; Moncaster and Song 2012). However, a number of authors (Dixit et al. 2012; Ibn-Mohammed et al. 2013; Pomponi and Moncaster 2016; Anand and Amor 2017) have identified that there are still multiple variations in methodological approach and that these still continue despite the publication of international and European standards. Greater harmonisation therefore has been, and remains, a key focus not just of Annex 57 (Lützkendorf et al. 2016) but also of a number of ongoing projects around Europe, as reported by Frischknecht et al. (2015).

Comparison between studies following disparate methodologies is also made additionally difficult, due to the frequent lack of transparency about the chosen methodological aspects and about the sources for the data used (Optis and Wild 2010; Dixit et al. 2012). These were therefore two important issues that needed to be addressed by Annex57 ST4 before meaningful conclusions could be drawn from the collection of case studies.

Before analysing the results, ST4 carried out two preliminary exercises. The first was the development of a template into which case study information could be transcribed, in order to show clear and comparable information about each while encouraging transparency and completeness of quantitative data. This ensured that the relevant methodological and data source information about each study used was known and stated.

The second stage was to identify the differences in approach across the case studies, in order to ensure that like was compared with like. This stage identified seven common variations in the approaches taken by the Annex 57 case studies, as:

- 'Purpose of study'
- 'Reference study period for the building'
- 'System boundaries'
- 'Scenarios for the building'
- 'Inventory level of detail'
- 'Background data'
- 'Performance indicator'

(Birgisdottir et al. 2016b)

The range of differences in approach found in the Annex 57 case studies is an important indication of the general range in published studies. As can be seen from Table 20.1, there were also some clear areas of omission in the life cycle stages considered by the case studies, with the majority focusing on the product (A1–3) stage and the replacement (B4) of those products during the reference study period.

The focus on stages A1–3 reflects widespread practice in published calculations of embodied carbon and energy. Figure 19.2 plots both the number of studies which included each life cycle stage within the 66 European Annex 57 case studies as shown in Table 20.1 (Birgisdottir et al. 2016a) and the numbers of studies which included each life cycle stage within the 102 case study publications reviewed by Pomponi and Moncaster (2016). While there are aspects in common between the two studies, Fig. 19.2 also shows that the calculation of the replacement (B4) stage was considerably more common in the Annex 57 European cases than in the review of international papers. This is discussed further later in this chapter. Stages A1–A3 were calculated by all but one of the Annex 57 studies, but only by 70% of the international studies, and C3 and C4 were also substantially more common in the Annex 57 studies, while A4 and A5 were more commonly studied by the international study.

Allowing for the multiple differences of approach, and concentrating on the life cycle stages for which there was the most information, Birgisdottir et al. (2016b) published a number of figures showing the wide range of embodied energy and greenhouse gas impacts of multiple case studies. The maximum and minimum values are given in Table 20.2. While the range of values conflate differences due to the building design and purpose with those due to the methodological approach taken by the modeller, further analysis in identifying the impacts of the latter made it possible to identify some general trends. These have been used to develop the mitigation strategies discussed in the next section.

Where calculated, the impact of replacing building components during the life of the building (stage B4) was as much as half the initial cradle-to-gate impact. This was particularly clear in cases where low-carbon energy technologies were installed

Table 20.2 Max and min values for A1–3 and B4 for Annex 57 case studies

		Min EC	Max EC	Min EE	Max EE
		kgCO$_2$e/m^2	kgCO$_2$e/m^2	MJ^a/m^2	MJ^a/m^2
New projects	(Figs. 9 and 10)	−10 (AT5)[b]	600 (CH6, NO9)	25 (AT5)[b]	13,000
A1–3 cradle to gate					(DE4)
New projects	(Figs. 11 and 12)	15 (DK3e)	300 (UK4)	300	4000
B4 replacements				(DK4d)	(UK4)
Refurb projects	(Figs. 9 and 10)	64 (AT4)	378 (CH13)	2350	4,670
A1–3 cradle to gate				(AT4)	(CH4)

Birgisdottir et al. (2016a, b)
[a]1 MJ=0.278 kWh approx
[b]This was a cross-laminated timber (CLT) building and the calculation assumed carbon sequestration

and where their whole-life impact (allowing for replacement during the building lifetime and end-of-life disposal) was calculated. These technologies are installed to reduce operational carbon, but it is important that their often significant embodied impacts are also calculated in order to give a fair estimate of the whole-life impact. Standard building services and fixtures and fittings also have a high impact on the B4 stage as they need replacing during the life of the building.

The extent of European construction works which are focused on refurbishment was reflected in the high numbers of case studies considering the impact of these projects within the Annex 57 studies, from six of the nine European countries who between them submitted 16 studies specifically looking at the impact of refurbishment projects. While none of the studies directly compared the impact of refurbishment with demolition and rebuild, Malmqvist et al. (2018) show that the 'product stage' figures for refurbishment projects are substantially lower than those for new build. Where these buildings are instead demolished in order to be replaced with new buildings, they will incur considerable impacts since they are often built robustly and with heavy materials; usually these impacts are not included in the analysis of the new buildings. It can be concluded therefore that within the European context, and dependent on the operational efficiency that can be achieved by refurbishment, considerable carbon savings are likely to be made through a strategy of refurbishment in preference to new build (Baker and Moncaster 2018).

The analyses also considered the relative impact of different construction materials. Concrete made the highest contribution to embodied energy and greenhouse gases in many case studies, but Birgisdottir et al. (2016b) note that concrete is often used in large quantities in foundations even of timber buildings. Steel and aluminium also have high impacts; however, in this case the report notes that the positive impacts which may be made by recycling are not accounted for in the A–C stages, but in the D stage, which is seldom included because it concerns the benefits of subsequent life cycles. Therefore, while there is strong evidence that timber buildings have a lower impact of embodied energy and greenhouse gases compared with other structural solutions (see next section), the report urges caution in the use of the exact figures calculated.

Of particular importance to note within the collection of case studies was the increasing focus on 'nearly zero-energy' or 'nearly zero-emission' buildings, particularly in more Northern European countries where heating loads tend to be high (Sartori et al. 2012). For zero-emission buildings in particular, the operational emissions can be balanced by the addition of low- or zero-carbon technologies to the building and by savings from the embodied emissions. Several of the cases collected were of buildings designed to have nearly zero emissions from operational energy and in some cases including embodied emissions for materials. As operational energy standards improve and resulting carbon emissions decrease, both the actual and the relative embodied carbon will increase. As over 50% of the total embodied impacts generally occur up to the end of construction (Moncaster and Symons 2013), this is likely to have the impact of increasing current emissions from construction in Europe. However, there were also several studies which considered the replacement of concrete or steel frames with timber structures as a way of minimising embodied impacts; if this trend increases, the current construction emissions will fall as a result. In this case, it will be important to support the European timber industry towards sustainable growth.

Approaches to Mitigation: Reducing Embodied Impacts of Buildings

There are a wide number of mitigation strategies proposed in the international literature. In their systematic review, Pomponi and Moncaster (2016) identify 17 separate strategies ranging from the specific ('use of materials with lower embodied energy and carbon') to the aspirational ('better design') to the strategic ('policy and regulations'). The range demonstrates that different strategies will be of relevance to, and actionable by, different stakeholders. Of particular note is that 19 papers suggest refurbishment as a mitigation strategy, although two papers suggest instead the alternative of demolition and rebuild.

The analysis of the Annex 57 case studies meanwhile also focused on strategies which could reduce embodied impacts compared with conventional buildings, specifically considering those that were actionable by designers and constructors. These are first divided into three main areas: substitution of materials, reduction of resource use, and reduction of construction stage impacts (p. 45, Birgisdottir et al. 2016b) which together include 17 sub-strategies (Malmqvist et al. 2018).

Table 20.3 maps the strategies identified by the two studies against each other, collating the individual strategies which fit into each of the three main areas, plus the additional strategies identified in Pomponi and Moncaster which relate to policy and management rather than design and construction. It is important to note that the two studies each collated and analysed information from a large number of independently published case studies.

As can be seen, there is considerable commonality between the two studies. Both Pomponi and Moncaster (2016) and the Annex 57 case studies show a strong focus on the first category, 'Substitution of materials'. In Pomponi and Moncaster, 'Use of materials with lower embodied energy and carbon' is the strategy most frequently identified, by almost half of the papers reviewed. Within the Annex 57 analysis, this strategy is further divided into four sub-strategies, from which the clearest evidence identifies reductions when replacing a reinforced concrete or steel frame with timber. Timber is grown across much of Europe and has been a traditional building material in the continent for centuries, with many timber-framed buildings still standing after several hundred years (see the examples discussed by Harris 1993). Modern timber structures occasionally still use large-section solid timber beams and columns but more often use smaller timber studs and increasingly 'glulam' beams and 'cross-laminated timber' (CLT) panels, sometimes with additional steel elements (Darby 2013). Traditional timber buildings

Table 20.3 Comparison of embodied carbon mitigation strategies identified by Annex 57 (Malmqvist et al. 2018) and by the systematic review (Pomponi and Moncaster 2016)

Malmqvist et al. (2018)	Pomponi and Moncaster (2016)
Substitution of materials	
Using timber structures	Use of materials with lower embodied energy and carbon
Using other 'natural' materials	
Using new, innovative materials/components	
Lightweight construction	
Reduction of resource use through building lifetime	
Flexible/adaptable design	Better design
Design for disassembly	
Design for low maintenance	
Optimising building form and design of layout plan	
Reusing building structures	Refurbishment of existing buildings instead of new built
Using recycled and reused materials/components	Inclusion of waste, by-product, and used materials into building materials
Building/service life extension	Extending the building's life
Reduction of construction/end-of-life stage impacts	
Minimising on-site waste material	Reduction, reuse, and recovery of EE/EC-intensive construction materials
Optimising energy use in construction process	More efficient construction processes/techniques
	Increased use of prefabricated elements/off-site manufacturing
Waste recycling in end of life	Increased use of local materials
Reducing end-of-life impact	
Policy and management	Policy and regulations (governments)
	Policy and regulations (construction sector)
	Carbon mitigation offsets, emissions trading, and carbon tax
	Carbon sequestration
	Decarbonisation of energy supply/grid
	Tools, methods, and methodologies
	People-driven change (key role of all stakeholders in the built environment)
	Demolition and rebuild

were limited to two or three storeys, but the Annex 57 case studies demonstrate that increasingly high-rise buildings are now possible using CLT; Annex 57 case study UK9, for example, is of an eight-storey building in London. The Annex 57 studies suggest that up to 77% savings in embodied greenhouse gas can be made by replacing a concrete or steel-framed building with timber and up to 24% savings by replacing a brick façade with timber (Malmqvist et al. 2018).

The Annex 57 strategies which consider 'Reduction of resource use through building lifetime' include several specific design approaches, including optimising the building form, design for flexibility, design for low maintenance and design for disassembly. Those that consider the impact of the later life cycle stages of the building are particularly important for the European context, where buildings often last for many decades and even centuries. These are combined by the Pomponi and Moncaster analysis of the international papers into a single sub-strategy of 'better design'. For the other sub-strategies in this category, there is a clear relationship between the two studies, with each including reuse of structures and buildings, recycling of materials and extension of the building life.

Within the category of 'Reduction of construction/end-of-life stage impacts', the sub-strategies found by the two studies map less closely onto each other; nevertheless, the reduction of construction stage impacts is generally reflected through minimisation of energy use and site waste, with similar strategies during deconstruction including a focus on waste recycling, bringing the focus full circle back to low-impact design with the ultimate aim of moving towards a genuinely circular economy. However, one strategy, 'Increased use of local materials', which was mentioned by 13 of the Pomponi and Moncaster international papers, does not appear in the Annex 57 study (Malmqvist et al. 2018). While the impact of transport to site is mentioned in a couple of the studies, there are no suggestions that use of local materials would reduce this. It is possible that the use of local materials depends on a local labour force, on labour being relatively cheap compared with materials and on lighter regulations which support small-scale operations. Therefore, the lack of focus on the use of local building materials as a mitigation strategy within the European Annex 57 cases may reflect again both the longevity of the buildings and the maturity of the construction sector in most European countries.

European Approaches to the Management of Embodied Impacts

While the analysis of the Annex 57 case studies focused mainly on the design and construction mitigation strategies, rather than on policy and management approaches, a number of such approaches across Europe were identified and discussed within the project. Chapter 5 of Birgisdottir et al. (2016b) divided these into deliberate interventions, such as policy instruments to reduce embodied carbon, and approaches that had unintentional impacts on embodied carbon such as construction cost and time savings. They were further divided by the contextual scale at which they acted, including international, national, regional and project level.

This section focuses on a few of these, discussing first the prevalence and impact of national standards and regulations and second the effect of professional initiatives, giving a specific example of an environmental assessment tool, BREEAM.

Standards and Regulations

At the international level, the publication of the 'TC350' standards by the European Technical Committee CEN/TC350 in 2011 and 2012 has provided a coherent framework for the calculation of whole-life impacts of buildings, marking a key defining moment for embodied carbon calculations in Europe. While non-mandatory, the advent of the standards EN 15978 and EN 15804 has been a significant step forward in the harmonisation of tools and methods. However, as discussed previously, there remain varying interpretations and applications, such as the sourcing of background data and the setting of assessment boundaries. These variations can lead to divergent results when comparing outputs from different compliant methods and can result in misguided decisions being taken on specification and life cycle impacts. For this reason, it is critical that the same methodological approach, product category rules (PCR) and data are used wherever direct comparisons are made or where outputs are aggregated.

The CEN standards define the life cycle of a building through a process-based method, rather than the input-output or hybrid approaches commonly used in many other regions (see other chapters in this section of the book). A process-based LCA is effectively a 'bottom-up' approach which understands buildings as a collection of products and processes which each have lifetime impacts. The standards define these into five main life cycle stages – product (A1–3), construction process (A4–5), use (B1–7), and end-of-life (C1–4) stages, followed by a stage beyond the building life cycle (D). Consideration of the embodied impacts is encouraged alongside the operational impacts (covered by B6–7) and at a building level.

While process-based environmental analyses have been carried out on buildings for many years before the new standards were published, the standards systematically set out the method for whole-life calculations. The TC350 standards also demonstrated that a whole-life calculation is necessary in order to understand the full life cycle impacts of a design or operation decision. As shown earlier in Fig. 20.1, few analyses in

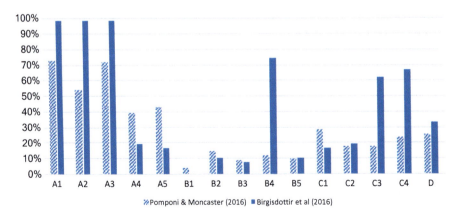

Fig. 20.1 Percentage of reviewed case studies including different life cycle stages, comparing Pomponi and Moncaster (2016) with Birgisdottir et al. (2016a, b)

practice do this. In some cases, the choice of life cycle stages calculated depends on the purpose of the assessment; for instance, where the purpose is to consider alternative structural materials, the assessment may focus on the relevant temporal and physical boundaries of the product stage only and the structural material only. In other cases, life cycle stages are omitted because of a lack of knowledge about the impacts. Indeed, one criticism of the standards is the level of detail included. Even for academic research, in which efforts are made to collect fine-grained information about a particular building, it is almost impossible to extend the analysis beyond the end of construction with any degree of accuracy, as there is considerable uncertainty about the future life of the building. For simple calculations at an early design stage, it is therefore necessary to adapt the standards and use a number of assumptions (Moncaster and Symons 2013).

At the national level, government building regulations are the obvious route for imposing measurement and then reduction of embodied impacts. However, a survey of the Annex 57 members, reproduced as Fig. 3, suggests that only the Netherlands currently requires measurement, and no country yet requires reduction of embodied impacts, in contrast to the focus on operational energy and carbon reduction (Birgisdottir et al. 2016b) (Fig. 20.2).

Questions: These have been slightly reworded for compactness.	Australia	Austria	Brazil	Czech Republic	Denmark	Finland	Germany	Italy	Japan	Republic of Korea	Netherlands	Norway	Spain	Sweden	Switzerland	UK
Do building regulations include embodied emissions?	x	x	x	x	x	x	x	x	x	x	~	x	x	x	~	x
Are there different requirements for domestic and non-domestic buildings?	✓	✓	x		x	✓	~	✓	✓	x	x	✓	✓	x	~	✓
Are there sustainability certifications specific to your country?	✓	✓	✓	✓	✓	x	✓	✓	✓	✓	~	✓	✓	✓	✓	✓
Do they include embodied emissions?	~	x	x	✓	✓		✓	x	✓	✓	✓	✓		x	✓	✓
Do other voluntary initiatives exist to measure embodied emissions?	✓	✓	✓	✓	✓	✓	✓		~	✓	✓	✓	✓	✓	✓	✓
Is there a construction LCA database for your country?	✓	✓	x	✓	x	~	✓	x	✓	✓	✓	✓	✓	x	✓	✓
Are there (LCA) tools to calculate embodied emissions in your country?	✓	✓	x	✓	✓	✓	✓	x	✓	✓	✓	✓	✓	✓	✓	✓
Are there any on-going initiatives to develop LCA tools?	✓	✓	x	x	✓	✓	✓	x	x	✓	✓	✓	✓	✓	✓	✓
Is it common for construction products to have EPDs?	~	~	x	~	~	✓	✓	~	x	x	x	~	x	x	~	~
Is there an EPD database for your country?	~	✓	x	✓	✓	x	✓	x	✓	✓	~	✓	✓	✓	~	✓
Are there any on-going initiatives to develop national databases?	✓	~	✓	✓	x	✓	✓	~	x	✓	✓	✓	✓	~	✓	✓

KEY: Positive answer ✓
Negative answer x
Ambiguous/complex answer ~
Question not answered (blank)

Fig. 20.2 Responses to the Annex 57 'Venice questionnaire' (Reproduced from Birgisdottir et al. 2016b, Fig. 28)

Professional Initiatives as Drivers

Several individual industry consultants have already developed their own tools to calculate life cycle impacts of buildings (Moncaster and Song 2012; De Wolf et al. 2017). However, many of these are commercially confidential and produce diverse results (Pomponi and Moncaster 2018). There have therefore been a number of recent initiatives in different countries to address these disparities, with industry consultants and academics working together to create greater harmonisation of approach; one of the aims of Annex 57 too was to create guidelines for the harmonisation of calculations (Seo et al. 2016).

In 2010, the UK government-funded body Innovate UK funded a number of projects to develop new design tools for sustainable buildings, several of which incorporated embodied impacts. A later funding stream looked at how these could be integrated into practice; one of these recent projects has worked with the Royal Institution of Chartered Surveyors (RICS) to write a professional statement – a mandatory code of practice for its multinational membership – on 'Whole life carbon measurement: implementation in the built environment' (RICS 2017). At the time of writing, this has only just been published, and so the impact on industry behaviour and the management of embodied carbon is as yet untested.

There are also a number of non-mandatory certification schemes in widespread use, such as the DGNB certification originating from Germany and also in common use in Austria and Denmark and GreenCalc from the Dutch Green Building Council. These encourage better environmental outcomes and best practice sustainable design against which the performance of buildings can be measured and independently certified.

One example of such schemes is offered by BREEAM, the environmental assessment tool from the UK Building Research Establishment, which is the most widely used environmental assessment method for buildings in the UK and is also used across Europe and other areas of the world. First launched in 1990, BREEAM's coverage of materials issues has evolved as awareness and availability of tools and data on impacts have increased within the industry and in order to keep abreast of current best practice. The BREEAM New Construction (NC) scheme currently assesses buildings for their performance across several sections: management, health and wellbeing, energy, transport, water, materials, waste, land use and ecology, pollution and innovation. Embodied carbon is primarily addressed within the Materials section, although there are also relevant aspects within the Waste section.

The scheme currently promotes more sustainable materials selection in several ways. Firstly, BREEAM NC takes account of embodied carbon through the use of LCA-based approaches to materials evaluation as part of its wider consideration of environmental impacts. It does this by awarding credits for the evaluation of separate, well-defined building elements, using the *Green Guide to Specification*. In this case, extra credits are awarded where materials specified are covered by independently verified EPDs. It also offers additional, 'exemplary-level' credits, which can be awarded for conducting a full-building LCA and using LCA-compliant software tools to measure the environmental impact of a building. Secondly, from a product and construction perspective, BREEAM NC promotes resource efficiency through encouraging the use of recycled aggregates and consequently the reduction of the demand for virgin material and through encouraging the effective management and reduction of construction waste. Thirdly, BREEAM NC encourages functional adaptability and design for durability and resilience, which aims to increase the lifetime of building elements, through the selection of durable materials and early consideration of the future needs of the building occupants. The accurate specification of material finishes by the building occupant also helps to avoid the unnecessary waste of materials.

The forthcoming versions of BREEAM will move away from the elemental approach towards a whole-building life cycle approach, following the example of LEED and DGNB. The new focus will be increasingly on encouraging designers and procurers to make decisions on the basis of robust and credible environmental LCA data through the use of EN 15804-compliant EPD and LCA studies regardless of provider. While it is recognised that there are still significant variations in the underpinning data, boundary assumptions and methodologies adopted by different EPD providers, there seems to be value in encouraging product and material manufacturers to publish credible data on their products. The forthcoming schemes will also encourage designers to use robust whole-building LCA tools in order to quantify and reduce life cycle impacts, thereby encouraging the use and development of such tools; BREEAM will differentiate between the use of simple/reduced scope tools and more accurate holistic tools.

It is hoped that the increased use of LCA and its documentation through BREEAM certification will further allow the benchmarking of building life cycle impacts due to materials specification to compare with local best practice. The use of comparable data and methodologies will be critical to the credibility of this benchmarking. BREEAM therefore aims to use peer-reviewed generic data which is regionally relevant and taken from a consistent data source, reward the refinement of whole-building LCAs through the use of proprietary data taken from third-party EPDs and encourage the sharing of LCA data to inform the future maintenance, and improvement, of the benchmarks.

Summary and Conclusions

This chapter has considered the current picture within Europe for measuring and reducing the embodied impacts of buildings. While this is an issue which has been considered for many years in this region, with increasing numbers of case studies published of individual buildings, there is clearly still a wide range of methodological approaches meaning that general reduction measures have been difficult to identify. The publications of the CEN TC350 standards in 2011 and 2012 have helped to develop a more harmonised approach to the measurement and identification of life cycle stages. Environmental certification schemes meanwhile are increasingly encouraging European designers to use LCA to measure and reduce embodied impacts of buildings. The combined approach of European-wide standards and national and international certification schemes is likely to mean that the embodied impacts of increasing numbers of new buildings are measured. This should produce data for benchmarking and lead to a rapid step change in design and construction approaches.

The work of Annex 57 has used a cross-case analysis of a substantial collection of case studies to clarify the most common approaches to measurement and mitigation of embodied energy and carbon, and to identify areas and life cycle stages for which further research is needed. The analyses suggest that substitution of materials is the clearest route to substantial embodied carbon reduction, including where appropriate replacing steel and concrete frames with timber structures and greater use of lightweight and recycled components and materials. There is considerable overlap with a systematic review of over 100 published case studies carried out by Pomponi and Moncaster (2016); the few differences reflect the fact that there are a high proportion of old buildings within the European building stock and that refurbishment and adaptation of existing buildings therefore account for a significant proportion of construction sector impacts. A key issue for Europe is the focus on brownfield development and the pressure to demolish existing buildings. The whole impact of demolition and replacement is seldom, to date, considered in industry practice, in spite of this being identified in the academic literature as a potential mitigation strategy. This area is one on which industry and academia should work together.

References

Aktas, C. B., & Bilec, M. M. (2012). Impact of lifetime on U.S. residential building LCA results. *The International Journal of Life Cycle Assessment, 17*, 337–349.

Anand, C. K., & Amor, B. (2017). Recent developments, future challenges and new research directions in LCA of buildings: A critical review. *Renewable and Sustainable Energy Reviews, 67*, 408–416.

Andresen, I. (2017). Towards zero energy and zero emission buildings – Definitions, concepts and strategies. *Current Sustainable/Renewable Energy Reports, 4*, 63–71.

Baker, H. E., & Moncaster, A. M. (2018). *Embodied carbon and the decision to demolish or adapt.* ZEMCH (Zero Energy Mass Custom Home) International Conference. Melbourne, Australia.

Baker, H. E., Moncaster, A. M., & Al Tabbaa, A. (2017). *The decision to demolish or adapt existing buildings on brownfield sites.* Proceedings of the Institution of Civil Engineers – Forensic Engineering, special issue: Forensic engineering in urban renovation, 170, Published online ahead of print 8th March 2017.

Ballarini, I., Corgnati, S. P., & Corrado, V. (2014). Use of reference buildings to assess the energy saving potentials of the residential building stock: The experience of TABULA project. *Energy Policy, 68*, 273–284.

Birgisdottir, H., Houlihan Wiberg, A., Malmqvist, T., Moncaster, A., & Nygaard Rasmussen, F. (2016a). IEA EBC Annex 57 ST4 Case study collection report ISBN: 978-4-909107-09-1.

Birgisdottir, H., Houlihan Wiberg, A., Malmqvist, T., Moncaster, A., & Nygaard Rasmussen, F. (2016b). Recommendations for reduction of embodied carbon from buildings. Report of Subtask 4 of IEA Annex 57: Evaluation of Embodied Energy & Embodied GHG Emissions for Building Construction. ISBN: 978-4-909107-08-4.

Birgisdottir, H., Moncaster, A., Wiberg, A. H., Chae, C., Yokoyama, K., Balouktsi, M., Seo, S., Oka, T., Lützkendorf, T., & Malmqvist, T. (2017). IEA EBC annex 57 'evaluation of embodied energy and CO2eq for building construction'. *Energy and Buildings, 154*, 72–80.

Bordass, B., Leaman, A., & Ruyssevelt, P. (2001). Assessing building performance in use 5: Conclusions and implications. *Building Research and Information, 29*, 144–157.

Chang, Y., Ries, R. J., & Lei, S. (2012). The embodied energy and emissions of a high-rise education building: A quantification using process-based hybrid life cycle inventory model. *Energy and Buildings, 55*, 790–798.

Darby, H. (2013). *A case study to investigate the life cycle carbon emissions and carbon storage capacity of a cross-laminated timber multi-storey residential building.* SB13, 24–26 2013 Munich, Germany.

Department for Communities and Local Government. (2017). *English Housing Survey, 2016: Housing Stock Data. [data collection]* (4th ed.). UK Data Service.

De Wolf, C., Pomponi, F., & Moncaster, A. M., (2017). Current industry practice in embodied carbon calculation. *Energy and Buildings, 140*, 68–80.

Dixit, M. K., Fernández-Solís, J. L., Lavy, S., & Culp, C. H. (2010). Identification of parameters for embodied energy measurement: A literature review. *Energy and Buildings, 42*, 1238–1247.

Dixit, M., Fernández-Solís, J., Lavy, S., & Culp, C. (2012). Need for an embodied energy measurement protocol for buildings: A review paper. *Renewable and Sustainable Energy Reviews, 16*, 3730–3743.

European Commission. (2010). Directive 2010/31/EU of the European Parliament and of the Council of 19 May 2010 on the energy performance of buildings.

European Committee for Standardization. (2011). *EN 15978 Sustainability of construction works—Assessment of environmental performance of buildings—Calculation method.* Brussels: European Committee for Standardization.

European Committee for Standardization. (2012). *EN 15804 Sustainability of construction works. Environmental product declarations. Core rules for the product category of construction products.* Brussels: European Committee for Standardization.

Frischknecht, R., Wyss, F., Knöpfel, S. B. S., & Stolz, P. (2015). Life cycle assessment in the building sector: Analytical tools, environmental information and labels, 57th LCA forum, Swiss Federal Institute of Technology, Zurich, December 2, 2014. *International Journal of Life Cycle Assessment, 20*, 421–425.

Hammond, G. P., & Jones, C. I. (2008). Embodied energy and carbon in construction materials. *Engineering Sustainability: Proceedings of the Institution of Civil Engineers, 161*, 87–98.

Harris, R. (1993). *Discovering timber-framed buildings.* Princes Risborough: Shire Publications.

Ibn-Mohammed, T., Greenough, R., Taylor, S., Ozawa-Meida, L., & Acquaye, A. (2013). Operational vs. embodied emissions in buildings—A review of current trends. *Energy and Buildings, 66*, 232–245.

International Energy Agency. (2016). *Annex 57 evaluation of embodied energy & carbon dioxide emissions for building construction [Online].* Available: http://www.iea-ebc.org/projects/completed-projects/ebc-annex-57/

Lowe, R. (2007). Addressing the challenges of climate change for the built environment. *Building Research and Information, 35*, 343–350.

Lützkendorf, T., Balouktsi, M., & Frischknecht, R. (2016). International energy agency: Evaluation of embodied energy and CO2eq for building construction (Annex 57) Subtask 1: Basics, Actors and Concepts.

Malmqvist, T., Birgisdottir, H., Houlihan Wiberg, A., Moncaster, A., Brown, N., John, V., Passer, A., Potting, A., & Soulti, E. (2014). *Design strategies for low embodied energy and carbon in buildings: Analyses of the IEA Annex 57 case studies.* World Sustainable Building Conference WSB14. Barcelona.

Malmqvist, T., Nehasilova, M., Moncaster, A., Birgisdottir, H., Rasmussen, F. N., Wiberg, A. H., & Potting, J. (2018). Design and construction strategies for reducing embodied impacts from buildings – Case study analysis. *Energy and Buildings.* Accepted 14 January 2018.

Mata, É., Sasic Kalagasidis, A., & Johnsson, F. (2014). Building-stock aggregation through archetype buildings: France, Germany, Spain and the UK. *Building and Environment, 81,* 270–282.

Moncaster, A. M., & Song, J.-Y. (2012). A comparative review of existing data and methodologies for calculating embodied energy and carbon of buildings. *International Journal of Sustainable Building Technology and Urban Development, 3,* 26.

Moncaster, A. M., & Symons, K. E. (2013). A method and tool for 'cradle to grave' embodied energy and carbon impacts of UK buildings in compliance with the new TC350 standards. *Energy and Buildings, 66,* 514–523.

Optis, M., & Wild, P. (2010). Inadequate documentation in published life cycle energy reports on buildings. *International Journal of Life Cycle Assessment, 15,* 644–651.

Pomponi, F., & Moncaster, A. M. (2016). Embodied carbon mitigation and reduction in the built environment: The evidence. *Journal of Environmental Management, 181,* 687–700.

Pomponi, F., & Moncaster, A. M. (2018). Scrutinising embodied carbon in buildings: The next (emission) gap made manifest. *Renewable and Sustainable Energy Reviews, Part 2, 81,* 2431–2442.

RICS. (2017). *Whole life carbon assessment for the built environment* (1st ed.). London, UK: RICS.

Sartori, I., Napolitano, A., & Voss, K. (2012). Net zero energy buildings: A consistent definition framework. *Energy and Buildings, 48,* 220–232.

Seo, S., Hajek, P., Birgisdottir, H., Nygaard Rasmussen, F., Passer, A., Chae, C.-U., Malmqvist, T., Houlihan Wiberg, H., Mistretta, M., Luetzkendorf, T., Balouktsi, M., Moncaster, A., Yokoyama, K., Yokoo, N., & Oka, T. (2016). Evaluation of embodied energy and CO2eq for building construction. Summary Report of Annex 57 to the International Energy Agency EBC Programme, ISBN: 978-4-909107-11-4.

Summerfield, A. J., Oreszczyn, T., Palmer, J., Hamilton, I. G., & Lowe, R. J. (2015). Comparison of empirical and modelled energy performance across age-bands of three-bedroom dwellings in the UK. *Energy and Buildings, 109,* 328–333.

Sunikka-Blank, M., & Galvin, R. (2012). Introducing the prebound effect: The gap between performance and actual energy consumption. *Building Research and Information, 40,* 260–273.

Thormark, C. (2002). A low energy building in a life cycle—Its embodied energy, energy need for operation and recycling potential. *Building and Environment, 37*(37), 429–435.

Treloar, G. J. (1998). *A comprehensive embodied energy analysis framework.* Melbourne: Deakin University.

Uihlein, A., & Eder, P. (2010). Policy options towards an energy efficient residential building stock in the EU-27. *Energy and Buildings, 42,* 791–798.

Yand, W., & Kohler, N. (2008). Simulation of the evolution of the Chinese building and infrastructure stock. *Building Research and Information, 36,* 1–19.

Yokoo, N., Oka, T., Yokoyama, K., Sawachi, T., & Yamamoto, M. (2015). Comparison of embodied energy/CO2 of office buildings in China and Japan. *Journal of Civil Engineering and Architecture, 9,* 300–307.

Zöld-Zs, A., & Szalay, Z. (2005). What is missing from the concept of the new European building directive? *Building and Environment, 42,* 1761–1769.

Chapter 21
Initiatives to Report and Reduce Embodied Carbon in North American Buildings

Catherine De Wolf, K. Simonen, and J. Ochsendorf

Introduction

Embodied carbon emissions during material extraction, production, transportation, building construction, and demolition are irreversible. Because of the threat associated with these emissions, targets have been set globally, including in North America, to keep global temperatures from rising above 2 °C.[1] To meet these targets, the Intergovernmental Panel on Climate Change (IPCC 2014) states that the building sector must be zero carbon by 2050. This goal is highly complex because carbon emissions include both embodied carbon as well as operational carbon from heating, cooling, ventilating, and lighting buildings. While standards and technologies exist to lower operational carbon, global and North American strategies to lower embodied carbon are critically lacking.

As demonstrated, this gap has primarily arisen due to a lack of aligned data and the unavailability of benchmarks that accurately assess the whole life cycle impacts of North American buildings. As a result, some firms have developed their own set of in-house tools to assess the embodied carbon of their projects (Thornton

---

[1]On June 1, 2017, the US president announced that the USA would withdraw from the 2015 Paris Agreement on climate change mitigation abiding the 4-year exit process. This decision is strongly criticized.

C. De Wolf (✉)
School of Architecture, Civil and Environmental Engineering, Ecole Polytechnique Fédérale de Lausanne (EPFL), Lausanne, Switzerland
e-mail: dewolf.catherine@gmail.com

K. Simonen
Department of Architecture, University of Washington, Seattle, WA, USA

J. Ochsendorf
Department of Architecture and Civil and Environmental Engineering, Massachusetts Institute of Technology (MIT), Cambridge, MA, USA

© Springer International Publishing AG 2018
F. Pomponi et al. (eds.), *Embodied Carbon in Buildings*,
https://doi.org/10.1007/978-3-319-72796-7_21

463

Tomasetti 2016; SOM 2013), which are not comparable across firms. Therefore, leading engineering firms (Kaethner and Burridge 2012; Yang 2014; SE 2050 2017) have called for a uniform standardized method for assessing embodied carbon.

This chapter presents data collection methods that allow the creation of new benchmarks that will hopefully motivate designers to assess and lower the environmental impact of their North American buildings. A lack of consistent data is not only responsible for the challenge of including embodied carbon in rating schemes, but it is also responsible for the scarcity of available benchmarks.

First, we review available life cycle assessment (LCA) datasets and tools in the region. The first part of this chapter indeed describes the available data in North America in terms of environmental product declarations (EPDs), LCA databases, and whole-building LCA tools.

Second, North American green building rating schemes are presented as potential incentives to reduce the environmental impacts of the construction industry. The second part of this chapter therefore evaluates the state of the art of embodied carbon credits in current rating schemes commonly used in North America. The Leadership in Energy and Environmental Design (LEED) rating scheme established by the US Green Building Council (USGBC) is widely used and currently constitutes a primary driver for the use of LCA data and methods by the building industry (USGBC 2016).

Third, we propose two consistent embodied carbon benchmark databases for buildings and building structures in North America. Hence, the third part of this chapter illustrates two benchmarking initiatives established by the authors of this chapter: the database of embodied Quantity outputs (deQo) developed at the Massachusetts Institute of Technology (MIT) and the Embodied Carbon Benchmark (ECB) Study developed at the University of Washington, both with the support of the Carbon Leadership Forum.

To collect data on the operational energy of buildings, the US Energy Information Administration (EIA) developed the Commercial Buildings Energy Consumption Survey (CBECS 2017). This database uses the energy use intensity (EUI) metric to express the building's operational energy consumption in terms of annual energy normalized by floor area. This is needed to establish benchmarks (Nikolaou et al. 2015). Similarly, the new embodied carbon benchmark databases established in North America presented in this chapter use the global warming potential (GWP) metric to express the building's embodied carbon in carbon dioxide equivalent emissions normalized by floor area ($kgCO_2e/m^2$).

Life Cycle Assessment (LCA) Data and Tools for Buildings

North American LCA data for building products and processes are available in various EPDs, LCA databases, and web-based tools. Given the low level of government support for data collection, there is a diverse range of data providers leading to the difficulty of aligning data and methods. The Carbon Working Group

(Webster et al. 2012) highlights the uncertainty and variability of the available data on common construction materials in the USA. Sources might not always use the same assumptions, resulting in the need for a more reliable and comparable definition of embodied carbon coefficients (ECC, expressed in $kgCO_2e/kg$) for building materials and components. The coefficients of construction materials can be found in EPDs, in LCA databases, or within whole-building LCA tools.

Environmental Product Declarations (EPDs)

An EPD is a verified and registered document that offers transparent and comparable LCA information about products. It reports the results of third-party-verified LCA calculations in a standardized manner. Multiple EPD program operators exist in North America. Nonprofit environmental organizations include Earthsure (2017) and the International Code Council Evaluation Service (ICC-ES 2017). Academic initiatives are carried out by the Carbon Leadership Forum (2017) among others. Standards bodies such as the American Society for Testing and Materials (ASTM 2017) and NSF Sustainability (2017) also provide EPDs. Industry trade organizations such as the National Ready Mixed Concrete Association (NRMCA 2014) publish EPDs of the corresponding products. For-profit organizations include Scientific Certification Systems (SCS Global Services 2017), Sustainable Minds (SM 2016), and Underwriters Laboratories Environment (UL Environment 2017).

In 2015, a group of EPD program operators developed the Program Operator Consortium (POC) to "provide more useful environmental product transparency solutions and to reduce complexity in the marketplace" (POC 2015). The USGBC collaborated with UL Environment to develop "enhanced EPD guidance" (USGBC 2017) providing a two-part framework with clarification on how to integrate requirements of ISO and European standards (ISO 21930 and EN 15804) within a North American context. Efforts exist to harmonize EPDs, but the nonhierarchical horizontal relationships of the many organizations working on them hinder consistent standardization and alignment of embodied carbon reporting. For the primary structural materials of concrete, steel, and wood, the adoption and use of EPDs are distinct and not aligned.

The concrete industry quickly saw an opportunity in using EPDs: they could differentiate "greener" concrete mix designs. The NRMCA developed their own EPD program in order to assist their diverse members to develop results (NRMCA 2014). The first company-specific EPD was issued by Central Concrete in 2013. In 2014, the US industry average EPD of concrete was issued for multiple standard mix designs in nine different regions in the USA. The top six concrete producers have adopted the use of EPDs to report the environmental impacts of their products. Over 2000 different mix designs or "products" now have EPDs. These are one of the primary mechanisms for design teams to get LCA credits in LEED. Concrete used in different areas of the building (e.g., foundations or floor slabs) can achieve the optimization material credit when selecting a mix that outperforms the industry average data. The second part of this chapter describes these credits in more detail.

The steel industry relies upon data collected from the World Steel Association ("Worldsteel") to report industry average life cycle impact assessment (LCIA) results of steel production. The industry has published global average data and a methodology report (Worldsteel 2011). The complexities of reporting the benefits and impacts of using recycled steel and future recyclability of metal products have made it challenging for the industry to reach consensus on methods and standards. In 2016, the first US industry-wide EPD for steel (cold-formed steel studs and track) was released concurrently with six industry average EPDs for fabricated steel products by the Canadian Institute of Steel Construction. The Steel Market Development Institute (2017) published a list linking to most current steel EPDs (2017). Three companies, ASC Profiles, CISC, and Gerdau, have issued company-specific EPDs for their products as published by the POC (2017).

The wood industry was an early adopter of LCAs and EPDs due to their interest in exploring how to quantify the perceived environmental benefit of wood products. Starting in 2005, life cycle inventories (LCI) of wood product manufacturing processes have been published via the Consortium for Research on Renewable Industrial Materials (CORRIM 2014) and are updated regularly. Funding from the US Department of Agriculture has helped to collect this data. Industry-wide EPDs for the primary structural wood products such as softwood timber, glue-laminated timber, and plywood were published in 2013, but it took until 2016 before two EPDs were created for specific products. Until more manufacturer (and ideally forest)-specific EPDs exist, wood product EPDs cannot be used to compare between different suppliers of similar wood products. The results of wood product EPDs are often integrated into whole-building LCAs to compare different system choices.

When considering CO_2 emissions related to wood products, there is a lack of consensus on how to account for biogenic carbon: the carbon uptake during growth and emission during combustion or decomposition of wood at the end of life. Disagreements exist on whether wood from North American forests can be considered sustainable and thus carbon neutral and how related emissions should be reported. This leads to difficulty interpreting wood LCA results when compared with other building products. Some practitioners exclude emissions related to burning wood waste from the LCA results as they assume carbon neutrality of biologically based materials from sustainably managed forests. Given the complexities of modeling the impacts of biogenic carbon, the comparisons of EPDs must be done with care. The LCA Practice Guide (in progress in 2018[2]) asks to report the following impacts separately in order to increase transparency and flexibility for the end user: contained biogenic carbon, combusted biogenic carbon from renewable sources used in production processes, combusted biogenic carbon from

---

[2]The progress and final LCA Practice Guide will be available on the Carbon Leadership Forum's website: http://carbonleadershipforum.org/2017/02/10/life-cycle-assessment-practice-guide/.

nonrenewable sources, and forest certification as discussed in ISO 21930:2017. This transparency should aid interpretation of future wood product EPDs.

In the most recent version of the LEED green building rating system, LEED v4, points are awarded for achieving different performance objectives including topics such as energy efficiency, water use, and material selection. Through the "Building product disclosure and optimization" credit, projects that can demonstrate the use of 20 products or materials that publish EPDs are rewarded. If the products demonstrate improvement beyond an industry average, the project obtains extra points. This credit has been the primary driver for manufactures to develop product and industry average EPDs. Prior to the establishment of the credit in 2013, the development of Product Category Rules (PCR), which define the rules and requirements for EPDs, was primarily driven by modest support of nongovernmental organizations (NGO) and sometimes governments (Simonen and Haselbach 2012). In 2012, PCRs for concrete, wood, cement, and windows were in progress, and only a few published EPDs existed for North American products. In the intervening 5 years, significant increases in PCR development and EPD publishing have taken place as organizations and companies look to ensure their products can be used to help achieve LEED points. Figure 21.1 outlines the number of products and materials that have EPDs (and by necessity, PCRs) as of April 2017.

LCA Databases

There are multiple LCA databases in the USA supported by different institutions and developed for different purposes. The US LCI Database was developed by the National Renewable Energy Laboratory (NREL 2017) starting in 2001 with the goal of being the primary source of US-based life cycle inventory data and remains a primary source for US-specific data on select materials and processes. The Building for Environmental and Economic Sustainability (BEES 2010) software and database was developed by the National Institute of Standards and Technology (NIST 2010) specifically for the building industry and contains product-specific LCA results. The LCA approach used in the BEES software is specified in ISO 14040, and environmental and economic performance are measured according to the ASTM standard for multi-attribute decision analysis with the classification of building products with UNIFORMAT II, an ASTM standard classification for building elements. This database was last updated in 2010 and unfortunately is not being populated with EPD results. The US Department of Agriculture funded the development of a data repository specific to agriculture (USDA 2017), the LCA Commons. The aim was to evolve to a Federal LCA repository housing the NREL database along with other government LCA datasets. The Canadian Raw Materials Database (CRMD 2017) is an initiative collecting environmental inputs and outputs of Canadian commodity materials based on LCI data.

Most LCA practitioners and tool developers rely upon proprietary datasets included in professional tools with North America-specific customization. Given

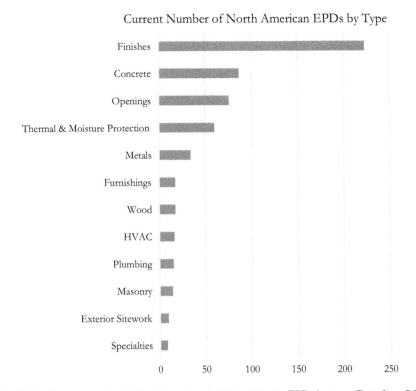

Fig. 21.1 Current quantity of North American building industry EPDs by type (Data from POC 2015; UL Environment 2017; NRMCA 2017)

that there are multiple data sources and that the customization to US applications requires some professional judgment, the LCI data of the baseline building can be difficult to compare with that of the proposed design. Indeed, the tools are not always transparent about assumptions and scope. Similar to other regions of the world, current LCA knowledge in North America is limited by a high level of uncertainty and a lack of standardized practice. Many architects and engineers, even in the USA, rely on the data included in the Inventory of Carbon and Energy (Hammond and Jones 2008) which is not North America specific but is free and transparent. This illustrates the lack of accessibility to regionally specific data that would enable building industry professionals to generate embodied carbon studies based upon their own material quantity estimates.

Privately funded, public access databases exist such as Quartz, a database which identifies environmental impacts of common building products based on a collaboration between experts in LCA, green building, and big data (Quartz 2016). Additional web-based tools help users search for data such as Mindful Materials (2017) and Pharos (Healthy Building Network 2017). As with the EPD programs, there is no national organization to standardize efforts resulting in a diversity of approaches to evaluating and reporting the environmental impact of materials.

Whole-building LCA Tools

A range of software tools exist to assist in evaluating LCA impacts, including the embodied carbon, of buildings over their full life cycle. Kieran Timberlake and Thinkstep released the Tally Environmental Impact tool (Tally 2017), which extracts data from 3D building models to calculate embodied impacts. In Canada and the USA, the Athena Sustainable Materials Institute has integrated LCI data into a free and building industry-specific tool, the Athena Impact Estimator (Athena 2009). These are streamlined and simplified tools that can be used by non-LCA experts. Conversely, LCA practitioner tools require experts to select the suitable datasets and calculation factors for different North American regions. An example of a dataset commonly used in the USA is the Environmental Protection Agency's (EPA) Tool for Reduction and Assessment of Chemical and Other Environmental Impacts (TRACI 2017). Both streamlined and LCA practitioner tools are developed based on the expertise of the individual organizations, leading to different LCA methods, LCI background data, or scenarios regarding transportation, construction, building use, and end of life.

Leading structural design and architecture firms have also started to develop their own in-house embodied carbon assessment tools. The SOM Environmental Analysis tool estimates the embodied carbon of design projects (SOM 2013). Thornton Tomasetti developed a calculation tool and started collecting data on material quantities in their projects (Thornton Tomasetti 2016). Arup developed their in-house database, called Project Embodied Carbon Database (PECD), populated with Arup buildings or projects from literature (Kaethner and Burridge 2012; Yang 2014). The authors initiated two additional databases in collaboration with industry. The database of embodied Quantity outputs (deQo) compiles structural material quantities from these and other sources to generate a database of embodied carbon results in building structures (deQo 2017; De Wolf and Ochsendorf 2014; De Wolf et al. 2016). The Carbon Leadership Forum has compiled the Embodied Carbon Benchmark database (Simonen et al. 2017a) for both structural and architectural studies.

Table 21.1 is a non-exhaustive list of EPDs, LCA databases, and whole-building LCA tools available to estimate the embodied carbon of construction materials and buildings in North America.

Because some LCA data are protected by intellectual property rights, available software can lack transparency. This "black box" effect leads to inconsistencies across regions and tools. To enable meaningful comparisons of buildings using two or more tools, a uniform methodology is needed.

Table 21.1 Non-exhaustive summary of the available EPDs, LCA databases, and whole-building LCA tools in North America

	EPDs	LCA databases	Whole-building LCA tools and databases
Nonprofit	ICC-ES	POC	Athena Impact Estimator
	Earthsure		
	USGBC		
Academic	Carbon Leadership Forum	Carbon Working Group	deQo
			ECB database
Standards	ASTM	USLCI by NREL	
Governmental	NSF Sustainability	BEES by NIST	
		LCA Commons by USDA	
		TRACI by EPA	
		CRMD	
Industry	NRMCA (concrete)	Worldsteel (steel)	SOM Environmental Analysis
		CORRIM (wood)	
			Thornton Tomasetti tool
			PECD by Arup
Commercial or privately funded	UL Environment	Quartz	Tally
	SCS Global Services	Mindful Materials	
	Sustainable Minds	Pharos	

Incentives to Reduce Embodied Carbon in North America

This part illustrates the main incentives for the North American construction industry to report and lower the embodied carbon of buildings. First, the most widely used rating scheme LEED developed by the USGBC is discussed. Version 4 of LEED includes a whole-building LCA credit that requires improvements in terms of embodied carbon reduction of a proposed building design compared to a baseline. In the credit, the term "proposed building" refers to the building design that is applying for the whole-building LCA credit in the LEED certification submission. The term "baseline building" refers to the reference building against which potential improvements are assessed. Other rating schemes such as Green Globes and Living Future are also described. Finally, industry-wide incentives such as Architecture 2030 and Structural Engineers 2050 Commitment Initiative are highlighted.

LEED Rating Scheme's whole-building LCA Credit

The LEED green building rating scheme is widely adopted in the USA and elsewhere. Some US market certifications and city or state legislations even mandate LEED for government buildings. Sometimes, users or building owners also expect it. The LEED Materials and Resources (MR) category includes embodied carbon within the whole-building LCA credit. Out of the 110 points a building can obtain for Certified to Platinum LEED levels, 3 points are allocated to "Option 4: whole-building LCA." This means that embodied carbon assessment only makes up 2.7% of all LEED building criteria. The whole-building LCA credit is applicable to new buildings and new additions to an existing building (USGBC 2016). A cradle-to-grave LCA of the building is required, i.e., including all life cycle stages as defined in ISO 21930.

To obtain the credit, the proposed design must improve its embodied carbon and other environmental impacts compared to a baseline building. However, this baseline building remains currently undefined due to a lack of benchmarks (discussed in the last part of this chapter). To compare apples to apples when looking at the proposed versus baseline design, it is crucial to unify existing embodied carbon measurement methodologies. The proposed and baseline building must have the same location, programmatic function, service life, size in terms of gross floor area, orientation, and operational energy. An operational energy model is not required for this credit, but the building must comply with the prerequisite minimum energy performance following the requirements of ASHRAE 90.1-2010 from the other LEED category Energy and Atmosphere (EA). It is important to use the same datasets compliant with ISO 14044 for both the proposed and baseline building.

Changing the baseline into the proposed building design has environmental consequences. The design changes include building footprint and shape, structural system choices, the selection of other products and assemblies, and the optimization of the structure. For example, a comparative assessment can show the environmental benefits or loads of shear walls versus columns (structural system). Varying column spacing and slab depths can also lead to improved embodied carbon (structural optimization). To apply for the credit, the submitted documentation must include an LCA narrative to describe the assumptions and scope as well as the analysis process for both the baseline and proposed buildings and an LCIA summary. The summary shows the outputs and the percentage of change of the proposed building compared to the baseline building for all impact indicators.

To obtain the whole-building LCA credit, five steps are necessary. First, the scope of the LCA must be clearly defined in terms of products and should include at least the structural and enclosure materials. Second, the appropriate tools and datasets for LCA must be selected and define whether an LCA expert is required. Third, the baseline building must be created and modeled. Fourth, the relevant impact measurement systems must be selected. Fifth, the LCA results must suggest design decisions that reduce the environmental impacts.

Table 21.2 Product scope in whole-building LCA credit (De Wolf 2017)

		Included	Excluded
Envelope	Complete envelope	✓	
Structure	Footings and foundations	✓	
	Structural wall assembly	Cladding to finishes	
	Structural floors and ceilings	✓	Finishes
	Roof assemblies	✓	
	Parking structures	✓	Parking lots
External works	Excavation and other site developments		✓
Equipment	Electrical/mechanical equipment and controls		✓
	Plumbing fixtures		✓
	Fire detection and alarm system fixtures		✓
	Elevators		✓
	Conveying systems		✓
Other	Interior nonstructural walls of finishes, etc.	May be included, earns no additional credit	

The minimum required scope has been limited to structure and enclosure until better LCA data is available for other materials. The material components taken into account include footing and foundation components, structural wall assemblies from cladding to interior finishes, structural floors and ceilings, roof assemblies, and parking structures. Table 21.2 gives an overview of the building products that must be included or excluded in the LCA's scope.

LEED imposes a service life for the LCA study of 60 years or more, to encompass enough replacement cycles in roof systems, curtain walls, and other envelope materials while not replacing the structure.

The selection of the type of LCA tool determines whether an LCA expert is required. Indeed, different tools require different levels of data control and interpretation. If streamlined and simplified LCA tools are selected, non-LCA experts and design teams can use them without customizing data. The calculation factors are specific to the region of the building, for example, with the Athena Impact Estimator (Athena 2009). If LCA practitioner tools are selected, LCA experts need to choose the suitable datasets, factors, and methodologies specific for the building's region on a product-by-product basis, for example, with TRACI (2017).

Out of the six impact categories defined by the whole-building LCA credit, the proposed building design must have a 10% reduction compared with the baseline building for the embodied carbon (also called GWP here) and at least two others. The other impact categories may not increase by more than 5%. The impact categories are illustrated in Table 21.3 and were selected due to their wide use. These factors are quantifiable environmental impacts evaluated with LCA tools.

Table 21.3 Impact categories considered in whole-building LCA credit (De Wolf 2017)

Impact category	TRACI 2.1	CML 2002	ReCiPe
GWP	CO_2 eq.	CO_2 eq.	CO_2 eq.
Ozone[a] depletion potential	CFC-11 e eq.	CFC-11 eq.	CFC-11 eq.
Acidification potential (land)	SO_2 eq.	SO_2 eq.	SO_2 eq.
Eutrophication potential (freshwater)	N eq.	PO_4^3 eq.	P eq.
Formation of tropospheric ozone[b]	NO_x eq.	C_2H_4 eq.	Kg NMVOC[c]
Depletion of nonrenewable energy resources	MJ	Kg or m^3 of raw material	Kg of oil eq.

[a]Stratospheric ozone layer
[b]Photochemical oxidant formation
[c]Non-methane volatile organic compound

Environmental impacts such as human health, ecological impacts, and land use are less easily measurable with common LCA tools and are addressed under other MR credits. Table 21.3 presents output units for each impact indicator for TRACI 2.1, CML (2002), and ReCiPe (2017).

Comparing a proposed building to the baseline can be time-consuming, as two designs and two LCA studies are needed to be able to apply for the whole-building LCA credit. However, as no benchmarks are defined yet, this is currently the best available method to assess the embodied carbon reductions of a new building design. A lack of benchmarks and uniform methodology has presented a challenge for the implementation of the credit.

Other Rating Schemes in North America

While LEED is the most widely used rating scheme in North America, competing rating schemes also require to compare the materials' performance of a proposed design against that of a baseline building. An example of a competing US-based rating scheme is Green Globes (2017). This rating scheme offers extra points for an embodied carbon assessment of the building. An example of an international rating scheme used in the USA is the Living Building Challenge. This rating scheme only has mandatory measures, and carbon emissions of a whole-building LCA study are one of them (Living Future 2017). Table 21.4 shows the rating schemes mostly used in the USA that have an embodied carbon component.

Local governments in North America are exploring the effectiveness of integrating LCA method data and standards into building policies and codes. An example is the state of Washington. Green building rating programs such as LEED are even mandatory in different jurisdictions. Optional whole-building LCA components exist in building codes in the USA, for example, the International

Table 21.4 Rating schemes and points to be obtained with whole-building LCA

Country	Rating scheme	Credit	Points/ maximum	Mandatory
USA	LEED	Whole-building LCA	3/110	No
USA	Green Globes	GWP, materials and resources	50–125/1000	No
International	Living Building Challenge	Embodied carbon footprint	1/20	Yes

Green Construction Code (IgCC 2012) and the California Green Building Standards Code (CALGreen 2011). These codes offer whole-building LCA requirements as a voluntary alternative to local prescriptive material requirements.

Architecture 2030 and Structural Engineers 2050

Other incentives for performing a whole-building LCA in North America include the Architecture 2030 Challenge and Structural Engineers 2050 Commitment.

Architecture 2030 is an environmental NGO that has led efforts to track and reduce the operational impact of buildings through the 2030 Challenge (Architecture 2030 2006) launched in 2006 to drive toward zero operational carbon by 2030. They developed the 2030 Challenge for Products aimed to motivate manufacturers to reduce product embodied carbon to 50% of current industry practice by 2030 in order to push for embodied carbon reductions on top of the operational carbon savings. The American Institute of Architects (AIA) developed the AIA 2030 Commitment (AIA 2030 2017) which provides a reporting framework for firms who decided to adopt the 2030 Challenge.

The Carbon Leadership Forum is developing the Structural Engineers 2050 (SE 2050) Commitment Initiative. The goal of this initiative is "to inspire structural engineers to contribute toward the global vision of zero-carbon buildings by 2050 and to provide measurement of progress toward that vision" (SE 2050 2017). Toward this goal, the initiative aims to enlarge the simple, straightforward, yet robust collection of structural material quantities on building projects to enable the determination of embodied carbon baselines. The SE 2050 Commitment Initiative asks that structural engineers commit to providing structural material quantities and key project information to deQo. This database will be illustrated in the following part of this chapter.

Benchmarking Databases

There is increasing interest in understanding and reducing the embodied carbon of buildings. Many programs to assess the environmental performance of buildings in the USA such as LEED require the comparison of a proposed design with a reference building as a baseline. To obtain the green building certification, the proposed design must achieve the performance targets set by the reference building (e.g., the proposed design must reduce its embodied carbon by a certain percentage compared to the reference building).

Rather than designing and assessing a baseline building every time a proposed building design wishes to obtain a better environmental rating, defining reference buildings would considerably save the assessor time. These reference buildings have comparable typologies with known embodied carbon. Defining them as benchmarks will pave the way to quicker carbon assessments and to lower carbon structural design. To address the need for benchmarks in embodied carbon, the authors created two databases: deQo developed at MIT (De Wolf 2017) and ECB developed at the University of Washington (Simonen et al. 2017a, b).

Database of Embodied Quantity Outputs (deQo)

Many previous benchmarking efforts only collect the end results of the embodied carbon in buildings or building structures. However, as noted, a lot of uncertainty and variability exist with regard to the ECCs of construction materials. Therefore, collecting the material quantities of buildings and building structures is a more robust way to respond to the need for embodied carbon benchmarks of buildings. Indeed, by collecting material quantities rather than embodied carbon end results, the collected material quantities in a database of existing building structures can be used to calculate and update new embodied carbon benchmarks when more reliable EPDs or LCA results become available (De Wolf 2014, 2017).

Efforts collecting structural material quantities such as Arup's PECD and Thornton Tomasetti's Embodied Carbon Database show that leading structural engineering firms are building in-house databases to this end. The aim of these databases is to increase the confidence in a project's environmental outputs with regard to the GWP expressing the embodied carbon in $kgCO_2e/m^2$ of the building.

In order to compare results across different companies impartially, the authors developed deQo at MIT to collect the structural material quantities (SMQ, expressed in kg/m^2) in buildings with data obtained from the construction industry. Due to intellectual property concerns, the results for the North American projects obtained from industry are presented in aggregated format in this chapter (Howard and Sharp 2010; Mathew et al. 2015). The projects are analyzed by building program type, main material for the structural system (concrete, steel, timber, masonry, composite), size (floor area), height (skyscraper versus low-rise), number

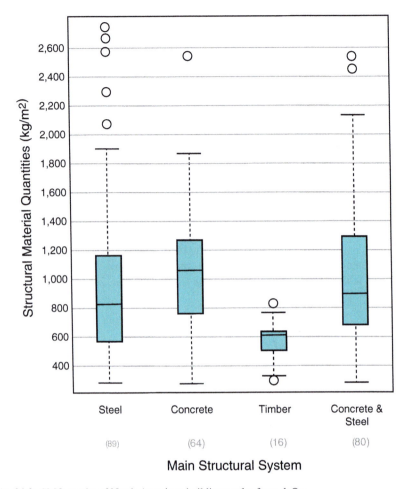

Fig. 21.2 SMQ results of North American buildings only, from deQo

of occupants, span, or rating scheme certification. The box-and-whisker graphical representation facilitates the visualization of the ranges, with median, lower and upper quartiles, minimum and maximum, and outliers (Tukey 1977).

Figure 21.2 shows the SMQ normalized by floor area and expressed in kg of material per square meter of floor area for North American buildings per main structural system. The number below the different main structural systems indicates the number of projects in that category. The outliers are illustrated by the gray dots.

Figure 21.3 shows the corresponding GWP for North American buildings per main structural system. Steel structures typically have lower or similar material quantities than concrete structures, but the higher ECC of steel leads to a similar GWP among typical concrete, steel, and composite concrete-steel structures. The results show that the best way to reduce the embodied carbon of structures is by

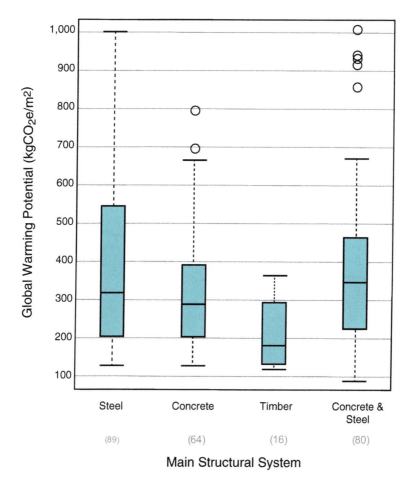

Fig. 21.3 GWP results of North American buildings only, from deQo

improving the material efficiency on a case-by-case basis, rather than by selecting a
particular structural system above another.

Typical buildings collected through deQo range from 650 to 1350 kg/m^2 for the
SMQs and from 200 to 550 kgCO$_2$e/m^2 for the GWP for all regions. North
American results show similar ranges. The SMQ of 95% of the collected buildings
is lower than 1900 kg/m^2. The GWP of 95% of the collected buildings is lower than
850 kgCO$_2$e/m^2. These results are a first step toward benchmarking the embodied
carbon of building structures. The main goal is to provide a baseline for comparing
a design to the range of existing buildings for similar structural systems and use,
rather than deciding if one is better than the other. The authors hope this will enable
engineers and architects to assess the position of their new design within the range
of existing similar building structures and improve its environmental performance.

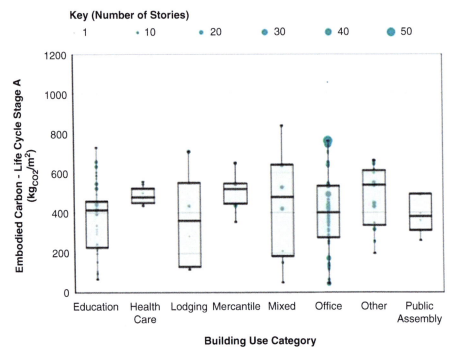

Fig. 21.4 Embodied carbon per m²: structure/foundation/enclosure/interiors

Embodied Carbon Benchmark (ECB) Database

In parallel to the deQo database, focusing on the structural part of buildings, the authors developed an ECB database illustrated by an Embodied Energy Data Visualization (Simonen et al. 2017b) at the University of Washington and the Carbon Leadership Forum (CLF). The data from deQo was added to this ECB database that not only includes building structure but also enclosure and occasionally interior finishes. The primary aim of this research was to understand the order of magnitude and range of building embodied carbon and to identify strategies to overcome the uncertainty in current knowledge. The project confirmed that the embodied carbon of buildings including structure, foundation, and enclosure is typically less than 1000 $kgCO_2e/m^2$. Figure 21.4 represents a clustered study of results of building structure, enclosure, and interior based upon building use type and number of stories (tall buildings shown as larger dots).

In addition to the need to develop more standardized methods to track and report building embodied carbon, the survey held by the University of Washington and the Carbon Leadership Forum (Simonen et al. 2017a) confirmed that there is a high value in developing databases that collect material quantities. Material quantity databases such as deQo track the material efficiency of building structures separately to their environmental impact. These material quantities can be used in the future when more accurate LCA data is available for North American construction materials.

Summary

While many of today's design efforts attempt to lower the operational carbon of buildings, embodied carbon is gaining increasing attention in North America as its significant contribution to whole-building life cycle carbon emissions is being more broadly understood. Because of the growing interest in LCA, practitioners continue to need a uniform and transparent method for estimating the embodied carbon of their projects. This chapter reviewed LCA data and tools such as benchmarking efforts as well as incentives for embodied carbon reduction such as rating schemes, available in North America.

First, the North American LCA data have been summarized. The main challenge with the available data and tools is the lack of transparency and consistency across different databases and applications. EPDs are available through a number of nonprofit and for-profit organizations, as well as academic, governmental, and industrial initiatives. The industry trade associations published several EPDs for concrete, steel, and wood in North America. LCA databases are also available through governmental agencies or privately funded organizations. Several engineering and architecture firms have developed their in-house whole-building LCA tools, but Canadian- and US-based organizations also offer free and commercial cross-company tools, including Athena and Tally.

Second, incentives to reduce the embodied carbon of North American buildings have been highlighted. LEED, a rating scheme of the USGBC widely used in North America and globally, in its latest version (v4) only attributes 3 out of 110 points to the embodied carbon assessment component, called whole-building LCA credit. To obtain the credit, the GWP and at least two other environmental impact indicators in the proposed building design need to be reduced, relative to a baseline building. EPDs can also be used to achieve material-specific credits. Other rating schemes include Green Globes and Living Building Challenge. The SE 2050 Commitment Initiative is also discussed as an incentive to report and reduce the embodied carbon of building structures in North America.

Third, two projects to collect data on the environmental impacts of buildings have been illustrated. Indeed, the authors of this chapter developed two databases to respond to the need for uniform and transparent benchmarks. The first database (deQo) collects structural material quantities and embodied carbon of building structures, and the second (ECB) collects the embodied carbon results of buildings from industry and academia. As these databases continue growing, they will create and refine benchmarks. This will hopefully allow designers to develop more confidence in embodied carbon assessments. Tools such as the Athena Impact Estimator and rating schemes such as LEED will then be able to implement these newly developed benchmarks in low-carbon design strategies.

Industry leadership and initiatives show that North America is moving forward in terms of embodied carbon measurement and reduction. The authors of this chapter propose tools and methods to record and report current practice and future progress (e.g., deQo or ECB), which can be used in emerging initiatives such as the SE 2050 Commitment, illustrating the aspirational challenges that North American professionals are increasingly taking on.

References

AIA 2030. (2017). *2030 Commitment*. American Institute of Architects. Available at https://www. aia.org/resources/6616-the-2030-commitment

Architecture 2030. (2006). *The 2030 challenge*. Available at http://architecture2030.org/2030_challenges/2030-challenge/

ASTM. (2017). *Environmental product declarations*. Available at https://www.astm.org/CERTIFICATION/EpdAndPCRs.html

Athena Sustainable Materials Institute. (2009). Impact estimator for buildings version 5.2.01. Available at https://calculatelca.com/software/impact-estimator/

BEES, Building for Environmental and Economic Sustainability. (2010). Available at www.nist.gov/el/economics/BEESSoftware.cfm

CALGreen. (2011). *A primer on California's green building code*. Presentation by M. Thomas, Brightworks.

Carbon Leadership Forum, University of Washington. (2017). Available at http://www.carbonleadershipforum.org

CBECS. (2017). Commercial Buildings Energy Consumption Survey. Available at www.eia.gov/consumption/commercial/

CML. (2002). *Tools and data*. Institute of Environmental Sciences. Available at https://www.universiteitleiden.nl/en/science/environmentalsciences/tools-and-data

CORRIM. (2014). *Carbon accounting*. Consortium for Research on Renewable Industrial Materials. Available at http://www.corrim.org/research/carbon.asp

CRMD. (2017). Canadian Raw Materials Database, Available at uwaterloo.ca/canadian-raw-materials-database

De Wolf, C. (2014). *Material quantities in building structures and their environmental impact* (MS in Building Technology Thesis). Massachusetts Institute of Technology.

De Wolf, C. (2017). *Low carbon pathways for structural design: Embodied life cycle impacts of building structures* (PhD in Building Technology Dissertation). Massachusetts Institute of Technology, supervisor: John Ochsendorf, June 2017.

De Wolf, C., & Ochsendorf, J. (2014). Participating in an embodied carbon database. *Structural Engineer, 1*(Feb Issue), 30–31.

De Wolf, C., Yang, F., Cox, D., Charlson, A., Hattan, A., & Ochsendorf, J. (2016). Material quantities and embodied carbon dioxide in structures. *ICE Journal of Engineering Sustainability, 169*(ES4), 150–161. https://doi.org/10.1680/ensu.15.00033.

deQo. (2017). *Database of embodied quantity outputs*. MIT, Available at deqo.mit.edu

Earthsure. (2017). Earthsure EPDs. Available at https://iere.org/programs/earthsure/earthsure-epds.htm

Green Globes. (2017). *The practical building rating system*. Available at www.greenglobes.com/home.asp

Hammond, G. P., & Jones, C. I. (2008). Embodied energy and carbon in construction materials. *Proceedings of the Institution of Civil Engineers – Energy, 161*(2), 87–98.

Healthy Building Network. (2017). *The pharos project. Web tool.* Available at https://www. pharosproject.net/

Howard, N., & Sharp, D. (2010). *Methodology guidelines for the materials and buildings products life cycle inventory database. Guidelines.* Sydney: Building Products Innovation Council (BPIC).

ICC-ES. (2017). *ICC-ES: The leader of technical evaluations of building products – Providing peace of mind to the construction industry.* Available at http://www.icc-es.org

IgCC. (2012). International Green Construction Code, International Code Council.

IPCC. (2014). *Climate change 2014: Synthesis report. Contribution of working groups I, II and III to the fifth assessment report of the intergovernmental panel on climate change. Synthesis report.* Geneva: IPCC.

Kaethner, S., & Burridge, J. (2012, May). Embodied CO2 of structural frames. *The Structural Engineer, 90*, 33–40.

Living Future. (2017). *Living building challenge.* Available at living-future.org/lbc/

Mathew, P., Dunn, L., Sohn, M., Mercado, A., Custodio, C., & Walter, T. (2015). Big-data for building energy performance: Lessons from assembling. *Applied Energy, 140*, 85–93.

Mindful Materials. (2017). *Web tool.* Available at www.mindfulmaterials.com

Nikolaou, T., Kolokotsa, D., & Stavrakakis, G. (2015). Review and state of the art on methodologies of buildings' energy efficiency classification. In T. Nikolaou, D. Kolokotsa, G. Stavrakakis, A. Apostolou, & C. Munteanu (Eds.), *Managing indoor environments and energy in buildings with integrated intelligent systems* (pp. 13–31). Cham: Springer International Publishing.

NIST. (2010). *BEES online: Life cycle analysis for building products.* Available at http://ws680. nist.gov/bees/

NREL. (2017). *U.S. life cycle inventory database home page.* Available at https://www. lcacommons.gov/nrel/search

NRMCA. (2014). *Sustainability and the concrete industry.* National Ready Mixed Concrete Association. Available at http://www.nrmca.org/sustainability/index.asp

NRMCA. (2017). *NRMCA/Sustainability.* Available at https://www.nrmca.org/sustainability/ EPDProgram/Index.asp#VerifiedEPDs

NSF Sustainability. (2017). *Environmental product declaration (EPD) publications.* Available at http://www.nsf.org/newsroom/publications/category/environmental-product-declaration

POC. (2015). Program Operator Consortium. Available at https://docs.google.com/spreadsheets/u/ 1/d/1XJtw9FN3AjTuMZiUENex6N7E7TawatNCCYxBEmrwdPM/pubhtml# and from http:// programoperators.org/goals/

POC. (2017). *Transparency catalog.* Program Operator Consortium. Available at www. transparencycatalog.com

Quartz. (2016). Available at http://www.quartzproject.org/p/CP168-a00/q/concrete#cp

ReCiPe. (2017). Available at http://www.lcia-recipe.net

SCS Global Services. (2017). *Environmental product declarations (EPD).* Available at https:// www.scsglobalservices.com/environmental-productdeclarations

SE 2050. (2017). *Structural Engineers 2050 commitment initiative, carbon leadership forum.* Available at http://www.carbonleadershipforum.org/2017/02/09/structural-engineers-2050-commitment-initiative/

Simonen, K. L., & Haselbach, L. M. (2012). Environmental product declarations for building materials and products: US policy and market drivers. *International Symposium on LCA & Construction 2012, Nantes, France*, (1), 197–204.

Simonen, K., Rodriguez, B., McDade, E., & Strain, L. (2017). *Embodied carbon benchmark study: LCA for low carbon construction.* Available at http://hdl.handle.net/1773/38017

Simonen, K., Rodriguez, B., & Li, S. (2017b). CLF embodied carbon benchmark data visualization, website. Available at www.carbonleadershipforum.org/datavisualization/

SM. (2016). 2016 SM Transparency Report™/EPD Framework: Aligning, and aligned with, the industry. *Sustainable Minds*. Available at http://www.sustainableminds.com/industry-blog/2016-sm-transparency-report-epd-framework

SOM (2013). *Environmental analysis tool™*. Skidmore, Owings and Merrill. Available at https://www.som.com/publication/environmental-analysistool-Tm

Steel Market Development Institute. (2017). *List of most current Steel EPDs/HPDs*. Available at http://www.buildusingsteel.org/why-choosesteel/product-transparency.aspx

Tally™. (2017). *Revit add-in real time environmental impact tool*. Kieran Timberlake Research Group, Available at http://www.kierantimberlake.com/pages/view/95/tally/parent:4

Thornton Tomasetti. (2016). *Embodied carbon and energy efficiency tool*. Available at *core.thorntontomasetti.com/embodied-carbon-efficiency-tool/*

TRACI. (2017). *Tool for reduction and assessment of chemicals and other environmental impacts*. United States Environmental Protection Agency. Available at https://www.epa.gov/chemical-research/tool-reduction-and-assessment-chemicals-and-other-environmental-impacts-traci

Tukey, J. W. (1977). *Exploratory data analysis*. Reading: Addison-Wesley.

UL Environment. (2017). *ULe, SPOT: The source for product sustainability information*. Available at https://spot.ulprospector.com/en/na/BuiltEnvironment

USDA. (2017). *Life cycle assessment commons/data and community for life cycle assessment*. Available at https://www.lcacommons.gov/

USGBC. (2016). LEED v4 – Building design and construction. MRc: Building life-cycle impact reduction. Full Reference Guide LCA text.

USGBC. (2017). *Part A and Part B guidance documents*. Available at https://www.usgbc.org/articles/advancements-product-category-ruledevelopment

Webster, M. D., Meryman, H., Slivers, A., Rodriguez-Nikl, T., Lemay, L., & Simonen, K. (2012). *Structure and carbon – how materials affect the climate*. SEI Sustainability Committee, Carbon Working Group. Reston: American Society of Civil Engineers (ASCE).

Worldsteel. (2011). *Life cycle assessment methodology report*. World Steel Association. Available at https://www.worldsteel.org/en/dam/jcr:6a222ba2-e35a-4126-83ab-5ae5a79e6e46/LCA+Methodology+Report.pdf

Yang, F. (2014). Benchmarking embodied impact performance of structure. In C. Griffin (Ed.), *Sustainable structures symposium* (pp. 131–146). Portland: Portland State University., April 17–18.

Chapter 22
Embodied and Life Cycle Carbon Assessment of Buildings in Latin America: State-of-the-Art and Future Directions

Francesco Pomponi and Liliana Medina Campos

Introduction

Central and South America[1] are characterised by rapid development and urbanisation and a steady population growth. They are also amongst the most urbanised areas on earth, with 80% of the population living in cities, and a prediction for a further 20% increase by 2025 (WGBC 2017). This region includes countries which, for the most, show a medium-to-high vulnerability and a poor readiness to climate change (ND-GAIN 2017). Additionally, most of Latin American countries regularly score in the top 20 of the most biodiverse countries in the world (Rankred 2015).

For all these reasons, the Americas Network of the World Green Building Council (WGBC) acknowledges that buildings, and their related environmental and social impacts, offer an unmissable opportunity to tackle such critical problems, mitigate risks and improve lives (WGBC 2017). For instance, in Colombia, the Government reports that industry is the second largest sector for potential mitigation of GHGs emissions in the country (ECDBC 2016). Given the rapid urbanisation and the developmental phase of the country, most of the industry is actually made of – directly or indirectly – the manufacturing of building materials

---

[1]Strict and widely agreed-upon classifications about which countries are part of the two regions do not seem to exist. For the purpose of this chapter, Central America comprises countries geographically situated between Mexico and Panama, whereas South America includes all remaining countries going southward. The two regions, i.e. Central and South America, will be referred to as Latin America.

F. Pomponi (✉)
REBEL (Resource Efficient Built Environment Lab), Edinburgh Napier University, Edinburgh, UK
e-mail: f.pomponi@napier.ac.uk

L. M. Campos
Universidad Colegio Mayor de Cundinamarca, Bogotá, Colombia

© Springer International Publishing AG 2018
F. Pomponi et al. (eds.), *Embodied Carbon in Buildings*,
https://doi.org/10.1007/978-3-319-72796-7_22

483

and construction products. Therefore, an acknowledged significant opportunity exists for GHGs mitigation from an embodied carbon perspective.

To this end, this chapter explores the current status of embodied carbon and life cycle assessment of buildings in Latin American countries. It provides an in-depth overview of available policies, existing data and current initiatives. It also identifies and discusses the main challenges for a more rapid and wider uptake of life cycle thinking when buildings are concerned. Additionally, given that data scarcity for the life cycle assessment (LCA) of buildings is a global problem – which is only exacerbated for developing countries – this chapter shows temporary mitigation measures to address such a problem that could significantly accelerate the spread of embodied carbon and LCAs of buildings in the region. The chapter concludes with a call for action, highlighting the roles that different stakeholders within the built environment can and must play in the immediate future if the whole region is to ever achieve truly a sustainable development.

Current Situation

This section reviews the state of the art in embodied carbon and LCAs of buildings in Latin America. Three key elements have been identified, which are dealt with in turn, and they are (1) current initiatives available at national or regional[2] levels, (2) existing policies that support a whole-life approach to the sustainability of buildings and (3) existing data on the embodied carbon and environmental impacts of buildings which might be available at both national and regional levels.

Current Initiatives and Existing Policies

The United Nations CEPAL (Comisión Económica para América Latina y el Caribe[3]) published in 2013 a document titled 'Estrategias de desarrollo bajo en carbono en megaciudades de América Latina' for low-carbon strategies in mega-cities of Latin America (CEPAL 2013). The report recognises that even if the LAC (Latin America and the Caribbean) area only contributes to 5% of the world's GHGs emissions, it is one of the regions which would suffer the most from the devastation of climate change (CEPAL 2013). Given the key role that cities and megacities play in energy demand, carbon emissions and waste generation

---

[2]In this chapter, regional is referred to a wider area than that of a single country. For example, Central America would be considered a region which is made of all individual countries (e.g. Mexico, Costa Rica, etc.).

[3]The reader will see that in many parts of the chapter, the original Spanish terminology has been given. This choice was made to allow the interested reader to search for those sources, links and documents on the web as they would not come up if searching for the English translation.

buildings, cities and urban areas are considered as the most effective starting point to consider low-carbon strategies (CEPAL 2013). The study focuses on economic, technological and regulatory means for climate change mitigation and adaptation around four main sectors (CEPAL 2013):

- Construction
- Urban waste
- Transportation
- Water and sanitation

The study concludes that no specific fiscal or economic initiatives exist aimed at promoting climate change adaptation and mitigation and a low-carbon development. It however highlights that individual megacities (Santiago, Mexico City, Buenos Aires, Lima, Bogotá and Sao Paulo) have included some elements of sustainable development within their political agendas and developed action plans to act upon them (CEPAL 2013). A key recommendation from the report is that LCA should be unreservedly and widely adopted throughout the whole region to analyse elements of and policies for sustainable development (CEPAL 2013).

While no policies on whole-life assessment of buildings have been identified in most LAC countries because the primary focus is still on operational energy, an exception is represented by Colombia. One of the key drivers for this has been identified as the Colombian Green Building Council (Consejo Colombiano de Construcción Sostenible – CCCS), which was founded in 2007 and has shown a great capability of translating elements of the global sustainability discourse about the built environment to the specific context of Latin America in general and Colombia in particular. For this reason, the following two sections review initiatives and policies available in Colombia, as a means to show the best practice available in the region.

Initiatives in Colombia

The Ministry for the Environment and Sustainable Development (Ministerio de Ambiente y Desarrollo Sostenible) created in 2005 the Group for Climate Change Mitigation (Grupo de Mitigación de Cambio Climático – GMCC) with the purpose of having a single entity responsible for the generation of Colombian policies and initiatives related to climate change.

In 2013, together with the Interamerican Development Bank (Banco Interamericano de Desarrollo – BID), a document was formulated (BID 2013) that allowed Colombia to implement a strategy for a low-carbon development (Estrategia Colombiana de Desarrollo Bajo en Carbono – ECDBC), which includes five components:

- Identification and formulation of alternatives for low-carbon development
- Design and implementation of polices, plans and measures

– Design and development of a monitoring, reporting and verification system
– Development of skills
– Creation of a platform of communication and collaboration

Within the first component, three major objectives were identified (BID 2013):

1. To build abatement cost curves in the different sectors
2. To formulate appropriate national actions of mitigation (Acciones Apropiadas Nacionales de Mitigación)
3. To formulate and strengthen documents of the national council for social and economic policy (Consejo Nacional de Política Económica y Social – CONPES)

The components and objectives enable the national government to develop and implement norms (through decrees, resolutions, etc.) aimed at the different economic sectors for the measurement and reduction of their carbon emissions.

The construction sector was first included through a study conducted in 2011 by the Universidad de Los Andes and the CCCS titled 'Estimación de curva de costos de abatimiento de emisiones gases efecto invernadero sector vivienda urbana en Colombia' (U.d.l. Andes 2012; Gamboa and del Pilar Medina 2013).

The study evaluated the state of the art in the housing sector, highlighting a significant issue around lack of information and data scarcity. However, the study allowed the identification of future actions related to improved designs and to new and lower-carbon materials even if the topic of embodied carbon was not specifically and thoroughly covered. The most notable numerical results from the study are:

- A 3.3% projected growth rate of the emissions of the housing sector, totalling 17 million tonnes of CO_{2e} in 2040 and 339 million tonnes as a cumulate figure for the period 2010–2040
- A significant mitigation potential in the cement industry – which is the top emitter within the construction sector – quantified at 127 million tonnes of CO_{2e}
- A potential saving of 28 million tonnes of CO_{2e} evaluated as a result of design strategies as well as appropriate choices of construction materials

To follow this up, in 2013, the Ministry of housing, cities and territory (Ministerio de Vivienda, Ciudad y Territorio) proposed a sectoral action plan, 'Plan de acción sectorial – PAS de mitigación para el sector Desarrollo Territorial y Vivienda', which included seven key strategic areas[4] to promote sustainability in the construction sector (Ministerio de Vivienda 2014). In particular, Strategic Area # 7 was related to efficient construction materials and revolved around three main actions:

---

[4]These are sustainable territorial development (1), sustainable cities (2), sustainable and efficient construction (3), productivity and competitiveness of the construction sector (4), liveability of new and existing dwellings (5), integrated energy management (6), and efficient construction materials (7).

1. Reduction of the carbon footprint of construction materials
2. Creation of projects of innovation and/or implementation for technological advancements in the manufacture of construction materials and building components
3. Development of methodologies to analyse the environmental impact of construction materials

All three actions are related to some extent to embodied carbon, but it is the first one which was more explicitly aimed at characterising the carbon impacts of buildings and their components from a life cycle perspective. In Colombia the three sectors which account for the most carbon emissions are cement, masonry and steel. Thanks to the PAS (Ministerio de Vivienda 2014), the cement sector began a journey to improve its environmental performance. For instance the leading cement manufacturer ARGOS S.A. set out to achieve by 2020:

- Thirty-five percent reduction of CO_2 per tonne of cement
- Sixty-five percent reduction of SOx per tonne of clinker
- Eighty-five percent reduction of particulate matter per tonne of clinker
- A maximum threshold of 1.35 kg NOx per tonne of clinker

The other sectors also followed similar paths to improve their environmental performance and reduce their emissions. For instance, the masonry sector adopted a Latin America specific programme for climate change mitigation (C.-S.-C. CAEM 2013).

Despite the undeniable challenges ahead in terms of emissions reduction and climate change mitigation, it is evident that Latin American countries have started their own initiatives which are often rooted in national contexts. The increase in environmental awareness and the desire to contribute to the global effort in combating climate change is even more evident in the following section which looks at the policies in the context of Colombia.

Policies in Colombia

The evolution of Colombian regulations for the construction sector and its associated impacts had a great acceleration after the creation of the Colombian Green Building Council in 2007. This was a privately led initiative born out of the will of different firms involved at all levels (design, consultancy and construction) in the building sector. As a first step, it was proposed to adopt the LEED (Leadership in Energy and Environmental Design) system to promote sustainable construction initiatives within the country.

An increase in the application of certification schemes for sustainable buildings, as well as the interest of governmental stakeholders for strategies of clean development (Mecanismos de Desarrollo Limpio) and their application to the construction industry, has allowed both the strengthening of existing policies and the development of new ones. The following table (Table 22.1) lists environmental policies related to the construction sector that have been promulgated since 2010.

Table 22.1 Regulations related to the construction sector promulgated in Colombia since 2010

Regulation	Year	Scope
Decree 3930	2010	Waste water management
Resolution 1115	2012	Construction waste management (segregation and recovering)
Decree 1640	2012	Aquifers planning, order and management
Decree 2667	2012	Water bodies for waste water disposal, quality conditions
Law 1665	2013	International renewable energy agency (IRENA) declaration approval
Law 1715	2014	Nonconventional energy sources integration to national energy matrix
Resolution 631	2015	Allowable maximum limits (quality waste water) disposal to municipal waste water system
Decree 1077	2015	Home, city and territory unique regulatory decree
Decree 1285	2015	Modifying decree 1077 regarding green building
Resolution 549	2015	Establishing criteria and guidelines to adopt energy and water efficiency for buildings
Decree 579	2015	Risks and climate change (2015–2050)
Decree 298	2016	Establishing climate change national system
Resolution 1283	2016	Promoting incentives to nonconventional energy sources and energy efficiency management
Resolution 472	2016	Integral construction and demolition waste management regulation
Resolution 330	2017	Regulation for water and sanitisation (enhancement)
Decree 926	2017	Carbon neutrality and carbon tax adoption

While not all of the above regulations are related to carbon emissions, it is clear that there is a significant increase in the attention towards the life cycle environmental impacts caused by buildings. The temporal trend highlights the significant role played by the PAS (Ministerio de Vivienda 2014) in promoting and supporting legislation across all seven strategic areas. At present, the regulatory framework for Strategic Areas # 4 and # 7 is still under formulation. Further, the Colombian Government is currently putting together a CONPES aimed at strengthening the regulations further for the construction sector through initiatives such as the stimulation in offer and demand of sustainable products under a mechanism labelled 'Compras Públicas Sostenibles y Eco-etiquetado'.

This initiative is led by the Ministry of the Environment and Sustainable Development in conjunction with other governmental bodies (Ministerio de Ambiente y Desarrollo Sostenible, en cabeza de la dirección de Asuntos Ambientales Sectoriales y Urbanos, en conjunto con el Departamento de Planeación Nacional). Not only will it create the need for the construction sector to reduce its overall emissions, but it will also promote the development of Environmental Product Declarations (Declaración Ambiental de Producto) related to the embodied carbon of building materials and components to hopefully end the data scarcity across the whole LAC region – which is discussed in the next section.

Available Data

No publicly available dataset on the embodied carbon of buildings and building products seem to exist for Latin America (De Wolf et al. 2017) but only a small number of environmental product declarations (EPDs) (UKGBC 2017). The creation and development of LCA databases in Latin America has been a subject investigated by the United Nations Environment Programme (UNEP) in recent years (Valdivia et al. 2015; UNEP 2016). It was recognised that more concerted efforts are necessary at regional (i.e. cross-country) level in Latin America to facilitate the creation and development of such databases, and for this reason a UNEP/SETAC session was held at the Pontificia Universidad Católica del Perú in 2015 (Valdivia et al. 2015).

This event highlighted an increase in available databases between 2005 (where no database existed in the whole of Latin America) and 2014. However, the optimistic view of the report does not seem to match reality. For example, the authors (Valdivia et al. 2015) report on the availability of the Mexican LCA database called Mexicaniuh, but a reality check seems to show that such database remains mostly an intention with the developer website still saying that '[translated from Spanish] at the beginning of 2012 it will be possible to have any LCA process in the new format [. . .] ILCD 1.1' (Centro ACV 2017).

Brazil seems to be more advanced in terms of LCA data availability and commitment to a wider uptake of life cycle thinking. This materialised in the International Week for Life Cycle Assessment held in Brasilia in 2016, where the Brazilian Institute for Information in Science and Technology (IBICT) launched the Brazilian National Bank of Life Cycle Inventories (SICV) (LCI 2017). As a follow-up, the Brazilian LCA community held a second forum in May 2017 to strengthen the role of life cycle thinking in Brazil and to support the wider adoption of LCA in the country (LCI 2017). At the time of writing, the SICV databases include 15 process datasets – half of which however have past their 'valid until' date, 795 elementary flow datasets, 252 product flow datasets and 40 flow property datasets (SICV 2017). The vast majority of such datasets however are not for construction products or building-related materials.

For Chile, UNEP inform that Fundacion Chile is the organisation in charge for the work on LCA databases, and they report the availability of inventory data on five building materials (Valdivia et al. 2015) although at the time of writing the authors could not retrieve such information. Chilean data appear instead to be available within the EuGeos datasets created and managed by the namesake company in the UK (EuGeos 2017) and available on the Nexus platform of the openLCA tool (openLCA Nexus 2017a). In fact the EuGeos database appears to have significant amount of data for all Latin American countries. However, a more in-depth check has shown that these datasets are in fact simply global datasets which are labelled under individual countries. For example, within the allegedly 306 datasets available for construction in Brazil, none is country specific, and they all fall into the global dataset category (openLCA Nexus 2017b).

The United Nations Environment Programme have praised the work done by the Iberoamerican LCA Network (supported by the UNEP/SETAC life cycle initiative) to promote life cycle thinking in the region (UNEP 2016). A further positive element of the collaboration in the region is the international conference on LCA in Latin America (CILCA - Conferencia Internacional de Análisis de Ciclo de Vida en Latinoamérica) held every other year in a different Latin American country. At the time of writing, CILCA 2017 is taking place in Medellín, Colombia, and most countries will report on their individual and joint efforts since the last UNEP/ SETAC meeting in 2015 (Valdivia et al. 2015). It seems widely agreed upon though that the main issue for the development of LCA databases as well as for the promotion and correct use of the methodology in sustainability decision-making in the region remains the lack of funding (UNEP 2016).

For this reason, the primary source of LCA data in Latin America is represented by academic publications. However, even the academic body of evidence is still very scant. Cerón-Palma et al. (2013) conducted an LCA to investigate the GHGs savings of a green sustainable strategy for a social neighbourhood in Mexico. Their results show significant CO_{2e} savings if the strategy proposed is adopted, but the underpinning dataset seems to be ecoinvent with no reference to country-specific data other than the use of the energy mix scenario of Mexico (Cerón-Palma et al. 2013).

Kashkooli et al. (2014) proposed a semi-quantitative framework for the life cycle assessment of building and applied it to a typical office building in Mexico. However, the underpinning data for embodied carbon are taken from the ICE database of the University of Bath, which clearly is not aimed to cover any country other than the UK – despite its wide but incorrect use across the world (Pomponi and Moncaster 2016, 2018).

Primary data for Mexico and, partly, for Chile can be instead found for waste flows in the work of Guibrunet et al. (2016). The authors (1) mapped the waste flows through the industrial cycle; (2) identified the stakeholders that produce, use, transport and transform waste; and (3) mapped the impacts related to the flows of materials (Guibrunet et al. 2016). Their work is not specifically focused on the construction sector; yet their results can be an important starting point to evaluate and assess the environmental impacts of both construction and demolition waste in Latin America.

Barbosa and Almeida (2017) developed a methodology to determine the relative weight of dimensions employed in tools to assess sustainable buildings. The authors did not contribute to any data creation or development in Brazil but highlighted that larger databases for the LCA of buildings and construction products are needed and the main barrier consists in the financial difficulty of securing funds for the creation of such datasets (Barbosa and Almeida 2017).

Ruschi Mendes Saade et al. (2014) proposed a set of lifecycle-based indicators to characterise material eco-efficiency of buildings. Their LCAs were partly based on average data from ecoinvent and partly on collected primary and secondary national data. They found that as few as 12 materials were sufficient to cover over 95% of both embodied energy and embodied carbon.

Castro-Lacouture et al. (2009) proposed an optimisation model for the selection of materials based on the LEED green building rating system in the context of Colombia. Their work seems however very much focused on obtaining a LEED rating as high as possible rather than on the reduction of the environmental impacts, and there is not yet convincing evidence that the two are strictly and directly related. Castro-Lacouture et al. (2009) say they have contextualised their work within the specifics of the Colombian construction market, but this does not emerge strongly in their research. To be fair, in 2009, the green building market in Colombia was just about to surface, and it must have been almost visionary to think of LEED-compliant buildings back then. The authors had an optimistic view about the future, believing that a higher maturity of the green building market in the country would have certainly improved the availability of high-quality data, as well as a wider uptake of LCAs in the construction sector (Castro-Lacouture et al. 2009). Nearly a decade has passed, but we are yet to see such positive transformation.

Ortiz-Rodríguez et al. (2010) performed the LCA of two dwellings, one in a developed country (Spain) and one in a developing country (Colombia). Their aim was to identify the differences of the environmental impacts of domestic buildings in the two countries, and they assessed construction, use and end-of-life stages (Ortiz-Rodríguez et al. 2010). However, all their LCA background data come from the ecoinvent databases, thus making their results questionable as not based on country-specific data. Yet, their work is to date one of the very few existing comparative LCAs between Europe and Latin America, and it is a shame that it has not been followed up in 7 years.

Oyarzo and Peuportier (2014) investigated the environmental impacts and the potential mitigation measures of housing in Chile through LCA. The authors acknowledge that LCA databases do not exist for Chile or even Latin America and have utilised ecoinvent data for their analysis 'rather than waiting for the development of such local data' (Oyarzo and Peuportier 2014, p.111). However, the authors have partly contextualised the ecoinvent data to the Chilean scenario by adopting the Chilean energy mix for electricity production and its use in buildings.

This section has shown that the availability of LCA data is a major impediment to carrying out LCAs of buildings in Latin America. The UNEP reported that most of the existing LCAs were carried out by academics in the region, but a review of the available scientific literature has shown that LCA data produced by academic studies do not contribute to advancing the data availability issue. Specifically only two studies (Guibrunet et al. 2016; Ruschi Mendes Saade et al. 2014) produced new data that could be used and incorporated in LCAs. The rest is mainly made of case studies of individual buildings where the lack of country-specific data has been solved by using existing databases (e.g. Bath ICE) that however cannot guarantee any representativeness of the Latin American context. Scholars say they do so because it is more useful to start applying LCAs based on existing data rather than waiting for pertinent databases to be developed. However, research efforts would perhaps be best placed in producing such missing data rather than utilising unrepresentative data to assess whole buildings without the necessary confidence in the numerical results produced.

Measures to Mitigate the Current Lack of Geographically Specific Data

Given how pressing a problem is the lack of data in the Latin American region, this section attempts to draft temporary measures that could enable more representative LCAs, while comprehensive and specific databases are being developed. Three main approaches are proposed, which correspond to the three approaches available to develop building life cycle inventories (LCIs) (Crawford 2011), namely, process, input-output and hybrid analysis:

1. Convert available process data on embodied energy of established production processes into embodied carbon data by taking into account the energy mix of the relevant country.
2. Utilise aggregated but country-specific data from multiregional, environmentally extended input-output tables as a trade-off between process-relevance and geographical appropriateness.
3. Combine the two in a hybrid LCA fashion.

The first two are explained through a worked example in the following sections, and a short section on hybrid LCA will point towards initial steps to apply the method.

Utilisation of Existing Data on Embodied Energy

It is reasonable to assume that commonly used construction materials are produced with relatively stable and established technologies across the world, with similar levels of inputs and production efficiency. With such an assumption, it is possible to translate data on embodied energy that are available for more developed countries into data on embodied carbon for developing countries.

In this section, it will be shown how this can be done for a material with a very standard production process: flat glass. It is produced with a technique by which the glass mix floats on a tin bath, which ensures perfect planarity. The technique was invented and developed by Sir Alistair Pilkington (2017), and it is now the standard production process across the world.

As a source for embodied energy, the ICE v.2.0 database will be used (Hammond and Jones 2011) due to its wide use in construction studies and its free availability. It should be noted that using embodied carbon values from the ICE database in context other than the UK does not ensure representativeness of the local context but the embodied energy – which represents the primary energy in a cradle-to-gate system boundary – can be a good starting point. Three Latin American countries have been chosen in this worked example – namely, Colombia, Peru and Chile.

Table 22.2 shows the different energy mix of the three countries, sourced for the year 2011(which is the same year of the latest version of the ICE database), from the

Table 22.2 Energy mix and carbon intensity of electricity in Colombia, Peru and Chile in 2011

Source	Colombia (2011)		Peru (2011)		Chile (2011)		Carbon intensity [kgCO$_{2e}$/GWh]	Carbon intensity [kgCO$_{2e}$/GJ]
	GWh	Percentage	GWh	Percentage	GWh	Percentage		
Coal	2300	3.8	622	1.6	19,619	29.9	53,100	14.75
Oil	145	0.2	2350	6.0	6355	9.7	52,380	14.55
Gas	7642	12.5	13,988	35.7	13,719	20.9	27,756	7.71
Biofuels	1999	3.3	675	1.7	4673	7.1	69,768	19.38
Hydro	48,878	80.1	21,587	55.0	21,009	32.0	7596	2.11
Wind	41	0.1	1	0.0	338	0.5	19,476	5.41

Table 22.3 Energy mix for the UK in the 2011 (IEA 2011d)

	UK (2011)	
	GWh	Percentage (%)
Coal	10,9397	29.77
Oil	3018	0.82
Gas	14,6190	39.79
Biofuels	12,238	3.33
Waste	3118	0.85
Nuclear	68,980	18.77
Hydro	8585	2.34
Solar PV	244	0.07
Wind	15,652	4.26
Tide	1	0.00

International Energy Agency (2011a, b, c). Table 22.3 shows the same information for the UK.

Carbon conversion factors have been sourced from Hill et al. (2011) for all nonrenewable energy sources and biofuels, from Thomson and Harrison (2015) for wind and from Akella et al. (2009) for hydropower.

The data in Table 22.2 allowed the calculation of a weighted average of the carbon intensity of electricity in the three countries, which is shown in Table 22.3 Energy mix for the UK in the 2011 – (IEA 2011d) (Table 22.4).

Once the average carbon intensity of energy is known, it is possible to calculate the embodied carbon based on the data on embodied energy available from Hammond and Jones (2011).

The results for the three countries are shown in Table 22.5. It is worth noting that this approach is not perfect, and the data it produces are affected by uncertainty like any other data for embodied carbon. However, the example shows that with little extra effort, it is possible to adapt available data on embodied energy to better represent the geographical context of Latin America.

It can be seen that the embodied carbon values calculated in Table 22.5 reflect a higher use of renewable energy sources in Latin America and a smaller percentage of electricity coming from fossil fuels.

Finally, good judgement should be used when applying this method across Latin America. For example, if the embodied carbon of a building in Honduras is to be assessed, it should be first evaluated whether Honduras has an internal production of – for instance – flat glass. If so, the above method can be applied, but if Honduras instead imports flat glass from neighbouring countries, the embodied carbon should be calculated for the country that produces the flat glass, and then a larger, realistic transportation distance should be used to estimate the embodied carbon of transport from the manufacturing site to the construction site in Honduras. Despite its limitations, the proposed method suggests an easy and alternative way to using, for Latin America, databases that were created for other regions of the world (e.g. ICE for the UK).

Table 22.4 Weighted average of the carbon intensity of electricity in the three countries

	Colombia	Peru	Chile
Weighted average carbon intensity [kgCO$_{2e}$/GWh]	13,988.66	19,260.63	34,203.60
Weighted average carbon intensity [kgCO$_{2e}$/MJ]	0.0039	0.0054	0.0095

Table 22.5 Embodied carbon of glass calculated for Colombia, Peru and Chile

	Embodied energy[a] [MJ/kg]	Embodied carbon [kgCO$_{2e}$/kg]
Colombia	62.00	0.43
Peru	62.00	0.52
Chile	62.00	0.77
UK	62.00	0.91 (ICE value)

[a]It is worth noting that a higher-than-average value was selected for the embodied energy of glass from the range within the ICE database as it better represents recent values developed from primary data (Pomponi et al. 2016). Thomson and Harrison (2015)

Utilisation of Aggregated MRIO Data

A further source of geographically specific data on individual countries in Latin America is represented by the input-output (IO) tables[5] that each country develops and makes available. A number of projects in the past few years have developed international databases that not only include available IO data but also link them with data on international trade, the embodied carbon emissions linked to international trade as well as other indicators. They are generally called multiregional input-output (MRIO) databases, and one of them is EORA (Lenzen et al. 2012, 2013), which was developed to map the structure of the world's economy. It is a widely and regularly used database, as it offers highly disaggregated data compared to other MRIO databases along with other benefits (Lenzen et al. 2012).

National data within EORA are available up to the year 2013, which is the one used in this worked example. In particular the 2013 Basic Price data was used (Lenzen et al. 2012, 2013). The EORA database includes all intersectoral links between all identified sectors of the Colombian economy, as well the transaction between each of those sectors with the other countries of the world. All such transactions are expressed in US dollars. In addition, for each sector of the economy, another matrix called 'satellite account' is available, which shows – for instance – energy inputs [TJ] and GHG emissions [Gg of CO_2, CH_4, N_2O] amongst others (Lenzen et al. 2013).

These two elements are expressed in IO LCA in matrix form. The first one is called transaction matrix and is square by definition because rows and columns refer to the same elements (sectors and industries); this is then transformed into a

---

[5]If the reader is unfamiliar with the IO concept it is suggested to refer to Leontief (IEA 2011d) for an introduction on how environmental emissions and economic structure are linked, and to Heijungs and Suh (Pomponi et al. 2016) for an explanatory introduction to its application in LCAs.

matrix of technical coefficients by dividing each input to its output (Heijungs and
Suh 2002), and in IO LCA is generally referred to as matrix $\mathbf{A}$. The second one is
called the satellite matrix and contains the environmental flows for each element of
the transaction matrix (Leontief 1970). The matrix $\mathbf{A}$ is linked to the final demand
vector $\mathbf{y}$ and the total output vector $\mathbf{x}$ of an economy through the famous Wassily
Leontief's equation:

$$x = (I - A)^{-1} y \qquad (22.1)$$

where I is the identity matrix and $(I - A)^{-1}$ goes under the name of the Leontief
inverse matrix (Lenzen 2001).

To calculate the embodied carbon of construction activities from IO tables, two
ways are possible – but only one is correct. The first, tempting but partial, way of
calculating it is simply by dividing the total emissions attributed to that sector
[Gg CO_2 [6]] by total output of the construction sector [1000 US $]. The IO table for
Colombia includes one sector out of m sectors that is labelled 'construction, and
construction and repair of buildings'. If we call this sector $\mathbf{S}$, the total direct
embodied emissions normalised to monetary unit can be calculated as follows:

$$\text{EC}_u^d = \frac{\sum_{i=1}^{n} \varepsilon_i}{\sum_{j=1}^{m} S_j} \qquad \left[\frac{\text{Gg } CO_2}{\text{USD}} \right] \qquad (22.2)$$

where ε_i represents the emission intensity of the construction sector across all n
different sources of CO_2 available in the IO tables (e.g. electricity, transport, etc.).
The method in Eq. (22.2) however leaves out all indirect emissions related to the
sectoral interdependencies that the construction sector has with other sectors in the
economy. The difference between the direct and indirect emissions is very significant, and it has been proven it can easily mislead decision-making (Lenzen 2002;
Wood et al. 2006).

To account for indirect emissions, a second way that accounts for all upstream
production layers need to be adopted. This is done through the composition of all
CO_2 embodied emissions into an infinite series of production layers. Let us assume
we have a functional unit vector f_0 which contains all zeroes other than the entry
related to the construction sector of the economy and that entry is 1 (which
corresponds to 1000 USD) so that we obtain normalised embodied carbon emissions like in the example in Eq. (22.2). The expansion into an infinite series then
works as follows:

---

[6]Only CO_2 is being used in this example, but data for all other GHGs are available within the
EORA IO tables. To obtain CO_{2e} values, the latest conversion factors between different GHGs and
carbon dioxide given by the (IPCC 2013; Lenzen et al. 2013) should be used.

$$\mathrm{EC}_u^{d+i} = \varepsilon \left(I + A + A^2 + \cdots \right) f_0 = \varepsilon f_0 + \varepsilon A f_0 + \varepsilon A^2 f_0 + \cdots \qquad (22.3)$$

where ε is the emission intensity vector of the whole economy. It should be noted that the more we expand the series, the smaller the terms get as they are powers of a matrix with values smaller than 1. Of course, an infinite series is not possible to calculate, but research has shown that there is a clear plateau after the analysis of about eight upstream production layers and stability is reached (Lenzen 2002; Wood et al. 2006).

This is method is applied to Colombia and Afghanistan to give an overview of two very different countries. Results are showed in the following figures where Figs. 22.1 and 22.3 show the cradle-to-gate emissions in kgCO2 per 1000USD of construction cost, while Figs. 22.2 and 22.4 show the emissions completeness against the supply chain production layers.

The figures show how the context of each country clearly influences both the absolute emissions value for a specific sector as well as the completeness of the emissions depending on the upstream production layers. At such, it is a compelling argument as to why utilising country-specific databases (e.g. ICE) in nations different from those for which they were developed can mislead judgements and decisions. In the specific cases above of Afghanistan and Colombia, it can be seen that the main contributor to the emissions of the Colombian construction sector is the construction sector itself. This explains why a 60+% completeness is achieved at the first production layer – clearly not the case for Afghanistan, which shows a rather steady 20% increase in the first few production layers.

It is worth highlighting that the use of MRIO data is not a perfect, nor a desirable long-term, solution. MRIO data suffers from aggregation error, and the use of specific, primary data should always be preferred. However, in a context where no country-specific data exist, MRIO databases could represent a viable short-term option to account for the peculiarity of each LAC nation and to avoid the truncation error typical of process-based analysis.

Combination of the Two Measures in a Hybrid-LCA Fashion

Hybrid-LCA has been proposed over two decades ago for use in the environmental and life cycle impact assessment of building (Treloar 1997). An explanation of the method goes beyond the scope of this chapter, but it basically brings together the strengths of process-based LCA (granularity and specificity of data) and IO LCA (comprehensiveness and inclusion of upstream production layers). Interested readers should familiarise themselves with seminal literature before applying the method (e.g. Heijungs and Suh 2002; Treloar 1997; Lenzen and Crawford 2009; Treloar et al. 2000 – amongst others).

It is worth noting that while a debate is still ongoing on which method should be used in LCAs (Yang et al. 2017), there is evidence that hybrid LCA offers more

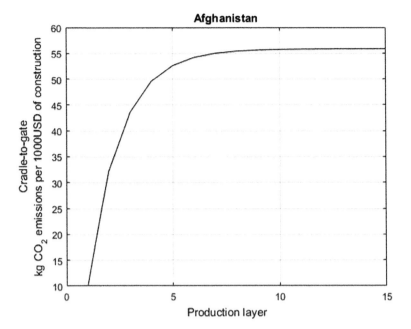

Fig. 22.1 CO_2 emissions per 1000 USD of construction (Afghanistan)

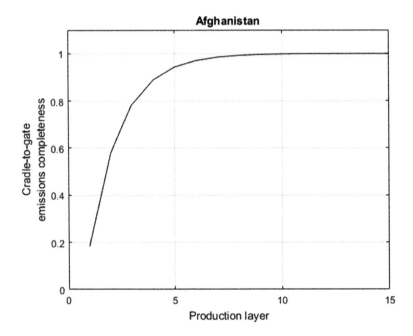

Fig. 22.2 Completeness of CO_2 assessment vs. upstream production layers (Afghanistan)

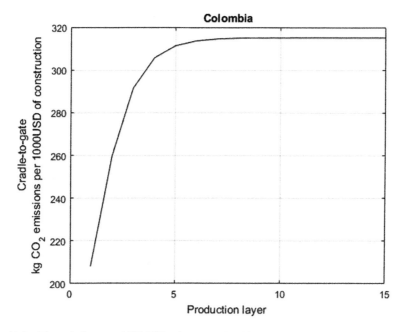

Fig. 22.3 CO_2 emissions per 1000 USD of construction (Colombia)

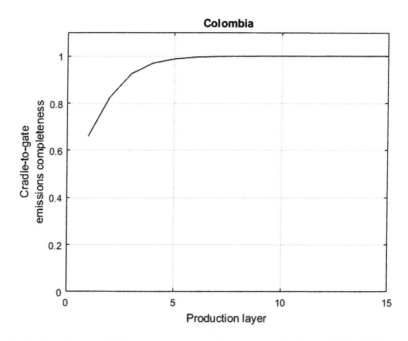

Fig. 22.4 Completeness of CO_2 assessment vs. upstream production layers (Colombia)

accurate results than both process-based and IO LCAs (Lenzen and Treloar 2002; Pomponi and Lenzen 2018). The main drawback of hybrid LCA is that it is a labour-intensive and time-consuming solution. However, newer methods are appearing to achieve an automated approach to compile hybrid life cycle inventories (Crawford et al. 2017).

If such new methods prove successful, it will be relatively easy to translate embodied energy data into embodied carbon-specific data for Latin America as suggested in this chapter and to combine them with available MRIO data in a hybrid LCA fashion. Given the advances in both technology and computational power as well as methodological clarity, Latin America is presented with an unmissable opportunity to quickly reduce the data lag compared to western societies and quickly gain a predominant role in both researching and promoting the sustainability of its built environment.

Concluding Remarks and Future Directions

This chapter has reviewed the state of the art of embodied and life cycle carbon assessments of buildings in Latin America. The task has proved very challenging due to the lack of data and information, and a relative novelty of the concept in the area, across both academics and practitioners.

The focus of this work revolved primarily around three elements that were identified as the most important for a comprehensive overview, which was the aim of this chapter. These are existing initiatives, existing policies and existing data. At a first look, several policies and initiatives seemed to exist across a number of LAC countries. However, when searching in depth, most of these seem to have remained good intentions that did not succeed in translating into practical implementations. A more virtuous example was represented by Colombia – which showed a good number of initiatives and policies aimed at a sustainable built environment, with also a focus on embodied carbon and the life cycle environmental impacts of buildings. A caveat here is necessary though because Colombia is also the context most known by the authors. However, there has been a genuine search across all major Latin American countries which yielded little to no results.

The main impediment in a wider uptake of embodied and life cycle carbon assessment of buildings in Latin America was once again identified in data scarcity – an issue that affects not just developing countries but also developed ones, with the exceptions of just a few. To address such a problem, while national databases are developed, three potential solutions have been proposed in this chapter, which relate to the three main approaches used in LCA, namely process-based, input-output and hybrid. The solutions proposed here represent neither a perfect nor a long-term solution, but they do present Latin American countries with the opportunity to 'catch-up' with developed nations in terms of embodied carbon and life cycle inventories. If this is done, the whole continent will be very soon able to gain a predominant role in both researching and promoting the sustainability of its built environment.

A moral responsibility remains with the many stakeholders that have a role to play for a sustainable built environment in Latin America. These include policymakers, practitioners, academics, professional bodies and NGOs – no one is exempt from playing their part. A very good opportunity does exist within the Americas Network World Green Building Council. In fact, individual national GBCs in Latin America could and indeed should share the effort of promoting a rapid diffusion of the concepts of embodied carbon, life cycle assessment and sustainable buildings from a holistic perspective.

References

Akella, A., Saini, R., & Sharma, M. P. (2009). Social, economical and environmental impacts of renewable energy systems. *Renewable Energy, 34*(2), 390–396.

Barbosa, M. T. G., & Almeida, M. (2017). Developing the methodology for determining the relative weight of dimensions employed in sustainable building assessment tools for Brazil. *Ecological Indicators, 73*, 46–51.

BID. (2013). Desarrollo de una metodología para la construcción de curvas de abatimiento de emisiones de GEI incorporando la incertidumbre asociada a las principales variables de mitigación. Banco Interamericano de Desarrollo. Departamento de Investigación y Economista Jefe. NOTA TÉCNICA # IDB-TN-541.

Castro-Lacouture, D., Sefair, J. A., Flórez, L., & Medaglia, A. L. (2009). Optimization model for the selection of materials using a LEED-based green building rating system in Colombia. *Building and Environment, 44*(6), 1162–1170.

Centro ACV. (2017). Centro de análisis de ciclo de vida y diseño sustentable (CADIS). Análisis de ciclo de vida – Inventario Mexicano. Available at: http://centroacv.mx/servicios.html. Last Accessed 19 Jul.

CEPAL. (2013). *Comisión Económica para América Latina y el Caribe (CEPAL) – Estrategias de desarrollo bajo en carbono en megaciudades de América Latina* [in Spanish]. Available at: http://www.cepal.org/es/publicaciones/36624-estrategias-desarrollo-carbono-megaciudades-america-latina

Cerón-Palma, I., Sanyé-Mengual, E., Oliver-Solà, J., Montero, J.-I., Ponce-Caballero, C., & Rieradevall, J. (2013). Towards a green sustainable strategy for social neighbourhoods in Latin America: Case from social housing in Merida, Yucatan, Mexico. *Habitat International, 38*, 47–56.

Crawford, R. H. (2011). *Life cycle assessment in the built environment*. London: Spon Press.

Crawford, R. H., Bontinck, P.-A., Stephan, A., & Wiedmann, T. (2017). Towards an automated approach for compiling hybrid life cycle inventories. *Procedia Engineering, 180*, 157–166.

De Wolf, C., Pomponi, F., & Moncaster, A. (2017). Measuring embodied carbon dioxide equivalent of buildings: A review and critique of current industry practice. *Energy and Buildings, 140*, 68–80.

ECDBC. (2016). *Estrategia Colombiana de Desarrollo Bajo en Carbon* [in Spanish]. Available at: http://www.minambiente.gov.co/images/cambioclimatico/pdf/Estrategia_Colombiana_de_ Desarrollo_Bajo_en_Carbono/FOLLETO_DE_PRESENTACION_ECDBC.pdf.   Last accessed 17 June.

EuGeos. (2017). *Sustainability services*. Available at: http://www.eugeos.co.uk/lifecycle_assess ment/openlca.html

Gamboa, C., & del Pilar Medina, M. (2013). Curvas de abatimiento para el sector construcción [in Spanish]. *Construcción Sostenible, 7*, 14–21.

Guibrunet, L., Calvet, M. S., & Broto, V. C. (2016). Flows, system boundaries and the politics of urban metabolism: Waste management in Mexico City and Santiago de Chile. *Geoforum, 85*, 353–367.

Hammond, G., & Jones, C. (2011). *Embodied carbon: The inventory of carbon and energy (ICE)*. *BG 10/2011*, University of Bath and BSRIA. https://www.bsria.co.uk/download/product/?file=znUADC5qYPE%3D

Heijungs, R., & Suh, S. (2002). *The computational structure of life cycle assessment*. Springer Science & Business Media.

Hill, N., Walker, H., Beevor, J., & James, K. (2011). *Guidelines to Defra/DECC's GHG conversion factors for company reporting: Methodology paper for emission factors*. London: DEFRA, Department for Environment, Food and Rural Affairs (DEFRA) and Department of Energy and Climate Change (DECC).

IEA. 2011a. Colombia: Electricity and Heat for 2011. Available at: https://www.iea.org/statistics/statisticssearch/report/?year=2011&country=Colombia&product=ElectricityandHeat

IEA. 2011b. Peru: Electricity and Heat for 2011. Available at: https://www.iea.org/statistics/statisticssearch/report/?year=2011&country=PERU&product=ElectricityandHeat

IEA. 2011c. Chile: Electricity and Heat for 2011. Available at: https://www.iea.org/statistics/statisticssearch/report/?year=2011&country=Chile&product=ElectricityandHeat

IEA. 2011d. UK: Electricity and Heat for 2011. Available at: https://www.iea.org/statistics/statisticssearch/report/?country=UK&product=electricityandheat&year=2011

IPCC. (2013). In T. F. Stocker, D. Qin, G.-K. Plattner, M. Tignor, S. K. Allen, J. Boschung, A. Nauels, Y. Xia, V. Bex, & P. M. Midgley (Eds.), *Intergovernmental panel on climate change. Working group I contribution to the fifth assessment report of the intergovernmental panel on climate change. Climate change 2013 – The physical science basis* (p. 1535). Cambridge: Cambridge University Press.

Kashkooli, A. M., Vargas, G. A., & Altan, H. (2014). A semi-quantitative framework of building lifecycle analysis: Demonstrated through a case study of a typical office building block in Mexico in warm and humid climate. *Sustainable Cities and Society, 12*, 16–24.

LCI. (2017). *Life cycle initiative*. www.lifecycleinitiative.org

Lenzen, M. (2001). Errors in conventional and input-output – Based life – Cycle inventories. *Journal of Industrial Ecology, 4*(4), 127–148.

Lenzen, M. (2002). Differential convergence of life-cycle inventories toward upstream production layers. *Journal of Industrial Ecology, 6*(3–4), 137–160.

Lenzen, M., & Crawford, R. (2009). The path exchange method for hybrid LCA. *Environmental Science & Technology, 43*(21), 8251–8256.

Lenzen, M., & Treloar, G. (2002). Embodied energy in buildings: Wood versus concrete – Reply to Börjesson and Gustavsson. *Energy Policy, 30*(3), 249–255.

Lenzen, M., Kanemoto, K., Moran, D., & Geschke, A. (2012). Mapping the structure of the world economy. *Environmental Science & Technology, 46*(15), 8374–8381.

Lenzen, M., Moran, D., Kanemoto, K., & Geschke, A. (2013). Building eora: A global multi-region input–output database at high country and sector Resolution. *Economic Systems Research, 25*(1), 20–49.

Leontief, W. (1970). Environmental repercussions and the economic structure: An input-output approach. *The Review of Economics and Statistics*, 262–271.

Ministerio de Vivienda. (2014). Plan De Acción Sectorial De Mitigación Parael Sector Vivienda Y Desarrollo Territorialestrategia Colombiana De Desarrollo Bajo En Carbono. Available at: http://www.minambiente.gov.co/images/cambioclimatico/pdf/planes_sectoriales_de_mitigación/PAS_Vivienda_y_Dllo_Terr_-_Final.pdf

ND-GAIN. (2017). *Notre Dame global adaptation index: ND-GAIN country index*. Available at: http://index.gain.org. Last accessed 7 Jul 2017.

openLCA Nexus. (2017a). Chilean data. Available at: https://nexus.openlca.org/search/query=chile!Database=EuGeos'%2015804-IA. Last Accessed 9 Jul 2017.

openLCA Nexus. (2017b). *Brazilian data*. Available at: https://nexus.openlca.org/search/query=brazil!Database=EuGeos'%2015804-IA!Category=F%3AConstruction.   Last accessed 19 Jul 2017.

Ortiz-Rodríguez, O., Castells, F., & Sonnemann, G. (2010). Life cycle assessment of two dwellings: One in Spain, a developed country, and one in Colombia, a country under development. *Science of the Total Environment, 408*(12), 2435–2443.

Oyarzo, J., & Peuportier, B. (2014). Life cycle assessment model applied to housing in Chile. *Journal of Cleaner Production, 69*, 109–116.

Pilkington. (2017). *The float process.* Available at: http://www.pilkington.com/pilkington-infor mation/about+pilkington/education/float+process/default.htm

Pomponi, F., & Lenzen, M. (2018). Hybrid life cycle assessment (LCA) will likely yield more accurate results than process-based LCA. *Journal of Cleaner Production, 176*, 210–215. https://doi.org/10.1016/j.jclepro.2017.12.119

Pomponi, F., & Moncaster, A. M. (2016). Embodied carbon mitigation and reduction in the built environment – What does the evidence say? *Journal Environment Management, 181*, 687–700.

Pomponi, F., & Moncaster, A. M. (2018). Scrutinising embodied carbon in buildings: the next performance gap made manifest. *Renewable & Sustainable Energy Reviews, 81*, 2431–2442.

Pomponi, F., Piroozfar, P. A. E., & Farr, E. R. P. (2016). An investigation into GHG and non-GHG impacts of double skin Façades in office refurbishments. *Journal of Industrial Ecology, 20*(2), 234–248.

Rankred. (2015). *17 megadiverse countries in the world.* Available at: http://www.rankred.com/ top-10-megadiverse-countries-in-the-world/. Last accessed 11 Jul 2015.

Ruschi Mendes Saade, M., da Silva, M. G., Gomes, V., Gumez Franco, H., Schwamback, D., & Lavor, B. (2014). Material eco-efficiency indicators for Brazilian buildings. *Smart and Sustainable Built Environment, 3*(1), 54–71.

SICV. (2017). SICV – Banco Nacional de Inventários do Ciclo de Vida. Available at: https://sicv.ibict.br/Node/. Last Accessed 15 June 2017.

Thomson, C., & Harrison, G. (2015). *Life cycle costs and carbon emissions of offshore wind power.* http://www.research.ed.ac.uk/portal/files/19730442/Main_Report_Life_Cycle_Costs_ and_Carbon_Emissions_of_Offshore_Wind_Power.pdf

Title Eficiencia Energetica En Ladrilleras – EELA. (2013). Avialable at: http://www.caem.org.co/ catalogo/docs/Avances%20EELA.pdf [last accessed June 16th 2017].

Treloar, G. J. (1997). Extracting embodied energy paths from input–output tables: Towards an input–output-based hybrid energy analysis method. *Economic Systems Research, 9*(4), 375–391.

Treloar, G., Love, P., Faniran, O., & Iyer-Raniga, U. (2000). A hybrid life cycle assessment method for construction. *Construction Management & Economics, 18*(1), 5–9.

Treloar, G. J., Love, P. E. D., & Holt, G. D. (2001). Using national input/output data for embodied energy analysis of individual residential buildings. *Construction Management and Economics, 19*, 49–61.

U.d.l. Andes (2012). *Estimación curva de abatimiento de gases efecto invernadero sector vivienda urbana* [in Spanish]. Available at: https://mesavis.uniandes.edu.co/Presentaciones%202012/ Mesa%20VIS%20Septiembre%2020.pdf

UKGBC. (2017). *Embodied carbon: Developing a client brief.* UK Green Building Council. Available at: [http://www.ukgbc.org/sites/default/files/UK-GBC%20EC%20Developing% 20Client%20Brief.pdf]. Last accessed 30 May.

UNEP. (2016). *Opportunities for national life cycle network creation and expansion around the world.* United Nations Environment Programm (UNEP) – SETAC Life Cycle Initiative. http:// www.scpclearinghouse.org/sites/default/files/unep-lci_mapping-publication-9.10.16-web.pdf

Valdivia S., Quispe I., Mila i Canals L. (2015). *regional stakeholder consultation on LCA databases in latin America. Session on 'regional roadmap towards the development of LCA databases'.* 14 July 2015. In conjunction with the CILCA 2015 Conference. Pontificia Universidad Católica del Perú (PUCP). Av. Universitaria 1801, San Miguel. Lima, Perú.

WGBC. (2017). *World green building council – Americas network.* Available at: http://www.worldgbc.org/our-regional-networks/americas. Last accessed 25 June.

Wood, R., Lenzen, M., Dey, C., & Lundie, S. (2006). A comparative study of some environmental impacts of conventional and organic farming in Australia. *Agricultural Systems, 89*(2), 324–348.

Yang, Y., Heijungs, R., & Brandão, M. (2017). Hybrid life cycle assessment (LCA) does not necessarily yield more accurate results than process-based LCA. *Journal of Cleaner Production, 150*, 237–242.

Index

© Springer International Publishing AG 2018
F. Pomponi et al. (eds.), *Embodied Carbon in Buildings*,
https://doi.org/10.1007/978-3-319-72796-7

Printed by Printforce, the Netherlands